SCIENCE, TECHNOLOGY, AND INNOVATION POLICY

INTERNATIONAL SERIES ON TECHNOLOGY POLICY AND INNOVATION

a joint initiative of
The IC^2 Institute, The University of Texas at Austin
http://www.utexas.edu/depts/ic2
and
the Center for Innovation, Technology and Policy Research
Instituto Superior Técnico, Lisbon
http://in3.dem.ist.utl.pt

The main objectives of this series are (1) to publish leading scholarly work representing academic, business, and government sectors worldwide on technology policy and innovation; and (2) to present current and future issues of critical importance for using science and technology to foster regional economic development and shared prosperity.

Forthcoming volumes: *Knowledge for Inclusive Development*
Systems and Policies for the Global Learning Economy
Towards an Inclusive Economy

Series Editors: Manuel V. Heitor, David V. Gibson, and Pedro Conceição

SCIENCE, TECHNOLOGY, AND INNOVATION POLICY

OPPORTUNITIES AND CHALLENGES FOR THE KNOWLEDGE ECONOMY

EDITED BY

**PEDRO CONCEIÇÃO, DAVID V. GIBSON,
MANUEL V. HEITOR, AND SYED SHARIQ**

Foreword by Diamantino F. G. Durão and Robert Ronstadt

INTERNATIONAL SERIES ON TECHNOLOGY POLICY AND INNOVATION
Manuel V. Heitor, David V. Gibson, and Pedro Conceição, Series Editors

IC^2 Institute, The University of Texas at Austin and
The Center for Innovation, Technology and Policy Research,
Instituto Superior Técnico, Lisbon, Portugal

QUORUM BOOKS
Westport, Connecticut • London

149847

Library of Congress Cataloging-in-Publication Data

Science, technology, and innovation policy: opportunities and challenges for the knowledge economy / edited by Pedro Conceição ... [et al.].; foreword by Diamantino F.G. Durão and Robert Ronstadt.
    p.cm—(International Series on Technology Policy and Innovation, ISSN 1528–1698 ; no. 1)
Includes bibliographical references and index.
ISBN 1–56720–271–3 (alk. paper)
    1. Technological innovations—Government policy. 2. Science and industry—Government policy. I. Conceição, Pedro. II. Series.
T173.8 .S38 2000
338.9'27—dc21   98–048944

British Library Cataloguing in Publication Data is available.

Library of Congress Catalog Card Number: 98–048944
ISBN: 1–56720–271–3
ISSN: 1528–1698

First published in 2000

Quorum Books, 88 Post Road West, Westport, CT 06881
An imprint of Greenwood Publishing Group, Inc.
www.quorumbooks.com

Printed in the United States of America

The paper used in this book complies with the Permanent Paper Standard issued by the National Information Standards Organization (Z39.48–1984).

10 9 8 7 6 5 4 3 2 1

Contents

PART IV: CHALLENGES FOR NEWLY INDUSTRIALIZED REGIONS

PART V: OPPORTUNITIES FOR CHINA

**PART VI: SUSTAINABILITY, ENVIRONMENT, AND BUSINESS:
POLICY AND STRATEGIES**

PART VII: BROADENING PERSPECTIVES

Foreword

Science and technology (S&T) are key global resources for wealth and job creation and for shared prosperity at home and abroad—challenges and opportunities are faced by corporations, entrepreneurs, universities, governments, federal laboratories, and research institutes alike. Success depends on educated management and employees, global networks and know-how, appreciation for the importance of both personal relationships and information technology in forming and maintaining partnerships, and a facility for successfully dealing with diverse regional, national, and international socioeconomic systems and cultures.

Governments, firms, universities, and research laboratories all take part in the process of building what has been conceptualized as national science and technology systems. The actions of these key players and interactions between them determine the impact of S&T activities on the well-being of nations and regions worldwide. Important challenges are to better understand and manage the complex processes that underlie world-class S&T research leading to successful technology commercialization and adoption. The impact of S&T on economic development and shared prosperity involves cooperation and competition among academia, business, and government at regional, national, and global levels of activity. As knowledge increasingly becomes a key strategic resource for economic development, there is a need to enhance our understanding of the barriers and incentives—in developed, developing, and emerging regions worldwide—for effective knowledge generation, transfer, adoption, and diffusion. Increasing interest in these processes has motivated creative and innovative research and practice across a wide range of businesses and academic disciplines—from management, marketing, entrepreneurship, engineering, and economics to government, public policy, area studies, sociology, history, and law.

In 1996 The Instituto Superior Técnico, Lisbon, Portugal and IC² Institute, The University of Texas at Austin, working with IC²'s Global

Research Fellows and partners, planned to initiate a series of international conferences on technology policy and innovation. Key objectives were (1) the exploration and analysis of unstructured problems using multidisciplinary perspectives for theory building and application research, (2) fostering world-class research and practice while helping to close knowledge gaps, (3) bringing together leading representatives from academia, business, and government worldwide to present and discuss current and future issues of critical importance, and (4) providing state-of-the-art and useful knowledge to decision makers in both the private and public sectors—including informed and effective education, business, and government policies and strategies for the global, knowledge economy.

This volume presents a select set of papers presented at the *1st International Conference on Technology, Policy and Innovation* that was held in Macau, July 2-4, 1997. The Conference theme was *21st Century Opportunities and Challenges for Asian Science, Technology, and Innovation Policy*. The goal was to present global perspectives and practice on the formation and impact of national and regional science and technology based systems leading to economic and social development. A total of 179 participants, from 27 countries, contributed 126 papers and 24 invited lectures. Of all the participants, 108 came from academia, 41 from the business sector, 16 from nonprofit organizations, and 14 from government. The Macau '97 conference brought together these scientists, engineers, managers, entrepreneurs, and policy makers to share knowledge and experiences on the role of science and technology in fostering sustainable economic development and social well-being.

Diamantino F. G. Durão
President,
Instituto Superior Técnico
Technical University of Lisbon
Portugal

Robert Ronstadt
Director, IC² Institute
The University of Texas at Austin
USA

Acknowledgments

The editors of this volume thank the organizers and sponsors of the *1ˢᵗ International Conference on Technology, Policy and Innovation* that was held in Macau, July 2-4, 1997. This list includes The Instituto Superior Técnico (IST), Lisbon, Portugal; The Calouste Gulbenkian Foundation, Portugal; The Portuguese Science and Technology Foundation; The Macau Foundation; The Macau Government; LNEC; The University of Macau; Banco Comercial de Macau; Banco Nacional Ultramarino, Macau; The RGK Foundation, U.S.; and IC² Institute, The University of Texas at Austin. We are grateful to the conference's International Organizing Committee for planning the conference and to those who so effectively managed the three-day event, in particular the staff of the University of Macau and the conference secretariat.

The editors thank the authors of the enclosed chapters for sharing their insights and country perspectives on a range of important topics. We especially thank the Luso-American Development Foundation and the Portuguese Fulbright Commission for their important and continued support of several Portuguese and U. S. scholars and contributors to this volume. Finally the editors are especially grateful for the excellent and timely work of Miguel Silveiro, Publications Coordinator of IN+ Center for Innovation, Technology and Policy Research, IST; and Masoom Khan, Publications Director; Heather Setzler, Program Coordinator; and Paula Correa all at IC² Institute, The University of Texas; at Austin and Katie Chase, Project Editor, Greenwood Publishing Group, for successfully bringing this volume to publication.

INTRODUCTION:
Knowledge, Technology, and Innovation for Development

Pedro Conceição, David V. Gibson, and Manuel V. Heitor

The importance of science and technology (S&T) for economic development is affirmed. Empirical evidence in support of this assertion has been mounting, in the form of both quantitative data and historical analysis (see, for representative surveys, Abramovitz and David 1996; Landes 1998). Accompanying the empirical efforts, the years since the mid-1970s were fertile in theories and empirical regularities that help us to better understand the process through which science and technology leads to economic and social progress. Current trends in engineering have also contributed to this process, both substantially in terms of adding new knowledge to emerging fields, and by creating demand for studies which require the intersection of many different disciplines, such as with environmental engineering. In economics, the new growth theories (summarized in Romer 1994), rekindled the interest in the process of long-run economic growth within neoclassical scholarship. Nelson (1997), in a critique of the new growth theories, provides an excellent survey of other contributions outside "mainstream" economics. From sociology we have seen the development and elaboration of the original Mertonian ideas into a better understanding of the functioning of the scientific and technological communities, and their interaction with the business and government worlds (Stephan 1996, although with an economics twist, summarizes some of the important contributions). Historians have continued to be a source of invaluable analysis, especially economic history (see the classic David 1985; Landes 1998). Other areas such as industrial archeology and business history are providing important contributions to the broad understanding of the process of technological change, innovation, and business development (recent examples include Hannah 1998; Lamoreaux 1998; Raff 1998). In law, the interest in intellectual property is soaring, and new questions are emerging, especially in highly competitive and dynamic scientific and technological fields such as biotechnology (see the recent controversy in the journal *Science*: Doll 1998; Heller and Eisenberg 1998).

Within this multifaceted community of scholars and practitioners, the series of International Conferences on Technology Policy and Innovation (ICTPI) aims at providing a forum for the discussion of recent research on the importance of science and technology for development, and of practices dealing with implementation of S&T policies. Benefiting from the rich contributions to the 1st ICTPI held in Macau in July of 1997, this book shares the reflections of a diverse group of scholars and practitioners. It is our aim in this introductory chapter to place some of the main insights of the chapters included in the book within the larger context of the existing literature on science and technology policy. The main outcome of this chapter will be the proposal of a research agenda to structure future efforts to enhance the understanding of issues relevant for policy making and scholarship in science and technology. Three main directions for research are proposed: (1) the development of a better conceptual understanding of the relationship between S&T and economic growth; (2) the construction of indicators associated with the immaterial aspects of the emerging knowledge-based economy; and (3) the study of the opportunities and threats faced by developing nations. The structure of this chapter reflects the organization of the research agenda. The three ensuing parts join the current literature with some of the contributions in the book discussing three directions for further research.

DEVELOPMENT OF A BETTER CONCEPTUAL UNDERSTANDING OF THE MECHANISMS OF KNOWLEDGE CREATION AND DIFFUSION

Research often builds upon a conceptual framework of the empirical reality being analyzed. Similarly, decision making is influenced by the understanding of the system upon which the decisions will have impact. Therefore, it is important to advance in the modeling of the processes through which science and technology interact with the economy and society at large. The relation between science and technology and economic growth is widely acknowledged to be very complex. Recent advances in the understanding of innovation have sharply criticized linear models of innovation (Kozmetsky 1993). Traditionally, successful commercialization of R&D was understood as being the result of a linear process, beginning with scientific research and then moving to technology development, before reaching the stages of financing, manufacturing, and marketing. No sustained connections among academic, business, and government institutions were explored, beyond the expectations that universities produce science and businesses commercialize technology. Today, it is understood, there is more to the relationships between technological innovation, economic wealth generation, and job creation than a one-dimensional flux from R&D to product. An integrated and interactive approach blending scientific, technological, social-economic, and cultural aspects is required to explain the dynamics of innovation.

At a more macro level, two traditional conceptual perspectives on how economic growth occurs have dominated scholarship and policy. Both have many variations, and many more models of development beyond these two exist. However, the seeds for the most active areas of scholarship on the relationship between technological change and growth are to be found in the following two theories:

1. The first was developed within the neoclassical models of growth, using a production function framework (basically, an input-process-output system). Inputs are called factors of production, and include labor and capital, which interact in a process of production of wealth that is limited by the current level of technology. The total maximum level of production, or total output of the economy, is given by $Y = F(K,L)$, in which Y is output, K is capital, L is labor, and F represents the process of transforming the factors of production into outputs. The idea that the pure accumulation of physical capital and labor is not sufficient to account for all the observed growth has been recognized since, at least, Adam Smith. However, it was Solow (1956, 1957) who first separated the effects of growth that went beyond the accumulation of physical capital and labor, which he called technological change (better than "the residual"). The Solownian theory of growth rests on purely neoclassical assumptions, in which resource allocation is mediated in free markets by pricing in a competitive environment that leads to market clearing. No process to account for the process of technological change was developed.

2. The second traditional perspective is associated with Schumpeter's economic theory. For Schumpeter, businessmen and firms are not passive elements merely adjusting the prices to the idiosyncrasies of the market. He argued that the expectations of profits would not only lead to price setting, but would also drive the "entrepreneur" to innovate. The entrepreneurial drive toward innovation was due to the temporary monopolistic position from which the innovator would benefit. Schumpeter regarded this position as temporary because the advantages from this privileged position would eventually "perish in the vortex of the competition which streams after them," since other firms would copy the innovator (Schumpeter 1934), following the process of *creative destruction*. Therefore, innovation appears at the forefront of economic progress, driving prosperity. In a later theory, Schumpeter

refined this earlier simplistic version of an entrepreneur in a market composed by a multitude of competing firms that destroy any persistent market advantage. In his final work (Schumpeter 1942), he acknowledged that some large corporations can sustain a market advantage by an institutionalization of the effort to innovate through the establishment of large R&D facilities.

Solow attributed the component of growth that went beyond the accumulation of physical capital to technological change, and Schumpeter to the process of creative destruction. However, implicitly at least, the underlying long-run economic growth mechanism is knowledge accumulation, through the generation of new ideas and new skills in people. In the Solownian framework, knowledge accumulation is regarded as an exogenous phenomenon, not formalized within the models of economic growth. Recently, endogenous (or new) growth theory incorporates in the Solownian formulation some aspects of the mechanics of technological innovation.

There are three basic types of new growth theories. The first was developed by Arrow (1962), who argued that, besides the accumulation of capital and labor, growth resulted from an increase in the knowledge used in production, due to experience gained in using capital. According to Solow (1997), this was at the root of new growth theories, and an example of a conceptual effort according to the same lines is due to Romer (1986). The second theory attributes growth to improvements in the level of formal education, and was proposed primarily by C. Lucas (1998). The third theory models Schumpeterian competition, and the impact of new discoveries that result from research efforts (Romer 1990; Grossman and Helpman 1991).

Despite the interest these theories have sparked within theoretical macroeconomics, Mankiw (1995), in a recent assessment of growth theory, has claimed that we should get back to the pure Solownian formulation, which conforms much better to the empirical facts. Pack (1994) surveys the lack of convincing empirical evidence. Nonetheless, new growth theories have yielded new insights concerning the historical patterns of growth (Romer 1996) and have provided new avenues for policy formulation (Romer 1993). Soete (1996) has blamed the empirical shortcomings of the new growth theory on the reductionism of the mainstream approach in economics, which looks for highly formalized and quantified evidence.

In fact, Soete has argued for the need to deepen more appreciative theories of the process of technological change. Freeman and Soete (1997) provide a comprehensive overview of the main findings and theories in the realm of existing appreciative theories. The distinction between these and new growth theories is hardly at the level of fundamental concepts. Both agree that knowledge accumulation (or, more traditionally, technological change) is endogenous, that new ideas and discoveries may entail processes of creative destruction, and so on. The main difference of the appreciative theorists is that reality is too complex to be modeled by a simple, stylized equation, and that differences across time and space are extremely relevant. It is, one could argue, primarily an epistemological difference.

Instead of the formalized production-function framework, the more appreciative theories have contributed to the concept of national innovation system, as articulated initially by Lundvall (1992) and Nelson (1993), and more recently by Edquist (1997). This concept reflects the perception that governments, firms, universities, research laboratories, the financial system, and a host of institutions take part in the process of knowledge accumulation, particularly as it is reflected in producing innovation. The actions of these key institutional players and the interactions between them determine the impact of innovation, and more generally of knowledge, on the well being of nations. The national innovation system perspective acknowledges the specificity of each country, defending country- and time-specific policies to harness the benefits of knowledge to economic growth.

The differences among these two perspectives are largely epistemological, and are not likely to be settled easily in the academic realm. However, from management and policy points of view, it is rewarding that the approaches briefly reviewed above have been able to coexist and prosper. Each of them provides important insights and implications for management and policy. If we try to find areas of consensus instead of looking at the points of disagreement, a fundamental aspect is the idea that knowledge accumulation contributes to economic growth. Therefore, from a management and policy point of view, an enhanced understanding of the processes of knowledge accumulation and diffusion is key for better policy analysis and policy practice. A clear example has recently been reported by Cooke and Morgan (1998) in terms of associational economies, in that the building-up of regional systems of innovation expresses a process of regional learning.

Consequently, Part I in this volume gathers five chapters dealing with the relationship of knowledge and economic development. Chapter 1, by de Maria y Campos and Machado, challenges us to look beyond developed countries. In fact, traditionally, there has been a bifurcation in the way economists address the problems of development. The academic discipline of "economic development" is focused on less-developed countries. It is assumed that in these countries, problems are fundamentally different from the problems faced by the "first world." Solow (1997) mentions that his models are not suited for most African countries, but rather for developed countries and countries in transition, such as Portugal (his example). However, regardless of the theoretical division that is made, the reality is, as de Maria y Campos and Machado show, that less-developed countries are not immune to the forces of globalization, concern for the environment, and, most important, new technology. This chapter is also a good motivation for some of the issues raised in Part VI of this volume.

In Chapter 2, Caraça calls for a renewal of science policy. In fact, Caraça defends a science policy that goes beyond the reactive and detached initiatives that abound in the context of the increasing liberalization of markets. Caraça begins by offering a conceptualization of knowledge that starkly illustrates the differences between this economic resource and the physical resources still located at the forefront of economic thought. If we accept the supremacy of knowledge, and see our societies as being "knowledge-based," then the demands of science policy are driven by the need to enhance the societywide ability to continuously learn.

Chapters 1 and 2 contrast with Chapters 3 and 4, in which the perspective of firm, rather than broad, policies is emphasized. Botkin, in Chapter 3, discusses the importance of sharing knowledge across firms and other institutional actors. The result of this communication process is the buildup of knowledge communities. Verdin and Van Heck, in Chapter 4, discuss the difficulties of building "knowledge communities," as defined by Botkin, within transnational corporations. They show that the successful integration of knowledge across different sites around the world depends crucially on the management of the integration process.

Concluding Part I (Chapter 5), Shariq argues for the need to actively construct a new disciplinary framework for knowledge management. This disciplinary framework, Shariq argues, should go beyond being a compass for research and scholarship, and must be used to benefit worldwide development. Therefore, the programmatic agenda within this new discipline should be set by engaging the academic, business, and government communities, and research objectives should be to seek immediate applications. Shariq focuses specifically on management issues, closer to the business perspective. Nonetheless, he emphasizes the idea of sharing and frames the need for a new discipline in terms of searching for economic and social relevance, particularly for businesses.

The university is a key institution in the processes of knowledge creation and diffusion. Part II focuses on universities in the context of the discussions in Part I. The role of the university in emerging knowledge-based economies has been discussed extensively in recent years (e.g., Ehrenberg 1997; Conceição et al. 1998b). Sullivan, in Chapter 6, challenges us with the claim that universities must transform themselves, in a context in which the creation and dissemination of knowledge is being carried out by many other institutions, especially since advances in information technologies and telecommunications have dramatically increased our ability to communicate globally. Sullivan presents the challenges of the digital age as opportunities for universities, despite their natural resistance to change. Konishi, in Chapter 7, foresees the future of the relationships between universities and industry in the digital age. In a context in which the difference between the creators and diffusers of knowledge is increasingly blurred, visions such as cyber classrooms and cyber laboratories are deemed to become reality within the context of "deinstitutionalization of universities and of the educational system" as a whole.

Dissenting somewhat from the two previous analyses, Conceição, Heitor, Oliveira, and Santos (Chapter 8), though agreeing that universities unquestionably need to change, argue that they should not undergo a transfiguration. To them, this transfiguration means the loss of their institutional integrity. Their argument rests on the uniqueness of the university in the current economy, especially as a provider of public knowledge. In Chapter 9, Jones-Evans describes interesting ways in which universities have changed without jeopardizing their institutional integrity. In a thorough analysis of a series of examples occurring in Europe, Jones-Evans shows how universities were able to reinvent themselves, acquiring an entrepreneurial role in fostering regional economic development. Jones-Evans launch a challenge, asking policy makers to understand that universities provide the engine and fuel for growth that coal mines, particularly in his native Wales, used to provide after the industrial revolution. However, Chapter 10, by Rogalev, shows the difficulties of going beyond concepts into practice. By discussing the efforts of a university in Russia trying to adopt the practices of developed and stable countries, Rogalev tells us that we need to put as much emphasis on principles for policy formulation as on ways to help us to implement what are regarded as being good policies.

Given the contributions in Parts I and II of this volume, we conclude that it is important to deepen interdisciplinary studies and to promote communication across scientific fields, in order to facilitate the understanding that learning is at the center of development. Clearly, work within specific academic disciplines is crucial, since it provides us with a better understanding of the knowledge-based economy from different perspectives.

IMPROVING THE QUALITATIVE AND QUANTITATIVE EMPIRICAL ANALYSES OF SCIENCE AND TECHNOLOGICAL POLICIES

The gathering of empirical evidence on science and technology policies, both qualitative and quantitative, is crucial to both validate and improve on the theoretical conjectures described in the previous section, and to diffuse best practices as well as major policy efforts to promote the learning society. The collection and analysis of quantitative data has been particularly problematic. Current accounting practices, at both the firm and national levels, make the gathering of quantitative information extremely difficult. National accounts, developed in the 1950s within the theoretical framework of Keynesian economics, are well suited to deal with material inputs and outputs of an economy. Only in the late 1960s, with the theories of human capital and the contributions of Solow (1956, 1957), did some attention go to measuring intangible assets such as human capital and R&D. Most efforts at the time focused on these assets as inputs, the outputs still being primarily material. Today, although we know that intangible assets are important as both economic inputs and outputs, our accounting framework has had difficulty in reflecting this reality (Edvinsson and Malone 1997).

Difficulties with the measurement of intangibles are at the core of the ongoing controversy of whether we are entering a new economic phase that relys increasingly on knowledge. While there is a general perception that today's developed economies depend on knowledge and information in an unprecedented way, Stevens (1996) mentions that the idea of a knowledge-based society remains more of a concept than a measurable entity. Still he makes the claim that in about half of the Organization for Economic Cooperation and Development (OECD) countries, gross domestic product (GDP) is now knowledge-based. Additionally, Greenspan (1996) said recently that America's GDP weighs today, in tons, as much as it did a hundred years ago.

But how can we measure these claims? So far, the evidence comes mainly from stylized facts, such as the growing incorporation of knowledge in products, the increasing value associated with software, as opposed to hardware, and the growing strength of services. Let us consider each of these in turn.

Manufactured products are increasingly characterized as "smart." For example, a typical car today has more computer-processing power than the first lunar landing craft had in 1969. Wyckoff (1996) mentions tires that tell the driver when the pressure is too low. Davis and Botkin (1994) describe how Massey Fergusson, a farm-tractor manufacturer, blended information and communication technologies to provide crop management tools to farmers (these are called knowledge-based farm tractors).

The incorporation of knowledge in products had led to the increasing importance of software, in comparison with hardware. A metric mentioned by Wyckoff (1996) reflects this fact: in the 1970s, 80% of the value of an IBM computer was hardware and 20% was the cost of the incorporated software. In the late 1980s the shares were reversed, and ever since have moved in the direction of increasingly valuable software. Finally, increased attention to knowledge production has led to a dramatic shift in employment in the last decade from the industrial sector to the services sector in most OECD countries, as documented extensively in a recent issue of *OECD Observer* (1996).

In terms of the measurement of "intangible" economic aspects, it is important to acknowledge the tremendous success of the OECD in establishing internationally comparable statistics and indicators for S&T, through the Frascati Manual, and in education, with the contribution of UNESCO. Recently, the OECD has been sponsoring similar ventures for innovation data. More advanced efforts include the measurement of innovation in the services sector by Sirilli and Evangelista (1998), and the development of indicators for the knowledge-based economy. In this volume, Parts III, IV, and V deal with empirical and conceptual studies of national and global practices on science and technology policies, with Part IV focusing on newly industrialized regions, and Part V focusing exclusively on China. There is a blending of quantitative and descriptive studies, illustrating the diversity of empirical work presented at the 1st ICTPI in Macau.

Part III includes Chapters 11 through 20. In Chapter 11, Iammarino, Prisco, and Silvani show how recent quantitative data gathered from Italian innovation surveys can yield fundamental insight into our understanding of national and regional aspects of the innovation process. The specific contribution of the chapter deals with the relationship between the concentration of innovative activity (both geographically and across sectors) and the concentration of production activity. The authors test the hypothesis that there is a self-reinforcing cycle between production and innovation in a specific region that leads to the emergence of distinctive "styles" of economic development.

In Chapter 12, Galbraith gives us a thorough and refreshing description of the industrial and S&T policies in the United States. Motivated by the rhetoric on the idea of competitiveness, Galbraith shows the divergence between the political discourse and the policy practice since the early 1990s. In fact, until the 1990s the discourse was emphatically against policy interventions in the industrial realm. However, policy practice contradicted this discourse, namely through heavy federal financial support to defense-oriented R&D. In the early 1990s, with the military threat vanished, the discourse bent toward interventionism—in order to assure the competitiveness of the United States in world markets. However, Galbraith argues, government support has been vanishing continuously.

Gibson and Stiles, in Chapter 13, go beyond the national perspective, and provide an overview of a globally oriented technopolis framework, perspectives on technology transfer, and the advantages that each of these perspectives would accrue from the establishment of a global entrepreneurial network. In Chapter 14, Gann argues that, within cities or regions, major capital projects can be of such a large complexity and scope that traditional views of innovation are inadequate to support the design of policies aimed at promoting innovation from the large investments involved in these projects. Telecommunications and information technologies are seen as major facilitators for the coordination of these projects, in the same way Gibson and Stiles see them as helping the formation of global entrepreneurial networks.

In Chapter 15, Reddy tackles a key issue associated with the establishment of global networks and investments of more developed countries in comparison with less developed countries. Invariably, there is an asymmetry involved in these networks and channels through which investment flows. More specifically, when do investments help the endogenous capabilities of less developed countries? Reddy analyzes this issue with a specific application to the investment decisions on R&D of transnational corporations. Looking specifically at the case of India, he discusses the drivers of the decisions to locate R&D activities in developed countries, and shows that an educated labor force is key to attracting this type of investment. Despite the fact that the location of manufacturing activities produces quicker and wider results, R&D facilities, if given the right incentives and policy context, Reddy argues, can lead to improved long-term development potential.

In Chapter 16, Archibugi, Evangelista, Perani, and Rapiti, draw on the wealth of information in the Italian innovation surveys. They characterize the patterns of expenditures, outcomes, and nature of innovation activities in Italy. This chapter is an important contribution to set a standard for the presentation and discussion of countrywide innovation data, now that more and more countries have adopted the practice of launching systematic innovation surveys.

The authors of Chapters 17 and 18 discuss some aspects of the development prospects of Hungary and Portugal, respectively. Inzelt shows the still-incipient institutional framework existing in Hungary, and argues for the need to promote and innovative society. Veloso and Felizardo focus on one of the most important industries in Portugal, the automobile parts cluster, and discuss the way in which this industry can act as a catalyst for development in Portugal and other medium-income countries.

In Chapter 19, Athreye challenges the view that full-fledged competition policies are the best to promote innovation. Following on the tradition of Schumpeter, she argues that better instruments to award monopoly power to innovators need to be developed. In the concluding chapter in this part (Chapter 20), Sarmento-Coelho discusses innovative aspects of the service sector, emphasizing the role of small- and medium-sized enterprises.

In Part IV, emphasis is given to studies on newly industrialized countries in Asia. In Chapter 21, Yuan gives an overview of the different policies to acquire endogenous technological capability in Asian countries. Chapters 22 and 23 provide in-depth analyses of different aspects of the development of Korea. Kim describes, from a macroeconomic point of view, industrial policies that helped Korea, while Sung, with a more micro approach, discusses different development strategies of cities across Korea. In Chapter 24, Chang, Chu, and Ting focus on the impact of a popular policy instrument, the science-based industrial park, on the development of Taiwan. Finally, the concluding chapter of Part IV (Chapter 24), by Dodgson, complements the initial chapter of this part by describing different innovation policies in East Asia.

Part V focuses on China, and includes six chapters which discuss different aspects of the technology and innovation development of China. Chapter 26, by Yingjian, describes the parallel development of China's adoption of information technologies and increased international cooperation. These facets are also key in the discussion of Chinese firms in Chapter 27, by Chu, which discusses the technological capability of Chinese firms in the context of the dominant strategies and organizational structures adopted by Chinese enterprises. Wang,

Qin, and Guan treat the issue of university-based technology commercialization in Chapter 28. On a related topic, Lin (Chapter 29) discusses the way a university can help to shape the industrial development of Hong Kong. As the economic structure of Hong Kong depends increasingly on services, Lin argues that a science and technology university can be important to retain the ability to perform sophisticated manufacturing in the region. The neighboring case of Macau is discussed in Chapter 30, where Martins and Iu address the challenges of science and technology policy for the 21st century in the territory. Concluding Part V, Chen provides us with a vision for the development of China as a regionally integrated economy, where networks of production and technological capabilities play key roles.

The bulk of this volume—Parts III, IV, and V—is devoted to the empirical analysis of science and technology policy across countries. Despite the quality of these contributions, we feel that more needs to be done in order to have better and more reliable indicators to both inform decision making and evaluate past policy decisions. Policy-making processes are very complex, and the aim of better indicators should not be to determine decisions, but to allow for a more informed public discussion. In particular, while S&T is, and will continue to be, important, in accordance with our earlier discussion, there is a need to move toward indicators of knowledge creation and diffusion. The OECD has been developing pioneering efforts in this area, and we hope this volume can illustrate both the potential and the need for further development of quantitative indicators.

THE NEED FOR AN "INCLUSIVE DEVELOPMENT"

A major theme of this chapter is that knowledge is a key resource for economic development. However, large regions fraction of the world lack the most basic resources. According to the World Bank, four-fifths of the world population lives in developing countries, and of these, two-thirds are poor. The strong participation of the United Nations Industrial Development Organization (UNIDO) in the Macau Conference was a wake-up call to the sometimes-closed section of the wealthy and developed world to the problems facing large portions of humanity. The lack of resources and infrastructure to deal with knowledge diffusion, storage, and use in underdeveloped regions is pervasive. In many regions of the world access to a telephone line seems to be an insurmountable obstacle.

Salomon (1995) discusses the one-way interaction among two civilizations: one which depends on the creation of knowledge and the other which passively accepts the knowledge created in the first. The emerging patterns of globalization and the increasing importance of regional sustainability, also a global issue, may lead to a change in the rationale for creating knowledge, towards a rationale that Salomon classifies as compassionate scientific curiosity.

The roots of the concept of "technopolis" — a utopian, radiant city based on science and technology — sprang from the humanistic mind of the Renaissance. The first example was T. Campanella's 17th-century vision of a "City of the Sun." This was a culturally diverse, poly-technical city — a showcase of science and continuing education. Here science was not just for an aristocratic elite. Philosophers, political planners and social prophets were intrigued by the idea of creating a "city of light," designed and ruled by wise scientists, where research and innovation were a way of life and invention and creativity were venerated. These ideal cities were to be the poles around which the economy of nations would grow and society would progress. The modern "technopolis" — "techno" for technology and "polis" Greek for city-state — combines scientific research and invention with the practical applications of technology through innovation. (Smilor, 1988; Gibson, 1992).

Part VI of this volume concerns the global challenges that are manifest across geography and levels of development. The perspective is that concern with the global environment goes far beyond the concerns with air quality or acid rain, and should be included in the broader perspective of achieving sustainable economic development. In Chapter 32, Barreto, Medina, Peitier, and Villas Bôas clarify the concepts associated with sustainable development, and propose scenarios and strategies for research. In Chapter 33, Peltier goes into more specific recommendations, describing different experiences on the usage of clean production methods. Ferrão and M. Heitor, in Chapter 34, propose a framework to integrate business strategies with environmental policy concerns. The ultimate goal is to provide a framework in which firms are able to translate the challenges of environment protections into further business opportunities and competitive advantages. A specific instrument for achieving what Ferrão and Heitor propose is described by Azapagic in Chapter 35, where the author elaborates on the potential of life-cycle assessment tools. Besides the asymmetries in development on a planetary scale, there are also differences in development within the so-called developed world. T. Heitor, in Chapter 36, notes that these asymmetries exist within cities, and proposes the analysis of the spatial morphology which can contribute to achieve localized sustainable development through urban planning.

The larger picture is that we must not only think in terms of sustainable development—that is, development that is continuous—but we must also consider present problems associated with the exclusion of countries, neighborhoods, and people. True sustainability requires full inclusion when approaching current and future challenges, and may require the strengthening of new research areas and the development of a new research agenda to deal with such inclusive development. Policies that might appear sustainable within a national or regional context might increasingly appear less so in an international context. This holds not only for traditional macroeconomic policy, but also for social policy, tax policy, social security policy, and other policies traditionally preserved at the national level.

The challenging and complex problems raised in the chapters of this volume emphasize the need to establish a research agenda to better devise policies and management practices in a global knowledge-based economy. Our purpose with *Science, Tecnology, and Innovation Policy: Opportunities and Challenges for the Knowledge Economy*, is to facilitate the development of a research agenda and action initiatives, by exposing the reader to international and multidisciplinary scholarship on science and technology policy across academic, business and government disciplines. To highlight our multidisciplinary emphasis, we close the volume with Part VII, in which Calado discusses the interactions between art and technology across time and space.

CONCLUDING REMARKS

The Macau Conference, as reflected in this book, provided a broad view of a multitude of perspectives to deal with the generic issue of technology policy and innovation. In our view, one important aspect evident during and after the conference was that, though being an important topic, technology policy and innovation studies are still lacking a disciplinary framework in which they can be grouped. Glancing at the various academic and professional backgrounds of the contributors to this volume is enough to conclude that the theme of the conference is still, at least academically, largely an undefined discipline.

The lack of a disciplinary framework represents both a threat and an opportunity. It is a threat in the sense that it may discourage contributions from people who work within well-established fields in the social, natural, and engineering sciences. On the other hand, it represents a vast opportunity to establish a forum of intellectual discussion of issues normally regarded as outcasts within academic boundaries. It is our hope that this volume will contribute to and raise the awareness of the importance of science and technology policy and innovation by exhibiting a diverse set of views thereby encouraging further research both within and across disciplines.

REFERENCES

Abramovitz, M., and P. David. "Technological Change and the Rise of Intangible Investments: The U.S. Economy's Growth-Path in the Twentieth Century," in *Employment and Growth in the Knowledge-based Economy*. Paris: OECD, 1996.

Arrow, K. "The Economic Implications of Learning-by-Doing," *Review of Economic Studies* 28 (1962) 155–173.

Arthur, W. B. *Increasing Returns and Path Dependency in the Economy*. Ann Arbor: University of Michigan Press, 1994.

Conceição, P., D. Durão, M. V. Heitor, and F. Santos. *Novas Ideias para a Universidade* [*New Ideas for the University*]. Lisbon: IST Press, 1998a.

Conceição, P., M. V. Heitor, and P. M. Oliveira. "Expectations for the University in the Knowledge Based Economy," *Technological Forecasting and Social Change* 58, 3 (1998b) 203–214.

Cooke, P., and K. Morgan. *The Associational Economy.* Oxford: Oxford University Press, 1998.

David, P. "Clio and the Economics of QWERTY," American Economic Review, 75, (May 1985), 332-337.

Davis, S., and J. Botkin. "The Coming of Knowledge-Based Business," *Harvard Business Review* (September/October 1994).

Doll, J. J. "The Patenting of DNA," *Science* 280, 1 (May 1998) 689–690.

Edquist, C. *Systems of Innovation: Technologies, Institutions and Organizations.* London: Pinter Publishers, 1997.

Edvinsson, L., and M. S. Malone. *Intellectual Capital.* New York: HarperBusiness, 1997.

Ehrenberg, R. G. The American University: National Treasure or Endangered Species? Ithaca: Cornell University Press, 1997.

Freeman, C., and L. Soete. *The Economics of Industrial Innovation.* Cambridge: MIT Press, 1997.

Gibson, D, G. Kozmetsky, and R. Smilor (Eds.). *The Technopolis Phenomenon: Smart Cities, Fast Systems, Global Networks.* Maryland: Rowman & Littlefield Publishers, Inc., 1992.

Greenspan, A. "Job Security and Technology," in *Technology and Growth* (J. Fuhrer and J. Little, eds.). Boston: Federal Reserve Bank of Boston, 1996.

Grossman, G. M., and E. Helpman. *Innovation and Growth in the Global Economy.* Cambridge: MIT Press, 1991.

Hannah, L. "Survival and Size Mobility Among the World's Largest 100 Industrial Corporations, 1912–1995," *American Economic Review* 88, 2 (1998) 62–65.

Heller, M. A., and R. S. Eisenberg. "Can Patents Deter Innovation? The Anticommons in Biomedical Research," *Science* 280, 1 (May 1998) 698–701.

Kozmetsky, G. "The Growth and Internationalization of Creative and Innovative Management," in *Generating Creativity and Innovation in Large Bureaucracies* (R. L. Kuhn, ed.). Westport: Quorum Books, 1993.

Lamoreaux, N. R. "Partnerships, Corporations, and the Theory of the Firm," *American Economic Review* 88, 2 (1998) 66–71.

Landes, D. *The Wealth and Poverty of Nations: Why Some Are So Rich and Some So Poor.* New York: W. W. Norton & Company, 1998.

Lucas, C. *Crisis in the Academy: Rethinking Higher Education in America.* New York: St. Martin's Press, 1998.

Lucas Jr., R. "On the Mechanics of Economic Development," *Journal of Monetary Economics*, 22 (July 1, 1988) 3–42.

Lundvall, B. A. *National System of Innovation: Towards a Theory of Innovation and Interactive Learning.* London: Pinter Publishers, 1992.

Mankiw, N. G. "The Growth of Nations," in *Brookings Papers on Economic Activity, Volume 1* (W. C. Brainard and G. L. Perry, eds.). Washington, DC: Brookings Institution, 1995.

Nelson, R. *National Innovation Systems.* Oxford, U.K.: Oxford University Press, 1993.

Nelson, R. R. "How New is Growth Theory?" *Challenge* 40, 5 (1997) 29–58.

Pack, H. "Endogenous Growth Theory: Intellectual Appeal and Empirical Shortcomings," *Journal of Economic Perspectives* 8, 1 (1994) 55–72.

Raff, D. M. G. "Representative Firm Analysis and the Character of Competition: Glimpses from the Great Depression," *American Economic Review* 88, 2 (1998) 57–61.

Romer, P. M. "Increasing Returns and Long-Run Growth," *Journal of Political Economy* 94, 9 (1986) 1002–1037.

Romer, P. M. "Endogenous Technological Change," *Journal of Political Economy* 98, 5 (1990) S71–S102.

Romer, P. M. "Idea Gaps and Object Gaps in Economic Development," *Journal of Monetary Economics* 32 (1993) 543–573.

Romer, P. M. "The Origins of Endogenous Growth," *Journal of Economic Perspectives* 8, 1 (1994) 3–22.

Romer, P. M. "Why, Indeed, in America? Theory, History, and the Origins of Modern Economic Growth," *American Economic Review* 86, 2 (1996) 202–206.

Salomon, J. J. "The Uncertain Guest: Mobilizing Science and Technology for Development," *Science and Public Policy* 22, 1 (1995) 9–18.

Sirilli, G., and R. Evangelista. "Innovation in the Service Sector: Results from the Italian Statistical Survey," *Technological Forecasting and Social Change* 58, 3 (1998).

Smilor, R., D. Gibson, and G. Kozmetsky (Eds.). *Creating the Technopolis: Linking Technology Commercialization and Economic Development*. Cambridge, MA: Ballinger Publishing, 1988.

Schumpeter, J. *The Theory of Economic Development*. Cambridge: Harvard University Press, 1934.

Schumpeter, J. *Capitalism, Socialism and Democracy*. New York: Harper Row, 1942.

Soete, L. "The Challenges of Innovation," *IPTS Report* 7 (1996) 7–13.

Soete, L. "The Impact of Globalization on European Economic Integration," IPTS Report 15 (June 1997) 21-28.

Solow, R. M. "A Contribution to the Theory of Economic Growth," *Quarterly Journal of Economics* 70, 1 (1956) 65–94.

Solow, R. M. "Technical Change and the Aggregate Production Function,' *Review of Economics and Statistics* 39, (August 1957) 312-320.

Solow, R. M. *Learning from "Learning-by-Doing": Lessons for Economic Growth*. Stanford, CA: Stanford University Press, 1997.

Stephan, P. E. "The Economics of Science," *Journal of Economic Literature* 34 (September 1996) 1199–1235.

Stevens, C. "The Knowledge-Driven Economy," *OECD Observer* (June/July 1996).

PART I:
KNOWLEDGE AND DEVELOPMENT

1

Renewing Technology Management and Policy: Innovation Crucial for Sustainable Industrial Development

Mauricio de Maria y Campos
and Fernando Machado

INNOVATION AND SUSTAINABLE INDUSTRIAL DEVELOPMENT

While industrialized countries, which already enjoy the benefits of world-class industrial capability, speak casually of deindustrialization and the evolution of their service economies, most developing nations remain absolutely convinced of the opposite. For them sustainable industrial development is the only hope they have of sharing in the social stability and quality of life that industrialized countries take for granted.

It is now over thirty years since developing countries succeeded in convincing the international community that industrialization was so important, and that the United Nations should set up a special agency to help them bring it about. Thus, the United Nations Industrial Development Organization was born, and its task is far from over. Today, three things are different. First, the process of developing country industrialization has to be sustainable—socially and economically as well as environmentally. Second, industrialization will have to take place in an increasingly globalized context in which international competitiveness is the major challenge. Third, globalization is here to stay: what matters is what kind of globalization will take place, and how less developed countries will be able to integrate into the world economy.

Above all, what concerns the whole international community, developed as well as developing countries, is that globalization itself will have to be sustainable—inclusive, participatory, equitable among and within countries—and people-centered. At the same time, the very success of the newly industrializing countries (NICs) shows that it does matter just how developing countries integrate into the world economy. Their integration cannot be indiscriminate; it has to be strategic.

At the same time, developing countries' future industrialization has to be very different than the patterns of the past. It has to be technology-led as well as sustainable. Bringing this about will be a joint task—of national governments who provide the appropriate economic environment, private firms that take the right technological risks, and an international community that is responsive to global environmental issues. In this context, technology will lead only if it is strategically innovative. And industrialization will be sustainable only if it not only protects the environment, but is also efficient and promotes social development.

This means that the world cannot afford a pattern of developing countries' industrial development that follows either their own pattern of the past four or five decades, or the pattern of industrialized countries over the past two hundred years. The first locks developing countries into permanent second-class world citizenship. The second leads to unsustainable demands for physical resources, energy, and waste assimilation capacity. Neither pattern is sustainable in the long run.

A third pattern, we suggest, can be found in how developing countries take up the main theme of Macau '97—technology policy and innovation—how they produce, diffuse, and apply knowledge, and how they manage technology through new approaches to policy and firm-level strategy.

OPPORTUNITIES IN DISCONTINUITIES

Elements of a possible new pattern for developing countries are opportunities deriving from discontinuities in present world development: (1) the unprecedented global concern over the environmental impact of continued high rates of economic growth; (2) the inability of industrialized countries to maintain their own long-term economic growth without accelerated and massive growth in developing countries; (3) the unlikelihood that developing countries will achieve technology-led development without radical changes in technology policies; and (4) a sea change in the role of universities as generators and providers of knowledge.

Taken together, these discontinuities are an unprecedented developing-country opportunity. To realize its benefits requires, however, the application of knowledge—technological innovation, in a pattern of industrial development in which they adopt a wholly new (and unaccustomed) relationship to technology—a relationship that makes them technology developers rather than merely secondhand users, and that permits them for the first time to close the technology and industrialization gaps between them and the developed world.

Four specific responses that will lead to a new pattern are:

- The immediate need to rethink developing countries' use of environmentally sound technologies (ESTs) as a means to sustainable industrial development.
- The crucial role of technological innovation, in particular strategic innovation—crucial for developing-country private firms' very survival, crucial for newly industrializing nations in closing the gap with developed countries.
- A new approach to technology policy in developing countries, an approach that places technology management and technological innovation at the center of their national development policies, especially those concerned with industrial development.
- The opportunities for developing countries in the technology-induced changes we predict for universities and industry-university relationships in the 21st century (see Chapter 7 by Konishi for more details).

TECHNOLOGY ACCESS CRUCIAL FOR SUSTAINABLE INDUSTRIAL DEVELOPMENT

One consequence of the increasing global concern for the environment, the destruction of the ozone layer and global warming in particular, is growing recognition that developing countries' development cannot sustainably follow the industrialization patterns of the now-advanced, rich countries. If they do it will destroy the planet. Nevertheless, their right to develop also has to be recognized. The new challenge therefore is to harmonize growth and job creation with the environment—a challenge they are unlikely to meet without significant external help.

The world's environmental burden is a function first and foremost of population and affluence. It is also a function of the technology applied by industry. Population in developing countries and affluence in the developed and the newly industrializing countries are both growing fast. Stabilization of the global population remains a distant and uncertain prospect. Developing countries' demand for affluence (and their ability to create it) is certain—and it is here today. The only remedy in the short and medium terms lies in environmentally sound (so-called cleaner) technologies that minimize materials and energy consumption and reduce waste. It is therefore crucial that developing countries' anticipated and necessary economic growth is based on technologies that satisfy their consumption aspirations in environmentally sound ways.

Will they be able to do it? Specifically, will developed countries fully play their part in providing "favorable access and transfer of ESTs to developing countries," as called for by chapter 34 of Agenda 21 for sustainable development? Frankly we are not sure.

At the time of writing, prospects for a positive outcome to the Special Session of the United Nations (UN) and the UN Conference on Environment and Development (UNICED) to review and appraise the implementation of Agenda 21 are not encouraging, particularly concerning a consensus text agreeing to concessional finance for transfer of relevant technology. Furthermore, we have serious doubts, because of the context in which it was framed, about the very ability of transfer of EST to achieve sustainable industrial development.

As defined by UNCED, ESTs are technologies that "protect the environment, use all resources in a sustainable manner, recycle more of their wastes and products, and handle residual waste in a more acceptable manner than technologies for which they were substitutes." In our view this concept is seriously flawed for several reasons. First, such ESTs focus on only one dimension of sustainable industrial development—namely, the environment. They neglect efficiency as a criterion of materials and energy use, and they ignore the need to promote equity— both intragenerational and intergenerational. Second, they lead to promotion of end-of-pipe technology rather than design choices that avoid the problem altogether (i.e., cleaner production through pollution prevention). Third, the practical emphasis on transfer of ESTs from abroad misses the fact that the main barriers to such transfer are indigenous to many developing countries.

In our view, to ensure that ESTs support all aspects of sustainable industrial development, developing country policies must change significantly in five ways:

- Environmental policies should assign a clear priority to cleaner production over end-of-pipe solutions. In China, for example, cleaner production in the form of preventive measures at a plastic additives mill cut the investment cost by 25% compared to an end-of-pipe solution. Total annual savings of US$750,000 meant a payback period of less than nine months.
- Technology policy should promote environmentally superior technologies, as urged, for example, by the World Resources Institute. This would recognize a hierarchy of technical change—from cleaning technology through process change, product redesign, and new infrastructure systems to fundamentally new technologies.
- Economic policy should explicitly recognize the multidimensional nature of sustainable development. China is proposing to do precisely this with its evaluation and adjustment of industrial policies for key industries to promote sustainable development.
- Institutional development should include an effective environmental regulatory program comprising discharge standards, permits, monitoring, and enforcement.
- Effective industrial extension services should support technological modernization and environmental change, while itself being supported by strengthened or newly created engineering research centers.

Given such changes, and with proper financial and technological support from the developed world, developing countries' industrial development need not add to the planet's environmental burden built up as a consequence of rich countries' industrialization.

PARTICIPATING IN THE TECHNOLOGICAL REVOLUTION

The second (and related) discontinuity concerns the way wealth is created and shared. Under the emerging paradigm, access to and application of knowledge are keys to further progress. Generation and application of knowledge, we predict, will redivide the world into haves and have-nots. With the right policies and strategies, developing countries can be among the haves.

New knowledge-intensive technologies will produce value-added products and services without unwanted by-products. Miniaturization, substitution of information for materials, and replacement of crude chemical processes with biotechnical ones already point the way. Cloned sheep and other large animals are already here, likewise genetically engineered crops that eliminate the environmental problems that come with insecticides and fertilizers. Plants that produce biodegradable plastics are just around the corner; the commercial use of nanotechnologies in health services, electronics, and other industries is but a few years away.

This unprecedented discontinuity for current technologies, markets, and businesses is potentially the biggest opportunity in the history of commerce and industry—truly an industrial revolution in the making. Within that revolution, application of knowledge in the form of technological innovation will become more important than access to physical and financial resources, achieving economies of scale and scope, or the ability to control markets and distribution systems. In essence, the levers with which rich countries and their large firms at present dominate world production and trade will pass into other hands.

This means an unprecedented opportunity for developing countries to do what development economists have long anticipated—namely, leapfrog in their use of technology. Leapfrogging is contingent, however, on policy commitment to radical improvement in technological innovation capabilities—in both qualitative and quantitative terms, and at government as well as company levels.

Typology studies of technological innovation distinguish the degree of management required to achieve such innovations and their impact on a firm's competitiveness and other success indicators. They show that solid technology management yields two types of innovation: incremental and strategic. Both are needed and it is important that they are not confused.

Incremental innovation leads to short-term gains in productivity, quality, and profitability—efficiency in resource use being a typical example. Such gains, however, do not guarantee sustainable competitiveness, ecological soundness, or returns on long-term investments. A company good at incremental innovation can be outflanked by a competitor with a better technology from fields completely outside the traditional business—witness the fate of the best of the 19th-century's saddle makers, who fell victim to bicycles and the internal combustion engine.

Strategic innovations result from a combination of technology foresight and market insight. Firms practice strategic innovation by positioning themselves to create the breakthrough new businesses and product categories thrown up by the new industrial revolution. They understand their role in shaping the future; they recognize complex trends, understand their implications, and develop the necessary core competencies that allow them to ride the new trends.

Management of the strategic technological innovation process is far from easy. It involves a blend of research, development, and technology transfer. It calls for multifunctional efforts that typically cut across organizational boundaries— boundaries within the company and between the company's suppliers and clients.

In this perspective, it is crucial that firms in developing countries learn to manage technological innovation, especially strategic innovation. They start, however, with major disadvantages in terms of costs, risks, and information gaps compared to their developed-country counterparts. As a result, the majority of small and medium-sized firms seriously underinvest in these aspects of their long-term survival. Essentially, they have to learn to learn, and market mechanisms are not going to teach them. Ensuring that private firms understand and practice technology innovation management is therefore a major agenda item of tomorrow's technology policies.

TECHNOLOGY POLICIES FOR TOMORROW'S WORLD

The third discontinuity concerns the inadequacy of most developing countries' present policies in the context of their need for technology-led development. Tomorrow's capacity for technological innovation will depend on today's technology policies. Today's policies must be informed by foresight and emphasize sustainable technologies, technology management, and people development.

The examples of South Korea and other Asian tiger economies show us how success hinges on having the right policies. Nevertheless, only recently have their policies focused on knowledge creation. In contrast, the policies of most developing countries are a long way behind—either not recognizing technology as a dynamic factor of production or lacking explicit policies to promote national technological development in any way. The challenge thus remains for nearly all developing countries to create and enhance policies and institutions that develop national firm-level capabilities not only to acquire and adapt good technologies, but also to capitalize on the growing discontinuity in the source of wealth by generating and applying new knowledge first.

New technology policies and industrial strategies must recognize threats as well as unique opportunities. Immediate threats to developing countries include their being overwhelmed by other forces shaping tomorrow's increasingly technological world. Globalization of industry will increase. Increased trade will follow World Trade Organization (WTO) requirements of lower tariffs and nontariff barriers. Production will increasingly conform to world standards of quality, environmental friendliness, and social equity—all heavily defined by developed countries. Technology itself will change even more rapidly with advanced technologies pervading the world even faster, challenging the competitiveness of even small firms that are geared only for domestic markets. Already by the year 2000, tariff levels on manufactured goods in most countries will be so low; these firms' protection against more efficient, lower-cost and better-managed competition will have all but disappeared. Developing-country governments and individual firms that do nothing in response face a bleak scenario of falling profits, pressure on wages, and eventual closure. Doing nothing on a national scale in developing countries will indeed mean deindustrialization—with the reversal of decades of industry-led growth and no thriving service sector to provide an employment and incomes substitute.

In this context, the role of the state, while reduced by deregulation and reform, remains crucial to technological development. An efficient government, capable of designing and implementing a consistent set of technology policies, and supporting industrial, educational, and technological institutions, is a prerequisite if the private sector is to remain competitive in the future. It will, however, require joint action by developing-country governments and their private sector: market mechanisms will take technology decisions involving risk; government programs in education, science, and technology will focus on closing the learning gaps that figure in the development and diffusion of new technologies in all countries, but which are infinitely larger in developing countries; any vertical programs—Malaysia's courageous promotion of its Multimedia Super Corridor and Korea's Highly Advanced National (HAN) R&D projects are examples—can only be developed in close consultation with the private sector, and with a risk-sharing approach.

FROM CYBER LABS TO CYBER CLASSROOMS

The final discontinuity is the breakdown of the traditional barriers between knowledge creators, distributors, and users. It proceeds with the information-technology-driven changes that we predict will be little short of revolutionary—deinstitutionalizing the universities and education systems of the future. When this happens, an impressive array of opportunities will open up for knowledge creators and incubators in developing countries.

Multimedia technologies, teleconferencing, computer networking, and related changes in the way knowledge diffuses through education and training already point the way. Their still more sophisticated successors and derivatives will allow researchers, teachers, and students to emulate present-day laboratories and classrooms, even to set up virtual research institutions, in cyberspace. Redefining the relationships between research and teaching, and cutting out the need to be in a certain place at a certain time in order to undertake research or learning, will vastly reduce the dominance of the great research and teaching institutions in industrialized countries. Such changes will encourage far greater involvement of the best of developing countries' scientists, engineers, and students in both basic and world-class applied research.

Developing countries that can establish the baseline requirements for participation will take advantage, as never before possible, of the deepening and broadening of accessible information. As a consequence they will participate in virtual laboratory work, for which they will be contractors as well as subcontractors. Their students meanwhile will profit from redefined learning systems featuring interactive learning and access to world-standard knowledge and teaching systems.

As the boundaries between knowledge creators, distributors, and users become porous, physical location and time will no longer determine accessibility to knowledge. Rewards will go increasingly not so much to the knowledge owners, but to those who use it most productively. Individual knowledge and specialized skills will assume greater importance than formal university education, which in turn will no longer be the sole qualification for such excellence. Industrial firms in developing countries will be able to become much more proactive in raising employees' individual knowledge and skill levels to meet their own productivity requirements. Their demands of universities will be less concerned with the flow, quality, and reliability of graduating students, and more with their provision of in-plant training and self-directed study such as distance learning. (Industry will also be looking to commercial providers of information and training, with emphasis on "just-in-time" training and skills development through new subcontracting schemes and on-the-job training.)

The universities themselves will emphasize thinking more than teaching and publishing. They will comprise bodies of researchers who are no longer tied to one entity or location. University-based research will be a platform for applications-oriented research at the private-sector level—in industry and at private (or privatized) industrial technology research institutes.

THE CHALLENGES OF OPPORTUNITIES

In summary, the four discontinuities, and the unique opportunities they open up for developing countries for sustainable development, technological and industrial leapfrogging, and education, are also enormous challenges for them, especially for the least developed countries of Africa and Asia. But even though they start with limited physical and institutional infrastructure and low and deteriorating skill levels, and now face the hardships of sudden globalization and trade liberalization, we believe they can share in the broad progress.

The crucial point is that developing countries and their private firms should take the alternative path of technology-led development, that they learn well the major lessons of the Macau international conference, and that they incorporate them in renewed technology policies and strategies that recognize that innovation is crucial for their sustainable industrial development. Then they will prosper and survive.

2

Toward an S&T Policy for the Knowledge-Based Society

João Caraça

THE EMERGENCE OF A COMMUNICATION PARADIGM, REVEALING AN "ARCHIPELAGO" OF KNOWLEDGE

We live in a complex world. The rate of change and the volume of communication in life today have no parallel at any period in the history of mankind. The growing development of industries based on information technologies, and the increasing importance of intangible or immaterial investment (R&D, software, education and training, organization, marketing, design) clearly indicate that the nature of the processes regulating economic activity is being profoundly altered.

We therefore need to rethink the economic analysis of human societies and their organizations, to introduce a new perspective that fully encompasses these intangible elements and their implications.

The basic regulators of information activities are not exchange (exchange used in a narrow economic sense: the exchange of tangible goods or the rendering of tangible services) and scarcity, but rather sharing and its limiting factor, misunderstanding. Following an "information transaction," both partners retain the information which was the object of the transaction (if, of course, the recipient was sufficiently capable of understanding it).

The increasing importance of sharing transactions reveals that the restrictions on the performance of modern economies are essentially inherent in the abilities of the operators themselves. Thus, endogenous (or accessible) knowledge potential, its form of organization, and the capacity to exploit it, are crucial elements for success and survival in the new economic environment. The main limiting factor encountered in this kind of transaction is not scarcity, but rather the effect of misunderstanding, of nonapprehension of communicated knowledge and information, which leads to an inability to generate relevant meaning at the receptor's level.

The economic impact of knowledge, arising from the globalization of the economy, forces us to reconsider ways of describing the space and dynamics of knowledge (Caraça and Carrilho 1994).

The system of classification we inherited from positivism—a pyramid with science at the top, aimed not only at the consecration of science as the model for all other fields of knowledge but also at establishing a corresponding hierarchy—is no longer adequate. A new factor has arisen from the emergence of the immaterial order in the realm of the hitherto material paradigm of progress and socioeconomic development.

Existing classifications of the fields of knowledge cannot be understood without reference to the societal context in which they were designed. In medieval times, when the concept of a central, finite space prevailed (the Earth at the center of the Universe), philosophy was put forward as the center of knowledge, surrounded by the seven liberal arts: grammar, rhetoric, dialectic, music, arithmetic, geometry, and astronomy. Later, enlightenment, corresponding to a highly advanced agricultural and commercial society, envisaged knowledge as a tree. The various fields developed as successive ramifications from a common stem, philosophy (three main branches of the tree were assumed: the science of God, the science of Nature, and the science of Man).

The success of industrial society, with the triumph of mechanics, railroads, and iron, brought with it a new rationale. In the positivist's pyramid, mathematics and the other "hard" sciences, in descending order, presided over philosophy, the humanities, and religion. This was the organization of knowledge which was transmitted to us and was undisputed until the 1960s.

However, from the standpoint of contemporary society, it has certainly become difficult to maintain rigid distinctions between different fields of knowledge, as we witness the emergence of a number of disciplines with composite names such as biochemistry and optoelectronics. Further, we have observed increasing internal complexity and autonomy in the various disciplines. Any attempt to think of contemporary fields of knowledge on the basis of their classical divisions and hierarchy is obviously risky, especially since important segments of contemporary knowledge, which are sometimes more innovative, may be excluded despite, or perhaps because of, the fact that they are unclassifiable in light of current criteria; examples of this are marketing and design.

In this sense, the metaphor of the archipelago of knowledge is useful and heuristically operative, particularly to the extent that it allows us to think, without any reductionism, about the linking of criteria and strategies that guide the internal thematization of its main areas. This approach leads us to a new descriptive understanding of the realm of knowledge, as an archipelago, suggesting a reticular form which does not postulate any common origin or accept any "natural" or functional hierarchy.

The loss of importance, if not of the aim itself, of arboreal or pyramidal conceptions of the fields of knowledge has been the most influential effect of the emergence of the immaterial paradigm. Further, it is a scheme that allows for and encompasses the creation of new disciplines. Such a network of disciplines carries clear implications for organizational behavior as well as for societal regulatory processes.

This arises from the need to take into account the profound changes which have marked the transformations occurring in economic activities—that is, the increase in intellectual investment compared to physical investment (Caspar and Afriat 1988). These have been accompanied by the growing role of complexity in the systemic framework (Nicolis and Prigogine 1989), which until recently was dominated by materiality, and by the emergence of *sharing* as the dominant process in communication.

The process of sharing, as it emerges from the immaterial paradigm, shows several specific characteristics typical of an activity which cannot be considered as an economic exchange, since it does not presuppose or imply ownership, and is a process which does not involve any loss. The nature of sharing stems from the fact that it is neither cumulative nor entails "expense;" from the fact that it is actually "enriched" the more it is exercised. Sharing is fundamentally a communicative-reticular process. It establishes networks. What this implies for society and its organizations is obvious: pure hierarchical-distributive interactions are no longer the sole legitimate modes of operation.

THE CENTRAL ROLE OF LANGUAGE IN COMMUNICATION

We now have to understand what knowledge is and where it is embodied. Knowledge can no longer be envisaged as a mysterious fluid, flowing in and out of countries and buildings, embedded in machines, encoded in documents, or even broadcast electronically. Knowledge resides in human bodies, like physical strength, and the whole issue revolves around the capacities of human bodies, organized in teams, firms, societies, or nations.

Unlike physical strength, however, knowledge is unfortunately not simply cumulative. Take two men of equal physical strength, and it is easy to understand that, given proper communication between them, they can combine their efforts to, say, pull a load of stones twice as heavy as one of them would be able to pull alone. What about the knowledge of those two men? Is it double the knowledge of one of them? Clearly not. Suppose they are both unskilled workers speaking the same language, then it is easy to see that the total knowledge of the "two-man system" may not differ much from the knowledge of each one of them.

Take as a further example a regiment of 10,000 soldiers. With proper training and coordination the physical effect of that regiment is very close to 10,000 times the power of a single soldier. What about the regiment's total military knowledge? Clearly, it is greater than that of the individual soldier, but equally clearly it cannot be obtained simply by multiplying by 10,000 the value corresponding to the soldier level. Here, we assume that the purpose of training, coordination, and command is to enable that regiment to function collectively (through the use of proper communication channels) with a "level" of knowledge much higher than that of individual soldiers. We assume that this level corresponds to the military knowledge of the general in command of that regiment.

Knowledge is not cumulative, unlike material things such as energy. Further, knowledge resides in each human body (with different levels and values depending on each individual person) and can be represented by the language, or languages, each human being employs.

It is tempting to ascribe a basic measure of the "quantity of knowledge" to the number of "words" each individual human being is able to use in his or her own daily and professional life. In this case, the military knowledge of a general is maybe only 100 times greater than that of each private soldier in the ranks (500 words for a soldier versus 50,000 words or "memory positions" for a general; these 50,000 words must necessarily encompass strategic options, visions of the future, and other ways of generating group cohesion). But it is this language advantage, provided communication channels (i.e., officers, sergeants, and corporals) function adequately, that allows a general to coordinate a physical power 10,000 times greater, supposedly, than his own. Historical examples of famous generals are very instructive. Not one of them neglected to carry out a personal inspection of the battlefront before launching an offensive. The history of battles and of their outcomes is surely a field of immense interest in this respect (Caraça and Simões 1995).

We thus see the power of knowledge, language, and communication. The role of language is not only that of a *medium* to communicate, enabling people to relate to the world around them, including their fellow human beings, but also that of a *repository* of knowledge, representing the capacity to force and direct humans' interaction with the world, and thus their ability to survive.

Knowledge thus arises as the outcome of successful communication with the outside world; it consists in a repository of *relationships* between subjects to which an individual attributes meaning, which is always linked not only with the contexts from which it emerges, but also with the web of beliefs of the intervening subjects. It is this situation that leads us to regard the creative process as a self-communicating one involving conjecture and verification—that is, a process in which from the ever-open repository of existing relationships, other relationships are proposed, which are capable of being conjecturally translated into new meanings. We can thus understand how the creative process can be accelerated by "imitation" or assimilation as a result of the socialization of communication, as well as how aspects which are rooted in the domain of the individual coexist with those which are relevant in terms of a perception of the collective.

The purpose of organizations, of institutions, is to enable (through communication) the creation of higher, more complex languages, and the "election" to command of a person who generates and manages a higher repository of knowledge. Obviously, the dominance relationship between power and knowledge frequently overshadows the clarity of this picture.

It has been pointed out (Caraça and Carrilho 1994) that sharing regulates the relationship of humanity with the world, according to three types of knowledge, which correspond to an increasing degree of complexity.

First, tacit knowledge, governing the relationship of an individual with the world as a whole (the outside world and his or her own group), experienced particularly as a confrontation of two orders, the "objective" and the "subjective." This is the level of organization of knowledge which corresponds to what can be defined as "common knowledge," the type of knowledge which we are not taught explicitly (Polanyi 1967), but which we learn by "exposure" in society or in our own groups. It also involves abilities and skills (Howells 1994). This level has evolved over historical time, but not in a simple linear progression.

Second, there is a level of *explicit knowledge* in which language emerges as its definite operator, through the creation of "specialized languages" leading to the affirmation of the identity and diversity of groups within a community, and corresponding to a growing level of complexity in the interaction between an individual and the world: that of "intersubjectivity." Explicit knowledge is associated with a regime of specialized information.

Third, emerging from the level of explicit knowledge through a constant process of increasing complexity in the relationship between an individual and the world, the densification and intensification of communication processes leads to intersubjectivity being replaced by a wider "interactivity," which in turn corresponds to languages of greater precision. This is the level of *disciplinary knowledge*, the context in which disciplines (e.g., sciences, philosophy, ethics, aesthetics) appear. Disciplines are associated with larger and highly developed repositories of meanings (Caraça and Carrilho 1996).

It is in this light that we must treat the effect of knowledge and communication in any area. For instance, if we are dealing with economic activity, we have to understand what the intellectual levels of the various sections of the population are, how the diverse institutions are organized, and which rules of overall coordination and operation of the economic system are being used. High-level repositories of knowledge are effective only if appropriate institutions are created, or are operating, which make full use of their specific meanings, values, and perceptions.

KNOWLEDGE-BASED SOCIETY IN NEED OF NEW SCIENCE POLICY INSTRUMENTS

The term "knowledge-based society" has become pervasive in contemporary literature in the socioeconomic field. It seems that no good definition, clear cultural contents, or behavioral modes have yet been fully established with regard to this new "object."

However, the expression "knowledge-based economy" has been somewhat validated and is now entering the vocabulary of respectable economists, new growth theorists, and information society "gurus." International organizations have been keen to investigate and launch activities designed to shed light on these new ways of operation in advanced economies (OECD 1996). The science and technology system is envisaged as carrying out key functions in the knowledge-based economy. Its performance and relevance will thus be under close public scrutiny. But new policies will have to be devised: let us understand why.

In the 1960s, a universal model for S&T was accepted that corresponded to an instrumental concept of scientific endeavor in relation to social and economic development. More investment in S&T was supposed to bring about a greater capacity to generate new wealth. Nowadays, however, we know that it is not possible to isolate research activities from the social context in which they are conducted; this is reflected in the growing "scientification" of the culture of contemporary societies, as well as by the increasing societal involvement of scientists and researchers and their organizations. A new need has been created: the need to ensure that public funds spent on S&T are used in a beneficial way for society—with evaluation performing the primary role of mediator in this process.

This is why the nature of science policies at the national level has changed from being mission-oriented (from the launch of strategic sectors to emphasis on the generation of technological innovations and support of national "champions") to being more diffusion-oriented, through enhancing the mechanisms of knowledge transmission and technology transfer and the exploitation of research results.

But the role of government in a modern economy is incompatible with conducting operations too close to the market. So, what is being left to government policy and action is the building-up of infrastructure (including the development of human resources—the notion of human capital); the support of networking activities (and hence the concept of human mobility); the financing of research programs in basic "pervasive" technologies (prompting the notion of generic technologies); and the provision of S&T services, education, and training at the national level (leading to the concepts of public understanding of S&T and of scientific literacy).

This is clearly too reactive and inadequate in terms of public S&T policy in this new era of an emerging epoch in which the new "knowledge society" is relentlessly replacing its main operative paradigm. Knowledge and learning will be the central resources and mechanisms of the new nations, communities, and organizations.

Science policy will have to be closely linked to policies in all other fields of knowledge, from the arts and humanities to the cognitive and social sciences. Further, implications of the free circulation of knowledge will have to be recognized and fostered: disciplinary knowledge (such as sciences, philosophy, aesthetics) can only evolve inside a strong communications framework which will enable the operation of sharing transactions to bear their full potential. And their relations with the body of explicit knowledge (technologies, specialized crafts, laws and regulations, management, fine arts) will need special attention in terms of the diffusion of new learning abilities throughout civil society, leading to the buildup of better capacities.

Learning will have to be seen as a genuinely life-long activity, from cradle to grave. Knowledge and the human condition will then enact a true marriage: till death does them part. The greatest barriers to this change stem from short-term vested interests and from misconceptions concerning the long-term interaction of knowledge and power.

Science has nurtured, enabled, and supported this transition. Let us now direct S&T policy, which is barely 50 years old, toward preparing the ground for the enlightened, humanistic perspective that will be crucial in guiding humanity through the next 50 years.

REFERENCES

Caraça, J. M. G., and M. M. Carrilho. "A New Paradigm in the Organization of Knowledge," *Futures* 26, 7 (1994) 781.

Caraça, J. M. G., and M. M. Carrilho. "The Role of Sharing in the Circulation of Knowledge," *Futures* 28, 8 (1996) 771.

Caraça, J. M. G., and V. C. Simões. "The New Economy and its Implications for International Organizations." Paper contributed to the European International Business Association Conference, Urbino, Italy, 1995.

Caspar, P., and C. Afriat. *L'Investissement Intellectuel.* Paris: Economica, 1988.

Howells, J. "Tacit Knowledge and Technology Transfer." Paper contributed to the European Innovation Monitoring Survey Workshop, Luxembourg, 1994.

Nicolis, G., and I. Prigogine. *Exploring Complexity: An Introduction.* New York: W.H. Freeman Co., 1989.

OECD. *Science, Technology and Industry Outlook.* Paris: OECD, 1996.

Polanyi, M. *The Tacit Dimension.* London: Routledge and Kegan Paul, 1967.

3

Corporate Knowledge: The Emergence of Knowledge Communities in Business

Jim Botkin

DRIVING FORCES

We are currently in the fourth era of economic life and already halfway through it. The previous economies were hunting and gathering, agriculture, and industry, and the current one, the information age, passed its halfway point in 1995, the year of the Internet and Windows 95. In 1995, more Web browsers were sold than word processors. We project the information age will be superseded around 2025 by the bioeconomy age. In the meantime, the professional lives of the members of the current generation will be determined for better and for worse by the search for knowledge trapped in information overload.

In this third quarter, lasting about a quarter-century, organizational learning will be the method, knowledge management the service, and smart products the offerings in an enterprise called "knowledge-based business." The structures emerging to meet the needs of knowledge-based business are called "knowledge communities." Over the next decades, knowledge communities will replace most hierarchical industrial models that characterize many companies today.

The success of knowledge communities, in turn, is highly dependent on connection. The ability to connect electronically and personally, things to things as well as people to people determines how effective a knowledge community is in meeting the challenges of the coming quarter-century.

When the infrastructure rumbles, everything else either changes or crumbles. The technology infrastructure is rumbling a lot these days, and it's changing everything. Changes in technology are resulting in smart products, increased competition, and radically changed cost structures.

Smith & Wesson makes a Smart Gun™ that will not fire unless its owner is wearing an electronic bracelet five inches from the trigger. In the United States, this may do more to reduce accidental shootings than any law banning handguns.

The chairman of Canada's premier bank, CIBC, met with the president of Intuit Software. The software president was asking the bank chairman for a listing of all the bank's customers for Intuit to link its home banking software. Understandably reluctant to release his customer list, the chairman was stymied when he realized his would be the only bank in Canada that would not provide this nationwide home banking service. CIBC suddenly had a new competitor, and it wasn't even another bank.

A man in Minnesota found a way to play music over the Internet. Usually it takes about $22 million to buy or start a new radio station. He started his for $200,000—1% the usual startup cost—and now operates Radio.Net with a worldwide audience of millions, no government regulation, no on-going costs, and he has advertisers. Wow, does he have advertisers!

Examples like these and countless others illustrate that the technological infrastructure is shifting from crunching (done by early computers) to connecting (available from advanced communications). If the three main laws of real estate are location, location, location, then the three main aims of the late information age are connect, connect, connect. In the early information age, the strategy was to overinvest in million of instructions per second (MIPS), or computer power. In the late information age, the strategy is to overinvest in bandwidth, or the power to connect.

KNOWLEDGE COMMUNITY: WORKING DEFINITIONS

What is knowledge? Knowledge is information put to productive use. Information is data arranged into meaningful sequences, and data are simply descriptions of things. To use an analogy, data are snapshots in time, information is a scrapbook, and knowledge is a blockbuster movie.

Linking knowledge to productive use is clearly not a rigorous or all-encompassing definition. Indeed, many universities pride themselves on creating knowledge for the sake of knowledge itself, whether it has any useful application or not. This is also the differentiation implied in pure versus applied research. Nonetheless, for the present, we are in a world and domain where knowledge is prized for its use, and the more broadly it is applied, the more highly it is prized.

New knowledge is often associated with science and with brilliant scientists. Albert Einstein and the theory of relativity come to mind, linking the notion of individual genius to knowledge creation. The desire to "anthropomorphize" new knowledge by associating it with an individual creator, however pleasing, is at variance with the facts. New knowledge today is seldom an individual act: it comes from building on the learnings and discoveries of others. Nobel prizes, particularly in medicine and science, are seldom awarded to individuals but rather to teams of colleagues.

This networking is essential for most professionals today because no one of them is capable of following all the developments that relate to any one specialization. Knowledge is said to double every seven years; this is not particularly new, as this represents a growth of only 10% a year. What really makes it hard to keep up is that the interconnections from one field to another have multiplied geometrically.

An analogy is that in the first half of the information age (1953–1995) we were concentrating on growing the knowledge base by crunching with our computers as individual researchers. In the second half we are concentrating our time on connecting one body of knowledge with others. Thus we find communities of research people working together.

What is a community? It is a group of people sharing a common interest or culture. Operative words are sharing, participation, and fellowship. Often a community lives in a common locale, but with our power to connect, virtual communities are becoming common. We speak of "the scientific community" or "the international business community" as people sharing a common profession. A "community of interests" is another way the word is used. Community problems or community facilities are yet others, sometimes implying society as a whole, joint ownership, or common goals.

The Institute for Research on Learning (IRL) in Menlo Park, California speaks of "learning" as occurring in "communities of practice." This perspective is a useful one that is helpful in conceptualizing why it is that knowledge, learning, and even technology are so difficult to transfer. The difficulty is that most communities have boundaries, which only "brokers and ambassadors" can cross. Cross-functional teams are popular in business because they transcend community boundaries. Teams are groups organized to work together. In sports, they are players on the same side. Teams imply more of a hierarchy than communities and a more limited focus, but they share some similar characteristics.

A knowledge community is a group of people sharing an interest in the use of knowledge for a specific purpose. The purpose may be to create new knowledge, as most research labs do, or it may be the application of knowledge, as many business communities want to do, or both. InterClass, the International Corporate Learning Association, is a knowledge community that does both. Its members meet every quarter to deepen their understanding about new knowledge and how to apply it in their work. Knowledge communities are often by necessity cross-functional. Many, like InterClass, are also cross-industry.

AT&T, the U.S.-based telecommunications giant, has a small team whose job is to create knowledge communities within AT&T. Their definition is more business focused: A cross-functional group of people accountable for business results within a defined market opportunity. They are geographically dispersed, encompass 70 to 700 people, and are cross-functional so that they include managers from sales, marketing, customer care, and others. It is the assignment of the knowledge communities team to create and launch eight such knowledge communities for AT&T in the next 12 months.

KPMG Peat Marwick, one of the premier professional services firms based in the United States, defines knowledge community in terms of its own business. It is market-smart professionals in the firm's five primary markets of x segments and y products and services, linked by intranet technology, which allows them to share resources, proposals, facts, and most important, learning.

The World Bank is exploring redefinitions of its ways of delivering services, which for the first 50 years of its existence focused on granting loans usually for large infrastructure projects in developing countries. In the process of evaluating loans, the Bank has acquired an immense repository of knowledge on best practices for economic development.

This knowledge is not always easily available to those who need it—inside and outside the Bank—when they need it, or in formats they find useful and accessible. As a result, the effectiveness of our service suffers. To address this problem, we will build a world-class knowledge management system throughout the Bank to capture and organize our knowledge, make it more readily accessible to staff, clients and partners and strengthen the knowledge dissemination and capacity building efforts of [the Bank]. In so doing, the Bank and its staff, clients, and partners can become a knowledge community.

TECHNOLOGY FOR KNOWLEDGE COMMUNITIES

Knowledge communities need information technology like fish need water. Usually the technology takes the form of an intranet, the Internet, or an extranet. AT&T and KPMG Peat Marwick use an intranet that connects the internal members of the corporate knowledge community to one another "7×24" or 24 hours a day, 7 days a week.

AT&T uses technology practices that are based on home pages, conference calls, symposia, and personal learning plans. KPMG looks for the technologies that can acquire, store, add value, and disseminate information as a first step in communicating knowledge. They have identified the key success drivers and have instituted a current rollout plan.

The advantage of an intranet is that it has higher speed and higher bandwidth. While it is protected by a firewall from outside viruses or unwanted participants, a disadvantage is that it is not able to incorporate knowledge from external partners or customers.

An extranet is technology that connects multiple intranets from partnering companies such as InterClass does for its 20 member companies. It may do so directly or more likely indirectly by using the Internet. So, for example, InterClass has a home page on the Internet and most of it is accessible only to its member companies in a password-protected "members only" section.

Remember when you were a child and had to repeat the password before your buddies would let you into the tree house or playhouse? It is interesting that passwords are back as the most practical way to determine who is a member of the community and who is not. But it is a pretty unsophisticated and childish way of making such determinations. Extranet-based communities walk a fine line between wanting openness to new ideas and new potential members yet enough closeness ("closedness?") to feel secure to engage in trial-and-error learning.

"If I know a competitor is in the room (virtual or actual), then I am only going to share the top 5% of my knowledge rather than all of it," is a statement typical of many members of the InterClass community. The problem with this statement is that the nature of competition keeps changing, and it is difficult to determine from day to day who the competitor is. AT&T for years considered Motorola a competitor. Since the recent voluntary trivestiture of AT&T, now the Motorola competition is with Lucent Technologies and the new AT&T sees Motorola as its largest supplier and a best customer.

What's an alternative to the password system? At the MIT Media Lab, scientists are developing a computer in a shoe (the Nike basketball sneaker!) that carries your personal and professional identification. This is part of the wearable computer initiative that crosscuts the Digital Life and Things That Think sections of the Lab. It is not so much the computer in a shoe that is important. It is that the information in the shoe can be transmitted via the human body to another shoe. It turns out that flesh and bones make reasonably high bandwidth transmission lines—100,000 bits per second. This means that you shake hands as a way of transmitting the contents of your Nike sneaker to the other person's sneaker.

While this may seem exotic now, it may seem less so when we realize that personal contact and trust are two important cultural aspects of a knowledge community. Perhaps in the MIT prototype may lie answers that marry technology and culture in important ways.

CULTURE AND KNOWLEDGE COMMUNITY

Any community takes on its own distinctive culture. A culture includes the norms, behaviors, and practices of the community members. As an example, the InterClass norms are confidentiality, learn from each other (benchmarking), learn by experience (failures as well as successes), truthful and thoughtful feedback (consultanices), learning by doing (site visits, interactive presentations), hard on ideas/soft on people, and trust.

The strong culture of openness and sharing can seem to conflict with trust and confidentiality. There is always a tension around "inclusion" and a seeming undue concern with "who is in the room." There is an inevitable degree of tension between the closeness of the core group and the need to include new participants and be open to new perspectives.

Time—or lack of it—is the greatest danger to the success of a knowledge community. Said more positively, active participation is one measuring stick of a successful community; a second mark of success is implementation of new information within a member's own organization.

Much has been made of the culture of "learning organizations" and how to achieve them (Senge et al. 1990). The InterClass learning culture can be described as:

- ask good questions
- be comfortable with not knowing and not finding solutions
- practice "hard on ideas/soft on people"
- value learning by doing—interactive presentations and minimal lecturing
- welcome brainstorming
- be open to sharing problems as well as successes
- be flexible—change what is not working
- be respectful of one another's time—no long monologues, leave time for others to share ideas too

Cultural implications for KPMG include the following: The company is dealing with management consulting, which may be the purest form of knowledge-based business, since it consists of transferring knowledge from one group to another. KPMG stresses the need to protect confidential information, a task that can be complicated by high turnover rates. Analytic knowledge, or the process of deriving knowledge, is the most important resource and the least accessible since it resides in people's heads and past experiences and training. There is a need to communicate across platforms, borders, and cultures, as well as to respond in real time.

AT&T has a two-part strategy: Increase Community Competency and Sell Offers that Delight Customers. Under the first would be learning, employee selection and turnover, access to information, and adoption of best practices. Under the second would be shared vision, aligned goals, and business process excellence.

The more information becomes an infinite resource, the more attention becomes a scarce resource. A good way to get attention is to appeal to emotions, or to share "emotional intelligence." In his recent book, Dan Goleman (1995) defines emotional intelligence as the ability to perceive your feelings as they are happening and to craft an appropriate response in real time.

Some corporations are more open to emotions than others, and the same can probably be said of country cultures. At InterClass, members believe that "make room for emotions" may be one of the most important vectors for 21st-century business and technology policy (see Davis and Botkin forthcoming).

REFERENCES

Davis, S., and J. Botkin. *Knowledge Communities: The New Generation of Smart Business*. New York: The Free Press, forthcoming.

Goleman, D. *Emotional Intelligence*. Bantam Books, 1995.

Senge, P. et al. *The Fifth Discipline* and *Fifth Discipline Field Book*. New York: Doubleday, 1990.

4

Knowledge Integration Across Borders: How to Manage the Transition?

Paul Verdin and Nick Van Heck

KNOWLEDGE, MANAGEMENT, AND INTEGRATION

While traditional strategy and management models concentrated on product market attractiveness and appropriate positioning of companies in the market and industry as a basis for company success, recent changes in the technological, political, and sociological environments of businesses have forced strategists to draw more attention to the "internal" view of business organizations. In this often more dynamic view of strategy, the foundation for sustainable competitive advantage lies in the current and future resource base, and especially the intangible and tacit resources, of the company[1] and its capability to grow and nurture them better and faster than the competition.

In light of this evolution, it is no surprise that research and framework development within the strategic management field has recently embarked on a "knowledge management route." Knowledge is called "the most strategically significant resource of the firm" (Grant 1996).

As a consequence, the demand for *innovative* and *integrative* capabilities in the company has dramatically increased. It is frequently said that top managers are concerned about the pressure for increasing sophistication and specialization of the company's knowledge. In addition, the ability to *share* and *coordinate* company-specific know-how, learning, and innovation between different parts of the organization is acknowledged to be of increasing importance. Not only is knowledge the crucial, or even sole, base for competitive advantage, the ability to share knowledge and learning, across functions and country borders, has grown to be a "strategic imperative" for any (international) company in the 1990s (Bartlett and Ghoshal 1989). A recent article, entitled "Making Local Knowledge Global," is a clear illustration of the complexity and difficulties, and the crucial importance of sharing learning and know-how across country borders (Cerny 1996). The role of the leader and manager of the organization has accordingly evolved toward "building a learning organization" (Senge 1995).

Company Integration

Bartlett and Ghoshal (1989) showed that in addition to the company's ability to exploit and leverage worldwide learning, innovation, and knowledge, global efficiency and local responsiveness are key success factors for international companies (see Figure 4.1). The art of international business is to *integrate* and *coordinate* the company's activities, resources, and knowledge, cross-functionally and across country borders, while staying or becoming locally responsive. It has frequently been argued that the pure multidomestic approach, with its widely dispersed and duplicated activities, resources, and knowledge, is completely outdated; an integrated and coordinated network approach to organization is required for success in the global information age.

Figure 4.1
Managing Integration: The Strategic Imperatives

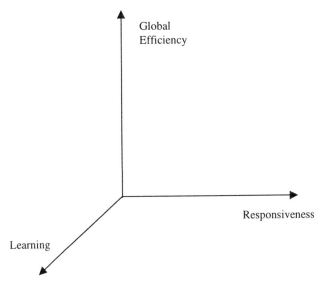

Source: Based on Bartlett and Ghoshal 1989.

In line with this view, today's business news is overwhelmed with articles about the integration challenges of various companies, especially in Europe. It seems that the recent pushes toward more deregulation, market convergence, and liberalization (especially within the European market: "Europe 1992," European Monetary Union, etc.), together with the overall globalization of the economy, have created new opportunities for increasing efficiency through coordination, and even centralization, of international activities and resources, and for worldwide and regional leveraging of innovation, learning, and knowledge.

While the specific drivers and requirements for internationalization and cross-border integration may vary by industry and business (as discussed below), it is clear these days that in most industries and companies, and especially those operating in Europe, more rather than less integration is required, and many companies have only just begun to make this happen in their organizations. Others embarked much earlier but found many pitfalls and frustrations. All of this is driven largely by lagging profitability (fragmentation; duplication of activities, resources, and knowledge; and NIH [not invented here] are significant causes of this[2]), increasing customer pressure (following the customer's internationalization or integration), and pressure on "time to market."

While various researchers have taken different angles and perspectives on this company integration debate and extensive discussions have taken place about why (not) to integrate and about the organizational requirements of an integrated company structure, relatively little research has been conducted on the management issues of company integration in a European or regional context. Even less attention has been paid to a related question, at the heart of the authors' ongoing research: how do companies achieve more and better integration? How can executives manage the transition from the existing organization to the new one? How can they manage the process?

Position of Research

In an earlier contribution (De Koning, Verdin, and Williamson 1997), the authors focused on the process management issue. One could indeed argue that their observations so far have focused especially on integration *per se* and less on the *learning* (which refers to the extent of worldwide or regional leveraging of innovation and knowledge, as indicated in Figure 4.1 on the "learning" axis). Nevertheless, it is frequently observed that opportunities along the "learning" axis have been the most important in driving cross-border integration in some cases, especially in what is often referred to as multilocal businesses. Various service businesses have embarked on explicit internationalization and integration, not because the key benefit for them was in reducing costs and/or eliminating duplication, but because primary opportunities and benefits for cross-border learning exist (e.g., the cases of Vedior International in the European temporary work services business [Van Heck and Verdin 1996b] and Eureko in the European financial services industry [Freeman and Verdin 1997]).

When the Integration (or Global Efficiency) axis was the most important dimension along which integration and internationalization was achieved, even then it turned out that initiatives to boost efficiency resulted in and/or were supported by improvement of worldwide learning. In that sense, one could consider the learning axis a dimension to push forward the integration for global efficiency.

While some of the objects we initially observed had only implicit (or at least less explicit) consequences for knowledge integration, the question remains whether the initial findings, mostly although certainly not exclusively, collected from integration cases with the objective of increasing global efficiency, apply when the explicit objective is to better integrate knowledge, innovation, and learning. In other words, we see knowledge and learning more broadly than solely as a dimension to push forward integration. The question under consideration in this contribution is: what can be learned from observations of company integration and management of the process, for making cross-border knowledge management and integration happen?

Although the observations were not limited to company integration in Europe, the issue seems particularly "hot" in a European context. Corporate Europe has been experiencing drastic changes in its business environment and has hence been exposed to various opportunities for cross-border leveraging of company knowledge, dispersed around European and worldwide subsidiaries.

Recent research has recognized strong regional (for example European) organizations as a key step in the globalization process (Malnight 1996). We would argue, however, that strong regional organizations have a role beyond the proverbial "transit stop" on the road to full globalization. As the world seems to be polarizing into ever stronger and more specific regional trading blocs, the importance of regional strategy and organization building has been recognized as an increasingly significant goal in its own right (witness the early calls made, for example, by Morrison, Ricks, and Roth 1991) and not just as some "transitory" state from national to global organizations (this point is further developed in De Koning, Subramanian, and Verdin 1997).

We used a longitudinal case study methodology to explore the managerial process of shifting from national subsidiaries to global or regional organizations. The core of the initial research project originated from in-depth field research at Procter & Gamble (P&G) Europe (Bartlett, De Koning, and Verdin 1997),[3] together with other integration examples like 3M (Van Heck and Verdin 1996a). The observations and findings from this were checked and deepened, making use of other case studies which had more emphasis on the *learning* and *knowledge* integration agenda: McKinsey and Company (Bartlett 1996) and Alcatel (Bonheure and De Meyer 1992; Bogaert, De Meyer, and Verdin 1997). All these served as the basis for the investigation and allowed a richer understanding to be built of the complex organizational changes that lead the way for the creation of integrated organizations, worldwide as well as in Europe.

Describing some of the insights gained and lessons learned from their ongoing research, we build a framework to help managers understand the ways in which their chosen process of knowledge integration influences the final outcome, to lay out the different options that are available, and to help them make choices both in terms of achieving their organizational goals and in identifying efficient and effective mechanisms for getting there. As the existing "old" sources of competitive advantage (like differentiation or cost leadership) fade, few managers need to be convinced of the importance of knowledge and the creation of it. Or as one said: "The belief is that the process by which knowledge is created and *utilized* in organizations may be the inimitable resource for creating sustainable rents" (Schendel 1996). Hence the interest in understanding how to make knowledge integration happen.

The Critical Role of Process: How to Get There?

Integration in global or regional blocs is often far less easily achieved than planned. Initial responses by many companies have led to the creation of global or regional headquarters, and integrating some parts of the value chain or introducing some kind of cross-border task forces (e.g., "Euro-teams," activity groups, or coordination centers; Schütte 1996). But in practice, few companies have achieved a high level of European integration simply by adopting these kinds of initiatives (Bleackley and Williamson 1995)[4] or have succeeded at all. Integration requires a great deal more than redrawing organizational boxes or creating new ones.

Some companies have chosen a slow but steady route to integration and have spent an impressive amount of time and energy to reach results, which could probably have been achieved much more quickly and more efficiently. Some decided, being forced by crisis, to drastically reorganize into cross-border structures. Others enthusiastically headed down the road to integration only to find out that they have become stranded halfway toward their goal, stuck in the middle as neither fish nor fowl. Still others emphasize the unexpectedly high cost of integration, in terms of the time and effort required. The general consensus is that actually capturing the benefits of highly integrated organizations turns out to be much more difficult than imagining how a shiny new global or regional configuration should look.

We believe that a large part of the problem can be traced back to insufficient attention to the planning and implementation of the process of integration as well as the overall framework within which it is to take place. The importance of the processes deployed to promote greater interaction has recently been flagged in the context of headquarters-subsidiary relations and the role of regional headquarters (Shütte 1996). Going beyond questions of the position and role of regional or global headquarters within multinationals, they find that the issue of what road should be traveled and how when changing the organization is critical in a broader context.

These findings are in line with research results on integration of mergers and acquisitions, which show how important preparation and execution of the integration process are for achieving the intended synergies and worldwide learning (Haspeslagh and Jemison 1991). In the broader field of corporate strategy, a growing literature on corporate transformation focuses on the importance, effectiveness, and requirements of alternative change processes (Strebel 1994;

Kotter 1995; Rumelt 1995; Chakravarthy 1996b). We agree with these researchers that change processes have "path-dependent qualities": in other words, that the pace, type, and style of initiatives have an impact both on what types of outcomes companies can achieve, and on their relative success in reaching their goals.

CASES

Described below is a "European integration" example in the *P&G Europe* case, while global knowledge integration is the central theme in the *McKinsey and Company* and *Alcatel* cases. Although more often the best insights can be gained from studying failures and pitfalls like Alcatel's struggle to integrate the company, the cases below set out to illustrate how the effective combination of various management aspects within some companies like McKinsey and Company resulted in a knowledge-focused and integrated organization. In addition, these cases illustrate the importance of knowledge management and integration in different kinds of industries. Illustrated by these cases, the key findings will be elaborated on later in this chapter. Complementary examples will be indicated in that section as well.

Integrating Europe at Procter & Gamble

Procter & Gamble (P&G) moved into European markets in 1932, starting with a U.K. acquisition. In the mid-1950s, P&G expanded into continental Europe. In these early stages of P&G's internationalization, each subsidiary was structured as a microcosm of P&G in the United States, including the full range of functions. All P&G general managers in Europe had a mission to adapt P&G's proven products for their local country market, and to use P&G's brand management approach to gain leadership in the local market. P&G had only a small European headquarters to overview the subsidiary activity. One major role of the headquarters was managing the trademarks and brand names for Europe.

By 1970 P&G had achieved a significant presence in most West European countries, although the relative market position and product range varied widely from country to country. However, P&G executives were not satisfied with these achievements. P&G Europe was under pressure from corporate headquarters in Cincinnati to improve financial performance: better economies of scale, speed to market, transfer of successful ideas, and overall effectiveness became important. What was needed was a new European perspective to solve these problems.

P&G decided to establish the European Technical Center (ETC) in Brussels, and initially the main focus of ETC was on common development programs. The first attempt for better European coordination, however, proved disastrous. Thereafter, P&G Europe hesitated to force the country managers to adopt new policies and, instead, a more voluntary approach to integration was adopted.

P&G's voluntary approach to integration was implemented by the formation of diverse project teams through which it created an ad hoc, matrix structure. R&D was the first function to take a clear step toward European integration. In 1977,

European Technical Teams were introduced with the goal of reducing development costs and leveraging the particular strong product capabilities owned by the local R&D departments. P&G continued to build on this initiative through the 1980s with a number of informal structures (such as a lead country responsibility for specific brands, etc.). Nonetheless, most of the R&D staff were located in the country subsidiaries and reported to the local country managers who paid their salaries.

From the marketing side, many European initiatives were started as well. Euro Brand Teams, drawing from the country marketing managers, were created to deal with specific issues. In the early 1980s, Euro Brand Teams (e.g., Vizir) membership was essentially voluntary and hence turnover in the teams very high. Using "center of excellence" logic, teams were led by a country-based marketing or country manager. Team decisions were all subject to ratification (and adaptation!) by the country organizations, which continued to keep P&G responsibility and remained the key arena for career development.

The project teams gradually included and were later led by the fast-growing cadre of ETC staff and senior management. By the end of the 1980s, team decisions had become less and less negotiable. As a next step in the integration process, P&G gradually centralized many functions, shifting reporting relationships from the country general manager to European management and executives (often moving through a matrix structure as an intermediate stage). This happened at different points in time for the different functions. The earliest steps were taken by the R&D function in 1977. By 1989 centralization was largely completed. The same process was used first in Purchasing, then in Manufacturing and Engineering, for all of Europe. Functional executives used this change process to leverage the new critical mass for greater effectiveness and efficiency. R&D, for example, grouped researchers by product categories, rather than countries and brands, to improve focus, learning, and productivity. None of these changes happened without pain or frustration, even if (as in R&D) the pressure for knowledge sharing was clear, the benefits obvious and "buyable" by all.

As a result, by 1990, the role of the general and other country managers had changed dramatically: with fewer functions and less autonomy, the general managers were given greater responsibility for public and government relations, as well as continuing their strong focus on the sales and marketing aspects of the business. Within a decade, they were forced to depend on a pan-European organization for product supply, product development, and consumer research, and also to absorb the allocated costs of the system. While P&L responsibility remained fully theirs for a long time, a system of shared responsibility and performance measurement was now being put in place.

Overview of Integration Process. The changes in P&G Europe and the country organizations represented a fundamental shift in P&G's structure and culture. Years later, the changes seem logical, yet executives and managers throughout P&G often recall the many doubts and passionate debates that surrounded the shift. At one level, the process of integration looks like emergent strategy and "muddling through." Yet, in retrospect, one can see a coherence in the many major and minor decisions that drove integration forward.

P&G's early organization in Europe was a loose federation of autonomous country organizations, within the International Division. The European perspective was virtually invisible: know-how and innovations were almost exclusively American. As P&G's leadership moved toward a regional organization, the vision was not fixed to dates or a specific structure. Two streams of decisions evolved from that point, one building a European perspective and cross-national coordinating abilities, and the other building a unified European infrastructure of information technology, finance, incentives, and other essentials. This infrastructure supported the broader coordination of business and knowledge and allowed managers to become adept at communicating across cultural barriers and to gain practical knowledge of the specifics of different markets. By beginning early, and holding a long-term vision, P&G was able to succeed in both aspects, while continuing to build revenues and profits.

The main vehicle for building the European perspective and capabilities among managers was the cross-national ad hoc structure of project teams. These teams began as voluntary projects, and over the years shifted to a standard and formal structure led by ETC management.

By then, sufficient time had passed to ensure that the new country managers had been groomed in the new system and the old country managers had either left or moved up and on to become part of the new integrated European organization.

Although the main challenge for P&G was to improve its position on the integration axis (see Figure 4.1) (e.g., through coordinating and centralizing purchasing, R&D, and others), some observations can be made of particular interest in the context of the current investigation. First of all, it turned out that the knowledge management or R&D function was not only the first function to be integrated but, with hindsight, can be considered key for the (success of the) overall integration process. One could say that some of the knowledge and R&D integration initiatives acted as "catalysts" (as discussed below) for integration in various other functions and fields. Conversely, it is clear that overall integration has affected P&G's position on worldwide learning as well. Although one could consider the improvements in worldwide or regional learning as a coincidental by-product of overall integration, it seems that in the process, the improvements in regional learning did not just pop up but have supported the overall agenda of company integration.

As a result of P&G's integration efforts, the time-to-market, quality of products, cost basis, and competitive position of the company improved. In addition, the flow of knowledge and innovation(s) has drastically changed over time: first, from one-way transfer of know-how from the American headquarters in the direction of Europe, to two-way transfer; and second, from local country-based know-how to shared and coordinated, pan-European learning.

Integrating Knowledge at McKinsey and Company

Since its beginning in 1926, McKinsey's image in the market had evolved from a company of "business doctors and efficiency experts" toward a highly respected and well-established consulting firm in the 1950s–1960s. Their international network had rapidly expanded in the 1960s and resulted in a solid presence in Europe and North America.

By the 1970s, however, McKinsey was observed to be an "elite firm unable to meet client demands." Their consultants were considered excellent generalist problem solvers but lacked the deep industry knowledge and specialized expertise necessary to meet the client's rising expectations. In addition, aggressive challengers like the Boston Consulting Group emerged in the consulting market. Pressure from the market and competition initiated a long process of knowledge management and integration within the company.

A committee of the most respected peers in the company was assigned to study the problems and make recommendations. In 1971, one of the first things they suggested was to position the consultants of the company in the future as "T-shaped" consultants. This meant that their broad general background would be complemented with in-depth knowledge and expertise in one industry or functional area.

Ron Daniel, who became managing director (MD) of the company in 1976, stepped up the pace of implementation of the committee's report and installed *Clientele Sectors*. These organizations centralized the company's experience and know-how in specific *industries* (like banking, consumer products, etc.) and acted in a sort of matrix structure with the traditional geographically organized offices. He also started initiatives for more formal development of *functional* expertise. He assembled working groups around two key areas, namely strategy and organization. Local experts were asked to lead those working groups—for example, Fred Gluck, from the New York office, was responsible for the strategy group. Throughout the company various concerns about these initiatives were raised: people did not want to compromise the local presence they had built up in the past.

By the early 1980s, Gluck had become the internal champion of the knowledge integration initiatives. The next step was the creation of 15 centers of competence around existing functional expertise (marketing, change management, etc.). The centers were headed by practice leaders and aimed to help develop consultants and to concentrate on continuous renewal of the intellectual competencies of the company. The centers were meant to complement the personal networks of the individual consultants, not to replace them. The message, widely communicated via endless meetings and discussions, was to emphasize knowledge creation, management, and integration within the company, and not only to leverage existing know-how in the market. The culture had to be changed from mere "client" development toward "client and practice development." In addition, the Practice Bulletins were introduced to facilitate the diffusion of important findings and ideas around the company.

Gluck soon wondered if no further organizational changes were to be made in order to support the process. A project team was started in 1987. It proposed a common database of knowledge within the company to be installed and to hire a full-time practice coordinator for each "practice area" (client sector and competence center). They would bear the responsibility for the quality and accessibility of the database. In addition, the team emphasized the importance of the specialist consultants and suggested enhancing their position within the company (in relation to the T-shaped consultants).

These recommendations led to the introduction of the Practice Development Network (PDNet), a computer-based assembly of documents representing the core knowledge found around the company; and the Knowledge Resource Directory, a sort of Yellow Pages, serving as a directory of all the experts and key documents. The key problem was to solve the issue of the specialist consultants' status.

In 1988, the same year Gluck became managing director, a Clientele and Professional Development Committee (CPDC) was established and took over Gluck's personal role in championing the practice development and knowledge integration agenda of the company. The committee observed that the original group of 11 sectors and 15 centers had grown out to "islands of activity" and "fiefdoms ruled by experts." The proposal was made to integrate the existing groups into *seven* sectors and *seven* functional capability groups. A lot of people interpreted this move as centralization and adding another organizational layer.

The CPDC concretized the suggestion for improvement of the specialists' internal position, through the introduction of multiple career paths within the company. Despite these initiatives, a lot of skepticism and confusion remained.

In 1994, Rajat Gupta took over as new managing director of the company. After listening to various comments on the knowledge integration initiatives and the status of the integration agenda, Gupta decided to push it one step further, through a combination of measures.

First, he committed the company to the investments made in the centers of competence and industry sectors. Second, after a successful tryout in Germany, he decided to organize worldwide *Practice Olympics* as a competition between different teams on the basis of their ideas presented to a jury of senior partners and clients. Third, he started diverse *special* initiatives, meant to let senior partners work on emerging issues within the management of companies. Last but not least, he expanded the *McKinsey Global Institute* as a research center focusing on the consequences of the global economy on business, leading to another center, the Change Center, in 1995.

Overview of Integration Process. The actions toward knowledge integration in McKinsey and Company were clearly initiated by customers and competitors. The commitment of successive managing directors within the company to this agenda, and the involvement of the individual consultants via various discussions and working groups, seem to have contributed to the successful integration of the worldwide offices' know-how and knowledge.

The knowledge integration process within McKinsey and Company was characterized by an initially slow and later on exponential institutionalization process: starting from a few initiatives of *ad hoc* working groups aiming to coordinate information flow, and moving toward various projects led by free-standing organizational units created and fully responsible for the exchange and coordination of company knowledge. Another characteristic of the integration process of the company was the slow but steady evolution toward a company culture and long-term career development focused primarily on knowledge creation and integration.

Crossing Borders at Alcatel

Alcatel NV is the result of the 1986 merger of CGE and ITT telecom activities. Despite attempts to present a European rather than French image, the company was from the beginning typically characterized by decentralized management with most decision making and power located in the national subsidiaries. In theory, Alcatel implemented a matrix organization in which product responsibility (development and marketing) was in the hands of five business units (basically product groups, like the network services group), while sales and profits were the responsibility of the national subsidiaries. However, it was widely known within and outside the company that the country managers were "mighty kings in their national kingdoms."

Product development responsibility was assigned on a "center of competence" basis and was totally unrelated to sales responsibility. Usually, the subsidiary that developed the product, ended up manufacturing it as well (e.g., due to different computer-aided design systems in the subsidiaries).

Alcatel had tried to address the issue of better cross-functional integration and coordination of R&D, manufacturing. The center was a coordinating mechanism that was intended to link product development and manufacturing. Communication and interaction between the marketing and product development side was still ignored, not to mention the problems with the overall responsibility for the development projects owing to a lack of project managers and management. It was clear that the issue of functional integration needed a better answer than the SDI center.

In 1988 the French subsidiary Alcatel-CIT announced the introduction of the concepts "Product Life Cycle" (PLC) and "Trio." PLC was a procedural information gathering and distribution system which spanned the entire life of a product in development, until it went onto the market. It described the flow of information and reports required; the people involved at the different stages, and especially the responsibility at various points in time. Especially for this latter aspect, the Trio was created. The Trio was a project management team consisting of three members with different roles to fulfill in the course of the development of the product.

Although it was meant to be implemented gradually and was supported by a number of pilot projects and formal training, it seemed that the implementation of PLC and Trio did not go as smoothly as hoped for. Some managers felt threatened and closely monitored, some others saw their job contents change significantly, and still others complained that the managers within the Trios lacked formal authority.

These problems were even more explicit in the international context. Taking the initial idea of PLC and Trios, Alcatel NV decided to implement the concepts globally within the Line Transmission Group (part of the Network Services Business Unit). But the French innovation did not address how the various national Trios could be coordinated and integrated. For example, one issue was that the national departments (R&D, manufacturing, and marketing) had started communicating through the Trio, but concerning international coordination, one could hardly expect the R&D department in country A to communicate with the sales department of country B. And even worse, it turned out that the various R&D departments seriously misunderstood each other in some situations. Despite the existence of a Central Product Manager (who headed one of the product lines of a Business Unit and brought the different national Trios together regularly), there was a clear lack of formal project coordination on an international scale. P&L responsibility remained with the country managers, who in the best cases lacked accurate information for taking optimal decisions and often had diverging priorities.

The consequence was that several development projects were seriously delayed. Due to early project failures, people were less and less excited about international coordination and blamed each other for the problems that showed up. Some complained that the numerous meetings with the different national Trios were time-consuming, not to mention the time spent on the reports that were supposed to be drawn up and the endless communication that was neither effective nor efficient.

The international Trio experiment was followed by some other cross-border integration projects, but basically the key issue remained the same: the country managers were still very independent and the attempts to break their power had by and large failed.

While earlier attempts to gradually tilt the matrix (as at Philips) had failed, the company went into a severe crisis and hence the need for a drastic reorganization was clearly felt. At the end of 1995, the company created business divisions (split up on the basis of technology) which became the key organizational units, with clear formal power and responsibilities. The regional dimension was not completely erased but clearly became less important. Some of the country managers became heads of business divisions.

Overview of Integration Process. Although the initial steps taken aimed to boost cross-functional knowledge integration, the Trio and PLC were later transferred to the international scene, in order to boost cross-border integration. The decision itself explains part of the problem encountered during its implementation: by copying the concepts to the international platform, the drawbacks of the system, as observed on a national level, were simply enlarged. The key problems were: people felt threatened and were not ready to take the steps toward integration; those who wanted to do it missed the required support (culture, human resources management) and formal authorization; the country managers remained very powerful and had only a national perspective. The initial implementation problems of the Trio in France already indicated the difficulties in making the system work internationally. Alcatel underestimated the formalization of the approach: the endless meetings and reporting attempted to formalize the integration, but it clearly did not happen just like that.

The Trio has initiated a change in the corporate thinking at Alcatel, which has resulted in maybe modest improvement in company integration, across borders as well as across functions. The key problem of the Trio was that it was completely isolated from the existing structure and organization. One could indeed wonder if the Trio was not a good first step (an informal working group trying to prepare the organization), which in the end failed because it was not followed by the necessary formalization of cross-border coordination. Under pressure from the market, Alcatel finally had to break drastically with its traditional geographically oriented organization and structure, since the smooth transition route seemed to have failed. The future will provide an answer to the question whether the current organization is sufficiently competence- or knowledge-driven.

KEY FINDINGS

In the following paragraphs, some tentative hypotheses will be put forward on how explicit attention to knowledge integration might affect the previous findings with regard to the integration process and the integration initiatives implemented within that process.

Pacing the Integration Process: Shock Therapy or Slow and Steady?

A growing stream of research on corporate transformation raises the question of whether transformations should be effected quickly or slowly. Working with a model of radical change they term "punctuated equilibrium," Tushman and Romanelli (1985) posit that corporate transformations, because they affect all the fundamentals of an organization, should occur quickly. Organization systems, they argue, are tight configurations of reinforcing patterns, and tweaking the system to achieve change simply does not work. In sharp contrast, others argue that a slower pace (as much as ten years) makes more sense. They contend that changing requirements for skills and, even more important, the need to build and retain trust, require a more patient approach (Kim and Mauborgne 1996).

Both approaches were observed in the companies we initially researched. P&G has followed a 20-year plan, building trust and organizational capability, and ensuring changes were positive with a substantial element of limited experimentation. By contrast, 3M Europe opted for shock therapy. The change was implemented quickly, even ahead of schedule, possibly even to the short-term detriment of employees and customers. Many employees had been involved in discussions about the problems that would arise if the company failed to integrate; nearly everyone in 3M Europe agreed that integration was the way to go. With widespread support for the changes, the general sentiment was that dragging out the awkward in-between stages would be too distracting for everyone. Arguably, both approaches were successful for the companies involved.

Applying this to the observations within McKinsey and Company and Alcatel, it seems that McKinsey explicitly decided to take the long route. They gradually built consensus about the need for knowledge integration and took one step at a time (although they stepped up the pace when consensus seemed to be established, see later).

The question remains, however, whether shock therapy (like 3M) is feasible and possible within the knowledge integration context. We have the impression that knowledge integration requires a long and steady process of building consensus, while leaving the option open for some "shock projects" in the meantime (the long route as a sequence of smaller shock projects or sprints). Although in some circumstances (e.g., near bankruptcy or severe shareholder pressure) slow change may not be feasible (Strebel 1994), the question remains whether pure shock therapy can be considered an option to implement better cross-border learning.

Despite being polar extremes, both the "shock therapy" and the "slow and steady" approaches share one thing in common: they minimize the trauma and confusion associated with fundamental change. In shock therapy, because the changes are implemented rapidly, people can settle down to learn the new systems relatively quickly (up to two years). The change may be cathartic, but confusion and trauma are short-lived. Under the slow and steady approach, most changes occur following discussion, debate, and experimentation. Relatively little resistance is created, and where it does arise, the organization can take the time to counter or bypass the resistance that might otherwise blossom. Within McKinsey and Company, the recommendations of the committee set up in 1971 were only slowly implemented from 1976 on when Ron Daniel took over the MD position. The five years in between had given the individual consultants the time to learn and live with suggestions for the more integrated approach that was suggested and required.

Which of these routes will be most effective in a particular situation depends on the pressure of the market (competitive or bottom-line pressure) and on how powerful the resistance encountered will be. The slow and steady approach is preferable for companies where the challenge lies primarily in natural inertia or resistance *in* the system (as is often the case for specific knowledge integration efforts), rather than active hostility. In this case, an initial program of incremental change may provide the basis for a more important or crucial change to take place later.

The companies opting for a "medium" pace of change seemed to experience all the trauma and confusion of short-term fundamental change, without the benefit of a quick shock to overcome organizational resistance. On the other hand, the change was too quick to allow true evolution of attitudes, responsibilities, and capabilities among managers. The result was a greater tendency to retreat to the old structure, making little or no progress toward integration. It is observed that these companies often got stuck in the middle between the two types of change processes because management had not explicitly made the necessary trade-offs and choices, and therefore were not able to manage the critical weaknesses inherent in either approach.

For example, for years Alcatel (and others like Philips and IBM) had been trying to "tilt their matrix" and foster more cross-border cooperation in Europe, without significant success. Alcatel expected the installation of Trios to be a one-time shot and did not plan to take additional steps: they were prepared for neither the quick route nor for the longer term integration process. It took some time (and failures of integration projects) and a severe crisis to finally make major inroads into the stifling power of the country baronies. Alcatel, like some of the other examples mentioned, seem to have become stuck in the middle for quite some time (see Figure 4.2).

Whichever path is chosen, it has to be carefully prepared and monitored. In the absence of such preparation, the quick route becomes dirty, while the slow and steady route involves high costs for little result, leading to lots of frustration and even more resistance to change in the future. This was the case in the Alcatel example: they had not carefully planned and prepared the organization for the integration process, which resulted in frustration and poor performance.

Figure 4.2
"Shock Therapy" vs. "Slow and Steady"

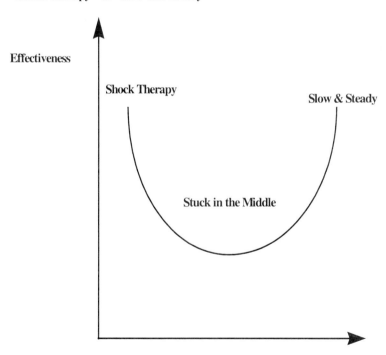

Build Capacity for Integration First

The next important question is how management can "prepare the way" for change as required, whether choosing shock therapy or the slow and steady route. The importance of the involvement of people, and the need to build a learning culture before implementing the changes, are important steps on the way to successful integration (See Figure 4.3).

Preparing the Scene: The More Involvement, the Less Need for "Implementation." First, we observed that participation in discussion before the actual change takes place is important. P&G, McKinsey, and 3M had a deep commitment to extensive discussion and tried to reach an overall shared purpose, even if individual decisions may have appeared haphazard. The approach mirrors research on Japanese change management, where it was observed that much more time was spent in discussions throughout the organization than in American companies, yet changes were implemented much faster. Overall, the Japanese had less resistance and more efficient changes. Likewise, P&G, McKinsey, and 3M have had significantly fewer problems adjusting than Alcatel, which initially involved a few people in the set-up of the Trios and PLCs but overlooked involving people (especially the country managers) in implementing the concepts internationally. The importance of this preparation phase for successful integration parallels that observed in successful mergers and acquisitions (Haspeslagh and Jemison 1991) or other strategy processes. It is important to generate a common perception and buy-in especially from those who will be most involved in producing the intended results or affected by them.

Figure 4.3
The Importance of Involvement in the Preparation Phase

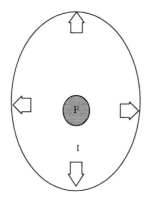
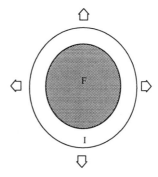

Deceptively rapid formulation, but slow or incomplete implementation.

Time invested in preparation, slow formation, but a more rapid result.

Circular area is proportional to time spent.

Creating the Culture: Learning by Doing. Second, the apparently ad hoc task forces and discussions ("soft" structures) put in place before implementing the integrated structures allowed managers within McKinsey, P&G, and 3M to develop the necessary skills and organizational capacity for the new structure. During the 1980s, for example, P&G marketing managers participated in numerous European projects, either as leaders or team members. During the long integration path within McKinsey and Company, several working groups have been called together to make suggestions and solve specific problems of knowledge coordination. 3M used European executives without line authority, based in Brussels, to encourage country managers to build consensus on strategies. These processes preceded formal restructuring in these organizations, and were necessary to build the capacity for cross-national perspectives and management skills. This is where Alcatel took a different approach: although they initially created what seemed to be *ad hoc* working groups (Trios) which could have been used as catalysts for further integration initiatives, they never formalized their initiatives and left these Trios out of the existing organization structure. The stages preceding restructuring are an important and necessary aspect of the path-dependent forces which affect the probability of successfully achieving integration, but are not sufficient. In addition, it was not easy for Alcatel to take the next step of formal restructuring, because frustrations and resistance had been built up due to a lack of attention to implementing and guiding the integration process (across both functions and borders).

In short, management should build the organizational capability and support required before and during the implementation of a shift to formal, integrated structures.

Start with Local Initiatives

It is observed that early local initiatives were important to long-term integration success. Local initiatives led by the staff and management of local units at P&G, McKinsey, and 3M, for example, began with ad hoc projects and later formal task forces or teams managed by various country units. Fred Gluck, from the local New York office, was in charge of the strategy team which consisted of various other strategy experts within McKinsey. The Practice Olympics organization in McKinsey is another example: they had been successfully organized in the German McKinsey office and were leveraged worldwide by the managing director. These projects led not just to the increased managerial capability noted above, but also allowed each local unit to build stronger ties to other units. Thus, individual nodes of the future network began by building stronger ties to other nodes, with both the worldwide and European center, and other local units. This has two types of benefits. First, it ensures that the process is better supported and more smoothly implemented. Second, it means that other linkages, especially those with real economic value (e.g., cost reduction via coordinated purchasing) are also likely to develop most actively. Initial successes motivate local units to push integration forward in these directions. As a result, the final outcome is likely to be superior, with the degree of integration aligned to demonstrated value-added, rather than a theoretical master plan (a similar result has been demonstrated by the procedural justice research within multinationals; Kim and Mauborgne 1996).

The failed integration processes usually started by trying to implement all the network links at once—for example, through major initiatives throughout Europe (Bleackley and Williamson 1995). This approach embodied a fundamental flaw: it required local units to contribute to a process whereby they would lose power or give up locally built-up know-how and innovations (see Alcatel), yet the benefits they would gain were both unproven and unclear. In other words: it might be true that you don't talk about Christmas to the turkey and you surely can't get it involved voluntarily!

In contrast to P&G, McKinsey, and 3M, Alcatel blindly copied the French Trio and PLC initiative in the complete Line Transmission Group and has never involved other national units in the process. The resistance from the national country managers was foreseeable, since they were supposed to let their people work with other national departments and hence give up some of their decision power, and yet were expected to evaluate them on their contribution to the local bottom line.

When initiatives cannot be generated locally, at the very least the benefits and gains of the integration should be widely and locally understood. We will add to this below in the finding of the importance of commonly perceived business benefits.

Choose Initiatives That Open New Options

Rather than seeing transformation simply as a shift from one structure to another, hopefully better, one, it has been argued that management must clearly envision the desired future company, and build the implications of that organization into the transformation process itself (Muzyka, de Koning, and Churchill 1995).

Our research suggests two important ways in which the final goal of integration should influence the process management adopts. First, in choosing an integration process, managers must consider the need to build managerial capacity for the new (still future) integrated company, so that once the formal organization changes, people within the new structure can work effectively. It is precisely the lack of capacity to deal with the more complex structure of regional organizations, the authors believe, that causes so many attempts at formally integrating to fail (Alcatel).

Second, managers need to take into account the impact their chosen integration process will have on the skill and knowledge base of the company. Some processes will result in skills being enhanced, other processes (such as closing down a particular function in a national subsidiary) will be skill-destroying. As a result, the integration process chosen by any company may either expand its future strategic options or close them off. We believe management must be sensitive to these long-term implications of the integration process they adopt. Moves that destroy too many skills risk boxing the company into a corner. This is especially dangerous given the considerable uncertainty about the rate of convergence imposed by external developments and the demand for internal learning capacity in newly restructured organizations.

Management at P&G and other companies also realized that their organizational choices would affect the outcome in other ways, beyond simply building capabilities and knowledge. The flexible commitments in R&D personnel assignments in P&G, for example, showed a preference for creating options for the future, rather than narrowly focusing the marketing strategy.[5] Creating options and trying to avoid unduly limiting future choices through the integration process are important considerations for management to be aware of. McKinsey clearly monitored and constantly checked their integration initiatives in the field. Key for them was their flexibility in pushing the integration one step further or in first letting people get comfortable with the current initiative or situation, whenever one of these was necessary and/or possible.

Pushing Integration to the Heart of the Business: A Multidimensional Perspective

Management faces the choice of a number of dimensions along which it can push its intended (knowledge) integration. Dimensions are defined as proxies of relevant cost savings and/or knowledge along which integration is pursued, in the hope of increasing cross-border efficiency and/or learning. Through reviewing the research and popular business press reports, the following options were identified:

- geography (e.g., adding a vice president for Europe, or a manager responsible for a group of countries)
- functions (integrating R&D, marketing, production, finance, etc.)
- processes (e.g., cross-border integration in the context of business process reengineering)
- activities (or parts of the "value chain" or "business system")
- products, product categories, brands (e.g., as in category management or Eurobrand management)
- customer key accounts (e.g., sales)
- customer-industry groups (e.g., industry verticals at IBM)

P&G Europe's approach clearly focused on functional integration in the first two stages. Later, they shifted to the product and product category dimensions. Over time, they took an eclectic approach to integration. IBM Europe, as part of a worldwide shake-up in 1994, chose customer-industry groups as a basis for integrating their European operations (internally referred to as industry solution units or industry verticals). A dual way was followed by McKinsey: simultaneous efforts on the functional and client industry dimensions were initiated. 3M Europe preferred to create European business units around product lines, grouped in business centers, inspired by their traditional U.S. structure. Alcatel in its December 1995 shake-up reorganized into Business Divisions (in combination with the geographic dimension). Other companies like Nestlé initially took a key account management approach, while maintaining national subsidiaries as strong local players (Parsons 1996).

In most cases we have observed substantial experimentation as integration pushed along only one dimension (e.g., on the basis of geography) got stranded. Therefore a lot of time and frustration could have been saved by carefully evaluating and using the different options available. The key issue is not only which dimensions for integrating provide the best results, but also which levers should be pulled in what sequence? How can more leverage be obtained? Our research so far has only begun to answer some of these questions.

Identifying the Dimensions for Long-term Integration Benefits. In selecting a dimension along which to push integration, management could first compare the relative strategic importance of each dimension for the business—that is, the long-term integration benefits it offers (even if these are sometimes difficult to quantify). When trade-offs between those dimensions with long-term benefits have to be made, those with immediate impact and those which act as "enablers" in laying a foundation for integration on other dimensions are suggested.

For example, in the case of accounting and information technology, integration is often necessary in order to provide the necessary support systems for integration on other dimensions. This category of initiatives, depicted as "A" in Figure 4.4, will act as enablers by facilitating cross-border comparison of information and coordination of activities. This category of integration initiatives may also offer substantial cost savings. But they are unlikely to have a broad-based impact on people's mind-set and skills, nor to fundamentally alter the strategic positioning of the company toward global or pan-European competition.

On the other hand, integrating on a critical dimension and changing performance measurements to highlight the change can have a powerful, direct impact on both mind-et and strategic positioning. These are termed category "B" initiatives (Figure 4.4). For instance, if a consultancy company like McKinsey observes that industry expert consultants will be key for future success, the establishment of Clientele Sectors will be a strong sign of the changes in "the way business is done." In fact, McKinsey chose client industry as a key dimension not only because the pressure for integration came from there, but also as a proxy for relevant knowledge, a key asset for a services business like consultancy, to be properly managed and integrated.

Ideally, one should begin by trying to identify those opportunities that will push integration along dimensions that offer substantial business benefits while at the same time acting as integration enablers (category C in Figure 4.4). When a trade-off exists (as in A vs. B) and cannot be overcome (by finding initiatives of type C or by combining both types A and B at the same time), it is important that management be explicit and realistic about its choice and what it entails. Given that fundamental (economic) benefits are the ultimate goal of integration, category B should be preferred over A. But there will be instances where unless category A initiatives are undertaken as enablers, the whole process will be impeded (e.g., Alcatel's attempt to coordinate activities and knowledge across borders initially failed partly because of a lack of overall structural support by HRM, etc.). This is the trade-off which is often hardest to resolve in reality. The situation is even more complex in a dynamic context, since the actual critical dimensions in the business may change over time as the industry or the company evolves. However, realizing what the terms of the tradeoff are in any particular case goes a long way toward managing it properly.

Maximizing Integration Spillovers Through a Multidimensional Approach.
Integration is rarely achieved in a single sweep, along one dimension of the
business. Building critical mass and creating spillovers of integration benefits are
the next requirement that is essential to achieve significant and lasting integration.
This means that integration initiatives may have to be sponsored on many
dimensions of the business simultaneously, guided by a common, long-term goal.

P&G, with their slow and steady process of integration, clearly showed how
initiatives along several dimensions in the business helped to create the highly
integrated organization of today. The benefit of combining dimensions were twofold.
First, by promoting initiatives on many dimensions, management built a critical mass
of strategic awareness and cross-national relationships. Thus, by electing to create
change along those dimensions, management built support for integration and avoided
needless battles. Second, the organizational capabilities required to cope with the
complexity of an integrated European operation are quite different than those needed
for a loose federation of country subsidiaries. By taking initiatives along many
dimensions of the business, management allowed themselves and others to learn
needed skills—before the final integrated structures were implemented. These kinds
of spillovers from specific initiatives were essential to the overall process. Progress on
any one dimension of integration is leveraged or reinforced by following a parallel
integration path along other dimensions.

Figure 4.4
Business Benefit vs. Integration Benefit: Trade-offs May Be Necessary

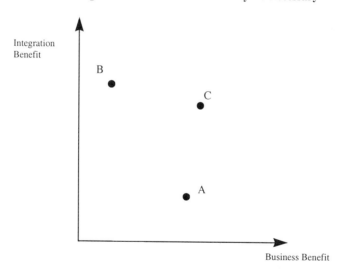

McKinsey and Company felt market pressure for increasing knowledge
integration along the client-sector dimension. However, from the very beginning, it
was clear that the functional dimension was not to be neglected. Gupta's comment

that the company should try all ways simultaneously clearly illustrates the multidimensional approach. The exponential integration process exemplifies the need to build up a certain critical mass before taking off at a higher pace.

Starting with Initiatives Giving Short-term Benefits

The multidimensional nature of the initiatives demonstrates that there are many paths to integration, but success is also driven by practical short-term concerns. In addition to assessing the long-term contribution toward integration of moving forward on a particular dimension, one should also evaluate initiatives on the basis of the extent to which they offer early, quantifiable benefits ("quick wins"). The findings here were in accordance with the change management and transformation literature, which has emphasized the importance of quick wins to rally support for the overall process (Kotter 1995).

For example, at McKinsey and Company, the pressure for more integration came from the customer who preferred industry specialists for its consulting projects. In light of this, the short-term benefit was especially evident on the client industry dimension. Another example in the same company was the local consultants who immediately experienced the benefits of the PDNet system. This computerized center of know-how had clear short-term benefits in the way consultants gathered information and know-how for their individual projects.

Alcatel's problems in implementing the Trio and PLC on a national level can be considered as an "early loss," since they only fed the conviction of the country managers that these concepts could not work (and especially not internationally) and were not optimal.

As part of the overall integration process, therefore, management should ensure they adopt at least some integration initiatives along dimensions that provide clear, ideally quantifiable, benefits (hard to accomplish for knowledge integration) that can be realized in the short term.

The demonstrable, short-term benefit does not necessarily need to be financial in nature. In many instances the organization can rally behind other commonly perceived business goals, as long as they are clear, represent a true challenge, and have a sense of urgency. Responding to a commonly perceived competitive threat, for example, may provide a strong impetus to the integration process (e.g., McKinsey and Company). In sum, management should give priority to integration initiatives built around focal points based on a commonly perceived, specific business need.

Given the limits and dangers of a purely short-term perspective, the importance of a longer term view is indicated below.

Top Commitment to Overall Long-term Vision

No matter how carefully management considers the above points, most benefits will not be apparent in the short term. It is clear that top commitment and a shared vision are as essential in order to make the kinds of fundamental changes to an organization's logic and functioning as those required to achieve integration.[6] Both of these prerequisites need to be long-term and sustained over a substantial period of time, especially when the slow and steady route to integration is chosen. But even if shock therapy is attempted, continued commitment and shared vision will be crucial in making the necessary behavioral and cultural changes after the drastic structural changes have been initiated. McKinsey and Company's internal champion of the knowledge integration process, Fred Gluck, illustrates the contribution of constant attention and support and empowerment for the integration initiatives.

Conflicts and resistance are bound to arise and top management must assume the delicate role of leading the change and empowering the key players or champions in the organization to push the changes through. Important adjustments will have to be made in reward systems, P&L responsibility, reporting lines, career development, and so on. For example, McKinsey adapted the career and appraisal systems for specialized consultants, while Alcatel's problems in the integration process were partly due to lack of support through P&L responsibility and reward systems. Hard business decisions, often involving trading off short-term benefits for long-term gains, will have to be faced. Here, as in any strategy process, top management will have to set and manage the context, and arbitrate or intervene clearly and decisively whenever conflicts or paralyzing ambiguities arise.

The nature of this commitment and top-level intervention may be somewhat different for the slow and steady and shock therapy routes to integration. Under the shock therapy approach, where the change is pushed through in a crisis context, a substantial change in top management itself is likely required, especially if the crisis is externally imposed, as was the case with IBM and Alcatel. If you can't change the people, move them!

The importance of shared vision and top commitment is all the greater whenever the benefits of integration are not immediate and easily quantifiable or demonstrable, as unfortunately is often the case for knowledge integration. But even if the benefits seem clear, the impacts of the required changes on the overall organization are likely to require top management attention to the process.

At one company (which had chosen a cross-border alliance route toward European integration), cross-border functional task forces, or activity groups which aimed at improving mutual cross-border learning, often got stuck despite initial enthusiasm as tough decisions and cross-border trade-offs had to be made. The problem was the absence of clear top management authority over the various country organizations, which de facto retained a high degree of independence.

CONCLUSION

Previous research has emphasized various aspects of integration management. Our initial research flagged the fact that the process by which a firm chooses to transform itself into a more integrated network organization proved important for two reasons. First, because the integration process a firm adopts significantly alters the probability that increased integration will be successfully achieved. Second, the final outcome itself is path dependent—that is, the type of integration a firm achieves depends significantly on the process (or path) it chooses to follow. It seems that applying this specifically to knowledge integration does not affect these observations.

Based on the findings on overall company integration, a framework is developed which emphasizes the options facing managers, relating existing organizational structure, change culture, and market conditions to the optimal or effective choice of integration processes. This chapter tests to what extent the process approach applies to explicit cross-border knowledge integration. It is observed that management of the knowledge integration process remains crucial for successful integration. The key conclusions are that long-term (slow and steady) processes are preferable to medium-term ones; that a preparation phase which builds integration capabilities remains crucial to success; and that integration efforts should combine initiatives that directly drive long-term success with those that can act as "enablers" to more fundamental types of integration. It is also suggested that the process should be begun with focused and quick-return initiatives, and that combining integration initiatives along multiple dimensions of the business increases the probability of success.

Some of the initial observations apply possibly even more to knowledge integration pushing cross-border learning than to overall integration efforts that push cross-border efficiency. The need to build trust and to involve people, the suggestion to prepare the company and build an appropriate coordination culture, and others, have played a key role in the observations of the success or failure of the integration initiatives. The question remains, however, whether shock therapy in making the transition can work for leveraging globally or regionally the learning and knowledge within the organization.

The question is most often not whether integration should be established either to increase global efficiency or to leverage innovation and know-how internationally. It seems that integration initiatives to boost the regional and global efficiency of the company are frequently in close interaction with cross-border knowledge integration attempts. In some businesses, especially multidomestic businesses, and companies, integration is particularly aiming at better overall leveraging of know-how, while in other businesses and companies the key focus is on global efficiency. It seems that for the first kind of industries and companies, integration along the knowledge dimension strongly supports or even initiates better overall integration; while for the second kind of industries and companies, the integration of knowledge is at least a by-product but often also a "catalyst" for overall integration.

Given the complexity and diversity of the markets, not to mention the rapidly changing technological and regulatory environment (e.g., European integration), companies are experiencing increasing pressure on their knowledge creation, management, and integration. Although a recent stream of research has focused on this evolution, the authors hope that their findings can contribute to the insight that the integration of company knowledge sources across borders is a complex process that should be monitored, planned and managed carefully.

ACKNOWLEDGMENTS

Research assistance and comments by Alice De Koning (INSEAD) and Venkata Subramanian (KULeuven) and support by Dirk Rijmenams (KULeuven) are greatly appreciated. Financial support from the Tractebel Chair for European Management Education and Development and from the Office of the Belgian Prime Minister (Federal Office for Scientific, Technological Cultural Affairs in the context of the Inter-University Attraction Poles) is hereby acknowledged.

NOTES

1. Although some put forward the internal view at the expense of the external view, some authors have tried to link the two sides and position them as complementary (Verdin and Williamson 1994; and others).

2. See article in *Financieel Economische Tijd* (May 23,1996) on the lagging profitability of European industry and rationale for European integration within multinational enterprises such as P&G, 3M etc.

3. The field research project at P&G Europe was conducted over a period of two years, and involved managers in many locations and levels of the organization. We interviewed senior executives assigned to the European Technical Center (essentially the European headquarters) in Brussels, including those who have since moved on to other parts of P&G, general managers of country organizations, and directors and managers of various functions and locations. One informant who no longer worked for P&G proved particularly valuable, partly because his memory was less "contaminated" by more recent events. In all, we interviewed about 20 top executives, currently filling managerial positions down to three levels below global executives in the organization. Several were interviewed more than once. Interviews were semistructured, with interviewees being explicitly informed at the beginning of each interview what information we were seeking, and why. This open approach allowed the interviewees to understand our broader agenda, and often led to the introduction of new information, documents, and contacts which greatly broadened our understanding of the historical evolution of the organization. In addition to the interviews, we also reviewed public sources and other published cases about P&G and competitors to fill in the external context of P&G's organizational evolution.

4. Similar findings have been reached at a global level, e.g. Ruigrok and Van Tulder (1995) found very few truly global companies.

5. The time line for 3M is much clearer, because the organization made a solid, massive shift in a very short period of time. P&G's more modulated approach meant that some departments had already completely centralized, while other functional areas or product lines were just beginning to explore the possibilities of a European approach. This varied approach created additional challenges of maintaining smooth cross-functional coordination and synchronization.

6. The original idea of the crucial importance of commitment in strategic choices has been elaborated in the context of business and corporate strategy by Ghemawat (1991); the notion of flexible commitment has been developed by Chakravarthy (1996a).

REFERENCES

Bartlett, C. A. *McKinsey & Company: Managing Knowledge and Learning.* Boston: Harvard Business School Case Study, 1996.

Bartlett, C. A., and S. Ghoshal. *Managing Across Borders: The Transnational Solution.* Boston: Harvard Business School, 1989.

Bartlett, C., A. De Koning, and P. Verdin. *P&G Ariel Ultra: First and Fast in Europe.* Harvard Business School-INSEAD Case Study in Progress, 1997.

Bleackley, M., and P. Williamson. "The New Shape of Corporate Europe." Unpublished Report, Conference Board Europe, 1995.

Bogaert, R., A. De Meyer, and P. Verdin. *Alcatel NV's International Organization.* INSEAD Case Study in Progress, 1997.

Bonheure, K., and A. De Meyer. *Product Development for Line Transmission Systems within Alcatel NV.* INSEAD Internal Case Study, 1992.

Chakravarthy, B. S. "Flexible Commitments," Strategy and Leadership 24 (1996a) 14–20.

Chakravarthy, B. S. "The Process of Transformation: In Search of Nirvana," *European Management Journal* 14 (1996b) 529–539.

Cerny, K. "Making Local Knowledge Global," *Harvard Business Review* (May–June 1996) 22–38.

De Koning, A. J., V. Subramanian, and P. Verdin. "Extensions to Malnight's Framework for Globalization: The Role of Cross-Border Regional Integration," KULeuven Research Note, 1997.

De Koning, A., P. Verdin, and P. W. Williamson. "So You Want to Integrate Europe? How to Manage the Process," European Management Journal 15, 5 (1997) 252–265.

Freeman, K., and P. Verdin. *Eureko Insurance Alliance*, INSEAD-KULeuven Case Study in Progress, 1997.

Ghemawat, P. *Commitment: The Dynamic of Strategy.* New York: The Free Press, 1991.

Grant, R. "Prospering in Dynamically-competitive Environments: Organizational Capability as Knowledge Integration," *Organization Dynamics* 7 (1996) 375–387.

Haspeslagh, P. C., and D. B. Jemison. *Managing Acquisitions: Creating Value through Corporate Renewal.* New York: The Free Press, 1991.

Kim, C. W., and R. A. Mauborgne. "The In-roll and Extra-roll of Multinational's Subsidiary Top Management: Procedural Justice at Work," *Management Science* 42 (1996) 499–515.

Kotter, J. P. "Why Transformation Efforts Fail," *Harvard Business Review* (March–April 1995) 59–67.

Malnight, T. W. "The Transition from Decentralized to Network-Based MNC Structures: An Evolutionary Perspective," Journal of International Business Studies 27 (1996) 43–65.

Morrison, A., D. Ricks, and K. Roth. "Globalization Versus Regionalization: Which Way for the Multinational?" *Organizational Dynamics* (1991) 17–29.

Muzyka, D., A. De Koning, and N. Churchill. "On Transformation and Adaptation: Building the Entrepreneurial Corporation," *European Management Journal* 13 (1995) 346–362.

Parsons, A. "Nestlé: The Visions of Local Managers: An Interview with Peter Brabeck-Lemathe, CEO-Elect, Nestlé," *The McKinsey Quarterly* 2 (1996) 5–29.

Rumelt, R. "Inertia and Transformation," in *Resources in an Evolutionary Perspective: A Synthesis of Evolutionary and Resourced Based Approaches to Strategy* (C. N. Montgomery, ed.). Norwich,: Kluwer Academic, 1995.

Ruigrok, W, and R. Van Tulder. *The Logic of International Restructuring*, London and New York: Routledge, 1995.

Schendel, D. "Editor's Introduction to the 1996 Winter Special Issue: Knowledge and the Firm," *Strategic Management Journal* 17, Special Issue (1996) 1–4.

Schütte, H. "Between Headquarters and Subsidiaries: The RHQ Solution." Paper presented at EIAB Conference, Stockholm, 1996.

Senge, P. "The Leader's New Work: Building Learning Organizations," in *The Strategy Process (European Edition)* (H. Mintzberg, J. Quinn, and S. Ghoshal, eds.). Hertfordshire, U.K.: Prentice Hall, 1995, 41–52.

Strebel, P. "Choosing the Right Change Path," *California Management Review* 36 (1994) 29–51.

Tushman, M., and E. Romanelli. "Organization Evolution: A Metamorphosis Model of Convergence and Reorientation," Research in Organization Behavior 7 (1985) 171–222.

Van Heck, N., and P. Verdin. *The 3M Company: Integrating Europe (A) (B) and (C).* (Adaptation of M. Ackenhusen, D. Muzyka, and N. Churchill 1994.) INSEAD Case Study, 1996a.

Van Heck, N., and P. Verdin. *Vedior International: The French Revolution.* KULeuven Case Study, 1996b.

Verdin, P., and P. Williamson. "Core Competence, Competitive Advantage and Market Analysis: Forging the Links," in *Competence Based Competition* (G. Hamel and A. Heene, eds.). New York: John Wiley and Sons, 1993.

5

Knowledge Management: An Emerging Discipline

Syed Z. Shariq

We need systematic work on the quality of knowledge and the productivity of knowledge—neither even defined so far. The performance capacity, if not the survival, of any organization in the knowledge society will come increasingly to depend on those two factors. But so will the performance capacity, if not the survival, of any individual in the knowledge society.

Peter F. Drucker (1994)

INTRODUCTION

As it begins to dawn on us that for the first time in human history, we may be able to see the impact of human decisions on the evolution of our society in our own lifetime. We become acutely aware of the unprecedented responsibility and the de facto power we all possess to influence the path of our evolution. Traditional disciplinary knowledge is limited in its ability to support the challenging decisions that lie ahead. Global stability in the future will depend upon our ability as a society to simultaneously address the three fundamental issues of prosperity, security, and sustainability.

THE CHALLENGE

Today, well over 3 billion people across the globe earn less than two dollars a day. As they move toward achieving prosperity they are likely to emulate our past overemphasis on material wealth, which has enormous consequences for the environment and global sustainability. At the same time, a vast number of citizens of the world remain illiterate, posing an enormous challenge for us in providing the necessary education and training to create global prosperity in the knowledge-based economy. We must find effective ways for capitalizing on emerging information technology to help the citizens of the world gain access to education and opportunities for adapting and prospering in the next century.

We are entering into an era where the future will be essentially determined by our ability to wisely use knowledge, a precious global resource that is the embodiment of human intellectual capital and technology. As we begin to expand our understanding of knowledge as an essential asset, we realize that in many ways our future is limited only by our imagination and ability to leverage the human mind. The future of the world lies in front of us unexplored and uncharted, and is full with great opportunities for expanding peace and prosperity in the next millennium through the responsible use of the knowledge.

As knowledge increasingly becomes the key strategic resource of the future, our need to develop comprehensive understanding of knowledge processes for the creation, transfer, and deployment of this unique asset are becoming critical. In the face of a globally expanding and highly competitive knowledge-based economy, the traditional organizations are urgently seeking fundamental insights to help them nurture, harvest, and manage the immense potential of their knowledge assets for capability to excel at the leading edge of innovation.

Primary and secondary schools, universities and training organizations (traditional suppliers of knowledge), and businesses and knowledge-based organizations in the public sector (growing users of knowledge) are in need of an integrative discipline for studying, researching, and learning about the knowledge assets—human intellectual capital and technology. This would be similar to the development of Operations Research as an integrative discipline during the Second World War.

An international society of knowledge professionals (as proposed in Appendix I to this chapter) can provide the necessary focus for fostering collaboration among the best minds and organizations of our time on study, research, and learning dedicated to the underlying disciplines and their integrative evolution in the emergence of Knowledge Management as a new discipline.

THE KNOWLEDGE MANAGEMENT DISCIPLINE

The task of developing Knowledge Management (KM) as a new discipline is a challenging endeavor. This new discipline must successfully respond to the diverse needs of knowledge-based organizations and knowledge professionals in a timely fashion. The pioneers of Knowledge Management must recognize that a great many of today's organizations are run largely on the basis of insights gained from the successes of the manufacturing-based capital-intensive industrial economy of the past. These organizations have fallen or are rapidly falling out of alignment with the evolutionary direction of the future as the economy transitions from the industrial economy to one that is rapidly becoming an intellectual capital and technology-based global knowledge economy.

A KM PROFESSIONAL SOCIETY

The problems of the future will be open-ended, complex, global, and adaptive in nature. We must look for a new synthesis of knowledge, integrating hard and soft sciences, to create the knowledge assets necessary for addressing the challenges in a rapid evolutionary era. The participation of experts from traditional academic disciplines (i.e., information technology, management, cognitive sciences, economics, finance, policy, law, social sciences), and business and government will be essential to the cohesion of an integrative body of knowledge leading to the formation of a Knowledge Management discipline and a community of scholars, teachers, and professionals associated with this new discipline. A professional society can serve as a home for enabling the development of the Knowledge Management discipline. However, it must be created as a hybrid, independent, and entrepreneurial organization with strong global participation and ties to leaders in academia, industry, and government.

In order to fulfill its mission the professional society needs to foster three functions: academic education, research, and advanced technology. These three interconnected sets of functions need to be pursued in a hybrid way by combining the strengths of the distinct institutional methods of academia, industry, and think tanks. The professional society may become an incubator for implementing new 21st-century models of operations for each of the three functions:

- An experiential learning-based academic environment
- A collaborative research community dedicated to life-long knowledge-based learning
- A multimedia and information technology-based knowledge-era tools development program for supporting the performance of the knowledge professionals and organizations

The professional society needs to foster the three functional areas by establishing a set of founding partnerships with innovative adult education universities, premier research universities/institutes, and a group of leading-edge industry and government collaborators from the United States and abroad. One of these founding partners needs to step up and become the founding home of the professional society, where the society's activities will be coordinated. Beyond the founding partners, opportunity for participation in the society should be open to all credible academic, government, nongovernment organizations and industrial organizations worldwide, including students, research fellows, and professionals across a range of disciplines.

EDUCATION

The professional society needs to consider the development of a comprehensive executive education program in Knowledge Management for leaders, executives, knowledge professionals, and policy makers. The program offering, as an executive master of science degree, should include state-of-the-art topics addressed through scholarly rigor to be able to earn accreditation with the support of members of faculties at the leading universities. This degree program, in collaboration with the leading-edge developers of multimedia and information technology-based knowledge delivery tools, can serve as an alpha site for testing and implementing state-of-the-art innovations in global instruction and learning delivered through distance learning technology (including learning and instruction delivered locally in real-life settings at a representative set of host organizations which are the leaders in applying Knowledge Management).

The postgraduate educational commitment should be explored as a means for testing and implementing network-based tools for delivering just-in-time learning to the graduates of the master's degree programs throughout their professional careers. (A Ph.D. program may also be developed and introduced.) The knowledge network consisting of students, faculty, and graduates across the globe can form the prototype of a knowledge community/university of the future where ongoing learning, resident in the network, will be the true competitive advantage.

The educational function, by convening annual real/virtual conferences addressing the topics at the frontiers of Knowledge Management, by serving as a knowledge resource network for sharing the latest research insights, and by providing professional advice and mentoring to instill the career self-management philosophy, can serve as a basis for facilitating the formation of a network-based global Knowledge Management professional society.

RESEARCH

The research function also needs to be carried out in collaboration with the university, government, and industry partners, and research fellowships/internships should be fostered to carry out basic and applied research into fundamental topics in Knowledge Management. These topics will aim to understand the processes and practices for generation, identification, assimilation, and distribution of knowledge as an asset for use by knowledge professionals and organizations, as well as to study the key knowledge-sector policy innovations at individual, organizational, and societal levels.

The economics of knowledge, addressing the accounting, valuation, and depreciation of knowledge assets, should be considered as one of the core themes on the overall research agenda. The initial research will lead to the formulation of specific and significant Knowledge Management research topics, some of which will be topics leading to MS theses, and Ph.D. dissertations. The results of research (in the form of best practices, case studies, papers, theses and dissertations) should be published and can be relevant inclusions in the instructional material used in the degree programs under the educational function.

Three preliminary research topics are presented below; they require further development and broader participation by the proposed Knowledge Management society:

1. A global change knowledge network, an experiment in the formation of a global community of students, scholars, and professionals to facilitate the co-emergence of prosperity, security, and sustainability in the 21st century. This concept is outlined in more detail in Appendix II.
2. A Knowledge Management program of graduate research, education, and implementation based on the new synthesis of hard and soft sciences. This program would include creation of a master of science degree in Knowledge Management and development of a professional Knowledge Management society. This goes well beyond the traditional inter-or multidisciplinary paradigm. The integrative synthesis will be a challenge unless we can get participation from most of our colleagues.
3. Research on global intellectual investment and knowledge assistance to create a culture of investment in intellectual assets. Traditional assets like money, land, and the like, can be given to anyone, whether deserving or not, but intellectual assets can only be earned by those who engage and invest their brainpower in learning. Accordingly, to be meaningful the paradigm of global aid and assistance must be transformed from monetary aid to one based on knowledge.

In order to carry out these programs, we should foster the creation of strategic nodes, with excellence in areas of complementary expertise, both within the United States and abroad, to create a virtual knowledge network, working within a common vision, addressing problems locally but from a global perspective (online resources are listed in Appendix III). The organization of the society must evolve to become an inclusive one based on shared stewardship and commitment.

ADVANCED TECHNOLOGY

The advanced technology program function needs to carry out cutting-edge technology development projects to maintain the educational functions for the Knowledge Management professionals at the state-of-the-art in educational technology, and thereby provide cost-effective enriched learning experiences to a diverse group of learners across the world. Each of these projects needs to be pursued through partnerships with the researchers and technology developers in the multimedia and information technology industry.

The use of advanced technologies and approaches, including those drawn from industry, government laboratories, and the intelligence community (such as artificial intelligence, data mining, knowledge representation, virtual reality, simulation, telepresence, teleoperations, Global Information Systems, distance learning, and cognitive sciences), in an environment open to exploration and learning, will lead to a unique development program for knowledge-era tools for supporting the needs of the knowledge professionals and organizations. In particular, the professional society is in a unique position to foster:

- the development of tools to support the strategic needs of the education function
- a cost effective just-in-time knowledge delivery capability
- the formation of a global knowledge network

CONCLUSION

The proposed Knowledge Management society will require a strong and sustaining commitment from the home institution for an initial incubation period, a visionary leadership, an enterprising group of faculty/scholars and dedicated staff, a strong advisory board and sponsors, and leading-edge partners and collaborators in the industry and government. Since Silicon Valley in California is recognized worldwide for its leadership in information technology-based innovations, the initial focus on Silicon Valley will serve as an ideal test bed for the professional society's home. A Silicon Valley location will be critical to the success of collaborations in all three function areas and therefore essential to the early development and acceptance of the society. However, from its inception the professional society will be international in its outlook and scope.

The professional society will serve as a beacon to knowledge professionals and organizations in the emerging knowledge-based economy. In order for the professional society to accomplish this challenging mission, it must be developed as a highly entrepreneurial, enterprising, innovative, and integrative catalyst organization. It must bring together the important components of Knowledge Management education and research, and the development of learning technology to significantly increase the productivity of knowledge-based performances and innovation (of professionals and organizations), to enable significant advancements in the capabilities for just-in-time knowledge delivery, and in knowledge-sector policy innovations.

The understanding of knowledge processes and its management will provide valuable insights for policy makers, learning organizations, knowledge professionals, individual citizens (including those in primary and secondary schools) and enable them to focus their investment in learning, unlearning, and life-long learning.[1] In short, the challenge is to help provide the decision makers at all ages, at all levels, and in all places with the necessary intellectual capital to be competent performers and proactive opportunity creators in the evolution of the global knowledge-based economy of the 21st century.

APPENDIX I: AN INTERNATIONAL SOCIETY FOR KNOWLEDGE PROFESSIONALS (ISKP)

Motivation

As we approach the dawn of the 21st century, we are humbled by the scale of challenges our world faces on the global scale. The unprecedented development and growth of knowledge during the 20th century notwithstanding, the evolution of a peaceful 21st century will depend on our ability to address three interdependent global challenges of prosperity, security, and sustainability. The new world order, though far from being fully defined and agreed to, already is beginning to point at some of the strategic threats and opportunities that would be determining factors in our ability to help shape a new century, which we can look back or as symbolic of the human spirit and its collective accomplishment at its best and most current stage in the evolution.

Knowledge Management affords knowledge professionals, those responsible for knowledge processes and its management, an opportunity to influence the policies and practices which will define the next generation of management principles and standards. Unlike its predecessors, an ISKP[2] is based upon bountiful resources, which—when shared—only increase in size and scope. Professionals across all functions, industries, and countries are embracing this new global agenda with a spirit of motivation, learning, and collaboration. Indeed, the changes are kaleidoscopic, the movement is pervasive, and the fundamental opportunity is to facilitate adaptive change to a stable world order in the 21st century.

Purpose

Each society member is an interactive knowledge node in a global network of knowledge professionals. Society members can play strategic roles in facilitating adaptive change to a stable world order in the 21st century by simultaneously addressing the three fundamental issues of prosperity, security, and sustainability through the advancement of state-of-the-art and the state-of-the-practice in the knowledge-based management of organizations and institutions across the world.

Objectives

- Create and communicate a shared vision of knowledge community that could foster stable world order in the 21st century through adaptive change.
- Catalyze the development of Knowledge Management as an integrative synthesis of traditional disciplines to address problems of the future.
- Link theorists and practitioners involved in the emerging community of knowledge practices to share successful strategic practices in the knowledge community and to help evolve the managerial standards for the knowledge economy.
- Initiate precompetitive collaborative projects on topics in knowledge management education, research, and technology development.
- Encourage and foster a spirit of volunteerism through ability to give the "gift of knowledge" on collaborative projects in the areas of critical needs.

Membership

Participation in the society is based upon professional responsibilities related to knowledge, learning, intellectual capital, innovation, and other declinations related to the creation and prosperous flow of ideas in support of the ISKP purpose. In addition to general membership, participants will have the opportunity to be organized according to functions, sectors, industries, and/or geography. Special focus on projects in education, research, and technology will be developed and coordinated according to particular topics of common concerns. In order to become a member of the society, professionals would need a recommendation from two current ISKP members and approval of the membership committee.

Proposed Next Steps

1. Preliminary announcement of the ISKP proposal.
2. Recruitment of founding members, the draft of the ISKP charter, and establishment of interim officers.
3. Incorporation of ISKP as a not-for-profit organization.
4. Solicitation of a start-up grant from a research or charitable foundation and "Design of Society" products and services.
5. Official ISKP launch announcement at the Knowledge Management 1998 conference in London.
6. Charter membership registrations and election of officers.
7. First International ISKP Congress, Germany (October 1999).
8. ISKP Global Millennium Summit, San Francisco (2000).

APPENDIX II: A GLOBAL CHANGE KNOWLEDGE NETWORK (GCKN)

The rising tide does not lift all boats. Three billion people live on the equivalent of $2.00 a day.

Ismail Serageldin (1997)

This is an experiment in the formation of a global community for facilitating the coemergence of prosperity, security, and sustainability in the 21st century.

The possibility for achieving future peace and prosperity is highly dependent on the creation of an emerging synthesis of knowledge as we know it today and our ability to engage in the creation of new knowledge through participation in an innovative Global Change Knowledge Network for research and experimentation. This "network" is an interactive community of knowledge professionals (across all knowledge domains who are responsible for knowledge processes and its management). Such a network, if successfully carried out, would not only be able to provide new insights addressing the intrinsic interdependence of the three critical global areas of challenges, but also may emerge over time as a trusted community, an intellectual asset, for effective communication and coordination of important ideas during critical times.

The principal charter of the GCKN will be an experiment in the formation of a global community for facilitating the coemergence of prosperity, security, and sustainability in the 21st century. The areas of research would focus on proactive assessment, development, and integration of knowledge in global change processes. Research areas would include:

- How are prosperity, security, and sustainability codependent in regional contexts and how may these interdependencies manifest into international or global conflicts?
- What are the intellectual capital (human skills) and institutional and infrastructure (both physical and technological) requirements for the region/nation to be able to achieve a sufficient level of prosperity in the global economy to ensure peace and security?
- What knowledge needs to be developed and integrated in the economic development process to ensure that the prosperity would lead to an environmentally safe and sustainable future?
- What research and developments need to be undertaken to bridge the gaps in knowledge exchange processes to identify and facilitate mitigation of emerging threats to security likely to be caused by environmental conflicts?
- How best can a Global Change Knowledge Network be formed to facilitate knowledge exchanges among members and provide a framework for an emerging knowledge-based international aid/assistance paradigm?

The regions of the world can be identified and invited to join the GCKN. In each selected region a university-based research group subscribing to a common charter with ability to carry out the integrative synthesis of knowledge for addressing the interconnectedness of the three challenges will be invited to become part of the network. A list of candidate members for the global change knowledge network would include recognized institutions from Brazil, Canada, China, India, Japan, Korea, Mexico, Portugal, Singapore, South Africa, Spain, Sweden, Taiwan, and the Unites States.

The GCKN member organizations will create a common framework for the development, exchange, and integration of knowledge in their respective regions. Under a joint GCKN vision, each member will develop an agenda for research, education, and knowledge dissemination programs that would incorporate the interests of scholars, students, regional organizations, and institutions in their region. The GCKN will be designed to electronically link the members through Internet, video conferencing, and distance learning capabilities. This will serve as the GCKN infrastructure and main communication mechanism for forming a virtual research community and a test bed for evaluating information technology-based concepts for developing global research and educational communities, as well as mechanisms for knowledge dissemination.

The GCKN members will develop research agendas by identifying their respective regional topics on prosperity, security, and sustainability, as well as incorporating research topics that would lead to development and sharing of a global synthesis. The GCKN community will provide opportunities for creating a virtual learning laboratory for scholars and students alike. It will also serve as a test bed developed and managed by students, for incorporating emerging information technologies to create a round-the-clock live virtual environment for the members.

One of the desired outcome of the GCKN, above and beyond developing a global community of scholars and students, would be for each of its members to serve as focal point for widely distributing the knowledge and insights resulting from the GCKN in their respective regions (to leaders, executives, policy makers, students, and the public at large) through the use of emerging information technology based tools and learning environments.

Development of a global community of scholars, leaders from the public and the private sectors, knowledge professionals and students, with commitments to build global trust and shared understanding is essential to ensuring the global security and stability over the 21st century.

APPENDIX III: ONLINE RESOURCES

American Productivity & Quality Center—publishes a newsletter, Knowledge Management in Practice, containing information on benchmarking, Knowledge Management, measurement, customer satisfaction, productivity, and quality.

Business Researcher's Interests—a searchable compilation of scholarly papers, articles, books, tools, and related topics on Knowledge Management.

The Complete Intranet Resource—magazine articles, case studies, software vendors, white papers, and so on, on setting up and using intranets.

Digital Knowledge Assets Offers—an on-line technology that links the knowledge needs of marketing professionals with the intellectual capital of academia and industry experts. Supports just-in-time learning rather than just-in-case learning by delivering filtered, on-demand information.

Electric Minds—an on-line virtual community of experts and colleagues discussing knowledge sharing.

Knowledge Inc.—site of an executive newsletter covering trends in information technology and knowledge strategy. Includes insights, articles, related links, and conference listings.

The MIT Organizational Learning Network—papers, case studies, and related articles on organizational learning.

Multimedia and Internet Training Newsletter—includes articles, resource lists, ROI studies, developer lists, software evaluations, and technologies to watch in the Web-based training industry.

Sveiby Knowledge Management—a "site devoted to creating business from knowledge." A collection of resources on measuring intangible assets.

Source: Forbes special issue on Intellectual Capital, April 1997.

NOTES

1. At the personal level, in the future knowledge professionals must learn to "slow down to go fast"—we need to create a personal and shared perspective within which we can operate as a community.

2. A proposed alternative name for the society is The International League of Knowledge Professionals (ILKP). The author is interested in hearing, via e-mail sshariq@mail.arc.nasa.gov from professionals interested in volunteering their time to help organize the formation of an ISKP.

PART II:
THE UNIVERSITY IN THE
KNOWLEDGE-BASED SOCIETY

6

Being Digital: The Unavoidable Transformation of Research Universities

Robert S. Sullivan

INTRODUCTION

During the past 20 years, using digital technology in higher education has been my avocation; what I do because I enjoy it. I see value and opportunity in it and I have capitalized, to some extent, on it. I have been a practitioner.

Some 13 years ago, in teaching graduate courses in operations research at The University of Texas, I worked with a wonderful doctoral student to develop software systems (even then, we referred to them as decision support systems). These systems were to help students focus on decision processes using structured mathematically based management models—we recognized that human decision processes were far more important than actually grinding through algebraic formulas by hand or calculator. It helped us to focus on creating more learning value for our students. On the opportunistic side, the software systems were widely adopted throughout the country. In fact, they still are widely used in many nations, including China.

During the period 1990–1995, when I was a dean at Carnegie Mellon University, digital technology became a central enabler for me and my faculty (and the faculties from other Carnegie Mellon schools) to experiment positively and proactively with the use of technology on our university processes, core functions, and even our mission of higher education. These experiments challenged the stability of what we did and how we achieved our mission.

Therefore, my former colleagues and I created new degree programs which centered on the use of digital technology. We experimented with the use of such technology for "enabling" and enhancing the learning process, and for intertwining education and research. We also used technology to distribute new learning opportunities to geographically dispersed communities both in the United States and abroad. That is, we assumed that digital technology would be central to redefining our "business" as a research university—a knowledge industry. This was interesting, fun, and enlightening, as well as valuable for understanding our profession as educators and scholars. These types of initiatives eventually were drawn together under the umbrella of a new center for innovation in learning at Carnegie Mellon.

I come from the perspective of a practitioner—one who views our profession and function in society as being ripe for experimentation and being appropriate as a laboratory itself. This requires discovery about ourselves. Digital technology has the potential to radically transform the mission and processes of higher education—and to do so in a very positive way. We need to better understand that potential.

For convenience, the chapter is divided into three parts. First I will provide a quick overview of a few common elements of digital technology. I will use these to demonstrate some existing technology-enabled programs. Second, I will give a sense of the future—focusing in on the opportunities associated with inevitable change. I will also present another scenario—an evolution that many of my colleagues in higher education fear—that universities might lose their "monopoly" as knowledge providers. Finally, I will return to the positive by posing the question: What initiatives must universities now take so that digital technology truly is beneficial, so that it "enables" us to achieve our important and evolving role in society?

DIGITAL TECHNOLOGY: COMMON ELEMENTS

What are we referring to when we talk about the impact of digital technology upon universities? Most generally consider the following basic elements:

- *Personal computers* have capacities and power which are expanding at astronomical rates. To repeat the well-known Moor's law, computing power continues to double every 18 months or so. We currently have computing power that exceeds the capacities of many for productive use.
- *Local area networks* link PCs and servers. These networks of PCs share software programs—word processing, databases, calendars, etc.—located on servers for general use.
- *E-mail* is viewed by many as a boon, and by others as the worst nemesis of interpersonal and group communications. Regardless, it is rapidly becoming a way of life.
- *The Internet* provides the communications highway for e-mail, and for the World Wide Web. It is growing in an explosive way, allowing individual and personalized

communications, collaborations, electronic business of all sorts, and access to unprecedented volumes of knowledge. It also is rapidly becoming a way of life.

- *Videoconferencing* (or video teleconferencing) has often been associated with the talking head. Now it is rapidly improving in quality, becoming less expensive, and emerging as a viable way to bring individuals and groups together in a personal and interactive way.
- *Individual and group application software programs* facilitate learning and decision processes. Many of these programs are interactive and can facilitate group interaction.

Additionally, there are the research applications—such as the use of high-performance computing for computationally intensive experiments. Here the technology has never been challenged and its need is increasingly essential. Because all of these elements are based on digital processing, they are rapidly converging—data, voice, and video are being distributed through common devices, with increasing capabilities and decreasing costs.

With this as a very simple framework, I will give one recent example from Carnegie Mellon University which demonstrates digital technology as an opportunity for us to better "enable" the learning process. The Institute for Technology and Management is a collaboration between Carnegie Mellon's Graduate School of Industrial Administration and Reuters. The Institute is a virtual university which delivers educational programs in advanced business subjects to executives around the world. By taking education to the learner, the Institute will aid industries in retooling their executives, as the role of the executive changes and is redefined. The Institute for Technology and Management brings together various technologies and ideas to create a highly effective virtual learning environment:

- Videoconferencing brings classrooms of students together from different locations, thus enabling the use of one instructor and multiple facilitators. In addition to the clear economies of scale, there are other, perhaps more important results. Students with different backgrounds and with different cultures interact and get to know and understand one another by working with each other. These behavioral and cultural benefits were not anticipated, but are clearly very important. Additionally, faculty from various locations are able to jointly plan, deliver, and discover during the sessions. Industry uses the term "global collocation" —being geographically dispersed, but also together. Technology allows this to happen, and offers tremendous opportunities for both students and for scholars.
- A local area network enables the creation/simulation of financial markets, with each student acting as a trader. This is experiential learning—the technological equivalent of on-the-job training.
- Additionally, there is the electronic textbook, an on-line decision support system, to facilitate the experiential learning process. The text automatically adjusts to the current simulated learning experience.
- Reuters provides real-time data from actual financial markets, which compliments the experiential learning. This is particularly important because the sheer volume of data now available can be overwhelming. The system allows for experimentation on

identifying and accessing the most relevant data, and processing it in a fashion that best contributes to decisions.

- Each location is linked together via the Internet to allow for the creation of a simulated global financial market, allowing students from different locations to make decisions that impact one another.

In the last two years, many new applications have come on-line—for globally dispersed student discussion groups, lecture notes and homework assignments on the Web, and direct e-mail access to faculty and facilitators. In a real sense, the only limiting factors are creativity and time.

Now I will provide another—more gloomy—perspective.

HERMITS SUFFER NO PEER PRESSURE

Those in higher education have been privileged, in a sense, because they have been entrusted with great traditions and responsibilities for society, as well as for the individual. In fact, many have noted that universities correspond to one of the most enduring forms of institutional organization ever created. Universities today remain similar to their original medieval form. Many of the existing European universities were already in place in the 15th century, and have endured despite devastating wars, religious turmoil, economic cycles, the industrial revolution, and many other transformations that destroyed and created innumerous institutions.

With this observation, why should we expect (or even suggest) that our universities, so well rooted in tradition, will change—even the suggestion of change, for some, borders on heresy. That is, digital technology is viewed by some as a demonic god about to violate the soul of our sacred institutions. This is a view put forth by Neil Postman, who refers to the god of technology, being preached on a pulpit of hype and cheerleading.

What is being challenged that evokes such negative reactions? William Massy of Stanford University and Robert Zemsky of the University of Pennsylvania suggest the following impediments to change:

1. The constraints of traditional academic values. Examples are the established institutional norms of teaching methods, faculty autonomy, and notions of productivity. In fact, faculty often cringe at the notion of measuring productivity—as if it is simply impossible or inappropriate for their unique world.
2. That faculty value having peers rather than technology. This tends to skew the appropriation of resources.
3. That faculty hide behind the "veneer" of collegiality, where the need for spontaneous, interpersonal dynamics among colleagues is purported to be essential for scholarship and discovery.

Consequently, there is a resistance to change in higher education—this fact generally is not contested. This resistance, along with the advantages of technology, has created opportunities for new nontraditional providers of knowledge and skills—who now threaten the long-established monopoly of universities. These profit-oriented providers can choose the perceived "best" instructors and scholars from around the world, in the most valuable or topical areas. They then use technology to deliver studio-quality learning to industry and individuals who have the capacity to pay. Learners no longer have to go to MIT or The University of Texas for certain types of learning and skills. The result is an erosion of the resource base of traditional universities, thus threatening the process of subsidy and cross-subsidy. These subsidies have enabled the maintenance of the breadth of inquiry and scholarship (even those areas that are not "popular"). It leaves the traditional universities with higher overhead costs, but with reduced revenues. IC2 Institute at The University of Texas at Austin has been involved in a strategic alliance with one of these nontraditional providers. Lessons learned from this alliance include:

- The company is able to capitalize on the cache of name-brand universities by using faculty from these institutions. In fact, the participating universities receive very little, if any, revenue from the programs. Generally, the faculty is paid directly by the company for their services.
- The company can assure the "best" instructors in various disciplines and guarantee the quality of the delivery process by cherry-picking from each university's specific qualities and expertise.
- They deliver multipoint interactive instruction to the learners at times convenient to the learners.
- They provide this learning experience at affordable prices (generally less than university executive programs). They are able to do this because of economies of scale, and because they do not have to conduct research or pay university overhead costs, and because they focus on educational areas of high perceived value.

This example is not just a business school phenomenon. There are similar initiatives in engineering, where graduate degrees are offered from colleges including MIT, Stanford, and others. There are also initiatives in healthcare and other professional programs. With all of these initiatives, the question is: How should universities respond to the challenge? Unequivocally, they must respond, and respond quickly. I offer the following suggestions.

A MODEST RESPONSE

We need to make digital technology our best enabler—a friend of higher education. I suggest two approaches that I believe are necessary (but likely not sufficient) for addressing the challenge of technology:

- Universities must proactively engage in a process of research, discovery, and experimentation on the core mission and processes of our profession. That is, we need to establish and literally become laboratories for better understanding the nature of work associated with learning and research. We need to understand the appropriate and evolving role of information technology; its impact upon individual learners, educators, and researchers; and we need to be willing to experiment with our process (and put in place incentives for such experimentation). We need to practice on ourselves that which we preach to others—the application of the scientific method of inquiry. In a sense, we need to systematically and continuously search for new ways to improve our quality and excellence. We must be able to experiment, discover, and implement in a timely fashion. Digital technology is breaking down the traditional barriers that defined the mission and processes of higher education. We must be willing to capitalize on these new opportunities.

- Universities must proactively and unabashedly attempt to understand the development and evolution of nontraditional providers of learning. They must understand the way they operate, the directions they are going, and the value they create. The brokering of highly visible scholars and education represents a new form of coalition or strategic alliance for the university, although it has been well-established in industry for years. Until these companies establish their own reputations, however, they must feed off the university infrastructure and reputation. Universities can remove the "middle man" by forming their own direct alliances—and by focusing on the perceived quality of education. This focus on quality is essential. Digital technology, which is being exploited by nontraditional providers, can be of even greater benefit to the "new" traditional universities—if they choose to seize the moment. Indeed, digital technology must be used as the enabler of our redefined roles.

Digital technology is opening up new and exciting opportunities for us. If we do not seize these opportunities, others will—as they rightfully should. We should not be afraid of the competition, especially since we cannot prevent it. We should welcome it to invigorate our own processes of reinvention.

I am confident of our future—confident that we will rise to the challenge as we seek to define new and exciting roles for universities into the 21st century.

7

Industry-University Linkage and the Role of Universities in the 21st Century

Yasuo Konishi

INTRODUCTION

The topic of industry-university linkage is not new. However, effective implementation of this concept to yield the expected output and outcome for both industry and university has been and continues to be a major challenge. The decline in volume of government-sponsored research programs, corporate downsizing, advances in information and multimedia technologies, and other global factors have redirected our attention toward the need to strengthen the relationship between industry and universities so as to create closer synergies, both technical and financial, to help produce mutually beneficial results.

In the United States, for example, the Morrill Act of 1862 helped set up the necessary legislative framework for the development of the Land Grant University System. This act, followed by the Hatch Act in 1887, helped establish centers for agricultural research, thus pioneering the concept of tying basic research undertaken at universities to commercial activities. Since then, many modalities for collaboration have evolved. It was not until the mid-1970s, however, that industry-university collaboration became a focus of debate in industrialized countries. Specifically, industrial productivity, the growing concern over global competitiveness, and restraints on public-sector finance pushed policy makers and private industry to seek partnerships with universities in which universities would take on a leading role, especially in the field of science and technology.

Perhaps the most impressive example of ties that evolved between industry and university can be found on Route 128 in Massachusetts, in Silicon Valley in Northern California, and at the Cambridge Science Park in the United Kingdom. During the 1980s, timing, combined with the prevailing university outreach and entrepreneurial spirit of the times, made it possible for industry-university linkage to take a foothold as an issue for policy consideration in the context of industrial competitiveness. While many enterprises that sprang up at these three technology sites have been hugely successful, these three examples are still considered an anomaly in a much larger universe of well-intentioned attempts to form partnerships between industry and universities.

In an attempt to replicate the successful industry-university partnerships found in the United States, the United Kingdom and elsewhere, governments, scholars, and private-sector managers have contributed to the development of a range of modalities for industry-university linkage. At the same time that new modalities are being developed, the environment in which data, information, and knowledge are imparted is changing rapidly as a response to the growing application of technology-based learning tools in both industry and academe. As multimedia technology rapidly permeates our daily lives, the way in which we acquire, assemble, and transform information into value will increasingly change. This transformation is expected to have a substantial impact on the role and significance of universities, both as a focal point for basic and applied research, and as the principal providers of education and training.

The objective of this chapter is to show how advances in information technology will contribute to the deinstitutionalization of universities and the education systems of the future, and the impact this will have on the dynamics between universities and industry.

MODALITIES OF INDUSTRY-UNIVERSITY PARTNERSHIPS

Enterprise-level collaboration with universities has yielded effective solutions in improving product development and manufacturing processes, enhancement of specialized skills, and the development and application of innovative technologies. While the type and degree to which various modalities for industry-university partnership are employed vary at the country, sector, and even at the enterprise level, an inventory of the various industry-university partnership activities suggests that the framework within which most partnerships are formed generally falls into one of three modalities: research, service/consulting, and education/training (Figure 7.1).

Figure 7.1
Modalities of Industry-University Linkage

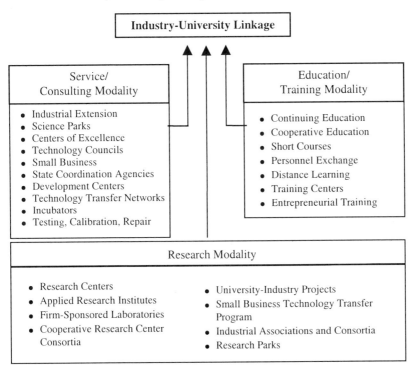

Research Modality

The origins of industry-university partnership are rooted in taking basic research conducted at universities and translating this knowledge into commercially viable goods and services. By doing so, the partnership creates value for the enterprise, which in turn helps fund further research for universities. Even now, the research modality continues to be the most relevant form of partnership for industries and universities. Research partnerships include a wide range of activities, from one-time product-specific research to much broader basic research sponsored by a consortium of interested investors. Generally, partnerships in the field of research fall into one of eight categories:

1.   Development and operation of *research centers.*
2.   Applied contract research undertaken by *applied research institutes.*
3.   Individual firms or groups of firms collectively commission both basic and applied research through *firm-sponsored laboratories* where researchers from university and private enterprise work closely together to achieve a specific objective.
4.   *Cooperative research consortia* are, by far, the most prominent form of partnerships involving groups of enterprises and sometimes multiple universities tackling both basic and applied research.
5.   On a smaller scale, *industry-university projects* involve small groups or individual firms working closely with university research laboratories, but with the added benefit of being partially subsidized through cost-sharing schemes with government.
6.   To assist small business during the preprototype stage of product development, some countries employ *small business technology transfer programs*, often subsidized by the government, to help translate concepts into realizable goods and services.
7.   *Industrial associations form consortia*, whether in-house, within a shared facility, or as a stand-alone research facility, primarily to conduct basic research in which universities take the initiative or help manage the process.
8.   Perhaps the most liberal interpretation of research partnerships falls into the category of *research parks*, where a cluster of related and unrelated enterprises congregate, often near a university, to address both basic and applied research concerns.

Service and Consulting Modality

Partnerships formed under the research modality are a useful launching point for developing concepts and ideas through basic and applied research. Once the technical platforms have been completed, the next step is to create value through the application and transformation of knowledge into actual goods and services. In order to do so, several types of partnerships have been devised to help enterprises translate existing or new technologies and techniques into productive use. It is important to note, however, that there are overlaps in the perceived comparative advantage of services offered by private-sector consulting firms and the types of consulting services most often associated with universities. Specifically, the focus of university-based consulting services places emphasis on design, testing, preproduction marketing, and assisting the innovation process of product development. In this context, partnerships between industry and universities take on a number of structures:

1.   Agricultural extension services are the earliest form of industry-university partnerships; these have evolved into other forms of *industrial extension services*, particularly to assist small and medium-sized firms in integrating their products and services into the market.
2.   Although there are a number of nonuniversity-based *centers of excellence*, most activities are part of a university, catering to specific technologies or servicing the regional needs of small and medium enterprises, particularly in areas such as prototyping, simulation, and scaling activities.
3.   *Small business development centers* are not necessarily a university-based function, but are often affiliated with a university, especially in providing business planning and other management-related services.

4. *Incubators* provide both the physical and financial infrastructure necessary to support start-up companies and help them become viable business entities, and some infrastructural support is effectively delivered by universities.

5. For small and medium enterprises in the process of launching a product on the market, access to low-cost *testing, calibration, and repair* services is considered a crucial first step to actual production of goods and services, and as such, universities often offer government-subsidized support to small and medium-sized enterprise (SME) sectors for such a service.

6. A hybrid between centers of excellence and small business development centers, *science parks* cater to enterprises with a distinct scientific and technology focus; universities are an integral part of the functions of the park, but do not necessarily share the same physical site.

7. Catering to regional and provincial needs of SMEs, universities often serve as a focal point to help identify the technology needs of enterprise through *technology councils.*

8. *State and provincial coordination agencies* are formed by governments to develop and promote technology assistance programs to SMEs, whereby universities often act as part of the support infrastructure for the delivery of services.

9. University administrations act as brokers of technology to industry by developing *technology transfer networks* through which know-how developed by the university is marketed to the private sector.

Education and Training Modality

Of the three modalities, the education and training modality is the most informal form of partnership between industry and university. Numerous attempts have been made to formalize the process of assessing the skill needs of industry and translating them into university curricula. To date, however, partnerships have tended to be targeted toward the education and training needs of specific individuals or have catered to the much broader knowledge requirement of the general public to help improve prospects for employability. In this area, partnerships fall into one of several formats for imparting knowledge:

1. *Continuing education* is a method by which educational services are offered to employees to help expand existing, or create new, skills and knowledge, where universities are often actively involved, but outside of the context of in-plant training.

2. Universities, particularly in delivering technical education and training, utilize *short courses,* another form of continuing education, whereby accessibility to education and training is improved by shortening the period in which information and knowledge are imparted.

3. Advancements in multimedia technology have allowed universities and vocational and technical institutions to increase the delivery of learning experiences by imparting information and knowledge outside the context of a traditional institutional setting.

4. Universities are often hosts to specialized *training centers* where specialized technical training is delivered to students and executives alike.

5. *Cooperative education* is a form of apprenticeship whereby the time is divided between education in a formal educational setting and on-the-job training.

6. As a way of imparting practical knowledge in a formal educational setting, *personnel exchange* programs allow experts from industry to become part-time faculty members at an educational institution.
7. Most recently, *entrepreneurial training* programs, where students are trained to evaluate and screen new ideas and processes through the experience of entrepreneurs, have gained in popularity.

LINKAGE MECHANISMS FOR INDUSTRY-UNIVERSITY PARTNERSHIPS

Integrating the highly technical and often theoretical undertakings of universities into practical applications at the enterprise level requires a complex web of formal and informal networks and agreements. While universities are seen as a focal point for imparting knowledge, in the context of industry-university partnerships, the flow of information is not one-way, but comprises a dynamic process of shared learning consisting of complex interactions between organizations, people, and ideas, eventually leading to a mutually beneficial outcome. Irrespective of the type of modality organizations select, nurturing both formal and informal linkages plays a crucial role in the success of an industry-university partnership.

A review of major industry-university partnership activities suggests that there are at least four types of linkage mechanisms, representing various degrees of commitment between the key players in an industry-university partnership.

1. *Contractual agreements* are the most formal and binding arrangement for industry and university to maintain a partnership with a clearly defined outcome.
2. A *Memorandum of understanding (MOU)* is an informal but binding agreement between two or more parties to undertake a shared activity.
3. *Grant-based agreements* are formed around contracts and grants sponsored through government programs or other funding agencies.
4. *Organization-based agreements* are formed as a result of the joint sponsorship, creation, or maintenance of an organization.

These forms of linkage serve as bridging mechanisms to bring together the interests and activities of enterprises and universities within formal and informal frameworks. In a formal setting, institutional networks are well-defined, the substantive interests and objectives of the group are well-integrated, fundraising is structured and often self-perpetuating, and organizational arrangements between enterprises and universities are structured. On the other hand, equally important is the informal structure within which partnerships exist. Informal structures are characterized by extensive use of personal contacts and networks, event- or activity-specific agenda; funding is often one-time and ad hoc, and organizational arrangements are fluid and unstructured.

In the previous section, the various forms of partnerships were described according to three modalities: research, service/consulting, and education/training. These three categories provide some insight into the most appropriate type of partnership according to the various stages in the development of a technology, product, or enterprise. At the same time, however, if the various types of partnership are regrouped according to the institutional arrangement as it relates to a university, we begin to see the relative importance of the university as a provider of institutional infrastructure for launching an industry-university partnership.

As is evident from Figure 7.2, a large majority of partnership modalities fall within the context of a university setting. At the same time, however, it is the know-how of the faculty, the high concentration of specialized expertise centralized in a single physical location that makes a university an attractive physical setting for establishing partnerships.

Figure 7.2
Partnership Modalities in the Context of Institutional Arrangements

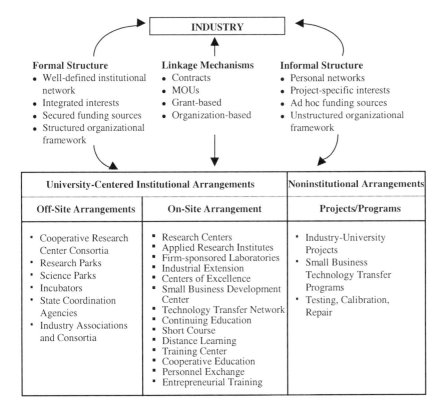

University-Centered Institutional Arrangements		Noninstitutional Arrangements
Off-Site Arrangements	**On-Site Arrangement**	**Projects/Programs**
• Cooperative Research Center Consortia • Research Parks • Science Parks • Incubators • State Coordination Agencies • Industry Associations and Consortia	• Research Centers • Applied Research Institutes • Firm-sponsored Laboratories • Industrial Extension • Centers of Excellence • Small Business Development Center • Technology Transfer Network • Continuing Education • Short Course • Distance Learning • Training Center • Cooperative Education • Personnel Exchange • Entrepreneurial Training	• Industry-University Projects • Small Business Technology Transfer Programs • Testing, Calibration, Repair

TECHNOLOGICAL ADVANCEMENTS AND THEIR IMPACT ON INDUSTRY-UNIVERSITY PARTNERSHIPS

Great technological advancements have been made to date, particularly in technologies readily adaptable to the educational and research activities often found in industry-university partnerships. The technologies most relevant to industry-university partnerships fall into at least four distinct categories:

Multimedia:	Compact disc (CD-ROM); and interactive video disc (CDI)
Teleconferencing:	Audio teleconferencing; audio-graphic teleconferencing; video teleconferencing (telephone-based); video teleconferencing (satellite-based); and desktop teleconferencing (telephone-based)
Computer Networking:	Local area networks; telnet; and file transfer protocols (wide area information system, Gopher, Veronica, WWW)
Courseware Applications:	Computer-managed instruction and computer adaptive testing; hypertext, hypermedia, and computer-based training; intelligent tutoring systems; electronic performance support systems; and simulation

The introduction of these technologies has had direct and indirect impact on both structural and substantive relationships forged between industry and university. Perhaps the most significant impact of technology intervention on the dynamic between industry and university is its influence in deinstitutionalizing the university as the bastion of learning and knowledge. The application of multimedia technology eliminates the need to "cram" learning into a limited time period or to congregate learners in a specific physical location or predetermined environment in order to bring about learning or experimentation. Experts no longer need to congregate in one physical location to conduct an experiment, but now have the freedom to simulate a collaborative experiment in a cyber lab.

In short, the intervention of multimedia technology in learning and research is creating opportunities for learners, teachers/facilitators, and researchers alike to emulate a classroom or a laboratory by redefining the relationship between time and space. Therefore, employing technology to overcome the limitations of physical space and problems of distance, to split space into time, has opened up a whole new dimension in the way industry and universities collaborate.

As evident from Figure 7.2, a number of research-related activities rely on universities as a physical entity. At the same time, however, if university researchers are no longer bound by the need to be located in a single physical location to conduct research, then the added value of the physical support infrastructure offered in a university setting is greatly diminished. In addition, the advent of multimedia technology in research allows developments to take place simultaneously in both physical and shared cyberspace. For example, the application of multimedia tools can create opportunities for a group of researchers located in different physical locations around the world to work on a single project in their own laboratories, while transmitting findings to others, whether in real time or with a time lag, so that research findings take on a pyramid structure in shared cyberspace.

In this respect, the role of the university as a large-scale, diversified research laboratory will decrease in importance. A probable outcome will be the emergence of a number of specialized independent laboratories or cooperative laboratory facilities where researchers can conduct off-site research either on a fee-for-use basis or as part of a larger private practice. A parallel example to this is the development of small- to medium-scale private medical practices in the United States. The growing trend has been away from large-scale hospital care, toward a network of small specialized private practices where a system of referrals is used through a managed care structure (the general practitioner helps to manage the medical needs of the patient).

In such a scenario, technical experts would participate in research projects on an individual basis as well as in groups, relying heavily on various modes of electronic communication to link them and their findings. This form of virtual laboratory work will most likely create various forms of subcontracting schemes which will evolve as a linkage mechanism to pool technical expertise to achieve a common objective. This suggests that cooperative research consortia-type institutional arrangements will lead the way in bringing together technical experts and managing research through cyberspace.

Similar changes are expected to take place in the field of education and training. Specifically, we can anticipate at least three major changes in the way information is created, imparted, analyzed, and translated into knowledge.

1. *Baseline requirements for learning to survive and compete effectively in the workplace:* Traditionally, basic employability skills included computation skills and communication skills (writing, reading, speaking, and listening). Given the demand for an understanding and command of the range of technologies incorporated in a workplace, information literacy, particularly within the context of information technologies and computer-related systems, has become an essential qualification for most jobs.

2. *Dimensional change in the depth and breadth of information available*: During the early stages of learning, the learning experience is often dictated by the volume and accessibility of information available to the learner. The gatekeeper of information, in turn, has traditionally been the teacher/instructor. Computer-based learning tools, however, have changed the traditional relationship between information and people. Thanks to the intervention of multimedia technology in education and learning, learners now have access to an unbroken stream of information and knowledge at any time and virtually anywhere. In addition, multimedia technology provides users with an opportunity to create new information and make it instantaneously accessible. While the volume of information flow has continued and will continue to increase, so too with the quality and validity of information. A number of implications are possible:

   a. A clear understanding of the differences between information/data and knowledge will become even more critical.
   b. The ability to translate and transfer information/data into knowledge will become a separate and independent skill.
   c. New methods will be required to manage and distribute information/data, making it possible to effectively create new knowledge.
   d. Having command over the mechanics of searching and retrieving information will be a prerequisite for learners and teachers/instructors alike to remain competitive and relevant in their own environment.

3. *Redefinition of hierarchy in the learning system*: Teachers and instructors are no longer at the center of a learning experience. With the advent of technology-based learning tools, accumulated knowledge, experience, and information through networks form the backbone of numerous learning opportunities. In this regard, students learn not only through teachers/instructors, but also along with them and other students through simultaneous interaction. Interactive learning is made possible due to the wide range of information now accessible through electronic media and networks. At the same time, access to information has created opportunities for learners to more effectively rebut or question teachers/instructors and learner:

    a. Teachers/instructors aid learners in the selection of appropriate information and learning tools.
    b. Qualitative and quantitative requirements for information are negotiated between teacher/instructor and learner based on the expected outcome of the learning opportunities.
    c. Learners are ranked or assessed according to the desired outcome and output of a learning experience.
    d. Performance of the learner is evaluated on the ability of learners to demonstrate originality and creative means of constructing knowledge.
    e. The ability of learners to cooperate and collaborate with others to achieve a desired output and outcome will weigh heavily in the evaluation of learners.
    f. The discipline of managing information, and the ability to command and organize a large body of information, will become a prerequisite for the interactivity of a learner with other learners, and the relationship formed between learner and teacher/instructor.
    g. The ability to integrate information/data into visual, audio, and text formats to create new knowledge will define the role of an individual within a group.
    h. Individuals will be simultaneously learners and teachers/instructors, thus creating a need to clarify the responsibility and accountability of individuals and what is presented to the public.
    i. "Providers" may become a new professional classification in the learning system, where a distinction must be drawn between providers and teachers/instructors.

WHAT CAN WE EXPECT IN THE FUTURE?

Before Microsoft, Netscape, and the Internet became household words, knowledge creation and dissemination, the business of packaging knowledge, the distribution of these packages, and users of knowledge existed as independent entities defined within the context of an organization, institution, or individual. In short, the compartmentalization of activities into a clearly defined institutional framework created physical boundaries between the various evolutionary phases of knowledge, with the chain of events beginning at knowledge creation and ending with wealth creation.

Thanks to the introduction of computers and the ever-increasing role of electronic media as a crucial life support system for the day-to-day operation of businesses and people's livelihoods, the clear institutional relationship and logical sequence of events that once defined the life cycle of information and knowledge has been gradually eroded, giving way to porous boundaries between knowledge creators, distributors, and users. In fact, wide acceptance of the Internet, and its inherent role as a petri dish for cultivating information, has created a new breed of thinkers, information-mongers, and consumers, where the creators of knowledge are simultaneously its packagers, distributors, and users.

The dynamics around which information and knowledge evolve are redefining the relationship between time, space, and resources associated with traditional education systems. Learning will be less concentrated in time, and physical location will no longer define accessibility to knowledge. Resource distribution will gradually shift away from assimilation of data toward stock in the productivity of knowledge. These changes threaten the future existence of traditional didactic methods and the role of universities as the Shangri-La of intellectual pursuit, while at the same time posing new opportunities for universities to create and deliver learning experiences once inaccessible to many.

The introduction of, and our dependence on, electronic media in our day-to-day lives will not diminish the importance placed on basic education, or on the ability of individuals to think, communicate, and take on basic workplace responsibilities. However, once these basic foundation skills are acquired, how we assimilate and transform information and know-how into productive means no longer hinges solely on formal university education. Similarly, access to electronic media and its effect of deinstitutionalizing universities is not expected to increase educational choices, but will have a tremendous impact on creating new opportunities to learn. At the same time, learning is likely to take place more and more on an individual basis, thus by definition introducing changes in the characteristics of the labor force, which will be composed of workers with increasingly individualized knowledge and specialized skills.

Likewise, proactive intervention by industry in direct learning opportunities aimed at tailoring individual knowledge and skills to meet its own internal requirements will set the pace for labor productivity, and will in turn define corporate performance. In this respect, demand for universities to deliver basic foundation skills and employability skills may shorten the time devoted to education in a formal educational setting. At the same time, however, supplementary skills training, whether within or outside the workplace, based on electronic media, will transform the ability of workers to perform specified tasks in the work environment. In this context, new forms of subcontracting relationships may evolve between industry and creators of knowledge—for example, professors, responsible for expanding the scope of employability skills that meet specified corporate objectives and culture.

As the traditional role of universities is redefined, the gap between skills training and formal learning becomes increasingly evident, such that more and more importance is placed on informal means of acquiring skills, know-how, and knowledge—whether through in-plant training, self-directed learning via distance learning schemes, or via commercial providers of information. While university education and accreditation are often viewed by managers of industry as a rite of passage in considering candidates for employment, industry will no longer rely solely on universities as a beacon to ensure the quality and reliability of its workforce. Likewise, it is expected that industry will take a greater interest in just-in-time training, and skills development will most likely be delivered through new subcontracting schemes and on-the-job training.

As educational systems become increasingly deinstitutionalized, the research functions of a university may take on greater significance, with universities becoming focal points for basic research, but within a less structured environment where researchers are not tied to a single physical entity or location. This would transform the university into a scaled-down think tank, where basic research and critical thinking can take place without the usual demands for teaching and publishing which professors regularly face. Concurrently, the outcomes of university-based research would then serve as the platform for application-oriented industrial research at the private-sector level.

8

On the Socioeconomic Context and Organizational Development of the Research University

Pedro Conceição, Manuel V. Heitor,
Pedro Oliveira, and Filipe Santos

THE UNIVERSITY CONTEXT: A HISTORICAL PERSPECTIVE

Created during the Middle Ages, the institution of the university soon developed its own identity and culture, which was not fundamentally altered until the 19th century. According to Boorstin (1983), the old European universities and colleges were not created to discover new knowledge, but to disseminate a heritage. The main purpose of these institutions was to rediscover and keep alive the cultural, philosophical, and religious heritage of classical times. The function of knowledge creation in society was thus mainly performed by the lone enlightened scientist, sage or artist, a typical character of the medieval and Renaissance period.

Later, in the 17th century, when knowledge creation became a more complex activity and the need for communication between scientists increased, informal networks were formed, which were later transformed into formal scientific associations. These associations created their own incentives for knowledge creation and distribution, particularly through mechanisms of dissemination and prizes, peer recognition being the most coveted of all. This was the birth of the modern academic system. Throughout this period, the role of universities in the process was not significant because the activity of knowledge creation was not considered an important element of the university's mission.

The industrial development of the 19th and 20th centuries broadened the employment base for qualified professionals, leading to a great development of universities, especially those connected with the exact sciences and industrial fields. The evolution of the university paradigm in the developed countries recognized that research is a fundamental activity of the university, equal in importance to education. These principles were first stated by Alexander von Humboldt, in the constitution of Berlin University in 1809. The basic principles of the Humboltdian research university were assimilated in the course of the twentieth century in almost all developed countries (Caraça et al. 1998).

The second half of the 20th century set the stage for an extraordinary development of higher education systems, which ensured the education of 20% to 30% of each generation cohort, whereas before this number was less than 5% (Gellert 1993: 17). The importance of the research function of universities grew considerably during this period (Rosenberg and Nelson 1996), and the traditional position of isolation of the universities began to be questioned. After the Second World War, universities were considered as promoters of socioeconomic development and cornerstones of the knowledge creation systems (e.g., Rosovsky 1990; Graham and Diamond 1997).

Strongly affected by the rapid expansion of the 1960s, the crisis of the 1970s, and the economic and technological changes of the 1980s and 1990s, universities are seeking creative responses to the new demands of society, by reforming the structure and organization of the activities inherited from the period of rapid growth (see, for example, Readings 1996; C. J. Lucas 1996; Rothbatt 1997; Ehrenberg 1997; Tierney 1998). Investing in closer links with the community, the university sets itself the task of examining the quantitative and qualitative requirements of its activities and new ways to exploit its scientific and technological potential, while striving to maintain its effective autonomy (OECD 1997). The impact of these efforts has led to the recognition that beyond the university's traditional roles in education and research, a wide range of other activities, usually grouped together under the heading of "provision of services" or "links to society," are now part of the university's mission (Roberts 1991; Rosenberg and Nelson 1996; Mitra and Formica 1997).

The contemporary university is thus faced with a twofold challenge: on one hand, society presents it with new and growing demands, while at the same time governments apply increasingly restrictive policies to the funding of its activities. The combination of these two factors is reflected in a growing diversity of funding sources and mechanisms (Caraça et al. 1998).

Given the above context, we contend in this chapter that the socioeconomic transformations occurring in the last quarter of the 20th century are strongly challenging the concept of the university as the center of knowledge creation in society. To sustain our argument, we first present some findings drawn from the emerging area of knowledge economics, discussing the characteristics of the innovation process and presenting the relation between the different kinds of knowledge and economic prosperity. Next we briefly discuss the mission of the university in terms of the current economic context, which may divert universities from their mission and harm their institutional integrity, because universities may be induced to compete with other types of institutions for research funding and technology application. The answer to this problem is a careful combination of public and institutional policies for R&D funding, organization, and knowledge exploitation. Then we set out the arguments for increased public funding of university R&D and develop a model for exploitation of universities' intellectual capital, also taking into consideration the necessary integration with educational activities. Finally we provide the rationale for a new organizational model of the research university, and we conclude by showing how this combination of public and institutional policies serves to protect the integrity of universities and ensure the added value of its activities for economic prosperity and sustainability.

TOWARD THE KNOWLEDGE ECONOMY

Empirical Evidence

There is a general perception that today's developed economies depend on knowledge and information in an unprecedented way (see, for example, the analysis of Conceição et al. 1997). Nevertheless, traditional accounting and statistical reporting methods, at both firm and national levels, make it extremely difficult to test any claims based on knowledge or intangible assets. Knowledge is not a traditional economic input, like physical capital or labor. For example, Stevens (1996) says that the idea of a knowledge-based society remains more of a concept than a measurable entity. Nevertheless, he also claims that about half of the OECD countries' GDP is now knowledge-based. Allan Greenspan, the Federal Reserve Board chairman, said recently in *The Economist* (1997) that the American economy's GDP weighed today, in tons, as much as it used to a hundred years ago. But how can we measure these claims? So far, the evidence comes mainly from stylized facts, such as the growing incorporation of knowledge in products, and the increasing value associated with software, as opposed to hardware. A figure mentioned by Wyckoff (1996) reflects this fact: in the 1970s, 80% of the value of an IBM computer was hardware, and 20% was the cost of the incorporated software; in the late 1980s, the proportions were reversed, and ever since they have moved in a direction where the value of software keeps increasing. Software, namely the knowledge incorporated in products, is more important not only in terms of product differentiation and value added, but also as a source of innovation. This led to the introduction of the idea of reverse "product life-cycle." This occurs when knowledge, say computer software, is developed in order to increase the differentiation of a product, or to improve the efficiency of a production process. Normally, the product or the process would be in a declining phase in terms of the traditional life-cycle. But then the computer software becomes in itself a new product or service, and is commercialized in its own right.

In addition to the facts mentioned above, it is also important to note that one of the manifestations of the increasing importance of knowledge-based activities is the tendency of the labor force to move to the service sectors (Conceição et al. 1998). This can be clearly argued on the basis of the structure of employment in the United States given by Wilson (1993), which quantifies the trend of new jobs in the service sector and indicates that job creation favors those with higher qualifications.

Conceptual Basis

Having discussed some of the empirical manifestations of the knowledge-based economies, we now aim to provide a conceptual basis for an economic perspective on knowledge. Using the traditional economic classification, economic goods can be classified according to their degree of rivalry and exclusion. Rivalry is associated with the scarcity and expandability of a good and reflects the idea that, if rival, a good can only be used by one person. Objects are typically rival goods. The consumption of an apple, for example, precludes others from consuming the apple. However, the knowledge contained in, say, a book, is nonrival. The fact that I am reading and enjoying a book does not preclude others from reading the same book. The same happens with music stored on a CD or with a software program. Excludability is associated with property rights over a good. A good is excludable if the owner has the legal power to prevent others from using it. Knowledge can be made excludable, through intellectual property rights. In the case of a book, the author holds the copyright, and may not wish people to read the book unless they pay a fee for it (buying the book), or for them to read it at all (taking the book out of print). Goods with high levels of both excludability and rivalry are designated private goods. In this case, there are private incentives for production because the producers can completely appropriate the benefits arising from the use of these goods by others. At the other extreme, goods with low levels of both rivalry and excludability are public goods. For these goods, such as national defense and public roads, there are few private incentives for production because it is impossible to completely appropriate the benefits of their utilization and existence. The production of these types of goods through competitive markets will not result in efficient resource allocation due to the absence of private incentives. Figure 8.1 presents a traditional matrix for the economic classification of goods in terms of their level of rivalry and exclusion.

Figure 8.1
Economic Classification of Goods

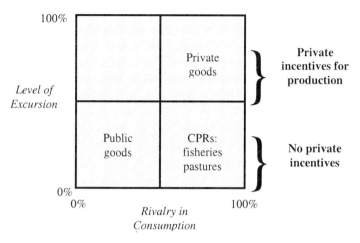

Nelson and Romer (1996) consider that all objects can be classified as hardware—material things that are nonhuman. Knowledge, on the other hand, is divided into wetware, the knowledge stored in each person's brain, and software, knowledge that is codified and is stored outside the human brain. The relevant distinction between these two types of knowledge is that wetware (more familiarly referred to as human capital) is a rival good, since it is linked to one human being and can only be used by its owner. Software, on the other hand, is nonrival, in the sense that a very large number of persons can benefit from the same codified knowledge. Nonrival software has a low marginal cost of reproduction and distribution (making it difficult to exclude people from its use) and is associated with high fixed costs of original production.

In the world of science and technology, there is a tendency to consider science as a public good and technology as a private good. Science rests on the public availability of scientific journals and is freely and rapidly disseminated throughout the scientific community and society at large. Technology is associated with more practical applications exploited by the firms that engage in its development and is protected by patents or other legal instruments for privatizing software. Nevertheless, being an embodiment of knowledge, technology is not truly a private good since it has a low level of rivalry in consumption, because the same technology can be used by many persons at the same time. Technology can then be considered as nonrival software as defined by Nelson and Romer (1996). Figure 8.2 illustrates this classification.

Figure 8.2
Technology and Science: Two Types of Software

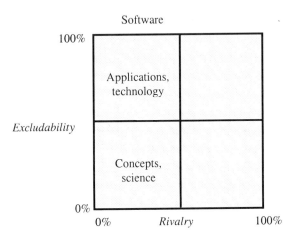

Software creation is the result of research. Most often, the software produced at universities is nonexcludable, being published in scientific journals that are freely accessible to the scientific community and to society at large. The public good characteristic of nonexcludable software creates the issue of individual appropriability. Since there are no private incentives for the production of this kind of knowledge, a market structure does not yield the motivation for scientists to produce. The answer lies in the rules of engagement of the scientific community, the result of a long and complex process of evolution. Stephan (1996), following the seminal sociological work of Robert K. Merton, described the functioning of the scientific community as being based on a "winner-takes-all contest" set of rules. This means that creativity is prized the most: the first scientist to achieve a result gets all the credit, and all similar ensuing results from other scientists are ignored. This type of work ethic has led to a type of appropriability that yields the necessary incentives for production.

Another way to solve the problem of private incentives is to make the software excludable by using intellectual property protection, mainly through patents. Patents are a way to allow the providers of innovations to benefit from a monopolistic position that yields them the benefits from the use of their ideas. However, the production of excludable nonrival software is not specific to universities. The privatization of research results has been the primary route of companies that engage in R&D. Naturally, companies are interested in receiving the payoffs of their research effort by protecting it and commercializing the resulting technology or new products. The production and patenting of new knowledge by companies has increased dramatically in knowledge-based societies. In part, this results from the dynamics of global competition, in which innovation and creativity are increasingly prized over more traditional competitive advantages such as costs.

The distinctive type of output that comes out of university research is public knowledge associated with basic and applied research. Naturally, a university deals with both science and technology, but the academic incentives and organizational structure are focused on the production of science, which has a more public good characteristic. Nevertheless, universities sometimes engage in aggressive and effective programs to protect their intellectual property. The motivation is clearly to derive financial benefits from the creativity of academic scientists. In effect, this action can be visualized as a shift upward in the matrix from nonexcludable to excludable software. The problem is that if the focus on excludable software is too high, the university will put its institutional integrity in danger. This issue will be further analyzed in the next section, after discussing the processes of knowledge production and accumulation.

The Process of Knowledge Production and Accumulation

In a discussion of the socioeconomic context of the university, it is important to note the consequences of the differences between software and wetware, with particular reference to their production and accumulation. The rivalry associated with wetware implies that, on the level of economic classification, it is similar to objects. As a consequence of this rivalry, it is clear who possesses a given object or ability, and simple to assign the corresponding property rights. On the other hand, objects and wetware are scarce, being limited by material and energy resources for the former, and by people for the latter. These two properties (ease of assigning property rights and scarcity) mean that the market functions as an efficient means of producing objects and wetware.

On the other hand, the nonrivalry of software and its typical low distribution costs mean, on the one hand, that it is very hard to assign property rights to it and to protect those rights, and on the other that there is no lack of software. Indeed, software tends to be abundant, especially given advances in information technology and telecommunications, which enable codified knowledge to be easily and inexpensively used and transmitted. Terms such as "the digital economy" and "the information economy" clearly reflect this. However, it is important to note that these terms are not synonymous with the wider concept of a "knowledge-based economy," which, as will be seen, has to do with the need for continuous learning processes, involving not only codified knowledge but also the skills needed to use that knowledge.

According to Solow (1997), the normalization of the process of economic development in the new growth theories follows the conceptual structure originally proposed by Arrow (1962). It is worth looking briefly at Arrow's analysis, as it contains the kernel of the reasoning behind the idea of economic development as a learning process. Instead of following the orthodox thinking of his time, which attributed to technological change the component of growth that could not be explained by the accumulation of labor and capital factors, Arrow argued that experience in the use of capital led to an increase in the knowledge used in production. In plainer terms, Arrow drew up a relatively simple model in which workers in a company learn by using the means of production, thereby increasing the company's productivity.

In this way, learning—that is, the accumulation of knowledge—appears as the driving force behind the increases in efficiency which lead to economic growth. It is interesting to note that Arrow chose an informal way of learning, learning by doing, as the basis for his reasoning. It should also be noted that in this model knowledge is accumulated only in the form of skills. The contribution of the new economic growth theories has been precisely to extend this reasoning to other types of learning, as well as to the accumulation of ideas, starting from when Romer (1986) showed the wider implications of Arrow's arguments.

Thus, R. E. Lucas (1988) also analyzed the accumulation of knowledge in the form of skills, but this time putting forward education as a formal learning process. In turn, Romer (1990) and Grossman and Helpman (1991) constructed models in which the accumulation of ideas results from effort put into research, another formal learning process. In this context, Table 8.1 summarizes how these contributions fit into a framework of possibilities which relates the accumulation of knowledge to the different kinds of learning that can lead to this accumulation. The construction of this table was also inspired by Foray and Lundvall's analysis (1996), in which they placed particular emphasis on the formation of networks of personal and professional contacts, which result from processes of social interaction, the fourth process in Table 8.1.

Table 8.1
Accumulation of Knowledge and Learning Processes in the New Growth Theories

Accumulation of:	Formal Learning Process		Informal Learning Process	
	Education	R&D	Experience (learning by doing)	Interaction
Software (ideas)		Romer (1990) Grossman & Hellpman (1991)		
Wetware (Skills)	Lucas (1988)		Arrow (1962) Romer (1986	

The analysis shows the very recent appearance of attempts to analyze the economic implications of learning processes that result from social interaction, particularly in the "information society." Indeed, this aspect puts forward a new vision of the university, notably with reference to the radical change from formal teaching to participatory learning, which is directly associated with continuous (lifelong) training and the need for the university to deal effectively with multiple demands and a multifaceted public. Furthermore, the fact that informal learning processes are shared between a varied range of institutions opens up new possibilities for the universities' ability to create and disseminate knowledge in the emerging economies.

ON THE MISSION OF THE RESEARCH UNIVERSITY IN THE KNOWLEDGE SOCIETY

Universities are diverse institutions across countries, and even within countries. It is always a great risk to refer generically to such a rich type of organization. Higher education occurs in many different institutional settings and is intended to serve different ends. In this context, there is a consensus that universities have the primary role of providing higher education and developing research activities. For a thorough discussion on the university's mission in the American context, see, for example, Christopher Lucas (1996), Readings (1996) and Ehrenberg (1997), while in the European setting see Caraça et al. (1998). In the scope of this chapter, the university function of teaching contributes to the accumulation of knowledge, specifically of skills, through the formal processes of learning through education, or "learning by learning." Given the aim of this chapter we will focus our attention on the research function of universities.

As we have seen in the previous section, there are two broad strategies that yield private incentives for software production. The first strategy relies on giving the scientist economic rights over his or her discovery, through the protection of intellectual property. The scientist may then release the results at will, and ask for a monetary payoff for external use of these results by others. The second strategy relies on the rules of engagement of the open science community. Results are made freely available to all, and the reward is based on the priority of discovery, which yields academic reputation and is reflected in promotions and income. In summary, the first strategy is one in which research results are privatized, and the second one in which they are public.

In the current context, under financial stress, a likely route for universities is to make the software they produce excludable, by issuing restrictive intellectual property rights, particularly through patents. The rationale for this would be to capitalize on the economic significance of the results, obtaining private benefits at the expense of making them freely available to society. This "privatization" of R&D results may have extremely negative effects, because the public good characteristic of this software is lost and because the institutional integrity of universities may be endangered.

By the institutional integrity of the university we refer to the idea proposed by several authors[1] that universities have, over the centuries, developed an institutional specialization by which they perform a unique societal role, by virtue of making publicly available the results of their research and education activities. This specialization has been accompanied by the emergence of institutions such as firms that have developed their own unique features, notably the fact that they strive for profit by privatizing the outcomes of their production processes. Although universities are important in creating technology, they are crucial in creating science, the nonexcludable portion of software. A threat to the institutional integrity of the university would mean, for example, that there would be less incentive to produce non-excludable software in comparison with the incentives to produce excludable software. This would mean in the first place that universities would have to compete with companies for the privatization of knowledge and, following Rosenberg and Nelson (1996), there is no reason to believe that universities would perform well in an environment in which decisions need to be made with respect to commercial criteria, and every reason to believe that such an environment would damage their legitimate functions. Additionally, there would be a lack of investment in the other type of valuable knowledge output—public, nonexcludable software. It is widely acknowledged that public university research results are necessary for long-term economic progress, and increasingly so in a knowledge-based economy. In terms of the innovation model proposed by Kline and Rosenberg (1986), this would mean that the links between R&D and the available knowledge pool would be severed. Consequently, firms would have to muddle through a pool of existing knowledge less vibrant, dynamic, and diversified than would exist had the universities kept the public good characteristics of their research.

In this context, the greater majority of the ideas that are generated in universities should be publicly available, this being the essence of the specific contribution that the university makes to the accumulation of ideas. In addition to this idea, and from a more pragmatic viewpoint, the university should respond to the needs of society, which include rapid and unpredictable changes in the structure of the employment market, and the need to furnish its graduates with new skills beyond purely technical ones, in particular learning skills.

The response to the first issue, relating to changes in the structure of the employment market, involves public policies designed to strengthen and preserve the institutional integrity of the university. The universities cannot actually be expected to foresee the demands of the employment market five or six years in advance. If they were to try, this would certainly entail jeopardizing their integrity. A solution to this problem is to develop a diversified higher education system, which would include various institutions with different vocations, in such a way as to promote a functional stratification of the system. This could be the way to ensure sustained flexibility capable of providing society with the instruments it needs to deal with instability in employment and, more generally, with the inevitable changes in technology, tastes, markets, and needs. This seems, moreover, to be the way to meet the challenge of maintaining excellence. The expansion of university education is obviously irreversible in the emerging society, but this fact cannot be allowed to stand in the way of creating centers of excellence. On the contrary, it should encourage their development, notably by means of the stratified system suggested above.

The American education system can give some pointers toward a possible path to follow. According to the Carnegie Foundation for the Advancement of Teaching, which produces a semiofficial classification of American higher education institutions, there are around 90 "research universities," being those which have generally been called simply "universities." These 90 institutions operate within a system of 3,706 institutions (not counting the 6,256 others that only provide vocational training), with a total of over 14 million students enrolled. In this way, the diversity and functional stratification of the system as a whole helps it to respond to rapid changes in the employment market, particularly through those institutions oriented more toward teaching and with shorter graduation times, without putting undue pressure on the universities.

A diversified and stratified system also presents advantages with relation to the second issue, the need to create and promote learning skills. This conclusion is reached by analyzing the function of university research. This function actually includes various subfunctions, not always clearly defined, but which should be the subject of separate public policies and forms of management, as follows:

- R&D (research and development) aims at the accumulation of ideas through learning processes and is associated with the codification of knowledge. This is the most common form of research, particularly in the context of economic development and from the standpoint of the relationship between universities and companies.
- R&T (research and teaching), in which research functions as a way of developing teaching materials, as well as of improving the teaching skills of the teaching staff.
- R&L (research and learning), in which the value of the research is not necessarily in the creation of ideas, but in the development of skills that enhance opportunities for learning.

According to the definitions in the previous section, R&D and R&T are learning processes, the purpose of which is the creation of ideas. In this context, selectivity is required in the choice of individuals with suitable skills for these types of activity. In turn, R&L is associated with a specific learning process, which seeks to develop learning skills through the experience of doing research. It is important to disseminate these opportunities, presenting research as a cultural factor.

In these circumstances a diversified system could respond effectively to the different demands made of it in the emerging economy, by being selective in R&D and R&T, and comprehensive in R&L. Indeed, in the context of the knowledge economy, the comprehensive nature of R&T should be extended beyond the university to cover the whole education system, as a way of promoting learning skills. In this situation, it seems essential to place renewed emphasis on education and, to a certain extent, to reinvent its social and economic role. Educational institutions must rethink their relationships with the individuals, families, and communities among which they find themselves, presenting themselves as vital providers of opportunities to develop formal learning processes, while at the same time encouraging a way of life that promotes learning through social interaction.

Among the challenges facing the university and the education system in general, we should also mention the need for lifelong learning. As an essential part of the knowledge economy and facilitated by the new information and telecommunications technologies, lifelong learning should also be seen by the universities as an opportunity to implement strategies that will help maintain their sustained flexibility, and confirms the need to diversify the system, as stated above.

In order to safeguard the role and institutional integrity of universities and ensure their impact in knowledge creation, a range of policies must be set in place, at both public and institutional levels. These policies will be discussed in the next section.

TOWARD PUBLIC AND INSTITUTIONAL POLICIES FOR THE UNIVERSITY

The analysis of the previous section provides the background to understand the relation between university research and economic growth, which, in turn, allows us to relate R&D to innovation. In the last decade, the traditional linear models have been subject to criticism based on both empirical evidence and conceptual grounds. As discussed by Kozmetsky (1993), the challenge of technology innovation requires the consideration of the entire process, from R&D in the laboratory to successful commercialization in the marketplace. Traditionally, successful commercialization of R&D was considered to be the result of an automatic linear process that began with scientific research and then moved to development, financing, manufacturing, marketing, and subsequent internationalization, without sustaining connections among academic, business, and government agents. Today, the relationship between technological innovation and economic wealth generation involves more than investments in financial and physical capital. It demands an integrated and interactive approach that blends scientific, technological, socioeconomic, and cultural aspects in rapidly moving environments.

Kline and Rosenberg (1986) argued that there are complex links and feedback relations between firms (where the innovation takes place) and the science and technology system. In their interactive model, innovation determines and is determined by the market. Myers and Rosenbloom (1996) extended this model to explicitly express organizational capabilities and the special characteristics of innovations. In this model, organizational capabilities are seen as the foundations of competitive advantage in innovation and include firm-specific knowledge, communities of practice, and technology platforms. In the context of this model, the main question is how research (notably at the university level) contributes most effectively to the profitable development of chains of innovation. Using the terminology of Myers and Rosenbloom (1996), "effective" research will contribute to the base of general knowledge, and "productive" research can be transformed into market value but requires the strengthening of organizational capabilities in companies.

Our thesis is that the university is a type of institution specialized in the production of "effective research" for public availability. In this sense, public funding of university research is a key factor to maintain, and possibly increase, the levels of public research results that are available socially. Moreover, there are huge externalities in terms of learning abilities for those who perform research at the universities. Engagement in research at graduate and even undergraduate levels (i.e., R&L above), leads not only to the production of knowledge, but also to training in search routines, inquiry, initiative, risk taking, and entrepreneurship. In summary, performing research enables the individual to learn, so that increasingly linking university research to education should be an important target for the university in the age of knowledge-based societies.

Nevertheless, given the complex and interactive nature of knowledge-creating activities and innovation, it is increasingly difficult to establish a clear demarcation between basic and applied research or, using Myers and Rosenbloom's nomenclature, between effective and productive research. So it is very likely that university researchers in the course of their activities may stumble into productive research with potential to be commercialized (i.e., R&D above). We claim that it is the responsibility of the university to overcome a possible market failure and ensure that these types of commercial opportunities are detected, protected, and transferred to an environment where they can be commercialized. In this sense, technology transfer should be explicitly acknowledged as a way to achieve the requirements of preserving the university's institutional integrity, while guaranteeing the impact of research discoveries.

Figure 8.3 conceptualizes a model in which technology transfer activities are valued in order to secure intellectual property rights, to assess the valuation of technological opportunities, and to implement transfer strategies. It is clear that it is in research, not in commercial design and development, that universities are expected to excel.

Figure 8.3
A Framework for the Interaction of University R&D and the Process of Technology Commercialization

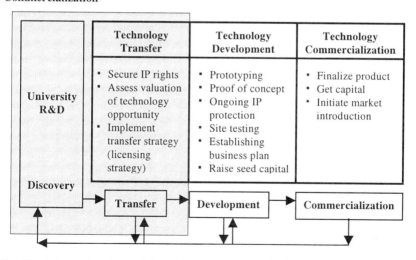

Note: The shaded area includes the priority activities on which a research university should concentrate.

The financial outcomes from technology transfer should always be marginal in comparison to the budget of universities, and should pay for the cost of the technology transfer process and give an additional reward or incentive to researchers. As an example, in the United States, in terms of technology transfer from universities to companies, the income from royalties does not exceed 0.2% of the average R&D budget and is about 2% in the most successful cases (AUTM 1997).

The next question we try to address is the type of organizational context in which university activities should be carried out and the kind of structure that will ensure the success of technology transfers.

THE ORGANIZATIONAL CONTEXT OF UNIVERSITY RESEARCH ACTIVITIES

According to the typology of organizations presented by Henry Mintzberg in his seminal work of 1979, the university can be included in the configuration of professional bureaucracy. This configuration is common in organizations which have a stable, but complex, set of activities to perform, demanding direct control by the professional workers performing those activities. Thus the structure of this type of organization is based on the decentralization of authority and standardization of the capacities of the professionals as a way to organize the activities (Mintzberg 1979). Coordination is achieved by the advanced training and education of the workers, giving birth to true professional organizations. This is also the typical organizational structure of hospitals and law or consulting firms.

The academic structure of the university is based on the disciplinary division of knowledge into formal departmental units and on the standardization of capabilities through the advanced education and training of academics. In a more dynamic environment this formal division is less effective because the organization cannot easily innovate and has difficulties in adapting to the rapid evolution of knowledge and to changes in the economy. This situation makes it more difficult to develop new education programs that are suited to the needs of the labor market, and also to develop new interdisciplinary research programs, which demand the cooperation and joint resources of several departments. This is one of the central problems of the modern university and the organizational solutions to this problem will probably reshape the very foundation of the university's organization.

The departmental organization of universities represents an evolution with relation to the traditional organization of universities in "chairs" and was a response to the growth and increasing complexity of the university's environment. The departmental structure nevertheless maintained some of the characteristics of the chair system and the concept of the chair still exists in present times. In a similar way, it can be conceived that the increasingly dynamic nature of the university's environment will force the creation of new organizational arrangements, which will extend the traditional concepts of discipline and departmental structure, concepts which have for decades been the basis of the university's organization and which have contributed to some extent to the erosion of the institution's unity (The Economist 1997).

The development of complex education and research activities, especially those of an interdisciplinary character, demands a new organizational evolution, with the creation of new structures, independent of departmental power. This independence is necessary to ensure the right incentives for decision making. As most educational programs are not entirely focused on one scientific area but instead share the resources of several departments, the decision of a program coordinator to increase the multidisciplinary standing of his or her program will probably be countered by the head of the main department involved in the program, who may see this movement as a threat to the development and growth of his or her unit. A similar reasoning can be developed for research programs, which have an increasingly interdisciplinary character and are geared to the solution of society's main problems. The scientific resources needed for a specific project may thus be dispersed among the various departments of the university, making it more difficult to gather resources if program coordination is not autonomous from departmental power. This analysis suggests the creation of units in the university responsible for the development of education activities and other units responsible for the development of research activities. These units should be independent of departmental power but would have to relate with the departments to obtain the scientific resources needed. This organizational principle is supported by recent corporate management practices, which tend to conceive the organization as a matrix, organized in terms of areas of competence versus areas of activity. In the case of the university we should speak of a double matrix because it performs two essential activities, research and education, as illustrated in Figure 8.4.

Figure 8.4 also illustrates the management styles proposed for the university. Knowledge is the foundation of the university, because the university exists to create and disseminate knowledge in a systematic and structured way. In this context, knowledge should be the source of authority in the university, as suggested by Rosovsky (1990). As the university is an organization of professionals, this naturally implies a bottom-up approach toward decision making, where the professionals at the operational level of the organization have autonomy to develop their activities in the way they see fit. To bring coherence and unity to the university's activities, a strong and visionary leadership for the university is thus necessary, based on an external board and embodied in the university leader, nominated by this board. Knowledge-based authority implies different management styles in the university:

- At the level of departments, which are the basic units responsible for the development of advanced knowledge in disciplinary areas, there should be a *democratic management* style, because the knowledge in each discipline is dispersed among all the members of the department.

- In terms of the management of research activities, it is essential that researchers have a high degree of autonomy because they hold the most advanced knowledge in their research fields. A *decentralized management* system is thus the most suitable for the management of research activities, which should function in a cluster of research teams and university centers geared to the development and management of research activities. Additionally, there should be a central unit, which would supply administrative services to facilitate and exploit research activities.
- In terms of the management of education programs, pedagogic knowledge is shared by all the participants in the education process, suggesting a *participatory management* style. Nevertheless, scientific knowledge is embodied in the faculty, and the complexity of the education process (management of schedules, students, faculty, curricula, and budget) suggests considerable intervention from the program coordinators.
- At the top management level, there should be an external board—a university council (which could also include university members detached from daily university affairs)—serving as a link with society and legitimating the power of the university leaders. The top management should develop a *cultural management* style, understood as the communication of the vision, ideals, challenges, and identity of the university in a top-down approach, and channeling the expectations of society to the university's members.
- At the support-structure level, including all administrative functions, there should be *professional management*, headed by the university leader, who should be appointed by the university council. This bureaucratic-type management should include job descriptions, formal units, and defined hierarchies.

The department still forms the building block of scientific competencies. Nevertheless, there should also be specific bodies to coordinate the different teaching activities and others to develop research activities, both types of bodies drawing their legitimacy from the university council and not from the departments. A coordinating mechanism, functioning by mutual adjustment and organized according to a set of clear and simple rules, ensures that the scientific competencies of the departments are efficiently allocated to the development of the university's activities. These new bodies do not need to be formal and rigid ones. One possible approach is the creation of a true *ad hoc* organization, as described by Mintzberg (1997), where the very boundary between the university and its environment would be blurred, and a network of individuals and teams would form relationships with outside organizations to develop and fund new research and education activities. This kind of *ad hoc* organization, also called a socially distributed knowledge-production system, should function in a very loose way but a set of very clear rules, management procedures, and coordination mechanisms should be devised in order to prevent conflicts of interest and protect the university's institutional integrity.

Figure 8.4
Proposed Organization and Management Model for the University

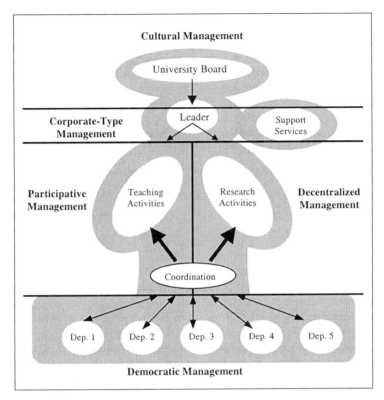

Source: Santos 1996

Based on these principles, an appropriate organizational model for the development of research activities is one based on research groups, independent from departmental power, which carry out their activities with a large degree of autonomy. Sometimes these groups join together for the development of more complex and interdisciplinary research programs. This association can be formal and lasting, thus creating the concept of centers or institutes, or can be occasional and decided on an ad hoc basis. Additionally, there should be a support structure for R&D activities, which should assure the infrastructural conditions for the development of research activities, should support the financial and legal administration of research contracts, and should guarantee the identification, protection and transfer of research outputs with potential for commercialization.

SUMMARY

The empirical evidence of the increasing importance that knowledge is assuming in economic activity in developed countries is discussed in this chapter based on recent conceptual advances in the understanding of the new dynamics of economic growth. These theories accord particular importance to the accumulation of knowledge by means of formal and informal learning processes. This accumulation takes place in the form of software and wetware, which have different economic properties but whose interdependence in a complex process of interaction requires a rethinking of the traditional role of the university.

The analysis shows that the preservation of the university's institutional integrity is essential in a situation of sustained flexibility, in which education, besides offering a specific qualification, should ensure the assimilation of learning skills. The signs of the knowledge economy, notably the expansion in university education and the need to manage multiple demands and to ensure participatory learning, point toward a diversification of the system, with reference to which it is particularly important to identify and understand the different components of the university's research function.

Our thesis has been formulated considering that one essential feature of technology transfer is the development of attitudes which accept personal responsibility for identifying intellectual property with commercial potential and then setting out to ensure that it is exposed to commercial scrutiny, that it receives adequate legal protection, and that finance can be provided so that the scope of the asset and its market potential can be assessed.

The analysis also shows that the successful positioning of the university in the knowledge economy is characterized by an increase in complexity and instability. The guiding principle for the management of the university should be knowledge-driven authority. This form of authority establishes different styles of management according to the different types of units found in the university.

NOTE

1. See, for example, Rosenberg and Nelson (1996), Pavitt (1991), and Conceição, Heitor, and Oliveira (1998).

REFERENCES

Arrow, K. "The Economic Implications of Learning by Doing," *Review of Economic Studies* 28 (1962) 155–173.

Association of University Technology Managers (AUTM). *Licensing Survey 1991–1995: A Five Year Survey of Technology Licensing (and Related) Performance for U.S. and Canadian Academic and Nonprofit Institutions, and Patent Management Firms.* AUTM, 1997.

Boorstin, D. *The Discoverers.* New York: Random House, 1994.

Caraça, J., P. Conceição, and M. V. Heitor. "On the Definition of a Public Policy Towards a Research University," *Higher Education Policy* (forthcoming).

Conceição, P., D. Gibson, M. V. Heitor, and S. Shariq. "Towards a Research Agenda for Knowledge Policies and Management," *Journal of Knowledge Management* 1, 2 (1997) 129–141.

Conceição, P., M. V. Heitor, and P. Oliveira. "Expectations for the University in the Knowledge-Based Economy," *Technological Forecasting & Social Change* 58, 3 (1998) 203-214.

The Economist. "Inside the Knowledge Factory: A Survey of Universities," Special issue (October 4, 1997).

Ehrenberg, R. G. *The American University: National Treasure or Endangered Species?* Ithaca: Cornell University Press, 1997.

Foray, D., and B. A. Lundvall. "The Knowledge-Based Economy: From the Economics of Knowledge to the Learning Economy," in *Employment and Growth in the Knowledge-Based Economy*. Paris: OECD, 1996.

Gellert, G. "Changing Patterns of European Higher Education," in *Higher Education in Europe* (G. Gellert, ed.). London: Jessica Kingsley Publishers, 1993.

Graham, H. D., and N. Diamond. *The Rise of American Research Universities: Elites and Challengers in the Postwar Era*. Baltimore: Johns Hopkins University Press, 1997.

Grossman, G. M., and E. Helpman. *Innovation and Growth in the Global Economy*. Cambridge: MIT Press, 1991.

Kline, S. J., and N. Rosenberg. "An Overview of Innovation," in *The Positive Sum Strategy: Harnessing Technology for Economic Growth* (R. Landau and N. Rosenberg, eds.). Washington, DC: National Academy Press, 1986.

Kozmetsky, G. "The Growth and Internationalization of Creative and Innovative Management," in *Generating Creativity and Innovation in Large Bureaucracies* (R. L. Kuhn, ed.). Westport: Quorum Books, 1993.

Lucas, C. J. *Crisis in the Academy: Rethinking Higher Education in America*. New York: St. Martin's Press, 1996.

Lucas, R. E. "On the Mechanics of Economic Development," *Journal of Monetary Economics* 22 (1988) 3–42.

Mintzberg, H. *The Structuring of Organizations: The Synthesis of the Research*. Englewood Cliffs: Prentice Hall Inc., 1979.

Mitra, J., and P. Formica. *Innovation and Economic Development: University/Enterprise Partnerships in Action*. Dublin, Ireland: Oak Tree Press, 1997.

Myers, M. B., and R. S. Rosenbloom. "Rethinking the Role of Industrial Research," in *Engines of Innovation* (R. S. Rosenbloom and W. J. Spencer, eds.). Cambridge: Harvard Business School Press, 1996, 209–228.

Nelson, R. R., and P. Romer. "Science, Economic Growth, and Public Policy," in *Technology, R&D, and the Economy* (B. L. R. Smith and C. E. Barfield, eds.). Washington, DC: Brookings Institution, 1996.

Organization for Economic Cooperation and Development (OECD). *Innovation, Patents, and Technological Strategies*. Paris: OECD, 1997.

Pavitt, K. "What Makes Basic Research Economically Useful?" *Research Policy 20* (1991) 109–119.

Readings, B. *The University in Ruins*. Cambridge: Harvard University Press, 1996.

Roberts, E. B. *Entrepreneurs in High Technology: Lessons from MIT and Beyond*. London: Oxford University Press, 1991.

Romer, P. "Increasing Returns and Long-Run Growth," *Journal of Political Economy* 98, 5 (1986) 1002–1037.

Romer, P. "Endogenous Technological Change," *Journal of Political Economy* 98, 5 (1990) S71–S102.

Rosenberg, N., and R. R. Nelson. (1996), "The Roles of Universities in the Advance of Industrial Technology," in *Engines of Innovation* (R. S. Rosenbloom and W. J. Spencer, eds.). Cambridge: Harvard Business School Press, 1996.

Rosovsky, H. *The University: An Owner's Manual*. New York: W. W. Norton, 1990.

Rothbatt, S. *The Modern University and its Discontents*. London: Cambridge University Press, 1997.

Santos, F., J. "The Organizational Management of Universities," Master of Science Thesis, Instituto Superior de Economia Gestão, Technical University of Lisbon, 1996.

Santos, F., Heitor, M. V., and Caraça, J. "Organizational Challenges for the University," *Higher Education Management* 10, 3 (1998) 87-107.

Solow, R. "Technical Change and the Aggregate Production Function," *Review of Economics and Statistics* 39 (1997) 312–320.

Stephan, P. E. "The Economics of Science," *Journal of Economic Literature* 34 (September 1996) 1199–1235.

Stevens, C. "The Knowledge Driven Economy," *OECD Observer* (June/July 1996).

Tierney, W. L. *The Responsive University: Restructuring for Higher Performance*. Baltimore: John Hopkins University Press, 1998.

Wilson, R. H. *States and the Economy: Policy-Making and Decentralization*. Westport: Praeger, 1993.

Wyckoff, A. "The Growing Strength of Services," *OECD Observer* (June/July 1996).

9

Entrepreneurial Universities: Policies, Strategies, and Practice

Dylan Jones-Evans

INTRODUCTION

The perception of universities as merely institutions of higher learning is gradually giving way to the view that universities are important engines of economic growth and development (Saxenian 1994), with increasing evidence that the university sector can undertake a variety of roles in developing the technological and industrial potential of a region (Cooke and Huggins 1996). This is not surprising, as regional and national governments view the high-technology sector as a source of direct and indirect employment opportunities, and universities are seen as crucial to facilitating the growth of the local high-technology potential, especially with regard to the small-firm sector. The development of a center of academic excellence in a certain field can also create or enhance a favorable public image and reputation and, as a result, additional jobs can be created not only in a university, but also in the wider community surrounding the university, because of its enhanced economic and social status (Malecki 1991). With evidence that the majority of jobs in high-technology industries have been created by small firms (Jones-Evans and Westhead 1996) there are opportunities, if the right mechanisms are in place, for this sector to benefit directly from the research carried out by university researchers.

This is because universities concentrate a large critical mass of scientifically sophisticated individuals who can generate new technologies, which, in turn, can lead to innovative ideas (and technological knowledge), which can be channeled and diffused by new ventures. High-technology companies thrive on state-of-the-art knowledge which they can harvest and apply to market opportunities, and many entrepreneurs establish their businesses near universities to profit in various ways from their creativity and technological output (Miller and Cote 1985). Indeed, there is increasing evidence to show that small innovative firms can benefit more from their links with universities than larger R&D organizations (Rees 1991; Feldman 1994), and various studies have recognized that a significant number of new technology-based businesses in both the United States and Western Europe have been established by scientists emerging from different types of academic-based organizations, such as nonprofit research institutes, government research centers, and universities (Jones-Evans 1996).

While the creation of spin-offs is one example of technology transfer into the small-firm sector, universities can also play a valuable role in directly transferring both technological expertise and knowledge—through either consulting activities or patenting—to existing local entrepreneurial ventures. Consulting by academic scientists and engineers can be the most versatile and cost-effective means of linking industry with the university sector, as it is a relatively inexpensive, rapid, and selective means of transferring information with few institutional tensions, and it rarely involves extensive demands on university personnel and material resources (Stankiewicz 1986). It is therefore recognized as the most effective two-way channel between university and industry, often leading to other forms of cooperation, as academic scientists and engineers who engage in consulting acquire knowledge about the needs of industry and can therefore identify how these needs can be met by the university sector (Mansfield 1994).

Another important means of technology transfer is the commercialization of research results through patenting and licensing, especially as it is becoming increasingly recognized that patenting contributes to the effective utilization of technology, in that a company is more likely to invest in new technology if it is protected by a patent (Vedin 1993). In addition, the technical equipment available in universities can also be used by existing local businesses to solve production process problems and to supplement their commercial advantage. This can particularly benefit small firms, who usually cannot afford the relevant technical equipment, especially as universities have computing, testing and analysis, and library facilities which are an incentive for small firms to engage in a university-industry based relationship (Westhead and Storey 1995). As a result, local firms can become more technologically sophisticated, thus enhancing their competitive performance and, in some cases, their survival.

Universities are also increasingly seen by high-technology businesses as a crucial source of skilled graduates whom they can employ after graduation. Universities are also involved in training, in both requisite quantity and quality, of the labor force of scientists, engineers, and technicians, who will provide the key ingredient for the growth of technologically advanced industrial centers. By providing graduates with the skills required by high-technology industries, they affect the overall level and focus of educational attainment, which in turn can affect a region's ability, especially in the small-firm sector, to adopt and exploit new technologies successfully (Gibson and Smilor 1991). Indeed, the presence of a large research university with its thousands of potential highly educated technical personnel can be a factor in attracting firms to a particular region, as firms can only easily recruit their personnel if they are already located in an advanced urban-industrial area. For technology-based firms being established within the region, the ability to build a local labor market of good-quality engineers and scientists is critical, and the university can play an important part in this through the supply of highly trained science and technology graduates. Moreover, students can also be used, through placements and assignments, by local small businesses to develop critical technical competencies which they could not otherwise afford.

Therefore, if economic growth in Europe is to increase in the near future, it is crucial to develop and commercialize technological knowledge into industrial success (European Commission 1993). In particular, the transfer of knowledge from academia to industry must be increased, and there is a need to focus on, and develop, the efficiency and effectiveness of the different forms of technology transfer in order to bridge the gap between university research, technological development activities, and the commercial market. Indeed, while it is widely recognized that it is increasingly important to successfully commercialize university research, it is also accepted that a greater focus on the development of technology transfer activities from universities to industry (especially small technology-based firms) is needed (Jonsson and Klofsten 1996), especially examples of "good practice" from existing successful initiatives.

METHODOLOGY

There have been a number of studies which have examined the contribution of the university to the technological development of industry from the viewpoint of the recipient firm. However, within Europe, there has been very little detailed examination (outside of specialist conferences for industrial liaison practitioners) of the proactive role that the university itself can play in developing strong linkages with industry, and the strategies that are undertaken to increase the process of technology transfer from academia into local indigenous business. A recent publication (Jones-Evans et al. 1997) describes the role of industrial liaison offices in generating specific policies for the development of closer academic-industry links. This chapter will build on this research by examining a number of different proactive approaches by universities in Europe in developing closer links with industry. As part of a European Commission project examining universities, technology transfer, and spin-off activities within peripheral regions, the research has attempted to examine specific cases of university initiatives to develop closer links with industry. The methodology utilized to examine these processes was the case study, as it enables an investigation to be made of a contemporary phenomenon within a real-life context (Yin 1994). In gathering the data, in-depth semistructured interviews were conducted with the main actors involved with each type of initiative, and supplemented with note-taking, observation, as well as relevant secondary documentation.

Although this is work-in-progress, the preliminary analysis of the research has identified a number of different types of initiatives developed by the university sector to work more closely with industry, including:

- adjunct professors (Finland)
- innovation centers (Ireland)
- networking activities (Wales)
- entrepreneurship stimulation programs (Sweden)

This chapter will briefly describe the main characteristics of the four specific types of initiative, and why the initiative has been seen to be successful and an example of "good practice" of university-industry collaboration.

THE ADJUNCT PROFESSOR MODEL (FINLAND)

Like many other universities in Finland, the Tampere University of Technology has developed a number of industrial links, mainly with large Finnish enterprises such as Nokia, VTT, and firms in the wood-processing industries. One of the most recent (and successful) types of industrial collaboration is that of adjunct professors from industry. The adjunct professors are selected from scientific or technological fields in which the university has active cooperation with industry, and share about 20% of their time with the university. These selected industrialists mainly undertake lectures for students in their particular technological area, thus providing up-to-date information from the "real world" to supplement the academic courses already on offer at the university. To date, this experiment has proven to be very useful for all the parties involved:

- the university maintains and deepens its collaboration and cooperation with industry
- the industrial partners get a ringside view of front-line technology R & D within the university
- the individual "adjunct professor" gains valuable experience by working on a challenging interface between industry and academia
- the students gain an important insight of the relevance of technology in industry

The financial structure of the model is mostly based on public or university finance, but since the commitment of the industrial professors is only 20% of normal working hours per week, the benefits easily exceeds the costs.

THE INNOVATION CENTER (IRELAND)

The Innovation Center at Trinity College Dublin (TCD) was founded in 1986 with the aim of seeking new ways to support and exploit Trinity's strong research resources and offer services on two levels: to both researchers and innovators. After consultation, the strategy adopted by TCD was one of hosting campus industrial laboratories and campus companies in order to enhance the interaction between the university and industry. For university researchers, this initiative has proved to be of use through the provision of information, intellectual property advice, and generally commercializing expertise to benefit the college. For campus innovators, there is the opportunity to maximize the potential of an innovation giving rise to spin-off companies and links with industry. Since its inception, the Innovation Center has created approximately 40 campus companies, which were formed from technologies originating from TCD's research income, which has risen to a steady annual level of £11 million. Highlighted cases/models of best practice include:

- *Magnetic Solutions Ltd.*—a private Irish company specializing in design and manufacture of advanced magnet systems. The company's core competence is based on the magnetic materials research group, which was recently awarded a European patent for their discovery of a new advanced permanent magnetic material, SmFeN.
- *IONA Technologies Ltd.*—This is a very successful indigenous software house which has developed and grown from originally being a campus company in TCD in 1991, and is led by a former lecturer in the Department of Computer Science.
- *The Institute for European Food Studies (IEFS)*—From a campus company formed by Professor Michael Gibney (Nutriscan), a proposal was put to industry to fund a new nonprofit institute to service European industrial needs on a sectoral basis. Recently the IEFS completed the largest ever pan-EU survey on consumer attitudes to food, nutrition and health.
- *UNIMED PLC*—This is a specialist healthcare R&D company. It has funded over £1.5 million in pharmaceutical research at TCD since 1994. The company is currently licensing its technologies to healthcare companies in the pharmaceutical and medical areas.
- *The Hitachi Dublin Laboratory* (HDL)—This lab employs 17 researchers and has links with 15 others. making it the largest element in Hitachi's European R&D operation.

The main incentives in forming a "campus company" at TCD as opposed to an SME outside of the university were not just financial. The location and site played an important particularly with regard to TCD being in the city center, with ease of access to student placements and graduate recruitment (which ensured that some of the most innovative minds were part of the business).

THE INNOVATION NETWORK (WALES)

The Innovation Network was set up by Cardiff University in March 1996 as an initiative to build bridges between the university and industry, by means of assisting industry in the process of innovation (the turning of creative ideas into commercial products). In response to this objective, the Network seeks proactively to build a network of contacts, between both the university and firms and between firms themselves, through which productive working relationships can develop.

The Network is jointly financed by the University of Wales Cardiff and, for a period of two years, by the European Regional Development Fund (ERDF) and has an annual budget of £90 million. To date, the Network has a list of approximately 1,000 firms who are involved in its activities. The Network performs a variety of activities in its aim to build a beneficial network of contacts between industry and the university, including:

- The net-line provides a point of contact for access to university expertise and resources in addition to access to advice and information that any of the Network staff, Steering Group, or Industry Advisory Board members can provide.
- Regular meetings/workshops by the Network on topics of key importance to industry and commerce are organized at which both academic and industrial actors give formal presentations on innovation issues of direct relevance to companies. The meetings are informative as well as offering an informal interactive opportunity for the guests. The success of the meetings relies on the usefulness of the topics to industry and on the willingness of the guests to interact with each other after the formal presentations.

The Network provides a variety of benefits for both the university and the firms involved. Establishing linkages with firms provides the university with invaluable opportunities with regard to research collaboration, technical and consulting services, as well as the opportunities it can provide for graduates/undergraduates with regard to employment prospects. The opportunity for interaction which the Network provides encourages both academics and industry representatives to overcome any cultural differences which may exist and to become aware of each other's working environment and professional requirements. The chance to share and experience each others' techniques and approaches to innovation are key concepts regarding the benefit of networking between the university and industry.

Therefore, the Network's activities are centered around being useful and relevant to industry with the indirect (though no less important) consequence of providing the university with the opportunity to develop closer links with the firms in the hope of establishing collaborative contracts. It is an initiative based on mutual benefit for all involved, which could be the reason for the Network's very successful first year. As has been previously discussed, the way forward now is to transform the contacts established through the Network into solid contracts between firms and academics at the university.

ENTREPRENEURSHIP STIMULATION PROGRAMS (SWEDEN)

The supporting network in the Linköping area of Sweden is highly conducive to the development of innovation within, and the growth of, new technology-based firms. During the last 10 years, an initiative has evolved which incorporates a close interaction between the small technology-based firms—represented by The

Foundation for Small Business Development in Linköping (SMIL)—and the university—represented by The Center for Innovation and Entrepreneurship (CIE)—an autonomous unit at Linköping University which is conducting activities (in cooperation with SMIL) intended to stimulate the growth and development of technology-based firms. A number of activities have been developed by CIE, with the main emphasis on examining the different problems which occur at various times during a firm's development. With tailored activities, firms gain access to resources relevant to their particular stage of development. There are three main activities undertaken by CIE for SMIL members:

1. *The entrepreneurship and new business development program.* The main aim is to solve the problems that can be encountered in establishing and managing a new firm, and to recruit students, researchers at the university, and people in established business who have both a plausible business concept and an interest in starting and running a business. The main structure of the program involves workshops on entrepreneurship and small business management, the development of business plans for the establishment of the new venture, the availability of financial resources to cover the costs of meeting customers and market surveys, as well as a process of mentoring for these fledgling entrepreneurs, which is carried out by one of SMIL's existing network of experienced businessmen. In addition, SMIL has secured cooperation with the local science park, which makes premises available for the new firms.
2. *The development programs and the management groups.* For the more established firms, various development programs and management groups are available. Here, the basic idea is to update the business expertise in the firm by working out solutions to certain known specific problems in their activities. Common problems can include internationalization, market positioning, professional board management, and quality assurance systems. While the development programs will be more general in nature (with different firms discussing a variety of company-related problems together), the management groups are more focused, and concentrate on the solution of just one specific problem common to all the participating firms. Another important difference is that the development program is targeted toward firms at an earlier phase of development than those which will partake in management groups. This concept is based on the fact that it is more suitable to first solve a firm's management problems on a more general level prior to concentrating on more specific problems. It is therefore an advantage if the participants in the management group have already taken part in one of the development groups.
3. *The club and networking activities.* The third function of CIE, in conjunction with SMIL, is the coordination of the club and networking activities whose main aim is to create a social network and exchange of information between firms in the SMIL group. Each month, activities such as pub-nights and various forms of seminars, where junior businesspeople can meet senior businesspeople and exchange experiences, are arranged. Other forms of network building can be found in the booklet "Ideas That Really Mean Business," where information such as firm addresses, lines of business, products, and markets are listed. The catalogue is distributed both in Sweden and internationally, and is a good marketing device for firms in the SMIL group.

The various stimulation programs have been successful for a number of reasons:

- *The ability to meet real needs.* The activities were grounded in the firms' experienced need of various kinds of stimulation. Through the activities, it was then possible to identify and put into concrete form the firms' real needs in order to then offer tailored solutions. As a result, the firms have been able to take appropriate measures.
- *The core group.* A management board of competent and committed people with different roles was available. The members of the board all have an understanding of small business, possess structural knowledge, and have access to an enthusiastic leader who can promote the activities of the group.
- *A clear focus.* Since the start, CIE-SMIL has focused on the development of the management of small, technology-based firms concentrating, in particular, on the executive group within the venture. Activities based on this principle have then been developed and carried out, which have resulted in an ability, by the management of these firms, to communicate effectively their plans for development to other actors, such as potential financiers.
- *Credibility.* The development of SMIL is characterized by a strong commitment by the majority of the firms. The trust they have in SMIL has depended on a well-functioning network with strong social dimensions. Participation in the stimulation activities has meant that firms, in many cases, must release information of a sensitive nature under an oath of silence. To date, nothing has leaked outside of the group.
- *Close relations between SMIL and the university.* Both institutions have complemented each other well through cooperation. SMIL can be thought of as being the eyes and ears of the marketplace where the firms' need for stimulation has been recognized, while CIE, as the university partner, has contributed a secretariat and financial resources, structural knowledge and credibility.

CONCLUSION

This chapter has introduced examples of university-industry relationships that have been implemented within four countries in Europe. The analysis of these and other cases of academic collaboration with industry in other European countries is currently ongoing. However, a preliminary finding of the study is that one of the key factors contributing to the success of these initiatives is the acknowledgment and incorporation of mutually beneficial activities for all partners involved and an awareness of the economy in which they participate (Manning and Cooke 1997). If an initiative involving universities and industrial collaboration is to succeed, it must fully acknowledge the economic implications for all parties and the kinds of approaches which would be most beneficial to all partners in view of their position and role within the economy. In addition to considering the relevance of the activities to each partner, it is necessary to consider what outcomes each of the participants expect to gain from the initiative. Only with both partners understanding their role in the project and appreciating the benefits that accrue from the collaboration can a university-industry technology transfer initiative succeed.

ACKNOWLEDGMENTS

The author would like to acknowledge the financial support of the European Commission DGXII Targeted Socioeconomic Research Program in undertaking this research. The chapter reports on the ongoing studies of the partners in the project and the author would like to thank the following individuals for their contributions: Professor Phil Cooke, Dr. Goio Extebarria, Professor Richard Harrison, Dr. Magnus Klofsten, Professor Antti Paasio, and Dr. Artur Da Rosa Pires. Thanks also to Dipti Pandya for her comments. The views in this chapter, while based on the work of the partners, are entirely the responsibility of the author.

REFERENCES

Cooke, P., and R. Huggins. *University-Industry Relations in Wales.* Working Paper, Center for Advanced Studies in the Social Sciences, UWCC, 1996.

European Commission. *Growth, Competitiveness, Employment: The Challenges and Ways Forward into the 21st Century (white paper).* Luxembourg: European Commission, 1993.

Feldman, M. P. "Knowledge, Complementarity and Innovation," *Small Business Economics* 6 (1994) 363–372.

Gibson, D. V., and R. W. Smilor. "The Role of the Research University in Creating and Sustaining the U.S. Technopolis," in *University Spin-Off Companies: Economic Development, Faculty Entrepreneurs and Technology Transfer* (A. M. Brett, D. V. Gibson, and R. W. Smilor, eds.). Savage: Rowman and Littlefield, 1991, 31–70.

Jones-Evans, D. "Technical Entrepreneurship, Strategy and Experience," *International Small Business Journal* 14, 3 (1996) 13–37.

Jones-Evans, D., M. Klofsten, D. Pandya, and E. Andersson. "Academic-Industry Links within Peripheral Regions of Europe: Policies, Strategy and Practice," presented at the Fifth International High Technology Small Firms Conference, Manchester Business School, May 28–30, 1997.

Jones-Evans, D., and P. Westhead. "High Technology Small Firm Sector in the United Kingdom," *International Journal of Entrepreneurial Behavior and Research* 2, 1 (1996) 15–35.

Jonsson, S., and M. Klofsten. *University-Industry Relations In Sweden: A Literature Review*, Center for Innovation and Entrepreneurship, Working Paper 96-1, Lund, Sweden: 1996.

Malecki, E. J. *Technology and Regional Development.* Essex, U.K: Longman. 1991.

Manning, C., and P. Cooke. *Academic Entrepreneurship in Wales: Third Interim Report for the EU-DG12 Fourth Framework "Targeted Socioeconomic Research" Program Project: Universities, Technology Transfer and Spin-Off, Academic Entrepreneurship in European Regions.* Cardiff: Center for Advanced Studies in the Social Sciences, 1997.

Mansfield, E. "The Contributions of New Technology to the Economy," presented at the American Enterprise Institute Conference on the Contributions of Research to Economy and Society, Washington, DC, October 1994.

Miller, R., and M. Cote. "Growing the Next Silicon Valley: The Right Strategy Can Trigger the Formation of a Regional High-Tech Cluster," *Harvard Business Review* (July–August 1985) 114–123.

Rees, J. "State Technology Programs and Industry Experience in the United States," *Review of Urban and Regional Development Studies* 3 (1991) 39–59.

Saxenian, A. *Regional Advantage: Culture and Competition in Silicon Valley and Route 128.* London: Harvard University Press, 1994.

Stankiewicz, R. *Academics and Entrepreneurs: Developing University-Industry Relations.* London: Frances Pinter Publishers, 1986.

Vedin, B-A. "Innovationer för Sverige," *SOU* 84 (1993) Näringsdepartementet, Stockholm.

Westhead, P., and D. Storey. "Links Between Higher Education Institutions and High Technology Firms," *Omega* 23, 4 (1995) 346–360.

Yin, R. K. *Case Study Research: Design and Methods.* 2nd ed. London: Sage Publications, 1994.

10

A Generation of Technology Innovation in a Russian Technical University in a Transition Economy: Challenges and Barriers

Nikolay Rogalev

INTRODUCTION

When looking at research and development activities at the Moscow Power Engineering Institute (MPEI) in terms of the generation of technological innovations, special attention should be paid to the research which the university considers advanced, because it is this research that produces results of greatest importance in terms of industrial applications.

Analysis of advanced R&D was carried out on the basis of the Moscow Power Engineering Institutes (MPEI) annual reports on research for 1985, 1987, 1989, 1991, 1992, 1993, and 1994, together with annotations from advanced R&D results. Figure 10.1 presents the yearly distribution of advanced R&D projects. There is a steady decrease in the total number of projects, which were considered advanced by MPEI. Figure 10.1 also shows the distribution of industrial contracts and advanced budget R&D. The beginning of reforms up to 1991 was characterized by an increase in industrial research contract work carried out in advanced fields, and thus by close relations with industry, without any significant decrease at the beginning of the structural reforms in the economy. However, 1992 and 1993 can be characterized by a break and discontinuance of this high-priority research work (10 industrial contracts in 1992 and 6 in 1993). Trends in state budget funding for the period considered show an inverse pattern: reduction in budget-funded research in 1985, 1987, and 1989, with an increase in 1992 followed by a decrease in 1993.

Figure 10.1
Number of Advanced Industry and State Budget R&D Projects, 1985–1993

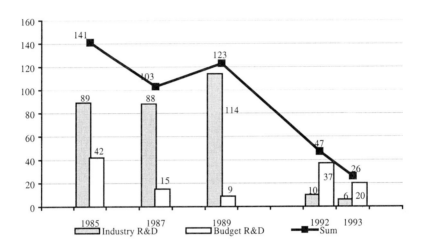

An important factor which determines the priority and basis for steady research is the length of research carried out within one contract. The length of a contract in this case determines the responsibility of both sides—the university and industry represented by particular customers—for the development of technologies and their industrial application.

Figure 10.2 presents data on the number of contracts lasting over three years. This criterion was chosen based on the assumption that three years is the average length of study for a Ph.D. or Sc.D. program, during which a specific field of science is investigated and a thesis is defended. This time is also considered long enough to study a problem, create all the necessary instruments for research in the form of displays or computer software, and obtain significant results.

The length of the contracts between 1985 and 1993 decreased dramatically. As compared to 1985, the total number of projects longer than three years fell in 1987 by 5.6 times, in 1989 by 17.8 times, and in 1992 and 1993, there were no projects at all that met this criterion. Further evidence of the fact that work in the field of advanced research is losing its significance is presented by the dynamics of publication, and protection of the intellectual property rights of research work. The total number of articles, monographs, theses, and research reports was used as a measuring parameter. As parameters of intellectual rights protection, inventor's certificates and patents were used. Figures 10.3 and 10.4 present these data graphically.

The negative trend seen in the total number of publications, inventor's certificates, and patents indicates cutbacks in scientific activity and in further development of advanced research in these fields. These factors reflect research efficiency: publications reflect scientific innovation, and patents represent the potential for their commercial industrial application.

Figure 10.2
Advanced R&D Contracts of More than Three Years, 1987—1993

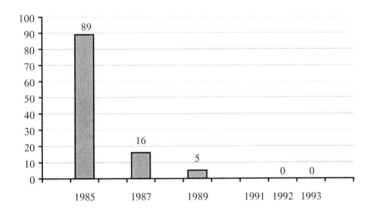

Figure 10.3
Publications within Advanced R&D, 1985—1993

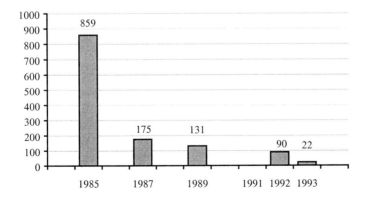

Figure 10.4
Certificates of Authorship, Patents, and Licenses, 1988—1993

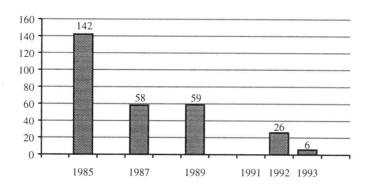

The data in question show, among other things, that in spite of a larger proportion of budget funding in the general structure of MPEI's R&D spending, the negative qualitative changes have not stopped and are in fact worsening. However, analysis of the results in the field of advanced research with respect to technology generation and its further industrial application would not be complete without considering the technological transfer level of the research in question.

Technology transfer implies four communication levels and four corresponding criteria for measuring success or efficiency of the technology transfer (Gibson, 1991; Gibson and Rogers, 1994). Technology transfer of the first level is considered passive, having weak communication between the participants, though researchers can work in teams or even communicate beyond their organizations or internationally. Success at the first level is measured by the quality and quantity of research reports and publications; and scientific significance, as a stronger point, is of greater importance.

The second level—the "acceptance" of technology—implies greater involvement of the technology transfer participants. The movement from level 1 to level 2 is complicated. Because this is a level of shared responsibility in the communication between technology emitters and technology acceptors, it is important to appoint key persons to collect the right information at the right time.

The third level of technology transfer, technology implementation, is characterized by the fact that its success depends on due and efficient application of the technology. To ensure successful technology transfer at this level, the users should possess the knowledge and resources required for its application. Technology implementation includes organizational steps such as product development, prototype construction, or approval of a concept for commercial application. Industrial power and industrial significance are the most important. At this stage, the customer—the technology user—contributes to the technology transfer.

The fourth level, technology application, corresponds to the stage of product commercialization. At this level the success of the previous three stages in achieving their goals is accumulated. In addition, a market advantage is necessary at this stage. Technology user feedback regulates the transfer process. The measures of success include capital revenue or share of the product market.

The movement from level 1 to level 4 is not a linear, progressive process. In the course of this movement, the complexity of the transfer multiplies. In general, the success of technology transfer at all levels is difficult to measure by traditional cost-benefit analysis, because: (1) it is difficult to estimate financial and other impacts of the technology over time; and (2) different persons involved in the technology transfer evaluate its costs and benefits differently depending on their personal ideas of advance and progress.

Figure 10.5 presents the results of the MPEI advanced R&D analysis in respect to technology transfer levels. The dynamics of change indicate a downgrading of technology transfer level, in both absolute and per unit relations, and a transition to lower levels, thus transferring the university's activity to the earlier stages of technological development with unclear commercial applications.

Table 10.1 gives the distribution of advanced research for the generalized knowledge fields traditionally considered in MPEI analysis. Electrical and power engineering includes all spheres connected with electrical technology development (electrical engineering, electrical machinery, electrical networks, etc.) and power engineering (boilers, steam and gas turbines, heat and mass transfer, technical hydro- and aerodynamics, etc.) The other fields of knowledge also are inclusive of various spheres. As can be seen, all the branches in question reveal a quantitative negative tendency. The distribution analysis for the technology transfer levels qualitatively reveals a general tendency of decrease for all research work. Another important result of the current situation is the weakness in, and even disruption of, contacts with industry (in respect to contracts) and the government (in respect to budget funded projects) with clearly observed reduction of cooperation and contacts.

Proceeding from the theory of technological transfer networks, let us specify the object of our study with respect to the university. First of all we shall consider the network formed within the scope of advanced R&D, which reduces the subjective element in identifying the boundaries of research and keeps us within the field, which is the most interesting in terms of technology generation and consequent transfer. Thus, the contents of the linkage network include licensing, technological transfer, technological exchange, and joint R&D. As particular participants forming linkages in the network there are industrial

Figure 10.5
Advanced R&D Projects by Technology Transfer Level

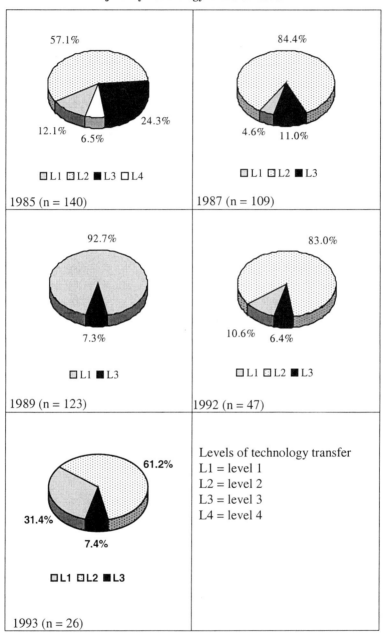

Table 10.1
Advanced R&D at MPEI by Major Knowledge Area, 1985—1993

Knowledge area	Number of R&D projects				
	1985	1987	1989	1992	1993
▪ Electroenergetics & Electrotechnology	29	28	58	10	4
▪ Thermo Power Engineering	32	23	38	10	7
▪ Electronics, Computers, Software and CAD	33	29	12	7	5
▪ Environmental Control	16	5	9	2	4
▪ Physics, Mathematics, Chemistry	5	8	6	3	3
▪ Other	25	16	0	15	3
Total	140	109	123	47	26

enterprises, universities (higher educational institutions), and organizations of the USSR Academy of Science (USSR AS), the Russian Academy of Science (RAS), the Russian Academy of Medical Science (RAMS), and industrial research centers. A linkage between the university and an enterprise is defined as technology transfer because there is a technology customer—industry—and a technology developer—the university. A linkage between MPEI and other universities and organizations of the USSR AS, RAS, or RAMS is considered to be technology exchange because in most research work this type of cooperation is observed. The majority of linkages with industrial research centers represent joint research and development ordered by industry and conducted by the university as a subcontractor or a co-executive partner on a number of defined problems. Research ordered directly by the government, ministries, state committees, and departments is considered separately.

In conformity with the objectives of our study we do not consider interorganizational linkages as a subject of analysis. Our study is aimed at defining the organizational set—namely, its size, density, network diversity, and linkage stability over time. The organizational level is the level of analysis. To estimate the network dynamics we shall assume that the number of associates (the network members) has remained the same since 1985—that is, we will run a comparative analysis of the networks based on the parameters at the beginning of the study period. This is possible because the university collaborated with large organizations that have preserved their functions even when their organizational forms have been transformed into joint stock companies or passed under the authority of another ministry, although some have remained the same. So, proceeding from these assumptions and with reservations on data accessibility as far as the most advanced research work is concerned, stability was analyzed as a time function of the amount of linkage change. Figure 10.6 graphically represents the change in the link density. As can be seen, a sharp decrease of all linkage types occurred up to

1987, which is likely to have been accompanied by a reduced number of contract and budget research projects, on one hand, and by weakened requirements to introduce research products to industry, on the other. By 1989, a recovery was observed in the fields of technology transfer, joint research and development with the industrial institutes, and technology exchange with universities. Linkages with academic centers, however, continued to decrease, which can be explained by reduced requirements on the part of the government to introduce research products to industry and by the orientation of academic centers toward fundamental research rather than toward commercial application. These types of ties with academic centers were minimized and came to an end by 1989. Cutbacks in the number of R&D projects financed by the state for the period are another important factor. However, the recovery in the number of linkages which had been observed in 1989 was later lost and after 1991–1992 the number fell dramatically. The disruption of all linkages and growing instability since 1992 can be accounted for by a number of factors external to the university on the national scale, particularly:

- a serious recession in manufacturing; a very difficult situation in an industrial sector striving for survival rather than profits; cutbacks in all types of R&D spending, including advanced research funding;
- reduced research spending in all organizations involved in scientific research; a falling number of research themes, conferences, contacts; a fight for budget funding from the Ministry of Science and the State Committee on Higher Education;
- disrupted linkages with associates from the former USSR; and
- a breakdown of Soviet Union programs and their transfer to Russian jurisdiction and participation in new programs.

The university's R&D policy has always been determined by the situation outside the university and has been aimed at achieving socially appropriate priorities and leadership both among universities and academic centers and in industrial cooperation.

The Moscow Power Engineering Institute's policy in the 1985–1989 period was characterized by the establishment of powerful strategic alliances with the academic community, the government, and industry. The alliance with the government is revealed in the university's participation in major research projects ordered by the executive bodies of the State Committee on Science and Technology (SCST), the USSR "Gosplan," the USSR Academy of Science, and the USSR State Committee on Education; and with industry in the direct contracts of the "MPEI-Ministry" type and its participation in industrial development plans. These contracts included about 60% of all industry R&D contracts, involvement in questions of postgraduate employment, and new product launches by the university. In general, as appears from the institute's R&D activity analysis, some 75–80% of research was ordered by industry. Cooperation with the academic community—higher educational institutions and academic centers of the USSR Academy of Science—included joint research programs, a number of which had MPEI as the head organization. In 1985–1989, research policy concentrated on broader cooperation, the strengthening of long-term contacts, and the establishment of leading positions. Thus, in 1985, MPEI signed contracts with four ministries, and in 1989 this number doubled.

Figure 10.6
Dynamic of University Linkage Density over Time, 1985—1993

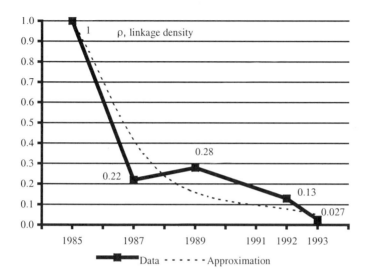

In spite of the somewhat reduced activity observed in 1985, by 1991 MPEI had recovered its contacts with industry, set up branch research centers as a consequence of economic decentralization, introduced a self-sufficient economy, and delegated rights to the university and its departments to sign direct contracts with overseas associates.

A number of negative trends related to the introduction of a self-accounting system, including run-off of young staff, separation of study from the research process, and a smaller number of contracts with MPEI, were reflected in R&D policy, particularly in the organizational structure of science, reduced overhead expenses, and the delegation of a great number of rights to the departments and research groups.

In 1990–1991, university policy remained the same, but after the events of August 1991 followed by the collapse of the Soviet Union, MPEI's position changed dramatically. Together with the 35 leading Soviet Union institutes directly subordinated to the State Committee on Science and Higher Education, MPEI was placed under the jurisdiction of Russia. Under these circumstances MPEI had to regain its position among the national universities, to establish new contacts with Russian government bodies, and to ensure participation in new Russian scientific and technical programs. At the same time the university continued to improve its research organization and cooperation with MPEI affiliates in order to avoid decentralization.

The policy during 1992 can be characterized as one of survival, with the following main criteria of efficiency: research funding, the integrity of the university research school, and preservation of research staff.

The "survival" policy, aimed at establishing a new position in Russian science, continued in 1993–1995. The efficiency analysis of this policy during 1991–1996 against a background of negative trends proved its consistency. Qualitative evidence of its efficiency is provided by the facts that MPEI kept its position among the leading technical universities subject to the former USSR Committee on Education (Moscow Aviation Institute, Moscow Institute of Physics Engineering, Moscow State Technical University) in terms of budget funding, preserved its leading role in a study of electrical engineering specializations, and had the function of head expert council in power and electrical engineering delegated to it. The efficiency of the university's R&D policy in organizing research and in preserving its scientific technical potential was proved by the results of a special social survey.

One can speak in general of a fairly successful research administration policy for the period considered, despite the dramatic changes that occurred in the country. This concerns both strategic external alliances with industry, the scientific community, and the government, and the university's internal policy. The influence of the environment, however, is a dominant factor which does not allow the development and implementation of a policy which will provide progressive results. The research analysis data show the same results in terms of their dynamics, as do the weakened technological transfer network and the transition to lower stages of technology transfer. Despite positive reactions to the scientific administration policy, 81.1% of the respondents stated that they would be unable to live without additional income from work that is not connected with their current job and profession.

The results of the study and the comparative analysis show that the destructive consequences of the policy carried out by the university prevent the situation from stabilizing and do not allow technologies on par with the West to be generated or developed.

REFERENCES

Gibson, D and E. Rogers. *R&D Collaboration on Trial.* Boston, MA: Harvard Business School Press, 1994.

Gibson, D and R. Smilor. "Key Variables in Technology Transfer: A Field-Study Based Empirical Analysis", *Journal of Enginnering and Technology Management,* 8 (December), (1991) 287-312.

Moscow Power Engineering Institutes. *Annual Report on the Institute's R&D Activities in 1987 and XI Five-year Plan.* Moscow: MEI, 1988.

Moscow Power Engineering Institutes. *Annual Report on the Institute's R&D Activities in 1987.* Moscow: MEI, 1985.

Moscow Power Engineering Institutes. *Annual Report on the Institute's R&D Activities in 1987.* Moscow: MEI, 1987.

Moscow Power Engineering Institutes. *Annual Report on the Institute's R&D Activities in 1987.* Moscow: MEI, 1988.

Moscow Power Engineering Institutes. *Annual Report on the Institute's R&D Activities in 1987*. Moscow: MEI, 1989.

Moscow Power Engineering Institutes. *Annual Report on the Institute's R&D Activities in 1989*. Moscow: MEI, 1990.

Moscow Power Engineering Institutes. *Annual Report on the Institute's R&D Activities in 1987*. Moscow: MEI, 1991.

Moscow Power Engineering Institutes. *Annual Report on the Institute's R&D Activities in 1991*. Moscow: MEI, 1992.

Moscow Power Engineering Institutes. *Annual Report on the Institute's R&D Activities in 1992*. Moscow: MEI, 1993.

Moscow Power Engineering Institutes. *Annual Report on the Institute's R&D Activities in 1993*. Moscow: MEI, 1994.

Moscow Power Engineering Institutes. *Annual Report on the Institute's R&D Activities in 1994*. Moscow: MEI, 1995.

Moscow Power Engineering Institutes. *Science of MPEI in Figures*. Moscow: MEI.

PART III:
NATIONAL AND GLOBAL PERSPECTIVES ON TECHNOLOGY AND INNOVATION POLICIES

11

Industrial Innovation in Italy: Regional Disparities and Winning Models

Simona Iammarino, Maria Rosaria Prisco,
and Alberto Silvani

INTRODUCTION

According to the results of surveys and studies, the spatial distribution of the innovative phenomenon in the manufacturing sector seems to be closely linked with the nature and disparities of the external social and economic environment. The identification of regional profiles of innovative activities allows national and international comparisons but cannot explain why and how a region comes to be recognized as a winner in this competition. Moreover, such identification is limited to an examination of results and performances. In an attempt to modify this assumption, we have introduced the concept of regional "style," defined as a combination of local factors, capabilities, and interactions. The aim of this chapter is a tentative approach to the problem of identifying a winning regional style among the four most significant "champions" as seen in the survey on industrial innovation conducted in 1990–1992. This has been done by exploring, on a measurable and quantitative basis, the relationship between the location of innovation activities in manufacturing and the geography of production. The hypothesis of the work follows that of "circular causation" between regional location of production, technological innovation, and international trade performance. After a presentation of the theoretical background to the work, we describe the general features of the Italian regional "champions" of technological innovation. Next the relationship found between the concentration of production and innovation at the regional and sectoral level is shown; attempts are made to shed some light on how innovative styles contribute to regional trade performance; and some concluding remarks and highlights future research lines are provided at the end of the chapter.

CIRCULAR CAUSATIONS AND REGIONAL STYLES

Theoretical research on technological innovation and its geographical dimension follows two main conceptual trends, respectively addressing the two phenomena of geographical agglomeration and diffusion of innovative activity.[1] Within the first approach—originating from Alfred Marshall's pioneering formulation of the concept of external economies—there is a further distinction to be made within the theory's objectives at different times in its development. Initial interest focused on identifying the advantages of geographical agglomeration and the ways in which innovation activities develop (Pred 1966, 1977). More recently, the investigation has moved toward analysis of the structural and environmental factors which shape regional capacity for innovation, a phenomenon which has been defined as the "regional system of innovation."[2]

Among the main structural elements producing the capacity for innovation there is, in the first place, the geographical and sectoral concentration of productive activity, lately the object of economists' attention as a consequence of the special consideration given by the "new economic geography" to the location of production.[3] Another crucial element is the "natural" tendency of innovative inputs to agglomerate, particularly with reference to R&D activity.

The importance of increasing returns and positive externalities in inducing the effects of polarization on economic activities is a well-known issue in economic analysis. The observation of such phenomena is the basic and common ground for the study of both agglomeration processes and the geographical diffusion of innovation.

In fact, the mechanisms and ways through which technological innovation spreads geographically are essential to the consolidation of a regional system of innovation. However, they evade the usual measurement criteria of economic analysis, especially if considered in light of those theories supporting the importance of tacit knowledge in learning processes (Pred 1966; Hagerstrand 1967; Lundvall 1996). Any attempt at measurement would turn out to be too partial, since it could only refer to the encoded elements of the transfer of technological knowledge, as in the case of structured and formal information.

This is the reason for limiting the object of the present analysis to the manifestations of the "external" outcome of the dynamics involved in determining a particular regional style. In other words, the aim is to measure, at a point in time, the effects of regional style, without detailing the specificities of its internal nature.

This means that the real nature and meaning of a "regional style of innovation" is estimated only on the measurement of limited and specific indicators. This study also attempts to capture some of the most relevant features of the geographical unit targeted in the process of opening up to foreign markets.

The role played by a country—with each of the units of which it is made up—in the international division of labor makes, in turn, a fundamental contribution to the outlining of national and local patterns of production and innovation, all the more so in the era of globalization of economic and social phenomena (see Garofoli 1991).

The hypothesis of this chapter follows that of "circular causation," which proceeds from the regional location of production activities, toward the location of technological innovation, hence affecting and interacting with the performance of the region on international markets. In other words, by considering the case of certain Italian regions, we will aim at verifying whether, and to what extent, such circular causation can be interpreted as a set of features that makes up a regional style. The latter is supposed to be sufficiently unique and strong to be differentiated from the national one, particularly when facing international competition. The basic assumption is that highlighted by the evolutionary interpretation of the globalization of economic relations. Such an approach attaches growing importance to the endogenous capabilities of a local system of production and innovation, emphasizing their geographical uniqueness and strong links with the local historical background. Moreover, the cumulative (over time and space) specificities of innovation constitute an interesting background, especially for a sectoral comparison between innovation and production activities.

Background to the analysis is the description of the regional innovation structure taken from the 1992 survey conducted by the Italian National Institute for Statistics (ISTAT) on technological innovation in the Italian manufacturing industry in the period 1990–1992 (Iammarino et al. 1995, 1996; Cesaratto et al. 1993). The initial interpretation of the "regionalized" data was meant to give a picture of the innovation phenomenon. We found a correlation between an assumed regional innovative structure—as gathered from previous analysis and from context variables related to innovation—and the most recent outcomes.

One of the possible developments of our previous work was to provide an in-depth analysis which could account for the increased level of geographical polarization of innovation, concentrating exclusively on some of the most significant regional cases. Such a line of reasoning is corroborated by some recent quantitative studies, which highlight the need for a specific interpretation of regional features within the general Italian process toward innovation (see, for example, Maggioni and Miglierina 1995; Ciciotti 1995). In fact, multiregional models of development would not, in this case, represent an appropriate tool for understanding the phenomenon under investigation.

To monitor the degree of geographical polarization of innovation capacity, the path followed has been suggested by economic geography—that is, comparison with the location of both industry and R&D activity. Such a comparison, in fact, although far from being exhaustive, could account for some inevitable factors when dealing with the geographical analysis of innovation (Krugman 1991a; Audretsch and Feldman 1994, 1995; Jaffe et al. 1993).

The results provide a rather complex picture, in which industrial location, technological innovation, and specialization only partially overlap, and even then to different extents according to the geographical context of reference. This situation produces regional styles with both similar and contrasting features, a peculiar outcome that aggregate analyses would not be able to elicit.

TECHNOLOGICAL INNOVATION AND REGIONAL "CHAMPIONS"

General Features of Regional Innovation

Some of the traditional categories of regional indicators have been used in this work, with the aim of verifying the assumption of "circular causation" of production, technological innovation, and export performance. These indicators were constructed from official statistics, integrated by the measurement of innovation obtained through the number of innovative firms surveyed by the ISTAT with reference to the period 1990–1992, and broken down by region.

In fact, notwithstanding all the limitations pointed out in previous works and in note 10, the number of innovative firms in each region, combined with further "territorialized" information, has already provided quite a suggestive description of the regional innovation profiles.[4]

Considering the strong polarization of the innovation phenomenon in Italy, the choice of the regions to be studied was somewhat conditioned. Therefore, our attention concentrated particularly on Lombardy, Piedmont,[5] Veneto, and Emilia Romagna, which have been labeled regional "champions" in the Italian case.

The 20 Italian regions indeed show large differences, not only in technological and scientific capacities, but also in relation to size, economic development, and overall contribution to the country's performance. Table 11.1 sums up these differences by means of some synthetic and comparative indicators. They confirm—also in macroeconomic terms—the consistency of the choice made of the four regional "champions" and the general relationship between innovation and economic contribution.

These imbalances are even more dramatic if we consider the results of the survey on technological innovation in the Italian manufacturing industry. The polarization of technological innovation in Italy is clearly represented in Table 11.2. Three-quarters of the innovative firms belong to the four "champion" regions, totaling 5,555 firms.

Table 11.1
Regional Macroeconomic Indicators, 1991-1992

REGIONS	Population 1991	Population/km² 1992	GDP* 1992	GDP (%) 1992
Piedmont-Valle d'A.	4,418,503	103	134,196	8.9
Lombardy	8,856,074	372	294,595	19.6
Trentino A.A.	890,360	66	28,852	1.9
Veneto	4,380,797	239	132,663	8.8
Friuli V.G.	1,197,666	152	36,671	2.4
Liguria	1,676,282	308	51,437	3.4
Emilia R.	3,909,512	177	126,320	8.4
Tuscany	3,529,946	153	98,650	6.6
Umbria	811,831	96	20,156	1.3
Marches	1,429,205	148	39,458	2.6
Latium	5,140,371	300	157,492	10.5
Abruzzo-Molise	1,579,954	96	35,397	2.4
Campania	5,630,280	417	103,136	6.9
Apulia	4,031,885	209	77,253	5.1
Calabria-Basilicata	2,680,731	100	42,101	2.8
Sicily	4,966,386	194	92,472	6.1
Sardinia	1,648,248	69	33,477	2.2
ITALY	56,778,031	189	1,504,326	100

REGIONS	Industrial employment/ Total employment (%) 1992	Industry value added* 1992	Industry value added (%) 1992
Piedmont-Valle d'A.	39.7	38,730	11.3
Lombardy	42.9	93,613	27.3
Trentino A.A.	26.0	4,799	1.4
Veneto	40.4	38,128	11.1
Friuli V.G.	31.1	7,543	2.2
Liguria	23.7	10,411	3.0
Emilia R.	34.5	33,328	9.7
Tuscany	34.3	25,142	7.3
Umbria	34.1	4,811	1.4
Marches	35.1	9,363	2.7
Latium	20.0	22,603	6.6
Abruzzo-Molise	28.9	7,201	2.1
Campania	24.4	14,104	4.1
Apulia	25.4	11,634	3.4
Calabria-Basilicata	22.2	4,468	1.3
Sicily	21.1	12,072	3.5
Sardinia	24.4	5,163	1.5
ITALY	31.9	343,167	100

(continued)

REGIONS	R&D expenditure 1992*	R&D expenditure (%) 1992	Fixed investments (%) 1992	Synthetix index of regional development 1992**
Piedmont-Valle d'A.	3,116,954	22.3	9.0	7.5
Lombardy	3,718,770	26.6	18.7	8.4
Trentino A.A.	58,430	0.4	2.5	5.9
Veneto	454,065	3.3	8.7	7.7
Friuli V.G.	209,163	1.5	2.5	6.1
Liguria	472,183	3.4	3.5	5.0
Emilia R.	819,510	5.9	7.3	7.3
Tuscany	658,532	4.7	5.5	6.5
Umbria	45,688	0.3	1.6	6.0
Marches	63,739	0.5	2.5	6.7
Latium	3,077,989	22.0	12.0	4.6
Abruzzo-Molise	211,960	1.5	3.1	5.1
Campania	509,254	3.6	7.2	3.7
Apulia	198,585	1.4	5.9	4.0
Calabria-Basilicata	81,515	0.6	2.9	3.3
Sicily	177,923	1.3	6.3	3.3
Sardinia	98,496	0.7	3.0	4.0
ITALY	13,967,756	100	100	5.8

*in billions of Italian lira; **i.e.: (2/3 x GDP/Pop.+1/3 x industr. Employ./Pop.) multiplied by 100.
GDP in current market prices; value added at factor cost (current prices).
Source: Our calculations on ISTAT data.

Considering now the differences between the four regional innovative "champions," it is useful to highlight the frequencies of the innovative subjects, weighted on the basis of three fundamental aspects of the regional dimension (Table 11.3): demographic size (resident population), geographic size (extent of the region), and relative economic potential (GNP).

While the index values for Lombardy, Veneto, and Emilia are fairly close, the ratio between innovation and dimensional variables is in all three cases definitely lower for Piedmont, where in fact the numeric incidence of innovative firms turns out to be the lowest. The presence of innovative firms is markedly higher than the regional economic contribution, especially for Veneto, which shows the highest economic potential (9%), and Emilia.[6] The contribution of the latter region to expenditure on innovation (fourth row of Table 11.3) is proportional to its contribution to GNP, while Lombardy and Piedmont show values higher than 1. Piedmont in particular—with an index of 2.5 and the highest average cost in the group—seems to regain predominance in the innovation dimension in terms of expenditure. Veneto is the only region showing a value lower than 1; in contrast to Piedmont, the relevance of regional innovation capacity turns out to be adequately represented by the number of firms, but not in terms of costs for technological innovation.

Table 11.2
The ISTAT Survey on Technological Innovation in the Italian Manufacturing Industry by Region, 1992

REGIONS	Surveyed Firms (#)	Surveyed Firms (%)	Responding Firms			Rate of Response (%)	Innovative Firms/ Responding Firms	Innovative Firms (Regional Share)
			Innovative Firms (#)	Non-innovative Firms (#)	Responding Firms (total)			
Piedmont-Valle d/A.	3,258	9.8	874	1,421	2,295	70.4	38.1	11.6
Lombardy	9,613	28.9	2,431	4,306	6,737	70.1	36.1	32.2
Trentino A. A.	546	1.6	175	233	408	74.7	42.9	2.3
Veneto	4,910	14.8	1,181	2,325	3,506	71.4	33.7	15.6
Friuli V. G.	859	2.6	224	431	655	76.3	34.2	3.0
Liguria	474	1.4	111	209	320	67.5	34.7	1.5
Emilia R.	3,705	11.1	1,069	1,729	2,798	75.5	38.2	14.2
Tuscany	2,687	8.1	419	1,313	1,732	64.5	24.2	5.5
Umbria	471	1.4	92	205	297	63.1	31.0	1.2
Marches	1,567	4.7	274	719	993	63.4	27.6	3.6
Latium	1,026	3.1	214	451	665	64.8	32.2	2.8
Abruzzo-Molise	879	2.6	109	395	504	57.3	21.6	1.4
Campania	1,159	3.5	144	498	642	55.4	22.4	1.9
Apulia	1,001	3.0	114	504	618	61.7	18.4	1.5
Calabria-Basilicata	266	0.8	26	113	139	52.3	18.7	0.3
Sicily	572	1.7	67	251	318	55.6	21.1	0.9
Sardinia	275	0.8	29	131	160	58.2	18.1	0.4
ITALY	33,268	100	7,553	15,234	22,787	68.5	33.1	100

Source: Our calculations on ISTAT data.

Table 11.3
Innovative Firms and Variables of Regional Size: Some Indices

REGIONS	Innovative firms/ Population*	Innovations/ GDP 1992**	Innovative firms/ Geographical area	Share of regional cost of innovation/ regional GDP	Average cost of innovation (in millions of Italian lira)
Piedmont-Valle d'A.	19.8	6.5	3.0	2.5	5,494
Lombardy	27.5	8.3	10.2	1.5	2,555
Veneto	27.0	8.9	6.4	0.8	1,359
Emilia R.	27.3	8.5	4.8	1.0	1,646
ITALY	13.3	5.3	2.5	1.0	2,861

* Percentage value multiplied by 1,000; ** Percentage value multiplied by 10.
GDP at current market prices
Source: Our calculations on ISTAT data.

With regard to this point, it is interesting to note the differences between the four "champions" concerning the composition of expenditure on technological innovation (Table 11.4). What is immediately evident from the overall figure is the strong concentration of total national expenditure—even greater than that found for the number of firms—of which more than 50% is accounted for in Lombardy and Piedmont. This percentage rises even further, to 60%, if we consider "aggregated" R&D expenditure. In this case, Lombardy, Emilia, and Piedmont show a share of research expenditure, of the total regional expenditure, which is higher than the national share (approximately 53%). Therefore, while Emilia shows values comparable to those of the two regions traditionally considered as leaders in the Italian innovative model—with a "propensity" to spend on research activities higher than Piedmont itself—Veneto, once again, makes a decisive difference in the group. In fact, even though it ranks as the fourth Italian region in terms of expenditure (7%), Veneto shows a lower scientific capacity and an innovative profile strongly directed to increasing production inputs, particularly physical capital, as is made clear from a share of investments of more than 67% of the total regional expenditure.

A further consideration should be given to the average size of firms. Even though the results of the survey confirm a local innovative structure traditionally based on small and medium-sized firms, we believe that some differentiation would be particularly useful to depict regional profiles. The percentage weight of the five classes of employees considered in the survey shows—in the four "champion" regions—the dichotomy in the Italian innovation system. In fact, in accordance with previous considerations of the concentration of expenditure on innovation, the increasing tendency to generate new technologies in-house and to commit to R&D

Table 11.4
Regional Breakdown of Expenditure for Technological Innovation, 1992 (in million of Italian lira)

REGIONS	R&D, design, trial production, marketing	Share of total regional expenditure	Regional share	Investment for technological innovation	Share of total regional expenditure	Regional share	Total Expenditure	
							Million	Share
Piedmont-Valle d/A.	2,747,063	57.2	24.0	2,055,170	42.8	20.2	4,802,233	22.0
Lombardy	4,118,897	66.3	36.1	2,093,630	33.7	20.5	6,212,527	28.7
Veneto	528,577	32.9	4.6	1,076,034	67.1	10.6	1,604,611	7.4
Emilia R.	1,066,485	60.6	9.3	692,977	39.4	6.8	1,759,462	8.1
ITALY	11,424,550	52.9	100	10,189,907	47.1	100	21,614,457	100

Source: Our calculation on ISTAT data

activities confirms the "Schumpeterian hypothesis" concerning Piedmont and Lombardy. On the other hand, smaller firms, usually involved in traditional sectors, innovating incrementally and—particularly in the case of Veneto—in production processes, show a degree of flexibility and a "catching up" ability that boost regional growth to very high rates, in spite of the lack of formal R&D in relative terms.

External Relations: Supply/Acquisition of Technology, Interregional and Intersectoral Relations

The extent and modalities for the transfer of technology from and to the outside have crucial relevance in determining the degree of openness of a region, especially if the aim is to identify its relative position within the national aggregate and in the wider global scenario.

With respect to the supply of technologies, the modalities showing the highest frequencies, both nationally and regionally, are the sale of equipment and the supply of qualified staff. With reference to the former, the Italian position in general—and of Piedmont in particular—confirms the predominance of one of the traditional technological strong points of the national and regional specialization model—that is, machinery and equipment. The prevailing geographical destination is in this case extra-European countries, presumably less developed ones. Piedmont, Lombardy, and Veneto indicate other European countries as their second market (mainly ex-planned economies), while for Emilia the national market is the second option.

Similar observations can be made for the supply of qualified staff, which is mainly directed toward less-developed extra-European countries. In connection with this, Emilia shows a greater degree of openness towards the other Italian regions, which is also confirmed by the highest proportion of firms transferring technology through cooperation with other firms to the rest of the national territory. The two modalities regarding formal research activities—that is, the granting of the right to use the firm's inventions and R&D carried out for other parties—show the low importance of outflows in all four cases. On the whole, the lowest level of internationalization is shown by Emilia, while Veneto registers relative openness toward the European Union (EU) for more formal research activities.

The acquisition of technology shows, as predicted, much higher frequencies. Also in this case, the most frequent modalities turn out to be the purchase of specialized equipment and the recruitment of qualified personnel. The regional picture, however, appears much more uniform from the viewpoint of geographical origin. In all four regions considered, innovative firms buy equipment mainly from the rest of the country, the second option being EU partners.

Piedmont also appears to be the most "open" region for technological inflows, while the lowest inflows in this regard are once more for Emilia. Also, cooperation with other firms and the acquisition of consulting services occur, in first place, within national boundaries. In second position, we again find the EU, which turns out to be the main geographical origin for R&D acquired from abroad. The high tendency to acquire EU technologies might be one of the effects of the European Union integration process, which reinforced the possibilities of transferring technology across member states and of assimilating European technological innovation, vis à vis that coming from other advanced countries (the United States and Japan).

In the survey questionnaire, the extent and the composition of expenditure on innovation were followed by the territorial attribution of the costs incurred. The relevance of introducing such a variable lies in the possibility of measuring the territorial dispersion of costs for innovation incurred by the firms of one region in all the others, and of testing the "attraction" capacity that a regional system exerts on the rest of the country. On this basis it was possible to produce a matrix, whose main diagonal indicates the degree of "self-reference" of the regional systems considered (Table 11.5).[7]

The highest "attraction" capacity, among the four regions considered, is shown by Piedmont, since approximately 30% of the costs attributed to this region come from external sources, particularly from firms in Lombardy. In the case of the latter, the main regions of origin are instead Veneto and Emilia, respectively, with 7.4% and 3.3% of the overall cost attributed to Lombardy.

More articulated, from the point of view of geographical origin, are Emilia's costs for innovation, attributed to the region by firms in different locations. It should be noted that the lower level of "attraction" exerted by the two Northeastern regions—relative to Lombardy and particularly to Piedmont—is correlated with a higher territorial dispersion of outflows in the case of Veneto. This region shows on the whole a greater intensity of interregional relations, expressed in terms of costs, compared to Emilia, which appears as the most "self-referring" among the regional "champions" of technological innovation.

A final element, useful in the identification of regional innovation profiles, is made up of sectoral interdependencies, which play a special role in verifying the connections between innovation and production activities. At the country level, 57% of the innovation produced by manufacturing firms is directed toward manufacturing itself—that is, it is reemployed in manufacturing processes. It is evident that the tendency toward intramanufacturing is definitely greater than average in Piedmont (67.5%) and Lombardy (63%), while the values for Emilia and Veneto are below the overall Italian figure (respectively 55.4% and 53%).

Table 11.5
Cost of Innovation—Matrix of Regional Distribution

REGIONS OF DESTINATION (j)

REGIONS OF ORIGIN (i)	Piedmont-Valle d'A.		Lombardy		Trentino A. A.		Veneto		Friuli V. G.		Liguria	
	% Column	% row	% Column	% row	% Column	% row	% Column	% row	% Column	% row	% Column	% row
Piedmont-Valle d'A.	71.3	89.7	1.2	1.6	0.2	0.0	1.0	0.2	0.4	0.0	23.0	5.8
Lombardy	10.4	10.1	80.2	81.3	1.6	0.1	2.7	0.5	3.3	0.2	0.6	0.1
Trentino A. A.	0.3	6.9	0.0	1.0	91.3	84.9	0.7	4.0	0.0	0.0	0.0	0.0
Veneto	0.1	0.5	7.4	29.0	4.5	0.6	86.0	66.8	1.7	0.4	0.2	0.1
Friuli V. G.	0.1	1.2	0.1	1.4	0.0	0.0	1.2	4.0	86.7	89.8	0.6	1.9
Liguria	0.1	1.0	0.1	1.8	0.0	0.0	0.0	0.1	3.9	4.2	25.1	83.6
Emilia R.	0.1	0.5	3.3	11.7	1.1	0.1	7.2	5.1	1.6	0.3	0.4	0.3
Tuscany	0.9	5.5	0.8	5.1	0.3	0.1	0.2	0.2	1.3	0.5	0.8	0.9
Umbria	0.0	0.3	0.5	19.5	0.0	0.0	0.0	0.0	0.0	0.0	5.5	40.7
Marches	0.0	0.0	0.2	4.4	0.3	0.2	0.2	0.8	0.4	0.5	0.0	0.0
Latium	1.1	4.5	0.8	3.4	0.1	0.0	0.0	0.0	0.2	0.1	3.8	3.1
Abruzzo-Molise	2.6	21.6	0.2	1.6	0.0	0.0	0.0	0.0	0.1	0.0	0.0	0.0
Campania	9.0	45.6	2.2	11.6	0.1	0.0	0.0	0.0	0.2	0.1	0.0	0.0
Apulia	0.6	4.8	0.5	4.6	0.2	0.1	0.0	0.0	0.1	0.0	39.1	65.3
Calabria-Basilicata	0.0	4.1	0.1	6.3	0.1	0.3	0.0	0.0	0.0	0.0	0.0	0.0
Sicily	3.3	48.7	0.7	11.1	0.1	0.1	0.0	0.0	0.2	0.2	0.9	2.7
Sardinia	0.0	0.0	1.7	57.2	0.0	0.0	0.7	4.8	0.0	0.0	0.0	0.0
ITALY	100.0	28.0	100.0	29.1	100.0	1.0	100.0	5.8	100.0	1.8	100.0	5.6

154

REGIONS OF DESTINATION (j)

REGIONS OF ORIGIN (i)	Emilia R.		Tuscany		Umbria		Marches		Latium		Abruzzo-Molise	
	% Column	% row	% Column	% row	% Column	% row	% Column	% row	% Column	% row	% Column	% row
Piedmont-Valle d'A.	1.6	0.5	0.4	0.1	0.4	0.0	3.7	0.2	3.5	1.6	0.3	0.0
Lombardy	4.9	1.3	2.6	0.4	7.1	0.1	3.5	0.2	14.7	5.2	2.0	0.2
Trentino A. A.	0.4	2.8	0.1	0.4	0.0	0.0	0.0	0.0	0.0	0.0	0.0	0.0
Veneto	1.1	1.1	1.3	0.7	0.6	0.0	0.7	0.1	0.3	0.5	0.2	0.1
Friuli V. G.	0.0	0.2	0.0	0.1	0.0	0.0	0.0	0.0	0.2	0.9	0.0	0.0
Liguria	0.0	0.0	0.6	1.5	2.0	0.4	0.1	0.1	1.0	5.8	0.0	0.0
Emilia R.	87.6	80.9	0.4	0.2	2.2	0.1	0.7	0.1	0.2	0.3	0.2	0.0
Tuscany	0.4	0.6	89.9	77.2	2.2	0.2	0.1	0.0	4.3	9.4	0.0	0.0
Umbria	0.0	0.0	0.1	0.5	83.5	38.3	0.1	0.3	0.0	0.3	0.0	0.1
Marches	0.2	1.1	0.4	1.2	0.5	0.1	84.7	87.0	00	0.3	2.8	4.1
Latium	0.6	0.6	3.0	1.7	0.9	0.0	0.1	0.0	58.3	86.3	0.0	0.0
Abruzzo-Molise	0.0	0.0	0.0	0.0	0.0	0.0	0.0	0.0	5.4	16.3	94.5	60.5
Campania	0.4	0.5	1.1	0.8	0.3	0.0	6.1	1.6	7.3	13.4	0.0	0.0
Apulia	1.9	4.2	0.0	0.0	0.2	0.0	0.0	0.0	2.2	6.7	0.0	0.0
Calabria-Basilicata	0.9	28.7	0.0	0.3	0.0	0.0	0.0	0.0	0.5	21.6	0.0	0.0
Sicily	0.0	0.0	0.0	0.0	0.0	0.0	0.0	0.0	0.1	0.7	0.0	0.0
Sardinia	0.0	0.3	0.0	0.0	0.0	0.0	0.0	0.0	1.8	21.7	0.0	0.0
ITALY	100.0	7.5	100.0	4.0	100.0	0.3	100.0	1.5	100.0	10.1	100.0	2.2

(continued)

155

REGIONS OF DESTINATION (j)

REGIONS OF ORIGIN (i)	Campania		Apulia		Calabria-Basil		Sicily		Sardinia		TOTAL	
	% Column	% row	% Column	% row	% Column	% row	% Column	% row	% Column	% row	% Column	% row
Piedmont-Valle d'A.	1.5	0.1	0.8	0.0	1.0	0.0	0.0	0.0	0.5	0.0	22.2	100.0
Lombardy	3.9	0.2	3.9	0.1	8.2	0.0	1.7	0.0	0.7	0.0	28.7	100.0
Trentino A. A.	0.0	0.0	0.0	0.0	0.0	0.0	0.0	0.0	0.0	0.0	1.1	100.0
Veneto	0.1	0.0	0.1	0.0	0.8	0.0	0.0	0.0	0.0	0.0	7.4	100.0
Friuli V. G.	0.4	0.4	0.0	0.0	0.0	0.0	0.0	0.0	1.0	0.1	1.8	100.0
Liguria	1.5	1.5	0.0	0.0	0.0	0.0	0.0	0.0	0.0	0.0	1.7	100.0
Emilia R.	0.4	0.1	1.6	0.1	0.4	0.0	2.1	0.2	0.0	0.0	8.1	100.0
Tuscany	0.3	0.1	1.5	0.2	0.4	0.0	0.1	0.0	0.0	0.0	4.6	100.0
Umbria	0.0	0.0	0.0	0.0	0.0	0.0	0.0	0.0	0.0	0.0	0.7	100.0
Marches	0.2	0.3	0.1	0.0	0.0	0.0	0.0	0.0	0.0	0.0	1.5	100.0
Latium	0.8	0.2	0.7	0.1	0.0	0.0	0.0	0.0	0.0	0.0	6.9	100.0
Abruzzo-Molise	0.0	0.0	0.0	0.0	0.0	0.0	0.0	0.0	0.0	0.0	3.4	100.0
Campania	90.1	25.8	2.0	0.2	16.2	0.4	0.0	0.0	0.0	0.0	5.5	100.0
Apulia	0.0	0.10	89.1	14.0	0.1	0.0	1.0	0.2	0.0	0.0	3.3	100.0
Calabria-Basilicata	0.4	2.8	0.0	0.0	73.0	35.6	0.1	0.2	0.0	0.0	0.2	100.0
Sicily	0.0	0.0	0.0	0.0	0.0	0.0	94.9	36.5	0.0	0.0	1.9	100.0
Sardinia	0.2	0.4	0.0	0.0	0.0	0.0	0.0	0.0	97.8	15.6	0.9	100.0
ITALY	100.0	1.6	100.0	0.5	100.0	0.1	100.0	0.7	100.0	0.1	100.0	100.0

CONCENTRATION OF PRODUCTION AND INNOVATION

While exploring the geographical and sectoral concentration of innovation activity as a whole, we can observe the location of production activity and its sectoral features within different regions. Following Krugman's example (1991a)—also applied by Audretsch and Feldman (1994, 1995)—the geographical and sectoral concentration coefficients have been calculated for innovation, expressed in terms of the number of innovative firms, for value added and for employment, in the 17 Italian regions and in the 10 manufacturing macrosectors for which a "homogeneous" database was available relative to the regional indicators considered.

The Gini concentration ratio has been calculated on the basis of a simple index of sectoral "intensity" of innovation at the regional level. In practical terms:

$$\text{ISRI} = \frac{I_{ij} / \sum_j I_{ij}}{I_{Ij} / \sum_j I_{Ij}} \qquad \text{where} \quad i = 1, \ldots \ldots 17 \\ j = 1, \ldots \ldots 10$$

where Iij is the number of innovative firms of the ith region[8] in the jth sector, and IIj is the total number of innovative firms in Italy in the jth sector (7,553 according to ISTAT). The above method of weighting Gini's coefficients allowed us—at least to some extent—to account for the remarkable differences in size of the territorial units targeted.[9] Not only does the index—which takes a value higher than, equal to, or lower than 1—represent the basis for the calculation of the concentration coefficients, but it has also proved a useful tool in the analysis of innovation profiles in the four regional "champions." In addition, it allows a direct comparison with the analogous index of revealed comparative advantage (RCA), used to define regional trade specialization in international trade.

Normalization, by sector and by region, for value added and employment has been carried out similarly to that of technological innovation, so as to obtain a comparative ranking of the concentration coefficients for both technological innovation and the variables related to manufacturing activity.

The analysis of Gini's geographical concentration coefficients—presented in Table 11.6—shows nonunivocal features. These can be summarized as follows:

1. The variance of concentration coefficients for innovation—on which they have been ranked—is much higher than in the case of value added and employment, confirming the strong polarization of the innovative phenomenon at the geographical level.

2. The ranking of concentration coefficients for innovation shows a stronger relationship with that related to employment than with the ranking on value added. However, we should bear in mind that the geographical correspondence between the production variables examined is *per se* rather uncertain in the Italian case, being biased by the technical developments of production activities. Not only does the employment rate vary appreciably within the same sector, but also the territorial trend for value added—as also indicated by the previous survey—turns out to have decreased in the most innovative regions (except Veneto) during the 1980s.

3.   The concentration coefficients of production variables are always lower in sectors showing a higher geographical concentration of innovation ("mining," "motor vehicles and other means of transport," "chemical and pharmaceutical products," and "textiles and clothing"). This is mainly due to two different reasons: first, the innovation phenomenon is particularly thin in some of the sectors considered (e.g., "mining"); second, because of the different unit of analysis used in the data collection (as stated, the unit considered for the geographical attribution of innovation is the firm, irrespective of the location of plants; the single production unit is instead considered for the measurement of production activities). Moreover, the list of macrosectors suggests that the different technological content at the intrasectoral level has been neglected—that is, even in high-tech sectors, such as "chemical and pharmaceutical products," a large variance is to be expected in the innovative efforts of various industrial activities. Therefore, production variables appear to be appreciably less concentrated compared to technological innovation;

4.   Conversely, the degree to which production is concentrated is often higher in sectors where the innovative phenomenon appears to be geographically scattered. The most obvious reason for this fact is the relatively low technological content in the majority of the products of the macrosectors examined (e.g., "minerals and ferrous/nonferrous metals," "minerals and nonmetallic products," "paper and publishing," "mechanical equipment and metal products"). However, the same cannot be said for "textiles, clothing, and leather." The "anomalies" of a higher concentration coefficient for the innovation variable than for value added (which is nevertheless one of the highest), and of a relatively lower position in the ranking of the employment coefficient, can be accounted for by the production modalities used in this sector. On the one hand, it is indeed a strong point in the national specialization model, boosting both innovative effort and wealth creation, even though it is geographically concentrated in some locations only. On the other hand, the sector is more affected than the others by a process of extra-national decentralization of production related to stages of production with lower value added, thus causing a lower concentration of employment compared to the other variables.

All this is in agreement with the points discussed in more extensive empirical investigations, according to which "while the location of manufacturing activity may explain the spatial distribution of innovative activity to some degree, it is certainly not the only factor" (Audretsch and Feldman 1994).

The most interesting aspect to help us distinguish possible regional styles among all the indicators examined, is in any case the sectoral concentration of production and innovation activities within the various regional contexts. The concentration coefficients are presented in Table 11.7 and, in spite of the caveat implied by this exercise, in this case also there are some interesting peculiarities to highlight:

1.   The greater variance shown by the concentration coefficients for innovation confirms what has been observed for geographical concentration.

2.   In the first few rows of Table 11.7 we find regions with the highest coefficients of sectoral concentration, but in which the innovation phenomenon has only a marginal role in economic performance (the low number of innovative firms itself explains the strong sectoral concentration of innovation).

3. Regions which have an intermediate ranking relative to the innovation variable exhibit rather similar values for the coefficients, but they are characterized by relatively different profiles, in terms of both average firm size and sectoral specificities.
4. The regions with a lower sectoral concentration of innovation are those with the strongest presence of the innovative phenomenon: its intensity, in fact, underlines the wider sectoral spread of technological innovation.

Table 11.6
Geographical Concentration of Manufacturing Activities (Gini coefficients)

SECTORS	Innovation	Value Added	Employment
Mining	0.7095	0.3955	0.3715
Motor vehicles and other means of transportation	0.4528	0.3326	0.3379
Chemical and pharmaceutical products	0.4085	0.2736	0.3175
Textile, clothing, and leather products	0.4016	0.3715	0.3067
Minerals and iron and noniron ore metals	0.3168	0.3707	0.3282
Food, drinks, and tobacco	0.2994	0.2408	0.2150
Minerals and nonmetallic products	0.2988	0.4042	0.2608
Wood, rubber, and other manufacturing	0.2804	0.2621	0.2070
Paper and publishing	0.2256	0.3073	0.2370
Mechanical equipment and metal products	0.1707	0.2054	0.1919

Source: Our calculation on ISTAT data.

In this latter group, precisely in the lowest places of the ranking, we found the four regional "champions," although, as already pointed out, they are not a homogeneous group. In Piedmont, the narrower sectoral diversification of the specialization of production gives rise to concentration coefficients for value added and employment which are markedly higher than for innovation. By contrast, in Lombardy, where the industrial spread is larger, the coefficients of the production variables are the lowest, with respect to that of innovation and to the other three regions. Also in this case, comparison between Veneto and Emilia clearly shows the different features of the two Northeastern champions. Veneto shows the lowest position in the ranking, not only because of the thinner presence of innovation relative to the other three case studies, but also because the region, as argued above, is a comparatively "open" system, acquiring a relatively higher share of technology from the outside. Emilia is instead characterized by fairly strong internal innovation, also connected to its robust "self-reference" in terms of innovation costs (a comparatively "closed" regional system).[10]

Table 11.7
Sectoral Concentration of Manufacturing Activities (Gini coefficients)

REGIONS	Innovation	Value Added	Employment
Trentino	0.5871	0.2827	0.2869
Sicily	0.5761	0.5212	0.4085
Sardinia	0.5731	0.4953	0.4297
Friuli	0.4913	0.4427	0.1222
Calabria-Bas.	0.4636	0.4030	0.3422
Liguria	0.4253	0.4426	0.40.82
Apulia	0.4011	0.3027	0.2365
Campania	0.3995	0.2647	0.2278
Marches	0.3972	0.3479	0.2989
Latium	0.3622	0.3490	0.2971
Abruzzo-Molise	0.3347	0.1526	0.1371
Umbria	0.3330	0.5096	0.3230
Tuscany	0.3056	0.2778	0.2609
Emilia Romagna	**0.2655**	**0.2983**	**0.2679**
Lombardy	**0.2461**	**0.2026**	**0.1880**
Piedmont-V.d'A.	**0.2394**	**0.3041**	**0.3160**
Veneto	**0.2145**	**0.2473**	**0.2387**

Source: Our calculations on ISTAT data.

Finally, to further corroborate the relationship between the concentration of innovation and that of production at territorial level—and thus, at least partially, our hypothesis about "circular causation"—Table 11.8 shows the correlation coefficients, for all 17 Italian regions, between the following variables: the sectoral Gini coefficient for innovation (Ginno); the sectoral Gini coefficient for value added (Gva) and for employment (Gocc); the regional share of national expenditure on R&D; and the regional degree of openness to international trade (relative to the whole country) (Open).[11]

The sectoral concentration of innovation is significantly correlated with all the variables considered. As expected, it is positively correlated with the concentration of both value added and employment, thus confirming the relationship between the sectoral localization of production and innovation activities.

Table 11.8
Correlation Coefficients (2-tailed significance)

	GINNO	GVA	GOCC	R&D
GINNO				
GVA	0.5825*			
GOCC	0.5057*	0.7876**		
R&D	– 0.5308*	– 0.3403	– 0.1890	
OPEN	– 0.6114**	– 0.4529	– 0.4477	0.3667

* indicates significance at the 1% level
** indicates significance at the 5% level

The negative coefficient observed with respect to R&D supports the fact that the greater the amount of resources devoted to R&D, the more the distribution of innovation is disseminated at the sectoral level.

The negative correlation shown by the variable Open seems to indicate that the regions with a relatively higher degree of openness are those characterized by an innovative capacity which is more homogeneously spread over manufacturing sectors. This suggests that the sectoral pervasiveness of innovation could lead to a more pronounced competitiveness of regional export performance.[12]

In conclusion, the analysis of data and coefficients for the four "champion" regions seems to give rise to a common paradigm—that is, a lower concentration of innovation is associated with a wider sectoral spread of specialization. This holds particularly for Lombardy, while in the case of Veneto and Emilia there are some incongruities, especially in the case of "paper and publishing" and "motor vehicles," respectively. Finally, to better understand the effects of circular causation and to give some indications on its cumulativeness, performance in international trade is examined in the next section.

INNOVATION AND INTERNATIONAL TRADE PERFORMANCE

The analysis of the sectoral structure of international trade performance in the four regions considered is a way to verify the "cumulative causation" process among technological innovation, production activity, and trade performance. The aim is to achieve a definition—even if only generic and limited to the external manifestation—of locally differentiated regional styles. The indices of trade specialization refer to 1994, a temporal lag of two years with reference to the introduction of technological innovations in the manufacturing firms surveyed by ISTAT.

Table 11.9 displays the shares of total Italian manufacturing exports, by sector and by region, and the sectoral composition of regional exports. Lombardy appears once more as the outstanding region, contributing 31% to total national exports. The highest values are to be found in the following sectors: "chemicals and pharmaceuticals" (45%), which however contributes to regional exports to a much lower extent (12%); "minerals and ferrous/non-ferrous metals" (40% of national exports), and "mechanical equipment and metal products" (38%). The latter sector also accounts for almost half of Lombardy's exports (44%), followed by "textiles, clothing, and leather products," providing 17% of the region's exports. Veneto is in second position with respect to the share of manufacturing products sold by Italy to foreign countries (14%). The most substantial contribution comes from "wood, rubber, and other manufacturing" (21%) and from "textiles, clothing, and leather products" (more than 19% of Italian exports), which is also the most substantial sector within the composition of regional exports (26%) behind "mechanical equipment and metal products" (over 33%). The strong specialization in the production of "motor vehicles and other means of transport" for Piedmont appears to be confirmed by its percentage of total Italian exports (31%) and by its significant contribution to the composition of regional exports (23%). Also "food, drinks, and tobacco" constitutes a good proportion of national sales abroad (20%), while approximately one-third of regional exports are accounted for by "mechanical equipment and metal products." Emilia provides 11% of all Italian exports (in particular, 36% of "minerals and non-metallic products" on the national scale). The composition of regional exports, however, turns out to be strongly concentrated in "mechanical equipment and metal products," accounting for 44% of manufacturing products exported by that region.

For the purpose of verifying the commercial specialization of our four innovative "champions," we used indices that are traditionally adopted in foreign trade analysis—the revealed comparative advantage index of Balassa (RCA)[13] and the normalized trade balance (NTB).[14] They are presented in Table 11.9 for the four regions and the ten sectors considered.

The RCA index shows a strong specialization of Piedmont, particularly in "motor vehicles and other means of transport," in accordance with the concentration of production and innovation of the above sector in that region. The export specialization also appears rather high regarding "food, drinks, and tobacco" and "paper and publishing." At any rate, in both sectors the regional contribution to innovation and production activities seems to be lower than the national one. For Lombardy, the export specialization is definitely in accordance with the regional contribution to innovation. In all three sectors of relative trade specialization—that is, "chemical and pharmaceutical products," "minerals and ferrous/nonferrous metals" and "mechanical equipment"—the value of the ISRI index on the number of innovative firms and on value added was also higher than 1.

Table 11.9
Regional Shares on Italian Exports and Product Composition of Regional Exports, 1994

MANUFACTURING SECTORS	Piedmont-Valle d'A.		Lombardy		Veneto		Emilia R.		ITALY	
	Shares	Prod. Comp	Shares	Prod. Comp	Shares	Prod. Comp	Shares	Prod. Comp	Shares	Prod. Comp
Mining	4.5	0.6	12.4	0.7	4.0	0.5	1.1	0.2	100	1.7
Mineral and iron and nonirong ore metals	12.1	4.1	39.7	5.9	10.2	3.4	4.3	1.8	100	4.6
Minerals and nonmetallic products	4.7	1.5	11.0	1.5	16.0	5.1	36.2	14.4	100	4.4
Chemical and pharmaceutical products	9.1	5.6	44.9	11.9	8.2	4.9	8.5	6.4	100	8.2
Mechanical equipment and metal products	13.2	35.1	38.3	44.1	12.9	33.2	13.4	43.7	100	35.6
Motor vehicles and other means of transport	31.3	2.7	17.9	5.6	6.7	4.7	9.1	8.0	100	9.7
Food, drinks, and tobacco	20.5	6.9	19.3	2.8	9.5	3.1	16.3	6.7	100	4.5
Textile, clothing, and leather products	8.4	11.4	28.2	16.6	19.4	25.7	6.9	11.5	100	18.2
Paper and publishing	18.0	3.0	27.0	1.9	15.2	2.4	6.9	1.4	100	2.2
Wood, rubber, and other manufacturing	11.0	9.1	25.4	9.1	21.3	17.0	5.8	5.9	100	11.0
TOTAL	13.4	100	31.0	100	13.8	100	10.9	100	100	100

Source: Our calculation of ISTAT data.

The relationship between regional intensity of innovation and export specialization is also clearly visible in Veneto and Emilia. In the former region, innovation helps to keep export performance above average in three of the four sectors of trade specialization ("wood, rubber, and other manufacturing," "textiles and clothing," and "minerals and nonmetallic products"). The only group of products in which the RCA is higher than 1 with an ISRI lower than 1 is "paper and publishing." However, the production specialization and the contribution to employment in this sector show rather high values. Finally, for Emilia we find a correspondence between "leading" sectors in exports and innovative contribution that is very similar to that observed in Lombardy. The comparative advantage of the Emilia region is identified by the RCA in "minerals and nonmetallic products," "food, drinks, and tobacco" and "mechanical equipment and metal products." This corresponds to a regional contribution of these sectors which appears to be higher than average in terms of both the number of innovative firms and of production capacity.

However, the RCA—based solely on export data—turns out to be a limited device for a more extensive comprehension of the regional position in international markets. Therefore, the normalized trade balance (NTB)—that is, the ratio between the trade balance and the total exchange—was also considered (Table 11.10).

We can first of all observe a lower international competitiveness for Lombardy compared to what the other indices indicated: the aggregate balance is the only negative one among the four regions and also in the whole country. Besides, the high specialization of exports, in those sectors for which the innovative contribution turns out to be relatively strong, cannot make up for the regional imbalance in the total foreign trade of manufacturing. The position of Piedmont appears more balanced, as the normalized balances turn out to be positive in all the sectors of export specialization, and the competitive success of "motor vehicles and other means of transport" is confirmed. Emilia shows a very high NTB in "minerals and nonmetallic products," in which it was also possible to observe the most striking comparative advantage in terms of the RCA index. Emilia's competitiveness in international markets, which for its "leading" sectors also turned out to be high on the national level, is confirmed for "mechanical equipment and metal products," with a strongly positive balance, but not for "food, drinks, and tobacco," another sector of relative export specialization in the region. It is however Veneto that stands out as the most balanced and dynamic region of the four as regards competitiveness. In all sectors of export specialization it is possible to note positive and high normalized balances, showing a surprising coincidence with the specialization pattern and with the strongest, most competitive sectors of the whole country.

Table 11.10
Regional Revealed Comparative Advantage (RCA) and Normalized Trade Balance (NTB), Indices 1994

MANUFACTURING SECTORS	Piedmont-Valle d'A.		Lombardy		Veneto		Emilia R.		ITALY	
	RCA	NTB	RCA	NTB	RCA	NTB	RCA	NTB	RCA	NTB
Mining	0.34	-0.83	0.40	-0.80	0.29	-0.64	0.10	-0.84	1.00	-0.70
Mineral and iron and nonirong ore metals	0.90	-0.21	1.28	-0.22	0.74	-0.55	0.40	-0.37	1.00	-0.30
Minerals and nonmetallic products	0.35	0.25	0.35	-0.03	1.16	0.48	3.32	0.80	1.00	0.43
Chemical and pharmaceutical products	0.68	-0.13	1.45	-0.29	0.60	-0.18	0.78	-0.03	1.00	-0.22
Mechanical equipment and metal products	0.99	0.31	1.24	0.08	0.93	0.59	1.23	0.59	1.00	0.27
Motor vehicles and other means of transport	2.34	0.38	0.58	-0.14	0.49	-0.46	0.83	0.41	1.00	0.01
Food, drinks, and tobacco	1.53	0.25	0.62	-0.43	0.69	-0.36	1.49	-0.21	1.00	-0.26
Textile, clothing, and leather products	0.63	0.35	0.91	0.38	1.41	0.47	0.63	0.47	1.00	0.44
Paper and publishing	1.35	0.13	0.87	-0.17	1.10	0.10	0.63	-0.24	1.00	-0.08
Wood, rubber, and other manufacturing	0.83	0.18	0.82	0.19	1.54	0.54	0.53	0.32	1.00	0.34
TOTAL	1.00	0.19	1.00	-0.03	1.00	0.20	1.00	0.34	1.00	0.08

Source: Our calculation of ISTAT data.

Thus, international trade data confirm, on the one hand, that innovative capacity, even in the case of the most innovative regions, does not result in stronger overall performance, probably because of the limited contribution of innovation to export competitiveness. On the other hand, a strong sectoral specialization has been observed in all four case studies, attesting that innovation capacity is used at least directly in the sector where it is generated, with relatively weak intersectoral linkages.

CONCLUSIONS: FROM REGIONAL PROFILES TO REGIONAL STYLES—THE ITALIAN EXPERIENCE AT A CROSSROADS

The research route followed in this work, aimed at exploring to what extent the geographical and sectoral concentration of innovation affects the overall economic performance of a regional system, has reached a first conclusion. On the one hand, in the four regions considered as "champions," organization and manifestation of innovation definitely show highly specific and differentiated profiles. This is particularly visible with regard to the sectoral spread of technological innovation in manufacturing, the relative size of innovative firms, and the degree of "self-reference" in terms of expenditure on innovation. On the other hand, it is hard to categorize the variety of regional profiles directly and explicitly into different regional styles. This is particularly relevant in the comparison between geographical concentration and sectoral concentration, where the role played by innovation seems to have had a greater influence in terms of the number of innovative firms, rather than in terms of "quality" and determinants of the innovation itself.

Taking into account the strong bias in the type of quantitative information available for innovation (especially the lack of any quantitative parameter to estimate the difference between actual innovation and that inferred on the basis of one firm-one innovation, geographically attributed solely to the headquarters of the firm), the sectoral concentration of technological innovation shows a strong correspondence with the ranking of the concentration coefficient for regional value added. This has been further confirmed by analysis of regional export specialization patterns.

The Italian pattern of economic development, therefore, seems to be at a crossroads. The contribution to the national performance comes from regions with more "mature" industrialization, as in the case of Piedmont, and from "leader" regions characterized by a high degree of extraregional decentralization of production, such as Lombardy. But a comparable contribution to national performance is also given by regional systems based on small firms, which show strong differences in technological content and sectoral specialization, as is the case of both Veneto and Emilia. It is possible to interpret these phenomena either as lacking a regional style winner compared to others, or as a restricted contribution of innovation to overall performance. The concentration of innovation, and its cumulative development over time and space, does not entirely explain sectoral performance, probably also because of the low homogeneity of technological content represented by the macrosectors considered.

However, a limited interindustry diffusion of innovation emerges, which affects overall economic performance and restrains the cumulativeness within each sector, rather than diffusing into the regional economic structure. It is thus necessary to improve the analysis along more than one possible line of reasoning, starting from a different weighting of innovative phenomena as depicted by the ISTAT survey and from a representation of regional systems wider than that provided by the variables analyzed here.

Two other final considerations can be added. First of all, at the sectoral level, the "circular causation" effect is generally confirmed, although it should also be tested in its evolution over time. Second, in the Italian case, the "power" of innovation in explaining production and trade performances seems to be rather limited. The regional analysis stresses, more than the competition between different styles, a substantial conformity to a national pattern which proved very difficult to model.

In other words, in an economic system potentially ruled by heterogeneous parts and subject to strong international pressures, such as the Italian system, the different regional styles are difficult to perceive. The difference in regional styles emerges as clear-cut only with reference to macroregions—that is, the most developed regions, like the four analyzed in this study, and the backward regions, as is the case of the whole Italian Mezzogiorno.

This rediscovery of the Italian dichotomy, however, does not correspond to the traditional model that is well-known in the literature. In fact, in the past, the social and economic functional relationships and interdependencies between the North and the South of the country gave rise to rapid growth and to the shaping of an Italian style. Today the globalization of production and innovation seems to jeopardize the agreement of interdependence and reciprocal interest between the different areas of the country, supporting the weakness of regional styles as a key for interpreting economic phenomena.

NOTES

1. For a clear and critical exposition of the two theoretical trends, see Breschi (1995).

2. See, for instance, Howells (1997). Among the studies of factors impacting on the local degree of innovation it is worth recalling the approaches based on the concepts of the *milieux innovateur* and the technological district (e.g., Aydalot 1986; Becattini 1987). More appropriate to our purposes, however, is the broader concept of regional systems, as it better represents the position and style of the region as a unit, setting aside the specificity of intra-regional relations.

3. See Paul Krugman (1991a, 1991b) who brought attention back to a research theme that was never completely set aside in economics, particularly in regional economics.

4. It is necessary at this point to recall briefly the main limitations of the data available. We suggest reference to our previous work for an exhaustive review of all the *caveats* concerning the general formulation of the survey and the method proposed to improve

the "regionalization" of the information gathered by means of questionnaires (Iammarino et al. 1995). In the first place, the "per firm" approach, though naturally apt to weigh the number of innovative subjects in each region, proves inadequate at explaining the phenomenon of innovation in all its complexity, as well as unable to provide any element of characterization of the different subjects. In fact, each firm "matters" as a unit, irrespective of production capacity and ability to produce innovation, as well as the quantity and quality of such innovation. On the other hand, the quality level of the replies given by the firms that defined themselves as "innovative" can be considered quite reliable, especially on the basis of a "check" section of the questionnaire which inquired about the costs incurred for innovation. This device allowed a substantial advance for the survey compared to the previous one, and provided a direct source of information on territorial features—a variable that does not appear in many of the other European questionnaires produced in the Community Innovation Survey (CIS) framework. In the second place, considering the case of multiplants firms, it should be noted that the geographical attribution of the innovative firm can be misleading in assessing the location of the innovation, as usually any innovation introduced is "officially" ascribed to the region of the headquarters. Finally, the survey addressed manufacturing firms with more than 20 employees, leaving aside a significant share of the innovation phenomenon throughout Italy, a country whose industrial structure is notoriously characterized by a substantial number of small firms.

5. The data of Piedmont and Valle d'Aosta collected by the ISTAT survey on technological innovation were not separable, therefore this aggregation was also kept for the other variables considered in this work. However, it is necessary to remember that this is not a particularly significant distortion, since Valle d'Aosta data taken alone would be rather thin for the majority of indicators considered.

6. It should be noted that polarization and imbalances encountered here are more or less equivalent to the outcomes of the previous survey carried out by ISTAT on technological innovation in the Italian manufacturing industry during 1981–1985 (for the "regionalized" results see Cesaratto et al. 1993), but unfortunately not directly comparable with the recent one. The most significant positive variations—in terms of the number of innovative firms—are those found in Veneto and Emilia Romagna, which perform far better than Lombardy and Piedmont. In particular, Lombardy had already recorded—in the analysis of the previous survey—a lower correspondence between innovation capacity and direct regional economic results. This disproportion had pointed up how the innovation capacity of the region did not seem to translate wholly into economic performance within the boundaries of the same region—at least not during the same period.

7. Obviously, this measurement—considering expenditure as a measurement unit—does not weigh the structural and dimensional diversities of regional systems. For details on the partiality of the information of data and for the construction of the matrix, see Iammarino et al. 1995.

8. We are referring to 17 regions, not 20, since the data available on the "minor" ones (so defined in connection with the phenomenon we are discussing—that is, innovation in the manufacturing industry) are aggregated to those of other regions, for statistical confidentiality reasons. This is the case of Valle d'Aosta aggregated to Piedmont, Molise to Abruzzo, and Basilicata to Calabria.

9. Clearly, the normalized shares expressed by the ISRI index (Balassa 1965) have been used both for the sectoral concentration in each region and for the geographical concentration in each sector.

10. The Gini coefficient for innovation has also been calculated using the number of local units (plants) to normalize the sectoral share of regional innovation. The exercise has led to a slight change in the ranking of the coefficients, but it has essentially confirmed the "uniformity" of the sectoral spread of innovation in the four regions under study, which kept their position in the lowest places of the scale.

11. See section entitled "Innovation and International Trade Performances."

12. It is worth recalling that, as explained in the next section, the relative degree of regional openness takes into account a lag of two years with respect to the introduction of innovation declared by the responding firms.

13. The RCA allows an easy comparison with what was observed in the innovative and production activities, since—similarly to the ISRI index used as a basis for calculation of the concentration coefficients—the index is given by:

$$RCA = \frac{X_{ij} / \sum_j X_{ij}}{X_{Ij} / \sum_j X_{Ij}} \qquad \text{where} \quad \begin{aligned} i &= 1,\ldots 4 \\ j &= 1,\ldots 10 \end{aligned}$$

where X_{ij} are the exports of region i in sector j, and X_{Ij} are Italian exports in the *j*th sector. A region is defined as specialized in exporting goods of a given sector—and therefore the index can take on values higher than one—if the weight of that sector shows a higher value of incidence on regional exports than on national exports.

14. The NTB is defined as follows:

$$NTB = \frac{X_{ij} - M_{ij}}{X_{ij} + M_{ij}} \qquad \text{where} \quad \begin{aligned} i &= 1,\ldots 4 \\ j &= 1,\ldots 10 \end{aligned}$$

where X_{ij} and M_{ij} are respectively the exports and imports of region i in sector j. The NTB—which was calculated by comparison for Italy as well—provides a synthetic measure of the degree of imbalance of a commercial exchange. It can take on values between -1 and 1, providing information on both the competitiveness of sectors (positive and high balances are found in those sectors showing competitive regional production in international markets) and the pattern of trade specialization (NB = 1 (–1) shows a complete specialization (despecialization) in the sector—that is, M = 0 or X = 0.

REFERENCES

Aydalot, P. (ed.). *Milieux Innovateurs in Europe*. Paris: Gremi, 1986.

Audretsch, D. B., and M. Feldman. "Knowledge Spillovers and the Geography of Innovation and Production," CEPR Discussion Paper no. 953, 1994.

Audretsch, D. B., and M. Feldman. "Innovative Clusters and the Industry Life Cycle," CEPR Discussion Paper no. 1161, 1995.

Balassa, B. "Trade Liberalization and Revealed Comparative Advantage," *The Manchester School of Economics and Social Studies* 33, 1 (1965) 99–123.

Becattini, G. *Mercato e forze locali: il distretto industriale*. Bologna: Il Mulino, 1987.

Breschi, S. "La dimensione spaziale del mutamento technologico: una proposta interpretativa," *Economia e Politica Industriale* 86 (1995) 179–207.

Cesaratto, S., S. Mangano, and A. Silvani. "L'innovazione nell'industria italiana secondo l'indagine Istat-CNR: tipologie settoriali e dinamiche territoriali," *Economia e politica Industriale* 79 (1993) 167–200.

Ciciotti, E. "Regional Dynamics, Innovation and SMEs in Italy," paper presented at the 12th Workshop "Changes in Industrial Organization, Small and Medium-Size Firms and Regional Local Policies," Rome, December 1–2, 1995.

Garofoli, G. "The Italian Model of Spatial Development in the 1970s and 1980s," in *Industrial Change And Regional Development: The Transformation of New Industrial Spaces* (G. Benko and M. Dunford, eds.). London: Belhaven Press, 1991.

Hagerstrand, T. *Innovation Diffusion As A Spatial Process*. Chicago: Chicago University Press, 1967.

Howells, J. "Regional Systems of Innovation?" in *National Systems of Innovation or the Globalization of Technology?* (D. Archibugi, J. Howells, and J. Michie, eds.). Cambridge: Cambridge University Press, 1997.

Iammarino S., M. R. Prisco, and A. Silvani. "On the Importance of Regional Innovation Flows in the EU: Some Methodological Issues in the Italian Case," *Research Evaluation* 5, 3 (1995).

Iammarino S., M. R. Prisco, and A. Silvani. "La struttura regionale dell'innovazione," *Economia e Politica Industriale* 89 (1996) 187–229.

ISTAT. *Rapporto Annuale*, Rome, 1996.

Jaffe, A. B., M. Trajtenberg, and R. Henderson. "Geographic Localization of Knowledge Spillovers as Evidenced by Patent Citations," *Quarterly Journal of Economics* 63, 3 (1993) 577–598.

Krugman, P. R. *Geography and Trade*. Cambridge, MA and Leuven, Belgium: MIT Press and Leuven University Press, 1991a.

Krugman, P. R. "Increasing Returns and Economic Geography," *Journal of Political Economy* 99, 3 (1991b) 483–499.

Lundvall, B. A. "L'economia dell'apprendimento. Una sfida alla teoria e alla politica economica," *Economia e Politica Industriale* 89 (1996).

Maggioni, M. A., and C. Miglierina. "Dov'è il motore del sistema tecnologico nazionale? Un'analisi spaziale dei flussi innovativi intersettoriali," in Regioni e sviluppo: modelli, politiche e riforme (G. Gorla and O. Vito Colonna, eds.). Milan: F. Angeli, 1995.

Pred, A. *The Spatial Dynamics of U.S. Urban-Industrial Growth, 1800–1914. Interpretative and Theoretical Essays*, Cambridge, MA: MIT Press, 1966.

Pred, A. *City-Systems in Advanced Economies. Past Growth, Present Processes and Future Development Options*, London: Hutchinson, 1977.

12

U.S. Industrial Competitiveness Policy: An Update

James K. Galbraith

INTRODUCTION

On February 22, 1993, the White House released a White Paper, under the names of President Clinton and Vice President Gore, entitled *Technology for America's Economic Growth, A New Direction to Build Economic Strength.* That paper set grand goals:

a growing economy with more high-skill, high-wage jobs for American workers; a cleaner environment where energy efficiency increases profits and reduced pollution; a stronger, more competitive private sector able to maintain U.S. leadership in critical world markets...(p. 1).

To achieve these goals, the White Paper demanded major changes in allegedly traditional American ways of business, and particularly of industry-government relations:

American technology must move in a new direction to build economic strength and spur economic growth. The traditional federal role in technology development has been limited to support of basic science and mission-oriented research in the Defense Department, NASA, and other agencies. This strategy was appropriate for a previous generation but not for today's profound challenges. We cannot rely on the serendipitous application of defense technology to the private sector. We must aim directly at these new challenges and focus our efforts on the new opportunities before us, recognizing that government can play a key role helping private firms develop and profit from innovation (Clinton and Gore 1993).

Industrial policy, in other words, seemed finally to have arrived in force on the American scene. The White Paper described specific actions that could form the core of a new policy. These included increased funding for "advanced manufacturing R&D" in consortia such as SEMATECH, a new "Agile Manufacturing Program," a national network of "manufacturing extension centers," "regional technology alliances," and support for expanded programs in manufacturing engineering education and "environmentally-conscious manufacturing." The White Paper also outlined programs for a "national information infrastructure," for policies to promote energy efficiency, and for a permanent research and experimentation tax credit.

The Clinton/Gore White Paper represented a departure in the rhetoric of public policy in the United States. But did departures in rhetoric also mean large departures in practice? If so, of what kind? In politics things are often not what they seem. I shall argue that the case of industrial policy is no exception to this rule. Indeed, though it is doubtful that the Clinton administration either willed or foresaw the course of its own tenure, one can argue that the Clinton-Gore years have seen a double irony: while moving to embrace industrial and sectoral policies in rhetorical terms, their administration has moved away from effective industrial interventions in practice, and has little to show for its efforts in this area.

This chapter presents a review of major policies affecting industrial sectors in the United States, alongside an assessment of their costs, effectiveness, and other consequences. The review begins with a discussion of macroeconomic, trade, and regulatory policies that have affected the composition and volume of industrial investment. It then turns to a range of sector-specific policies that have, in the past, affected the pattern of development of industry, agriculture, and services. The chapter concludes with an assessment of changes under the Clinton administration.[1]

MACROECONOMIC, TRADE, AND REGULATORY POLICIES

Macroeconomic Policies

The effects of U.S. macroeconomic policy on the pattern of American industrial development could be, and indeed have been, the subject of a book (Galbraith 1989). Most of macroeconomic theory abstracts from these effects, perhaps because they are specific to the position of the United States in the technological hierarchy of the world economy, and therefore not terribly useful as grist for textbook models. But the effects are present in the real world, and it is useful to begin with a summary.

In brief outline, the period from 1968 through 1992 can be divided into four major phases: the Vietnam War phase (1968–1972), the oil shocks (1973–1979), the Volcker era (1979–1986), and the Greenspan era (1986–). From 1968 through 1973, the United States experienced inflationary growth fueled by the Vietnam War, which in the context of fixed nominal exchange rates led to a progressive real overvaluation of the dollar and a declining trade balance. This, coupled with a development and commodities boom around the world at that time, began a process of globalization in the manufacturing sector that has marked U.S. economic development ever since.

Globalization has two essential industrial effects. First, it has fostered American world dominance in a wide range of specialized, sophisticated manufacturing and service activities, each of which pays high wages to comparatively small numbers of skilled employees. As globalization progressed, particularly in the late 1970s, the relative importance of these activities in the American economy increased. Second, globalization has eroded employment and wage standards for less-skilled American workers in industries ranging from automobiles to garments, cutting into what was formerly the basis of worker affluence in North America. The tension between these two forces—one pushing upward and the other down—goes far to explain many of the twists and turns of American economic policy, and the politics surrounding it, in the years since globalization began.

From 1973 through 1979, policy instability accompanied profound industrial adjustment. The dollar fell. The recession of 1974–1975 weakened labor and domestic capital in major industries such as garments, steel, and automobiles. Import penetration intensified, particularly in consumers' goods. Yet by the end of the decade worldwide development had demonstrated the continuing resilience and competitiveness of U.S. industries producing aircraft, computers, construction and farm equipment, industrial machinery, and advanced intermediate goods. The characteristic patterns of U.S. trade, as an exporter of advanced goods and an importer of labor-intensive manufactures, became much more pronounced in this period, even as trade overall remained roughly in balance. The globalization of U.S.-based manufactures became, for practical purposes, irreversible. This was, however, not a good position for what lay ahead.

The arrival of Paul A. Volcker at the Federal Reserve in early 1979 and the 1980 election of Ronald Reagan brought on a fundamental policy shift. Macroeconomic policy in the early 1980s tried, in effect, to reestablish a primarily domestic basis for stable, noninflationary economic growth, independent of foreign demand. It was a throwback policy, suited to an era of hegemony that no longer existed. And it was derailed, in the end, by international consequences unforeseen either in private or in public by the architects of the policy itself, as memoirs later revealed[2] (see Stockman 1986; Niskanen 1988a).

Tight monetary policy, beginning in 1979 but greatly tightened in 1981, raised interest rates sharply—which was foreseen—and drove up the value of the dollar by around 60% in real terms, which was not. The resulting rise in the real value of dollar-denominated debts led to the eruption of financial crises throughout the world in mid-1982, beginning in Mexico. External markets for U.S. industrial equipment and intermediate goods collapsed. Only then, a series of tax reductions and increases in military spending brought the American economy out of recession on the strength of growing domestic demand and a vast increase in private and public debts.

When recovery came, the combination of expansionary fiscal and tight monetary policies, and the cheap imports associated with them, did restore growth without inflation for a time. But the dollar remained overvalued for five years, and advanced export industries entered a prolonged depression. The domestic recessions of 1980 and 1981–1982 had already precipitated severe cutbacks in steel, automobiles, engines, and industrial and construction equipment, all of which were ripe for heavy competitive challenge, mainly from Japan, and also to garments, textiles, and labor-intensive manufacturing, which were vulnerable to market penetration from low-wage countries. The accompaniment to recovery was therefore a flood of manufactured imports, while advanced U.S. exports continued to stagnate and many areas of technological advantage were lost. Together, these effects drove the current account deficit toward $150 billion by the middle of the decade and signaled the unsustainability of the high-dollar policy. They also produced a revolt in the advanced technology sector, eventually leading major corporate players to align themselves with the Democratic Party for the first time since 1964.

After 1985, the high-dollar policy was progressively abandoned. But the damage it had done to the competitive position of U.S. industry could not be undone, and this made a high level of domestic demand inconsistent with reasonable trade balance. Instead, in a fourth phase lasting through the Bush years, U.S. administrations sought to reconcile a falling dollar and improving trade position with low inflation, mainly by tolerating low rates of economic growth.

Slow growth turned to recession in 1990, and the recession proved unusually long and stubborn. Fiscal policy was immobilized by powerful forces of political gridlock and parliamentary budget crisis. President Bush effected a modest stimulus by executive order affecting the timing of income tax collections, and through customary manipulations of military and public works spending. But more ambitious proposals—for example, a tax incentive for first-time homebuyers—met defeat in Congress. Easier monetary policy proved unable to end the recession on its own.

The 1992 election brought an immediate surge of public confidence and a bootstrap revival of animal spirits; in addition the delayed effects of the surreptitious Republican stimulus measures were finally felt. Meanwhile the Federal Reserve cooperated with the new administration and long-term interest rates fell steadily until President Clinton's inauguration. The new administration also accepted some early currency adjustment, especially with Japan, as an inevitable consequence of a deteriorated competitive position and precondition to a further recovery of exports. In early 1994, in a move widely misinterpreted as a tightening of policy, the Federal Reserve began a sequence of increases in short-term interest rates. The practical effect was to push the banking sector, which in the previous era of high spreads had become a sink for government bonds, toward the resumption of commercial and industrial lending. The expansion from that point was financed by private debt accumulation, and public deficits started shrinking.

There followed a long and essentially stable period of slow growth without rising inflation. The expansion was marked by reduction of public deficits and an increase in private debts, with the result that by 1997 private financial portfolios had become dangerously unbalanced. The characteristic symptom of this, a rapid run-up in common stock prices, was not long in developing. Declining commodity prices beginning in late 1996 and a small but indicative increase in the short-term federal funds rate in March 1997 drove real and relative American interest rates up, beginning an inflow of capital funds to the United States and a rise in the value of the dollar relative to the yen and to European currencies, further fueling the domestic stock boom. A collapse in the currencies of those Asian countries most dependent on U.S. portfolio investment followed; this was the so-called Asian crisis—in fact a crisis of a system based on globalized portfolio investment.

By mid-1998 it was beginning to look as though the Clinton administration had re-created the essential macroeconomic conditions of the middle Reagan years without the high budget deficits of that time: high employment, high real interest rates, a rising dollar, falling export competitiveness, and a rapidly increasing deficit in the balance of trade. Global macroeconomic stabilization had become a pressing priority; should it fail, then no doubt the hard choices between protection and contraction would shortly reemerge (as they did in the late 1980s). The pressing question was whether the instruments for an effective stabilization—which must include a sharp reduction in interest rates, taxation, and other limits on capital movement, and a return to balance between public and private sources of economic activity—could be mustered in the prevailing political climate in Washington. Given the peculiar mixture of triumphalism about free-market capitalism in that precinct, combined as it is with low partisan warfare on all fronts between the Democratic president and the Republican Congress, the chances for this seem, at the time of writing in 1997, to be poor.

Trade Policy

In the early 1990s trade policy stood out as the major arena for discourse over sector-specific policies in the United States, well in excess of its actual powers. And trade policy was debated and indeed employed to the point where spokesmen from many advanced industries argued that it bore an excessive share of the burden of competitiveness policy in the United States (Council on Competitiveness 1993). By the end of the decade, in sharp contrast, free trade orthodoxy was in complete control of American political rhetoric; suggestions of a contrary or divergent policy had all but disappeared.

The broad evolution of U.S. trade policy since the Second World War has been to promote the progressive multilateral reduction of barriers to trade, through the successive rounds of the General Agreement in Tariffs and Trade (GATT) and on to the World Trade Organization (WTO) and the North American Free Trade Agreement (NAFTA), with exceptions made for a specific but limited set of industries enjoying particular patterns of economic vulnerability and political clout. Thus, oil import quotas protected domestic petroleum extraction until the early 1970s. Beginning in 1962, the textile industry came under the Multi-Fiber Agreement (MFA), establishing quotas that were intended to slow the penetration of U.S. markets by inexpensive foreign cloth and clothing. In the 1970s, the Carter administration intervened repeatedly to forestall, with trade relief, the impending collapse and restructuring of United States basic steel production. Even tighter restrictions have applied, with some gaps, to dairy products (a Wisconsin political fief), sugar (ditto for Louisiana), and citrus (ditto for Florida). A well-known measure also restricts the use of foreign-built vessels in coastwise trade and fishing (Commission of the European Communities 1990: 11). The United States also restricts technology exports, a policy justified on national security grounds but most obviously in the interest of retarding the development of competition in advanced industries overseas.

Countervailing duties and antidumping measures are the major instruments of U.S. trade policy. They are based on the principle, consistent with the GATT code, that relief is due when injury is caused by unfair trade practices, whether dumping or subsidy, so that domestic producers may enjoy protection from damaging actions by overseas competitors similar to that provided against unfair competition originating at home. Thus "predatory pricing," or sale below cost with intent to drive competitors out of a market, is prohibited to U.S. companies under the antitrust laws; corresponding relief when similar actions originate overseas comes under the rubric of antidumping. On the other hand, procedures for obtaining relief in the two circumstances necessarily differ, and therefore so may the distribution of effective relief. An administration that is philosophically opposed to rigorous antitrust enforcement may be more open to antidumping sanctions, creating a situation in which domestic predators (or price-fixers, for that matter) achieve both impunity and protection.[3]

During the Reagan administration free market rhetoric combined with a strong move toward protectionism in practice.[4] William Niskanen, a member of the Council of Economic Advisers from 1981 to 1985, has written: "U.S. trade policy turned sharply protectionist during the Reagan years. Moreover, *all* of the new trade restraints were initiated or approved by the administration, despite a general endorsement of free trade in its public rhetoric" (1988b).

Beyond the increased use of countervailing duty and antidumping measures, two types of protective trade regime were expanded during the Reagan years. Where products are widely available from many sources, complicated systems of import management and country quotas are required. This is true of textiles under the Multi-Fiber Agreements, established in 1961 and renewed and broadened under Reagan in 1981. It is also true of sugar, for which a system of quotas that had been suspended in 1974 was revived in 1981, and later, according to Niskanen, "progressively tightened and later extended to the import of food products containing even minimum amounts of sugar" (Niskanen 1988b: 36).

In instances where the threat to a major industry arises from just a few national competitors, the Reagan administration took the step of negotiating "Voluntary

Export Restraints" (VERs). Steel came under such a regime from 1984 to 1992 (Seebald 1992). A VER negotiated with the Japanese for the automobile industry in 1981 led Japan in 1985 to a self-imposed limit on annual automobile exports to the United States (Niskanen 1988b: 34).[5]

In recent years, innovation in trade policy has shifted toward the high-technology arena. The pathbreaking event in this area was the U.S.-Japan Semiconductor agreements of 1986, revised and renewed in 1991 (see Flamm 1993). The range of issues addressed by trade policy in the high-tech arena is quite wide, and includes dumping, market access, pricing in third-country markets, foreign subsidies and investment, and foreign government procurement practices. In a useful paper, the Council on Competitiveness (1993) has surveyed eight cases, including the imposition of antidumping duties on flat panel displays in 1991, the U.S.-Japan Semiconductor agreements of 1986 and 1991, the U. S.-Japan conflicts and negotiations over market access to the Japanese cellular telephone market, the Uruguay Round negotiations over R&D in the GATT Subsidies Code, commercial aircraft subsidy negotiations between the United States and the European Community, the U.S.-Japan Supercomputer Procurement Agreements of 1987 and 1990, the review of the sale of Semi-Gas Systems, Inc. to Nippon Sanso KK in 1990, and the debate over support of High Definition Television (HDTV) in the United States. Some of these cases are also treated in detail in an important book by L. D. Tyson (1992). One conclusion of both studies is that while trade policy can and has played an important and constructive role in certain cases of conflict over advanced technology, the role of trade policy must be supplemented by more effective domestic technology and competitiveness policies, and by more effective coordination between branches of government dealing with trade and those dealing with other areas. The incoming first Clinton administration seemed determined to come to grips with this issue of coordination.[6]

The extreme salience of trade policy in U.S. political discussion was vividly illustrated by the debate in 1993 over NAFTA. From a reading and evaluation of the agreement itself, it would be impossible to understand the vehemence and depth of bitterness that characterized this discussion. The flow of manufactured imports into the United States from Mexico was, after all, already essentially free; NAFTA proposed reducing average U.S. tariffs over time from a base level of about 3%. U.S. investment into Mexico was also already a fact of life, with some 2,000 foreign-owned factories in operation along the U.S.-Mexican border. And the reduction of Mexican trade barriers and remaining investment barriers viz. the United States was, after all, a Mexican policy decision, reflecting changes that could be made unilaterally—and that would have been made unilaterally by Mexico if the free trade agreement had failed in Congress.

So what was the fuss about? Evidently, NAFTA became an issue that symbolized class differences in the United States and the pervasive insecurity felt by working Americans about the security of their employment and incomes. These differences and insecurities are real, and were reflected in a brutally real mistrust of the workings of American government, as the NAFTA debate showed. The details of what NAFTA "really would" or "really wouldn't" do faded from the discussion, which centered most heavily on competing estimates—all artificial and most demonstrably inflated—of the jobs that NAFTA would create or destroy. In this way a discussion about job creation that did not occur over macroeconomic policy in the first months of the year, when it might conceivably have been useful, reemerged over an issue to which it was only peripherally related.

The real significance of NAFTA was that it embodied an implicit commitment by the larger partner to the stability and growth of the smaller one, a point perfectly evident to high officials in Mexico but virtually missed in the U.S. discussion (Galbraith 1993b). But in this respect, NAFTA did underpin the interconnections between trade, competitiveness, and macroeconomic policies. By providing a seal of creditworthiness to the current direction of Mexican policy, NAFTA was intended to reassure foreign investors, to help stabilize Mexican macroeconomic performance, and therefore to contribute indirectly to Mexican demand for advanced U.S. capital goods and intermediate exports.

The events in Mexico in NAFTA's opening days—the assassination of the likely PRI presidential candidate, Colosio, the rebellion in Chiapas—might be thought to call into question this linkage, but in fact they illustrated it. The Mexican stock market rose in the early days following ratification, and the Mexican government capitalized on it to borrow extensively on the American market—in this way effectively financing the reelection of the Institutional Revolutionary Party (PRI). A severe financial crisis followed, but so too did an American rescue program, something that had been wholly lacking in the crisis of 1982. The result was that the Mexican economy recovered from the peso crisis relatively quickly: 1996 and 1997 were years of growth, and in 1997 Mexico was not instantly afflicted by the Asian flu, but was, rather, viewed by investors as a close American dependency and therefore an insider of the North American bubble. No doubt the prospective extension of NAFTA to other countries in the Americas can likewise be viewed in terms of access to (the perception of) United States financial protection. In practical terms, however, it remains doubtful whether such a perception would be meaningful. Because of the immigration issue, the position of Mexico with respect to the United States is special in a way that the position of South America is not.

Regulation

The study of the U.S. regulatory system is not usually conducted—in the United States—as an extension of discussions about sectoral and industrial policies. However, it is not unnatural to consider regulation in this light, a fact perhaps best seen by observing how American analysts look at regulatory systems in other nations. Critics of Japanese trade practices, for example, often point to the high effective barriers posed by the pattern of Japanese business practices, business structures (notably, vertical integration, and *keiretsu*), and government-business relations, including regulations on health, safety, and other matters. To name one prominent example, Tyson (1992: 57-58) surveys these considerations, and writes:

In technology-intensive industries [differences between the United States and Japan show] up most clearly in procedures for product standards, testing and certification....Whereas the US system of more lax standards and certification places heavy reliance on the common-law system of product liability, the Japanese system places heavy reliance on elaborate certification standards and administrative rulemaking.[7]

This language, curiously enough, finds its echo in the words of a 1990 report from the Commission of the European Communities (1990: 15):

According to U.S. sources, as of 1989, out of 89,000 standards used in the U.S., only 17 are directly adopted from ISO (International Organization for Standards) standards. No IEC (International Electrotechnical Commission) standards have been adopted. The Federal Government refers to about half of these standards in its technical regulations, thereby making them mandatory. This situation is difficult to reconcile with the GATT Standards Code [under which the United States] is obliged to use international standards as a basis for its own technical legislation....The U.S. Federal government is also obliged to take such reasonable measures as may be available to it to ensure that private standardizing bodies and states use international standards. None of this seems to happen in practice.

There are also cases where U.S. regulations have played a strong role in patterns of industrial development. In an article that touches on a striking example, F. M. Scherer reviews the history and effects of regulatory practice in the pharmaceutical industry, a case where American "procedures for product standards, testing, and certification" are surely not to be sneezed at (Scherer 1993). Here, the relevant statute is the 1962 Kefauver-Harris Act, which extended the regulatory ambit of the Food and Drug Administration to cover the therapeutic efficiency of new drugs. Scherer writes:

Prior to that time...the average expenditure required to develop a new drug through the regulatory approval stage was approximately $6 million (converted to 1990 price levels)....By contrast...achieving a successful 1980s-vintage drug entailed research, development and testing outlays...averaging $53 million during the clinical (human testing) stage plus $73 million at the preclinical stage....The average time required to bring a drug from the start of clinical trials to FDA approval exceeded eight years. Thus, drug development has become a high-stakes, high-risk game (p. 99).[8]

Despite—one is tempted to say because of—these costs, as Scherer documents, the U.S. pharmaceutical industry enjoys high and stable profits,[9] and an export-import balance in the front ranks of American manufacturing. It can be argued that regulations have helped, rather than impeded, this development. High regulatory barriers to entry in the U.S. market, alongside patent protection for new drugs, foster both a monopoly structure in the domestic industry and a reputational advantage in foreign markets for firms that pass through the American process.

A high-technology example can be found in the recent case of High-Definition Television. Here, United States policymakers resisted pressures to invest directly in particular technologies, even while European and Japanese advances in the field, heavily subsidized by their respective governments, were being reported and strong pressures were being brought in the United States for a direct response. The United States instead relied on a regulatory agency, the Federal Communications Commission, to set standards for the final product, as well as on the Defense Advanced Research Projects Administration (DARPA) to fund "display, digital signal processing and data compression technologies." In the end, advances in digital technology by an American firm trumped the emerging European and Japanese systems, leaving the technological lead in HDTV firmly in American hands.

Two significant developments on the regulatory front have marked the Clinton years, though their full implications remain to be seen. One is telecommunications reform, and the other is antitrust action, including the newly launched case against Microsoft. Both were advertised as procompetitive measures. However, in the telecommunications case there is not the slightest evidence that increases in competition resulted from the 1997 telecommunications bill, which appears instead to have fostered increases in concentration and control through mergers—something that suggests that the underlying purposes of the bill may have been to strengthen American telecomm companies rather than to assist the American consumer. Prospects for the Microsoft suit are, perhaps, somewhat better, but the case is very new.

Some insight into the Clinton administration's approach to antitrust can be gleaned from the 1998 *Economic Report of the President*, which devotes an entire chapter to this topic (United States Congress 1998). The *Report* boasts of an aggressive approach to antitrust in general: over $200 million in fines have been levied under this administration, and investigations have pursued targets as diverse as Archer Daniels Midland, Toys 'R' Us, Staples, and Office Depot.

The Justice Department prides itself on taking a nuanced approach to antitrust enforcement. Bigness *per se* is not the criterion—it is doubtful that it ever was, except in the rhetoric of those opposed on principle to antitrust enforcement. Rather, the department pursues specific cases and allegations of anticompetitive behavior, and brings antitrust charges where detailed evidence of such behavior can be brought to bear. The mission is not to achieve a particular market structure, but to "safeguard the competitive process," which presumably means to foster standards of procompetitive conduct in the corporate sector by imposing stiff penalties on those who do not uphold such standards.

This is necessarily a decision-making process requiring judgment, particularly as antitrust seeks to cope with the rising tide of mergers and acquisitions:

In evaluating these mergers and deciding which ones to challenge, the enforcement agencies must strike a fine balance. A merger may yield significant cost savings, but it may also threaten to increase industrial concentration…and stifle competition, allowing the remaining firms to increase prices and reduce output. The impact on concentration and competition is particularly difficult to evaluate in the many industries currently experiencing rapid technological change (United States Congress 1998: 198).

Thus the department looks for specific activities that threaten competition or that are the hallmarks of anticompetitive practice. Bundling is one such practice—the compulsory packaging of television cable channels or computer software in a single package. Concerns about bundling led the authorities to stipulate that consumers could not be forced to purchase both HBO and certain Turner Broadcasting channels when Time Warner was permitted to acquire TBS.

Similarly, the department has targeted airline price-fixing; the proposed merger of Microsoft and Intuit, maker of the popular financial management program Intuit; and price-fixing among market makers on the NASDAQ stock exchange. In each of these cases, the government's focus has been on conduct and behavior, rather than on market presence as such. The new Microsoft case, which alleges a series of anticompetitive practices including the bundling of a Web browser into the Windows 95 operating system, is in this mold. There is apparently no objection to Microsoft's quasimonopoly position in operating systems software as such; the issues turn on specific practices that appear to violate competitive standards.

EXPLICIT INDUSTRIAL AND SECTORAL POLICIES

The United States does not have an "industrial policy." There is no Department of Industry, no Directorate for Technology, and no Ministry of Civilian Investment. In official doctrine until 1993, public investments were presumed best restricted to generics, such as support for basic research and development and the provision of education and infrastructure, the latter not especially tied to high-technology endeavor.[10] Private investment decisions were to be the sole arbiters of the direction of economic development, mitigated only by the wide range of public concerns with which private profitability must be tempered: national security first and foremost, energy security, rural development, public transport, occupational and consumer safety, public health, and environmental protection.

Yet as these widely accepted, politically legitimate public concerns[11] cumulate and overlap, it becomes possible to ask whether, and to what extent, have they influenced the structure of the American economy taken as a whole. And if they do exert a strong influence over that structure, are they not properly to be considered as a form of industrial policy in themselves?[12]

In this section, we begin the process by reviewing selected literature on industrial and sector-specific policies. In the next, we attempt an evaluation of cumulative effect.

Broadly speaking, U.S. sectoral policies can be classed under five major headings. *Technology policies* have aimed at promoting research and development in a wide range of advanced industries, often as the collateral effect of a particular defense-related or scientific mission. Under this heading, we will note literature covering aerospace, communications, and electronics, especially computers, as well as emerging efforts in the underlying areas of advanced materials. *Energy policies* have often been presented to the public as a matter of national economic security and independence. Nevertheless, their effects have had strong sectoral implications, in particular in efforts over the years to advance the civilian uses of nuclear power. *Agricultural policies* have long been celebrated as a central example of successful government intervention to promote productivity gains. While the most salient triumphs in this area lie in the past, the pattern of agricultural policy underlies much of the conceptualization of modern industrial policy in the United States to this day. *Health policies* have led to significant federal sponsorship of epidemiological and pharmaceutical research, strongly influencing the pattern of private investment in drug research and treatment strategies. And finally, *small business policies* have long sought to preserve and protect a stratum of American business which might otherwise have been sharply reduced in size and political importance by technological developments and the forces of economic competition.

Technology Policies

As economic historians from Frank Bourgin to Gavin Wright have documented, U.S. technology policies are not new (Bourgin 1989; Wright 1990, 1992). Indeed, Wright argues that technology policy and the technological developments they produced themselves have often defined the nature of U.S. economic development, mainly by reducing the relative price, at different times, of alternative methods and types of natural resource exploitation.[13] From this perspective, the descriptive problem is less one of finding important examples of technology policy, than of choosing the most salient cases from a wide array.

Aerospace. Government support for commercial aviation dates back to the First World War, and covers the range from technology development (including airframes, propulsion, and navigation systems), provision of infrastructure in the form of airports and air traffic control, uniform safety standards and inspection procedures, and, until the early 1980s, a tightly regulated domestic market. In 1970 Congress narrowly defeated proposals to extend the technology frontiers to supersonic travel, and in 1974 had to bail out a failing manufacturer, Lockheed, ostensibly in order to preserve that company's role as a supplier of military and space systems. In 1980, of course, Congress also deregulated entry to the airline business, a step that has had profound implications for the technology and market structure of air travel—implications that were largely unforeseen at the time and which may not have yet fully played themselves out.

Mowery and Rosenberg (1989) have provided a concise history of U.S. government policy affecting aviation. As they show, the research budget in this industry has been overwhelmingly financed from governmental coffers. From 1945 through 1982, the federal government spent over $86 billion (in constant 1972 dollars) on aeronautic research—$77 billion of that military and $9 billion explicitly civilian. Private industry spent $17 billion on research and development over this time. Moreover, the annual variation in these patterns of spending is not large: in 1982, governmental appropriations for aeronautical R&D came to about $2.4 billion, private to $732 million; both figures are within 10% (in real terms) of the values prevailing in, say, 1969, and within 25% of the figures for 1953 (Mowery and Rosenberg 1989: 179).

Were federal R&D expenditures for aerospace "limited to support of basic science and mission-oriented research"? Mowery and Rosenberg document that this was not their effect, insofar as private companies could approach the execution of government-financed military research programs with future civilian spin-offs explicitly in mind. Thus, in the case of power trains:

From the Pratt and Whitney Wasp of 1925 to the high-bypass turbofans of the 1980s, commercial aircraft engine development has benefited from and frequently has followed the demands of military procurement and military-supported research. The development of the first U.S. jet engine was financed entirely by the military. More recently, military-supported research on turbofan engines for the C-5A transport influenced the development of the high-bypass engines that power the latest generation of commercial transports (1989:185).

Another author comments that the Boeing Corporation's development of the 707, far from being a "serendipitous" spin-off, actually represented an illegal diversion of federal R&D monies to commercial ends:

During the early 1950s, Boeing faced an effective tax rate of 82%. Instead of permitting its funds to return to the government through renegotiation, Boeing diverted funds to development. The prototype for the 707 was offered to the air force as a jet tanker, but Boeing's "tanker" was painted like an airliner and had passenger seats. The air force accepted the craft as a prototype and subsidized the costs of setting up production facilities and working out the bugs in design and production. In 1962, the U.S. tax court found Boeing guilty of diverting funds and ordered the firm to reimburse the government. By this time, however, Boeing had established itself in the airliner market and never looked back (Hooks 1990).

Electronics, Communications, and Computers. Direct government financing of research and development was important throughout the development of the microelectronics sector, though not as dominant as in aerospace. Mowery and Rosenberg show that as late as 1971 federal expenditures on research in communications equipment and electronic components still exceeded private spending, by $1.5 to $1.1 billion (constant 1972 dollars) (1989: 142). By 1981, industry had taken the lead, $2.1 billion to $1.3 billion for government-financed R&D.

In the case of microelectronics, however, comparative R&D expenditures do not tell the full story. Rather, the prospects for military procurement of large volumes of privately developed product long provided the principal spur for private investment in technological advance. Mowery and Rosenberg write that in the case of semiconductors, procurement-pull probably outweighed research-push:

The military services were intensely aware in the 1940s of the potential military applications of semiconductors and followed developments in this technology closely. "Followed," however, is the operative word. The major scientific and technological breakthroughs in semiconductors were achieved in the private sector with private funds, and not, in the most important instances, with military R&D support. Yet much of this privately funded work on semiconductors was motivated by an awareness that military electronic equipment was plagued by equipment failures that stemmed in part from the systems' complex circuitry and reliance on vacuum tubes. The continual lure and the eventual reality of vast procurement contracts drove much of the R&D effort in this sector (1989: 144).

Nor was the eventual role of the federal government in providing demand for semiconductor output a small factor in the growth of production in the industry: "In the first year of integrated circuit production, the federal government purchased the entire $4 million of output...It remained the largest buyer for the first five years, although the government share declined rapidly" (Mowery and Rosenberg 1989: 145).

Hooks takes an even stronger position. He argues that aggressive programs of military-directed research were required to overcome the technological conservatism of the private industrial firms. Without the assertive meddling of procurement bureaucrats, it almost seems, we might yet be mired in the age of vacuum tubes:

The semiconductor was "so radically different...in the way it worked, in the way it could be manufactured and sold, and in its apparent potential, that it could not be comfortably accommodated within the existing electronics industry without changes that that industry was then unwilling or unable to make,"....The diversified firms were equally resistant to integrated circuits....As late as 1962, most firms shied away from revolutionary integrated circuits, concentrating instead on evolutionary changes (1990: 386).

From code-breaking and bomb-making, and later from semiconductors and integrated circuits, there sprang the revolution of the computer. Kenneth Flamm has created a compelling portrait of the computer industry and the role of government in its creation and development (Flamm 1988). Flamm contributes important new history to an understanding of the government role in early computer development, both through the military during and after the Second World War and, albeit abortively, through the work of civilian agencies, notably the National Bureau of Standards. He also describes the changing role of the government as the private-sector market for computers grew:

In the 1950s, before a large commercial market existed, the government role was pervasive. However, as a commercial industry matured in the mid-1960s, the government role switched to one of sponsoring basic research and infrastructure and what might be called leading-edge technological projects in which R&D was divorced from shorter-term commercial benefit (1988: 254).

In other words, a policy of support mainly for basic research, far from being the approach of a "previous generation," was actually the condition into which industrial policy affecting computers evolved, *after* a policy of massive government sponsorship of the new industry had proved successful.

In modern circumstances, moreover, the metaphors of "spin-off" and "serendipity" may cloak, in electronics as we have previously seen in aerospace, an underlying reality of intertemporal interdependence between the design of a government-sponsored research program and its eventual commercial application. The fact that new techniques and technologies can now, *in general*, be adapted to mass markets is sufficiently familiar so that rational companies should come to expect it, and therefore to plan for it. Flamm summarizes the scene:

The computer networks that link today's mini- and microcomputers to other machines, the fancy graphics, the "mouse" and graphics tablets used to draw complex designs, the modems that bring computer users together over the public telecommunications network—all have at least some roots in expensive and exotic research projects funded by the taxpayer in past decades.

And products that are now at the leading edge of the commercial market—systems and languages using the concepts of "artificial intelligence," special computers designed to run this programming efficiently, new and different kinds of computers with "parallel" processing capabilities, ever more powerful supercomputers—are just emerging from research programs fueled by government spending (Flamm 1988).

The creation of SEMATECH in 1987 signaled, perhaps, the beginning of the emergence of a commercially oriented industrial policy in electronics from its closet. SEMATECH is a federally sponsored research consortium financed on a 50/50 matching-fund basis by the government and private industry, and dedicated to the improvement of manufacturing in the U.S. semiconductor industry. As the authors of major new study of the relationship between civil and military uses of advanced technology write,

Although Sematech gets nearly half its annual budget from DARPA, the goals are strictly commercial: to rebuild U.S. capabilities in microelectronics processing so that American merchant firms can compete with their Japanese rivals. Congress channeled the government share of Sematech's budget through DoD [Department of Defense] largely for lack of alternatives: no other agency had the ability and experience to oversee public spending on a complex technical undertaking of this sort (Alic et al. 1992).

A 1990 Congressional Budget Office (CBO) study of the enterprise gives a limited endorsement to the concept of federally supported R&D consortia in general, while conceding that the motivations not only are but have to be commercial:

The study finds that federal support for R&D consortia can be a useful, if limited, tool to support commercial innovation. To be effective, such consortia must develop from private-sector interests, thus delegating a relatively passive role for the federal government (Congressional Budget Office 1990: ix).

Explicit acknowledgement of the role, if not paramountcy, of commercial objectives in technology policy is spreading. In 1991 the Bush administration undertook a High Performance Computing and Communications (HPCC) program. In a new study of these efforts, CBO writes:

Although one reason for this legislation was to improve the performance of federal missions, an important consideration was the benefits such technology might yield for private industry. Such benefits can be realized only if the technology becomes widely used in industry—in short, if it is commercialized (1993: ix).

Energy Policies

The history of energy policy in the United States covers many issues, products, and processes. Those surrounding oil and coal, in particular, transcend mere questions of technology and industry, and encompass the full range of environmental, economic, political, strategic, and even imperial concerns.[14] Policies affecting oil occupy more political space than those concerned with any other commodity—from the import quotas of the 1950s and 1960s through the allocation mechanisms and price controls of the oil crisis period in the early 1970s,[15] even to the Gulf War of 1991. Certain constants run through the history: protection of the oil majors overseas and of independent refiners at home, preservation for the consumer and automotive sector of access to inexpensive gasoline, preservation for the petrochemical industry of inexpensive feedstocks, and struggle over the environment.

The consequences, in the deformation of the U.S. industrial structure toward energy-intensive applications, from the size of our cars to our reliance on highways and aircraft to the competitive strength of our petrochemical sector, are apparent to the naked eye. For this reason, and perhaps also because the technological questions are not novel and because these consequences resulted as much from purposive governmental inaction in the presence of powerful private interests (failure to tax gasoline, notably) as from autonomous state initiative, analysts of industrial policy have tended to treat energy policy as a distinct analytical category. Major recent surveys (e.g., Mowery and Rosenberg) often do not cover it, and a vast proliferation of writings on the energy problem *per se* petered out after 1980. While certain features of fuel policy bear the markings of "industrial policy" writ narrow, they are minor in the scheme of things. The synfuels program of the late 1970s came and went, victim of changing oil prices and its own technical problems. Solar energy research, more promising, fell victim to political pressure and bureaucratic disinterest at the Energy Department in the early 1980s. Continuing federal support for the development of methanol-gasoline blends is a minor feature of the petroleum scene, and more a concession to agricultural and distilling interests than an assault on energy problems.

Insofar as it intersects unmistakably with industrial policy, the modern history of energy policy in the United States is dominated by the drive for nuclear electric power, perhaps the largest and most concentrated use of state power for an industrial objective on record. The story is well-told in the literature (Camilleri 1984), and need not be repeated here. All the instruments of classical industrial policy were brought to bear, on a gargantuan scale: from state finance of the basic and applied research (the Manhattan project and the nuclear weapons laboratories), to government procurement of a major share of power plant output (for naval nuclear propulsion), to the creation of a unified regulatory body dedicated to the advancement of atomic energy, at whatever cost (the Atomic Energy Commission), and a joint congressional committee (the Joint Atomic Energy Committee[16]) intended to streamline the processes of legislative review. To this was coupled a strategy of utility regulation that guaranteed profits and facilitated essentially unlimited capital investment.

The Atomic Energy Commission was in no doubt as to the commercial character of its mission. As William J. Barber wrote in a study of energy policy in the 1960s,

the AEC report of 1962 was hardly an analytic study. It was written in the style of a promoter's prospectus. Nuclear power was depicted as the solution to a host of problems. In a letter transmitting the final version of the AEC report to the President, Chairman Seaborg [wrote] "nuclear power promises to supply the vast amounts of energy that this Nation will require for many generations to come, and it probably will provide a significant reduction in the national costs for electrical power" (1981: 329).

As the 1960s unfolded, the United States did in fact embark on a nuclear power program, beginning with contracts for light water reactors in Connecticut and California in 1962. As James L. Cochrane reported,

by 1966 orders were booming. Twenty-one contracts for nuclear power plants were awarded in 1966 and thirty the following year. By the beginning of the Nixon administration, utilities had ordered over sixty-five large nuclear power plants (with capacity in excess of 500 megawatts), thirty of which were already being built. The rush to a nuclear future was on, not as a result of present or future policy but as a result of an economic click, the culmination of past policies and programs. Of course, it all turned out to be based on extraordinarily optimistic estimates of construction costs (1981b: 366).

From this point, disaster was not far off. The energy crises of the early 1970s revealed the vast overoptimism in utility projections of energy demand, while in the meantime cost overruns in construction destroyed the promise of unlimited power on the cheap. By the mid-1970s the nuclear industry was in deep financial trouble— difficulties that were to be made much worse by the emergent understanding of environmental and waste hazards at the end of the decade. As Cochrane sums up: "by 1979 it had been a very long time since any power company in the United States had ordered a new plant based on nuclear energy" (1981a: 597). Despite repeated and sustained efforts, the Reagan administration proved unable to overcome public resistance to further nuclear adventures, as projects like the Clinch River breeder reactor met defeat in Congress and efforts to start up new power plants met frustration in the courts. In the 1990s, the principal issues surrounding nuclear power are the unresolved ones of decommissioning obsolescent plants and disposing of radioactive wastes.

Thus, while the history of aviation and computers reveals the wide range of economic benefits to which the successful co-optation of state power in the service of accelerated technological change and industrial development may lead, the same, alas, cannot be said for the American experience in the energy sector.

Agricultural Policies

U.S. agricultural policy is a field of specialization unto itself, and defies characterization in simple terms. Farm policy has also been the prototype of state management in all phases of technological development and diffusion, the exemplar of industrial policy as later practiced in other fields.

A 1986 book by Alain Revel and Christophe Riboud outlines the complex history of policies that have led to American dominance of world markets for agricultural exports. They point out that (in 1981) the United States was the world's leading producer of wheat (17% of total production), barley (16%), tobacco (17%), cotton (46%, excluding the Communist countries), corn (63%), and soybeans (63% and growing despite output expansion in Brazil and Argentina). Revel and Riboud assess these numbers in the following way:

Except for corn and soybeans, these figures do not appear especially dramatic. Nonetheless, they reflect a formidable dominance because they could be increased, even doubled if necessary. Recent calculations...have shown that by using all available arable land, the United States could feed 4 billion people more than 3,000 calories each per day. In other words, the United States could feed the entire world's population a higher energy diet than currently available (on the average 2,400 calories per day) (1986: 43).

U.S. advantages in production are magnified when export markets and stocks are considered. America produces only 13% of the world's wheat, yet controls stocks that

have never constituted less than 30 percent of the stocks of the world's principal exporting countries....The U.S. share of world wheat exports is almost as great today (35.8 percent for the average 1984–86) than it was in 1949 (37.8). For corn, the United States has exported more than 70 percent of the world exports since 1980, as compared to around 35 percent in the 1950s. For the total feed grains the U.S. share of world exports was 59.3 percent for the average 1984–86. The same is true for the U.S. share of the soybean and soymeal market, which appears to have stabilized at around 70 percent (Revel and Riboud 1986: 43–44).

How was this remarkable dominance achieved? Revel and Riboud cite the continuous application, in continually varying forms since 1933 or even 1862, of a wide range of policy instruments, including production controls, support prices, direct payments to farmers, food aid both internal and external, export credits, and a national system of agronomic research.[17] Of these, only the last is immediately relevant to the debate over industrial policy in other domains.

The agricultural research effort in the United States is not very large. Revel and Riboud count 6,000 research professors in 53 agricultural universities, 17,000 Agricultural Extension Service agents, and 3,000 researchers at the Federal Research Service of the U.S. Department of Agriculture. They also note the presence of 16,000 "mid- and high-level" employees engaged in private agronomic research (Revel and Riboud 1986: 151). Agricultural research funds accounted for only 2% of the federal research budget in the first half of the 1970s.

Whence, then, the success? Revel and Riboud attribute it to persistence:

The success of the American agricultural sector, therefore, can be attributed more to its exemplary continuity for over a century than to its annual access to government funding. The United States is the only country in the world to have so consistently accumulated such a vast amount of knowledge and above all to have transmitted it so rapidly to its farmers and food industries. Its success is also due to the professionalism of American researchers, who display an exceptional combination of training and ability to identify objectives and make available financial, material and technological resources (1986: 153).

It is not surprising that certain multiple aspects of the drive to legitimize industrial policy as such in the United States, notably proposals for civilian research laboratories and for a "manufacturing extension service," are lifted directly from the repertory developed in agriculture.

Health Policies

Another area of active state intervention along lines that have strong consequences for industrial development occurs in the medical and pharmaceutical sphere. We have already noted the effects of federal regulatory policies on the market structure of the pharmaceutical industry. It remains to observe that the federal government also funds a vast array of research endeavors aimed at the conquest of diseases, from cancers to AIDS. As of 1985, the basic research budget of the National Institutes of Health alone was nearing $3 billion, and expenditures on all forms of R&D in the health area exceeded $5 billion. This compares with, for example, $3.1 billion for space research, $952 million for transportation, and $778 million for all agricultural R&D—though $42.4 billion for defense (Mowery and Rosenberg 1989: 139). By the mid-1980s, the share of medical research in nondefense research was over 30%, up from about 17% in 1971. It also goes without saying that the state-supported system of medical cost reimbursement, incomplete though its coverage is, has fostered a rapid diffusion of new (and sometimes inordinately expensive) medical equipment, contributing to the burden of health care on American taxpayers but also to the strong position of U.S. industry in the world market for advanced medical supplies.

Small Business Policies

Finally, an explicit interventionism in patterns of economic development has long characterized the attitudes of the American state toward small business. A Small Business Administration, sporting an annual budget of $458 million in fiscal year 1994, caters to small business needs. Regulatory policies of all kinds exempt small business from paperwork and other compliance burdens. And in the workings of the private-market health care, an accepted obligation of larger businesses, is not widely provided by the small. The result of all these factors is a generalized difference in wage rates, known to labor economists as the "size-wage differential," in favor of workers employed in large establishments at the expense of those employed by the small. While the public policies that contribute to this stratification of the American industrial and service-sector scene by business size are not strictly "industrial policies" as such, their effects are nonetheless conspicuous.

There was in the 1980s a celebration of small business as a backbone of American society, alongside a considerable volume of moderately scholarly work purporting to show the essential contribution of small business to employment creation. It would not be too cynical, I think, to suspect that the politics of keeping small businesspeople happy may have had something to do with this effort. The findings, however, turned out to be incorrect: most new jobs are created in enterprises that are not small, and even where jobs are created in newly created enterprises, it develops that a large share are created in enterprises that are already larger than small businesses at birth. The literature in the latter part of the decade tended to debunk the small business myth: most small businesses are less efficient and pay lower wages than larger ones operating in the same fields.

In the 1990s, interest in small business development as such seems to have declined sharply, and there is no evidence that the Clinton administration has launched any significant initiatives in this area. This may be in recognition of the political reality that small business organizations, such as the National Federation of Independent Businesses, form the intransigent backbone of the hard right of the Republican Party on business issues, and therefore small business courtship by a Democratic administration has no political payoff. The sole exception concerns minority set-aside programs, which cater to a specifically Democratic clientele and are, of course, regarded as anathema by conservative interests. Correspondingly, considerable time and attention have been devoted to defending against legal challenges to these programs.

THE CLINTON RECORD ON SECTORAL AND INDUSTRIAL POLICIES

The Clinton administration's bold 1993 initiative can now be seen in its true light: not as a change of policy, but as a frank embrace of the long-standing pattern of public and private cooperation in technology and industrial development in the United States. The remaining question is, Having come out of the closet, did industrial policy prosper? The answer, of course, is that it did not.

A key assessment of critical technologies was published by the White House Office of Science and Technology Policies in March 1995 as the *National Critical Technologies Report*. The report in some ways epitomizes the Clinton administration's approach. On the one hand, the ambition is large; on the other hand, policy development has been practically nonexistent.

The *Report* itself consists—for 27 distinct technological areas, ranging variously over energy, environmental policy, information, manufacturing, and avionics—of a simple assessment of whether the United States continues to lead in each area, and whether the lead is diminishing or growing. In summing up, the report admits to not a single case where European or Japanese technologies lead those of the United States. But, lest complacency set in, our lead is deemed substantial in only two cases (information management and "human interface" in transportation technology), and slight leads are considered to be diminishing in ten cases. Yet the *Report* does not discuss whether any new policies are required.

Four recent documents provide insight into the Clinton policies in the industrial and technology areas. They are *The Emerging Digital Economy* (Department of

Commerce 1998), *Science and Technology: Shaping the Twenty-first Century* (Office of Science and Technology Policy 1997), *National Security Science and Technology Strategy* (Office of Science and Technology Policy undated), and *Technology in the National Interest* (Office of Science and Technology Policy 1996). These are all beautifully produced, glossy documents replete with photographs and, as one might expect, reassuring sentiments from the president himself: "We are, in a way, a whole Nation of inventors and explorers...We believe in technology and we are determined to pursue it in all of its manifestations. I do believe that the 21st Century can be a Golden Age for all Americans."

The new documents celebrate past achievements, including republication of such famous photographs as that of Earth from the Moon, taken in 1969. They list, one supposes comprehensively, the achievements of more recent technology endeavors, including in one case a $50,000 experiment in the production of the Einstein-Bose concentrate, the proof of Fermat's last theorem (by a solitary professor working alone in his attic), and an advanced technology award to a two-person start-up company in "ultrafine ceramic and metal powders... for applications ranging from skin-care [mud facials?] to high-performance engine parts."

The reports include much more, most of which can perhaps best be described as descriptive trivia. American winners of the Nobel prize are duly listed in one document, even in economics (Office of Science and Technology Policy 1997:14–15). One of the reports also includes an analysis of the performance of the U.S. educational system in comparative perspective.[18] Important work on nuclear security cooperation with Russia is mentioned, as are biodiversity research and global warming. A small business success story, the development of electronic labels for freight containers, is duly noted; so too is the inspiring example of an Internet-based gardening supply store.

But what the documents do not contain is any statement of the goals for new technological development, nor any assessment of the resources to be mobilized in this endeavor. The essence of actual policy—which is to say budgets and objectives—is missing and unaccounted for. None of the initiatives mentioned first in 1993—from a "national information infrastructure" to "regional technology alliances" to "environmentally conscious manufacturing" seem to remain on the agenda today.

The reasons for this are no doubt complicated, and great responsibility must fall on the Republican Congress, which has launched savage assaults on all civilian science and technology ventures, including on all environmental laws and research, the abolition of Congress's own Office of Technology Assessment, and threats to the existence of the Department of Commerce.

Still, the effect, in the end, is a little sad. The Clinton administration may have succeeded in bringing frank acknowledgement of the role of government in the great technological achievements of modern times, including those usually attributed to the private sector. It has, however, failed entirely to mobilize the creative energy and the political will required to think of something substantial to do next.

NOTES

1. This chapter was written for the United Nations' Economic Commission on Latin America and the Caribbean (ECLAC). The paper was developed under a project on Technological and Competitiveness Policies jointly organized by the Industrial and Technological Development Unit of ECLAC and the Chilean Ministry of Economic Affairs. The chapter draws heavily on my 1993 essay, "Sectorial Policies in the United States: An Overview."

2. Including by this author and in the public statements of the Federal Reserve itself.

3. Schott (1990: 24) alludes to this situation. See also Messerlin (1990).

4. Messerlin (1990: 110) shows that U.S. antidumping and countervailing duty actions rose from 35 in the recession year of 1980 to 208 in the recession year of 1982. These actions continued to average more than 100 per year in 1984–1986, over twice the late Carter rate, until dropping in 1987–1988.

5. The limit was set at 2.3 million cars from 1985 until April 1, 1992, when it was lowered to 1.65 million.

6. It is not surprising that several key players in this work became high-ranking officials in the Clinton administration, notably Tyson herself, who served as chairman of the Council of Economic Advisers; Kent H. Hughes, who moved from the Council on Competitiveness to the Commerce Department, and Kenneth Flamm, who became an Assistant Secretary of Defense.

7. For a survey of Japanese regulatory practices, Tyson cites Vogel (1992).

8. Scherer attributes half of observed cost increases to the effect of market forces, and half to tougher regulation.

9. The exact magnitude of these is a matter of dispute, though Scherer ends by citing a calculation that U.S. drug prices could have been reduced during the 1980s "by 3.4 percent without driving profits to subnormal levels" (1993: 113).

10. Except of course in aviation, where it is the foundation of the world-beating U.S. domestic airplane market.

11. One might also mention the questionable use of national security grounds to restrict federal procurement to domestic suppliers. The Commission of the European Communities "Report on United States" (1990: 37–38) cites instances from supercomputers and pancarbon fibers to ball bearings to coal and coke for U.S. troops in Europe.

12. The OECD report on "Industrial Support Policies in OECD Countries" lays heavy emphasis on tax preferences and indeed counts these as providing 86 percent of all U.S. industrial support in 1988 (p. 40). Unquestionably tax preferences represent a major tool of policy in the United States, with effects that distort resource allocation. However, many tax preferences are available irrespective of sector, and it is quite difficult to compare their effects with those of policies organized explicitly on a sectoral basis. Hence this array of policies is not dealt with here.

13. Examples go back to the building of transcontinental railroads, and include such developments as the effects of New Deal hydropower policies on the aluminum industry and therefore on the production of aircraft.

14. The literature is vast, but a major source remains Blair (1976).

15. A good survey of energy policies in the Nixon administration is provided by De Marchi (1981).

16. A true anomaly, the JAEC was the only joint House-Senate committee in the history of the Congress to be given legislative (as opposed to advisory or mere internal legislative-branch housekeeping) powers. It was finally abolished in the congressional reforms of 1974.

17. See the chart in Revel and Riboud, pages 74–77, which details 24 major pieces of legislation and executive action from 1933 to 1985.

18. One list of countries with test scores significantly lower than those of the United States mentions the name of "Columbia," [sic], which may explain why Professor Robert Barro chose to remain at Harvard (Office of Science and Technology Policy 1997: 119).

REFERENCES

Alic, J. A., L. M. Branscomb, H. Brooks, A. Carter, and G. L. Epstein. *Beyond Spinoff: Military and Commercial Technologies in a Changing World.* Boston: Harvard Business School Press, 1992.

Barber, W. J. "Studied Inaction in the Kennedy Years," in *Energy Policy in Perspective: Today's Problems, Yesterday's Solutions* (C. Goodwin, ed.). Washington, DC: The Brookings Institution, 1981.

Blair, J. M. *The Control of Oil.* New York: Pantheon, 1976.

Bourgin, F. *The Great Challenge: The Myth of Laissez-Faire in the Early Republic.* New York: Harper and Row, 1989.

Camilleri, J. A. *The State and Nuclear Power: Conflict and Control in the Western World.* Seattle: University of Washington Press, 1984.

Clinton, W., and A. Gore. *Technology for America's Economic Growth, A New Direction to Build Economic Strength.* Washington, DC: Office of the President, 1993.

Cochrane, J. L. "Carter Energy Policy and the 95th Congress," in *Energy Policy in Perspective: Today's Problems, Yesterday's Solutions* (C. Goodwin, ed.). Washington, DC: The Brookings Institution, 1981a, 597.

Cochrane, J. L. "Energy Policy in the Johnson Administration: Logical Order versus Economic Pluralism," in *Energy Policy in Perspective: Today's Problems, Yesterday's Solutions* (C. Goodwin, ed.). Washington, DC: The Brookings Institution, 1981b, 366.

Commission of the European Communities. "Report on United States: Trade Barriers and Unfair Trade Practices, 1990," mimeo, 1990.

Congressional Budget Office. *Using R&D Consortia for Commercial Innovation: Sematech, X-Ray Lithography, and High Resolution Systems.* Washington, DC: Congressional Budget Office, June 1990.

Congressional Budget Office. *Promoting High-Performance Computing and Communications.* Washington, DC: Congressional Budget Office, June 1993.

Council on Competitiveness. *Roadmap for Results: Trade Policy, Technology and American Competitiveness.* Washington, DC, June 1993.

De Marchi, N. "Energy Policy under Nixon: Mainly Putting Out Fires," in *Energy Policy in Perspective: Today's Problems, Yesterday's Solutions* (C. Goodwin, ed.). Washington, DC: The Brookings Institution, 1981, 395–473.

Department of Commerce. *The Emerging Digital Economy.* April 1998.

Flamm, K. *Creating the Computer: Government, Industry and High Technology.* Washington, DC: The Brookings Institution, 1988.

Flamm, K. "Semiconductor Dependency and Strategic Trade Policy," *Brookings Papers on Economic Activity: Microeconomics I.* Washington, DC: Brookings Institution, 1993, 249–333.

Galbraith, J. K. *Balancing Acts: Technology, Finance and the American Future.* New York: Basic Books, 1989.

Galbraith, J. K. "The NAFTA and Labor: A Short Report," *Economic Development Quarterly.* 7, 4 (November 1993a) 323–327.

Galbraith, J. K. "What Mexico—and the United States—Wants: What NAFTA Really Means," *World Policy Journal* X, 3 (Fall 1993b) 29–32.

Hooks, G. "The Rise of the Pentagon and U.S. State Building: The Defense Program as Industrial Policy," *American Journal of Sociology* 96, 2 (September 1990) 382.

Messerlin, P. A. "Antidumping," in *Completing the Uruguay Round: A Results-Oriented Approach to the GATT Trade Negotiations* (J. Schott, ed.). Washington, DC: Institute for International Economics, 1990, 108–129.

Mowery, D. C., and N. Rosenberg. *Technology and the Pursuit of Economic Growth*. New York: Cambridge University Press, 1989.

Niskanen, W. *Reaganomics*. New York: Oxford University Press, 1988a.

Niskanen, W. "U.S. Trade Policy," *Regulation* 3, 34 (1988b).

OECD. "Industrial Support Policies in OECD Countries."

Office of Science and Technology Policy. *National Security Science and Technology Strategy*. Undated.

Office of Science and Technology Policy. *Technology in the National Interest*. 1996.

Office of Science and Technology Policy. *Science and Technology: Shaping the Twenty-first Century*. April 1997.

Revel, A., and C. Riboud. *American Green Power* (trans. by Edward W. Tanner). Baltimore: Johns Hopkins University Press, 1986.

Scherer, F. M. "Pricing, Profits and Technological Progress in the Pharmaceutical Industry," *Journal of Economic Perspectives* 7, 3 (Summer 1993) 91–115.

Schott, J. *Completing the Uruguay Round: A Results-Oriented Approach to the GATT Trade Negotiations*. Washington, DC: Institute for International Economics, 1990.

Seebald, C. P. "Life after the Voluntary Restraint Agreements: The Future of the U.S. Steel Industry," *George Washington Journal of International Law and Economics* 25, 3 (1992) 875–905.

Stockman, D. *The Triumph of Politics*. New York: Harper and Row, 1986.

Tyson, L. D. *Who's Bashing Whom: Trade Conflict in High Technology Industries*. Washington, DC: Institute for International Economics, 1992.

United States Congress. *Economic Report of the President*. Washington, DC: Government Printing Office, 1998

Vogel, D. "Consumer Protection and Protectionism in Japan," *Journal of Japanese Studies* 18, 1 (Winter 1992) 423–444.

Wright, G. "The Origins of American Industrial Success, 1879–1940" *American Economic Review* 80, 4 (September 1990) 651.

Wright, G. "The Rise and Fall of American Technological Leadership: The Postwar Era in Perspective," *Journal of Economic Literature* 30, 4 (December 1992) 1931.

13

Technopoleis, Technology Transfer, and Globally Networked Entrepreneurship

David V. Gibson and Christopher E. Stiles

INTRODUCTION

Strategies for building and sustaining successful high-technology regions have been proposed and implemented worldwide ever since leaders from business, government, and academia began to take notice of the wealth-creation potential and technology-spurred growth of such pioneering "technopoleis" as Silicon Valley in California and Route 128 in Boston (Rogers and Larsen 1984; Botkin 1988). However, outside of a few select and important visionaries, such as Professor Frederick Terman at Stanford University in California, these initial and perhaps most successful technopoleis were not planned nor were they managed as strategic regions. They were primarily fostered by entrepreneurial behavior within regional universities and businesses, entrepreneurs that at times fostered spin-out, fast-growth companies.

Japan was one of the first nations to engage in long-term planning for managed high-tech growth. In 1983 the passage of Japan's technopolis law resulting in 20-year economic development plans. In May 1986 the Japanese government approved the Ministry of International Trade and Industry's (MITI) Regional Research Core Concept, which called for the establishment of 28 research centers or technopoleis. The program passed by the Japanese Diet promoted four types of research facilities (Tatsuno 1986):

1. experimental research institutes for joint industry/academic/government research;
2. new research training and educational facilities;
3. the creation of conference and exhibition halls, and data base systems for improved access to technical information;
4. venture business incubators.

Ample funding and meticulous, long-term plans led to the emergence of Tsukuba (about 70 miles north of Tokyo) and Kansai (outside of Osaka).[1] While attractive buildings have been constructed and beautiful parks landscaped, the creation of wealth and high-value jobs has not been as dramatic as expected. For example, a ten-year survey (completed in 1998) revealed there have been no spin-outs from resident companies and R&D facilities located in Kansai. While basic research has been a primary stated objective in these "science cities," issues of science and technology commercialization and return on investment (ROI) are becoming more pronounced.

Smilor, Gibson, and Kozmetsky (1988) suggest that four factors are fundamental to the development of a region as a technopolis leading to the creation of wealth and high-value jobs: (1) the achievement of scientific preeminence in technology-based research, (2) the development of new technologies for emerging industries, (3) the attraction and retention of major technology companies, and (4) the creation and nurturing of home-grown technology companies. Many scholars, practitioners, and government leaders would also suggest that three underlying phenomena are critical and necessary to achieve these four factors:

1. A world-class research university with programs in emerging technology areas to train the needed talent and to research new and emerging technologies.
2. A "smart infrastructure" or the managerial, entrepreneurial, legal, financial, manufacturing, sales, and distribution talent and infrastructure needed to commercialize emerging technologies and innovative business ideas.
3. A high quality of life to attract and retain talented people.

A traditional challenge faced by technopoleis has been to foster regionally based collaboration across academic, business, and government sectors. On the one hand, competition in the pursuit of institutional excellence within, but not across, institutions, has been the norm. On the other hand, new kinds of relationships/alliances/partnerships between the public and private sectors are having far-reaching consequences on the way people think about and implement regional economic development.

IC² Institute researchers offer the framework of the Technopolis Wheel to assess the impact of critical components in regionally based high-technology development. The sectors are as follows: quality education with an emphasis on the research university; large companies; start-up and spin-out companies;

federal, state, and local government policies; and support groups that include "smart" infrastructure, professional associations, and organizations that foster entrepreneurship. Emphasis is placed on the importance of regional collaboration as well as competition within and across these sectors.[2] This regional collaboration is fostered by first and second level influencers from academia, business, and government.[3]

Figure 13.1
Expanding the Technopolis Wheel Framework for the Global Economy

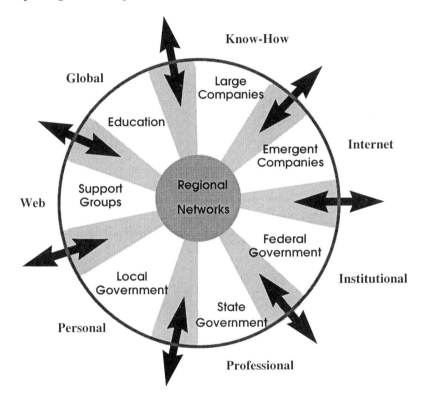

While this chapter supports the importance of a regional focus, it also emphasizes the fostering and leveraging of global linkages through regionally based education initiatives; large and small companies; local, state, and federal government; and support groups (Figure 13.1). For the more established as well as emerging worldwide technopoleis, wealth and job creation in a sustainable environment increasingly depends on globally linked public/private collaboration. A major challenge facing high-tech regions, and the firms that reside in these regions, is how to effectively and efficiently acquire, transfer, and commercialize science and technology that is developed locally and globally at research universities, federal laboratories/institutes, and consortia. Often these R&D facilities are physically and culturally separate from organizations that seek to apply and commercialize the technologies.

TECHNOLOGY TRANSFER AND APPLICATION

Two kinds of technology transfer—from research to commercial application—directly impact the creation of wealth and high-value jobs (Figure 13.2): (1) spinning-out technologies into start-up companies (the dashed line), and (2) the transfer of creative and innovative technologies to established firms (the solid line).

Spin-out technologies may originate in the private sector, government labs, universities, and consortia. These spin-out companies may or may not be nurtured by an incubator. The United States is a successful role model for much of the industrialized world regarding spin-out technology transfer leading to fast-growth firms. The latter part of this chapter emphasizes accelerating the growth of spin-out and start-up companies through Globally Networked Entrepreneurship.

Figure 13.2
Two Basic Forms of Technology Transfer Leading to Wealth Creation

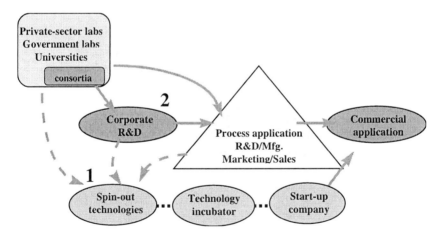

With regard to the transfer of technologies to established firms, IC² Institute research on technology transfer suggests four levels of collaborative activity and four correspondingly different definitions of technology transfer success (Figure 13.3). Moving from Level I to Level IV is *not* a linear, step-by-step process. Multidimensional collaboration is required and complexity increases significantly as the technology and perhaps the technology developers move from Level I to Level IV.

At Level I, researchers conduct state-of-the-art, precompetitive research and transfer these results by such varied means as research publications, students graduated, and personnel transfers. Technology transfer at this level is often viewed as a largely passive process that requires little collaborative behavior among the source and receivers of the technology, although the researchers themselves may work in teams or across organizational or even national boundaries.

Level I success is commonly measured by the quantity and quality (usually determined by peer review) of research reports and journal articles. Technology transfer plans and processes are commonly not considered very important. *Research strength* is most important. The belief is that good ideas sell themselves; pressures of the marketplace are all that is needed to drive technology use and commercialization.

Figure 13.3
Technology Transfer at Four Levels of Involvement

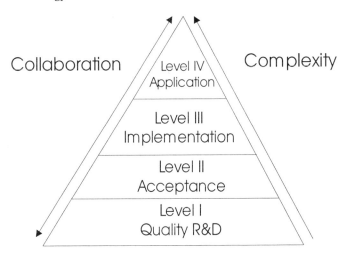

Level II transfer, technology acceptance, calls for the beginnings of shared responsibility between technology developers and users. Success occurs when a technology is transferred across personal, functional, or organizational boundaries and it is accepted and understood by designated users. Moving from Level I to Level II technology transfer is an especially difficult task when the organization conducting the research is at "arm's length" from the organization that is the prospective user of the technology. A Level II perspective encourages the belief that successful technology transfer is simply a matter of successfully getting the right information to the right people at the right time. The mistake of limiting involvement in technology transfer to this level has led to the demise of many prestigious R&D organizations worldwide.

In Level III transfer, technology implementation, success is marked by the timely and efficient implementation of the technology. For Level III success to occur, technology users must have the knowledge and resources needed to implement, or Beta test, the technology. Technology implementation can occur within the user organization in terms of manufacturing or other processes, or it can occur in terms of product development, such as building a prototype or proof of concept for commercial application. *Industrial strength* is required. It is at this stage where the receptor organization provides value added to the transferred technology.

Level IV transfer, technology application, centers on product commercialization. Level IV builds cumulatively on the successes achieved in attaining the objectives of the three previous stages, but *market strength* is required. Feedback from technology users drives the transfer process. Success is measured in terms of ROI or market share. Here, we take a longer-term view. It is with respect to Level IV technology transfer that universities, government labs, R&D consortia, incubators, and science parks are increasingly being judged.

Overall technology transfer success in terms of Levels I to IV is difficult to measure by traditional cost-benefit analyses, since (1) it is often difficult to quantify financial and other impacts of a technology over time, and (2) different persons involved in the process are likely to evaluate costs and benefits differently, depending on their unique perspectives. Different participants commonly hold different expectations as to which level of technology transfer they value. Some are happy with research reports, while others want market-strength products. However, it is becoming increasingly apparent to managers in government, academia, and business that Level I, Level II, or even Level III measures of success, however impressive, will not motivate much less sustain technopolis development.

GLOBALLY NETWORKED ENTREPRENEURSHIP

No matter what field you are talking about—electronics, medical, education, the environment, entertainment—the global marketplace opens up more opportunities than I have seen in my lifetime. Very few generations throughout history, perhaps not since the Renaissance, have been accorded the opportunities this period provides. It is a profoundly different world.

Dr. George Kozmetsky, Chairman of the IC² Institute
Advisory Board, 1997 IC² Annual Report.

The remainder of this chapter focuses on using personal and Internet networks to accelerate the growth of spin-out and start-up technology companies through Globally Networked Entrepreneurship (GNE). Technology entrepreneurship as described by IC² Institute in the early 1990s focused on regionally linking talent (entrepreneurs), technology, capital, and business know-how (see Figure 13.4). GNE focuses on having small and emerging high-tech companies globally linked through personal networks and Computer and Information Technologies (CIT). Ideally such globalization occurs as firms are launched and as they grow and not after they are established and large.

Entrepreneurial talent results from the perception, drive, tenacity, dedication, and hard work of special types of individuals—people who make things happen. Where there is a pool of such talent, there is opportunity for economic growth, diversification, and new business development. Talent without ideas is like seed without water. When talent is linked with technology, when people facilitate the push and pull of viable ideas to commercialization, the entrepreneurial process is under way. Every dynamic process needs fuel, and here the fuel is capital. Capital is the catalyst in the technology-venturing chain reaction. Know-how is the ability to leverage business or scientific knowledge by linking talent, technology, and capital in emerging and expanding enterprises. It finds and applies expertise in a variety of areas, making the difference between success and failure. This expertise, or "smart infrastructure," involves management, marketing, finance, accounting, production, manufacturing, sales, and distribution, as well as legal, scientific, and engineering skills.

GNE argues for the importance of globally sourcing and linking of talent, technology, capital, and know-how as firms are launched and as they grow. In short, firms do not wait until they dominate local or domestic markets before they go global. Start-up firms with specialized technologies use networks to globally seek out niche markets and partners. For example, GNE suggests that while a regionally based start-up may have an exciting technology that is locally competitive and the company may benefit from exceptional entrepreneurial talent, it may lack sufficient capital and business know-how to market globally.

Figure 13.4
Critical Success Factors for Regionally Based Technology Entrepreneurship

Barriers to a global strategy for small start-up firms include:

- The small, entrepreneurial staff being preoccupied with local markets and challenges
- Limited personnel, resources, and time
- Limited tolerance for the added challenges a global perspective brings, such as intellectual property protection, trade and labor issues
- The challenge of assessing foreign opportunities and the ignorance of possible benefits as well as potential challenges to a global perspective

GNE is made possible through globally networked teams that include technologists and entrepreneurs as well as managers, finance, legal, sales, and distribution expertise. International collaborations are seen to be important for overcoming challenges of local language, culture, legal, and policy differences, challenges that can quickly overwhelm and defeat a small company. International GNE teams are seen to be a way to rapidly provide local expertise and knowledge. The objective is to solve regionally based problems faster, shortening learning curves and minimizing the threat of costly business and marketing mistakes.

Benefits to GNE can include:

- Niche market dominance and growth
- Access to needed talent, technology, capital, and know-how
- Minimizing costly mistakes and misspent resources
- Shortening the learning curve and solving problems faster

Incubators have traditionally been a way to foster public/private collaboration at the regional level—to spur economic development, fill vacant office space, train entrepreneurs, and create high-value jobs. At their best such incubators have acted as "lightning rods" at the regional level, linking talent, technology, capital, and business know-how to market needs. And the incubator can serve as a "learning laboratory" for local colleges and universities as well as regionally based professionals. Globally Networked Entrepreneurship seeks to foster the global linking of incubators. As a regionally-based incubator can facilitate local public-private collaboration it can also serve as an effective bridge across national boundaries. Globally networked incubators can act as catalysts to:

- promote partnerships between large companies and entrepreneurial firms regionally and globally;
- foster collaborations across academic, business, government sectors globally;
- promote and support entrepreneurial vision and leadership;
- develop innovative ways to regionally and globally leverage capital, talent, technology, and "know-how" resources; and
- leverage R&D to achieve early success, and contribute to newer industries and smart infrastructures for the 21st century.

Figure 13.5 depicts how the regional and global know-how networks of Globally Linked Incubators cross the technology transfer gap between research strength and market applications leading to wealth creation. The primary drivers are entrepreneurs and technology, which come from the private sector, universities, federal laboratories, and R&D consortia nationally or globally.

The challenge is to foster the global linking of cutting-edge research and technology with venture financing and the realities of the international marketplace. Whereas technology reports, patents, and technology licenses are often the output of R&D environments, they are considered inputs to the due diligence and business plans required for the Globally Linked Incubator. GNE strives to shorten product development cycles by broadening an entrepreneurs' global know-how in market research, finance, advertising, quality issues, management, sales, and service. The GNE culture emphasizes the importance of intangibles and networking (e.g., business know-how and learning) over tangibles (e.g., physical assets) (Figure 13.6).

Figure 13.5
Crossing the Technology Transfer Gap with Know-How

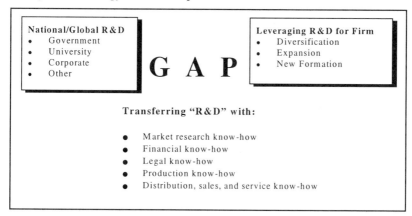

Figure 13.6
Globally Networked Entrepreneurship (GNE)

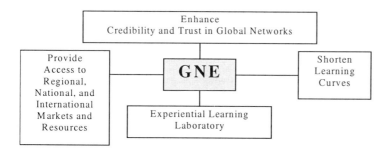

FOSTERING VIABLE GLOBAL ENTREPRENEURIAL NETWORKS

While computer and information technologies are key to GNE implementing these enabling technologies is not sufficient for a viable global network. IC^2 Institute's most successful global collaborations are based on personal relationships with locally respected and active champions. Long-term personal relationships based on trust and mutual benefit are considered essential. Short-term relations are not sufficient because to realize sufficient win-win scenarios collaborators often must overcome, in the short term, lose-win and win-lose situations. Short-term gains and loses, for all parties in the collaboration, must be balanced with longer term visions that all parties can support.

On-site visits need to occur despite frequent technology-mediated communication. Mutual trust is essential and is facilitated by a realistic understanding of the "real world" constraints that face the global partners, constraints that often inhibit "ideal" behavior and outcomes. Such "partnering" requires an appreciation of regional norms, values, and policies in terms of formal as well as informal business environments.

IC² INSTITUTE'S STRATEGY FOR FOSTERING GNE

Since its founding in 1977, IC² (Innovation, Creativity, and Capital) Institute has fostered global partnerships and alliances through its associated Fellows who work in academic, business, and government sectors. These networks have centered on both research and education or "think" activities as well as enterprise development, or "do" activities centering on the commercialization of science and technology. International and multidisciplinary teams work on "unstructured problems" in such areas as technopolis and regional economic development, entrepreneurship and technology venturing, and documenting critical success factors for technology incubators and capital networks.

IC²'s locally based learning laboratories included the award winning Austin Technology Incubator (ATI), The Capital Network (a computer-mediated network linking Business Angels and entrepreneurs), and the fast developing Austin technopolis. Based on the lessons learned innovative concepts and management techniques regarding technology commercialization and regionally based economic development have been transferred nationally and globally.

The challenge for GNE is to foster and maintain viable entrepreneurial and professional networks for accessing and leveraging global talent by specialization; technology by area of expertise; capital by type (e.g., seed, venture, business angels, banks, government); and know-how (e.g., marketing, legal, sales and distribution, management, manufacturing). In this regard an on-going IC² research project is to complete the matrix (shown in Figure 13.7) for specific technologies with regard to specific global partners.

In 1996 IC² Institute, The University of Texas at Austin launched an MS Degree Program in Science and Technology Commercialization. A main goal of this program is to effectively leverage the Institute's national and global "learning laboratories" for educational and research advancements while fostering regionally based economic

Figure 13.7
Globally Networked Entrepreneurship Partners Linked by Talent, Technology, Capital, and Know-How

GLOBAL LOCATION	TC	Talent	Tech.	Capital	Seed	V.C.	Establ.	Other	Know-How	Marketing	Sales/Distr.	Legal	Mgmt.	Production	Prototype	Mass
Brazil																
Curitiba																
China																
Shanghai																
Wuhan																
Hefei																
Russia																
Moscow																
Sarov																
Ukraine																
Kiev																
Taiwan																
Hinschu																
South Korea																
Taejon																
Japan																
Osaka																
Portugal																
Lisbon																
Chile																
Santiago																
Romania																
Iasi																
Mexico																
Monterrey																

development. Classes for this innovative one-year degree program are held simultaneously in Austin; Houston, and Dallas, Texas; Monterrey, Mexico; Australia; and in Europe through the Instituto Superior Técnico in Lisbon. Two-way interactive video and e-mail are used to link "technology commercialization teams" in these geographically separate sites. Current plans are to offer most courses over the Internet while using two-way video to facilitate virtual teaming. Students enrolled in the program have come from Brazil, China, Japan, Mexico, Portugal, and Russia in addition to the U.S. This global classroom, to a certain extent, exists in cyberspace where international teams of entrepreneurs and managers from a range of backgrounds collaborate to access needed talent, technology, capital, and know-how. Early on it was realized that the benefits of bringing a functionally and "situationally" diverse group of professionals together virtually overwhelmed the challenges of teaching across different academic and work backgrounds as well as cultural and geographic distances.

In conclusion, while this chapter supports a regional focus in technopolis development it also emphasizes the fostering and leveraging of global partnerships across academic, business, and government sectors. An overview of two basic forms of technology transfer emphasizes the importance of spin-out and entrepreneurial activities to regional economic development. Finally, Globally Networked Entrepreneurship seeks to use Internet and Web technologies as well as personal networks to (1) globally link talent, technology, capital, and know-how, and (2) shorten learning curves and accelerate the growth of select small and mid-sized enterprises in emerging, developing, and developed technopoleis worldwide.

NOTES

1. Some technopolises are the result of long-term planning and varying degrees of public/private collaboration, such as Tsukuba and Kansai, Japan; Bari, Italy; Sophia-Antipolis, France; Hsinchu, Taiwan; Taejon, Korea; Raleigh-Durham, North Carolina; and Austin, Texas.

2. "Coordinate" means to bring to proper order or relation; to harmonize; to adjust. "Cooperate" means to act or work together with others for a common purpose; to combine in producing an effect. "Collaborate" means to cooperate treacherously with an enemy; to be a collaborationist or to cooperate in a literary or artistic manner. "Synergism" is the simultaneous action of separate agencies which together have a greater total effect than the sum of their individual effects (Webster's New World Dictionary).

3. First- and second-level influencers are identified in the communication literature as (1) cosmopolites (individuals who have a relatively high degree of communication with a system's external environments), (2) opinion leaders (individuals who are able to influence other individuals' attitudes or over behavior), and (3) liaisons (individuals who connect otherwise separate communication networks). The personal communication networks of first- and second-level influencers tend to be outward looking and global, as opposed to closed and provincial. The success of Global Networked Entrepreneur is to a large degree dependent on the networking activities of such influencers from different sectors of the technopolis wheel.

REFERENCES

Botkin, J. "Route 128: Its History and Destiny," in *Creating the Technopolis: Linking Technology Commercialization and Economic Development* (R. W. Smilor, D. V. Gibson, and G. Kozmetsky, eds.). Cambridge: Ballinger, 1988.

Gibson, D. V., G. Kozmetsky, and R. W. Smilor, eds. *The Technopolis Phenomenon: Smart Cities, Fast Systems, and Global Networks.* Savage: Rowman & Littlefield, 1992.

Gibson, D. V., and E. R. Rogers. *R&D Collaboration on Trial: The Microelectronics and Computer Technology Corporation.* Cambridge: Harvard Business School Press, 1994.

Gibson, D. V., and R. Smilor. "Key Variables in Technology Transfer: A Field-Study Based Empirical Analysis," *Journal of Engineering and Technology Management* 8 (December 1991) 287–312.

Rogers, E. M., and J. K. Larsen. *Silicon Valley Fever: Growth of High-Tech Culture.* New York: Basic Books, 1984.

Smilor, R. W., D. V. Gibson, and G. Kozmetsky, eds. *Creating the Technopolis: Linking Technology Commercialization and Economic Development.* Cambridge: Ballinger, 1988.

Tatsuno, S. *The Technopolis Strategy.* Englewood Cliffs: Prentice Hall, 1986.

14

Technology Policy: An International Comparison of Innovation in Major Capital Projects

David Gann

INTRODUCTION

Conditions determining the development, construction, operation, and use of large capital projects are changing rapidly. Many new projects are being constructed in the swiftly developing economies within China and South East Asia, and these often involve firms with specialist technical know-how from Europe, North America, and Japan, which compete in international markets for new work.

Few studies have focused specifically on innovation in major capital projects. The particular technoeconomic and socioinstitutional conditions found in markets for these long-lived, fixed capital assets mean that conventional, mainstream theories of innovation are not always applicable for understanding processes of change (Hobday 1996). This chapter draws upon the findings of an international collaborative research program on the management of innovation in large, complex capital projects: the International Program on the Management of Engineering and Construction (IMEC), formed in 1995. IMEC is sponsored by large owner/operators of major capital projects and governments. It involves a network of researchers led from Canada, and including the United Kingdom, the United States, France, and Japan. The University of Sussex Science Policy Research Unit (SPRU) contribution to IMEC is managed through an Innovative

Manufacturing Initiative/Royal Academy of Engineering Chair in Innovative Manufacturing for Construction, which is also involved in a number of related studies on innovation in large projects (Gann et al. 1996; Groák et al. 1997). The work also relates closely to that of the new SPRU/CENTRIM Complex Product Systems (CoPS) Innovation Center, discussed in Andrew Davies' paper at the First International Conference on Technology Policy and Innovation in Macau, July 2-4, 1997 (see, for example, Davies 1996; Hobday et al. 1995; Hobday 1997; Hobday and Brady 1996; Brady 1995).

This chapter begins with a definition of major capital projects, set within a framework for the analysis of complex product systems derived from the work referenced above. It discusses the importance of these projects to both developed and rapidly developing economies and societies. The chapter explores a framework for understanding innovation in these projects through a synthesis of a number of different streams of literature relating to technical and organizational change and the need to understand innovation within complex, multivariate environments. The discussion of innovation is developed first by focusing on organizational questions relating to the supply chain and literature on networks of innovators, and second, through a discussion of the role of new technologies in facilitating communication, coordination, and control within supply networks. The benefits of new ICT (information and communication technologies) in supporting the development of knowledge and creating new relationships between actors and new decision-making processes are then presented. The chapter goes on to outline technology policies for the adoption of new procedures in the management of innovation in large capital projects, with a particular focus on strategies of project-based firms. The chapter concludes with a few remarks regarding international comparisons of the different rates and directions of innovation in major capital projects.

THE IMPORTANCE OF MAJOR CAPITAL PROJECTS

Major capital projects are engineering-intensive, high-cost products and systems which embody large numbers of highly customized, interactive subsystems and components. These may include projects in the following areas:

- power, oil and gas, off-shore, and process plants
- transportation infrastructure
- utilities infrastructure
- "intelligent buildings" such as hospitals, R&D laboratories, control stations, and some offices

These types of projects are important because their production, maintenance, and adaptation are often significant contributors to GDP. This is less so in developed economies where there is already a large installed base of major fixed capital assets, but these nevertheless need to be replaced or undergo extensive refurbishment from time to time. In some respects, because of their long life with often unpredictable changes in patterns of use, these facilities are never quite completed, and some are in almost continual cycles of change. For example, many European countries are currently investing heavily in new railway infrastructure, airports, and air-traffic control systems. In rapidly developing countries such as those in East Asia, the production of these projects plays a very significant role in terms of contributions to GDP. Furthermore, the use of major capital projects underpins the competitive performance throughout the rest of the economy. For example, these facilities create the basis for the production of goods by high-volume, mass-production industries and in the supply of modern services. In essence, they provide the infrastructure which makes a modern economy and society viable.

While these projects provide facilities we could not live without, their production and use are destroying the natural environment and causing problems for existing and older forms of social cohesion. Finally, their development and use often raises transnational as well as regional and local issues about the governance of technology.

THE NATURE OF INNOVATION IN MAJOR CAPITAL PROJECTS

The construction and operation of these facilities usually involves long-term business-to-business interactions with feedback between users and producers. Production is triggered in response to user needs, and, in this sense, projects are demand-derived rather than the result of arm's-length market transactions, which typify consumer-goods industries. Production occurs through project-based activities rather than batch or mass-production methods. Operation usually involves the management of facilities as part of larger technical systems and infrastructures. Thus individual projects are often designed within constraints defined by existing systems, creating technological legacies and path dependency.

In this environment there is a need for integrity of information between suppliers, designers, systems integrators, engineers, constructors, clients, and end users. The choice of technology and management of innovation takes place under conditions where it is not usually feasible to test full-scale prototypes. This usually involves the transfer of information within complex networks of suppliers and includes a large number of interactions between many different specialists. Moreover, it includes the need to deal with complex technical decisions in which the interdependency between components and subsystems creates the need for the interchange of technical know-how across a range of professional and engineering disciplines.

Design and engineering processes often occur concurrently and these are increasingly affected by events which had previously been considered to be exogenous to engineering decision-making processes. For example, the structure and timing of project finance has introduced new institutional decision makers in project planning, and this may have implications for the procurement of technologies, path-dependency, and technological "lock-in." Demands concerning environmental protection, made by what were once considered to be "external" pressure groups, are also shaping the environment within which technical decisions are made.

The need to innovate is driven by pressures on the demand and supply sides. Owner-operators of large facilities are exerting pressures to improve the ways in which complex engineering and construction projects can be delivered on time, within budget, and to specified quality. The role of project finance and new financial institutions is becoming an important stimulant of change. Firms in engineering and construction compete in dynamic environments in which they need to manage technological innovation and uncertainty across organizational boundaries, within networks of interdependent suppliers, customers, and regulatory bodies. Knowledge is differentiated and distributed throughout these supply networks. And in these conditions, the management of technical know-how has become a significant strategic consideration for suppliers and operators.

In our research we have begun to construct a conceptual framework for understanding these patterns of innovation. We have drawn upon a number of distinct bodies of knowledge, including:

- *Systems theory*—von Bertalanffy on open systems—biological sciences; Wiener and Stafford Beer—feedback and automation on cybernetics; games theory; Jay Forrestor on simulation
- *Complexity and chaos theory*—Brian Arthur on socioeconomic systems and path dependent processes; Stuart Kauffman with reference to biological systems
- *Innovation in systems*—Nathan Rosenberg and Sahal on interdependencies, complementarities, and bottlenecks in technical systems
- *Sociotechnical systems*—Thomas Hughes on the economic and institutional shaping of technologies; Chandler and Davies on the economics of systems
- *Evolutionary economics*—Giovanni Dosi and Chris Freeman on co-evolutionary processes of technological change and socioeconomic development.

NETWORKS OF INNOVATORS AND THE ORGANIZATION OF PROJECT-PROCESSES

The types of knowledge required by managers change over time as projects proceed from prefeasibility and initial conception through design and execution to use and decommissioning. The ability to utilize new Information and Communication Technologies in planning, design, engineering, and management is becoming an important condition for the successful management of innovation in other technological systems which form component parts of major projects. Pressures to develop new methods and tools for managing uncertainty, systems integration, and changes in specification of large complex engineering construction projects are resulting in the need to assess and change business processes, and firms' competitive strategies. These result in a set of issues which condition the ways in which technology policies can be developed and managed.

Perhaps the most important findings from our research to date concern the organization of project-processes. Projects vary greatly in the ways in which they are organized and this has major consequences for coordination and communications and whether decisions concerning the choice of technology are made sequentially or concurrently. We have identified three basic forms of organization:

- relatively stable partnerships between firms involving long-term interorganizational networks in which it is possible to use integrated information systems—for example, Japanese, East Asian, and some French approaches;
- unstable market-relations between temporary coalitions of firms in which there is greater uncertainty; risk is managed in hierarchical networks where problems are *interdependent,* but people, methods and organizations are *independent* (Tavistock Institute 1966)—for example, in the U. S. and in many U.K. approaches;
- hybrid forms of organization, which combine a mixture of market-based transaction approaches to organization, with partnering; partnering itself may take a number of forms from tactical to strategic (Barlow forthcoming)—for example, in some U.K. and North European approaches.

In the United States and the United Kingdom, projects tend to be organized in a way which epitomizes the fragmented form of hierarchical network organization, in which contracts are highly specialized. The approach relies upon price-based contracting methods (market transactions). This often results in adversarial relations between participants, in which information is concealed and information flows are disrupted. This is particularly the case when projects are carried out by temporary coalitions of firms working in geographically distributed and often concurrent processes.

THE IMPORTANCE OF INFORMATION TECHNOLOGY (IT) IN COORDINATION AND DECISION MAKING

Large projects involve the integration of many technical subsystems and components in a context which often involves complex and sensitive economic, social, political, and environmental decisions. The flow of information from early debates about project concepts through to strategic project definition, engineering, procurement, construction, operation, and decommissioning is therefore extremely important if life-cycle costs are to be reduced. This is difficult to achieve because the nature of decisions and types of decision makers change considerably over project life cycles. Moreover, there is a lack of historical data about decisions relating to the choice of technologies, which could prove useful in improving the quality of information provided at different stages of the life cycle.

New information systems to support decision making in the procurement, construction, and operation of major capital projects are being developed and deployed rapidly, particularly in the United States (Gann et al. 1996). These systems offer several possible advantages:

- the *integration* of information flows
- *automation* of routine information processing and communication activities

- generation of new information on processes and systems integration—*informating* processes (Zuboff 1988)
- providing new levels of *transparency* about processes
- increasing capabilities for knowledge acquisition, feedback, and learning

Such systems are being used to involve different interest groups in early phases of planning and design, increasing the potential for participation in decision making, which ultimately affects the choice of technology. They are also being used to involve clients in decision making throughout the design and construction process. For example, Bechtel is using Virtual Reality systems to share information with its clients in order to reduce risk and uncertainty and improve predictability in design decisions. This enables clients to modify design decisions at little expense. These systems have been found to save time and reduce travel costs.

Some firms have adopted an approach to sharing project information aimed specifically at extending the market for their services. For example, Parsons has created opportunities to provide clients with new value-added services, extending its market for engineering, procurement, and construction, into early project decision making and downstream facilities management. To achieve this, Parsons is developing its existing Computer-Integrated-Engineering IT support systems to form new Computer-Integrated-Project systems. These are supported by a variety of technologies such as Geographical Information Systems. The adoption of this approach resulted in the need for internal business process changes and new relationships with suppliers and design organizations.

The notion of decision making and nature of decisions themselves are changing due to the implementation of new technologies relating to management of information in major projects. Successful implementation of IT-based decision support systems in leading U.S. construction organizations demonstrates that emerging construction processes are quite different in character from conventional approaches. The use of IT systems is resulting in fundamental changes to the timing, sequencing, and hierarchy of decision making. The most important aspects of change are:

- the speed and concurrence of decision making
- the ability to make information readily available when and where it is required
- increased visibility of decision making processes, including access to other people's decisions.

There are significant regional differences in the development and implementation of some IT-based decision support tools. These differences can be seen most clearly in comparing activities in the United States with those in Europe and the Far East. While most firms participating in large international projects can gain access to IT systems for decision support, the United States is probably the world leader in most areas. This is because it has many of the world's leading IT firms, which are providing enabling technologies that form the infrastructure and applications for decision support systems: much of the R&D is carried out in the United States. U.S. firms are generally advanced in their level of business networking and use of IT systems. Some of the world's leading research institutions work in close proximity with companies, providing a fertile environment for innovation.

Europe is possibly more advanced than the United States in the development of elegant concepts about the use of IT, but markets are more fragmented and Europe lacks powerful large software firms to develop these ideas into applications technologies. Japan and the Far East are thought to lag behind the United States and Europe in some areas of development and implementation.

TECHNOLOGY POLICY: A FRAMEWORK FOR UNDERSTANDING AND STIMULATING PERFORMANCE IMPROVEMENTS WITHIN FIRMS

Rapid innovation underpins competitiveness in the development and use of major capital projects. Deregulation and internationalization expand and change business opportunities for supplier firms. Moreover, they need to manage innovation in response to changes in software engineering, information and materials technologies, rising costs, and the need to deal with increasing complexity due to social and political, as well as technical circumstances. Governments and international agencies, financial institutions, and insurance organizations all have a part to play in developing a new framework for the governance of technology in this dynamic environment. These issues will be considered in subsequent papers.

At the level of the firm, technology policy and the strategic management of resources concern questions of how firms develop and resource their core technical competencies. In the production and use of major capital projects, this relates directly to issues of integration in planning, design, and construction. Supply firms in project-based industries trade on their reputation, based on their performance track record from previous projects. Their ability to leverage reputation to win new orders and thus gain further experience in the deployment of technical expertise is of crucial importance to firms' long-term competitiveness. It also has a direct impact on profitability and therefore on the ability to invest in new generations of technology and business process changes.

The extent to which technical competencies are specialized and located in different places within and between organizations affects how they can be deployed, ultimately affecting project performance, the ability to deliver value to clients, and firms' profitability.

An initial model indicating the interactions within which technical support activities takes place is shown in Figure 14.1. This illustrates in-house support and external research and technical support services bought in by firms.

Technological capabilities take several forms and are located within different functions in the firm. These capabilities may be tangible, codified, transferable assets, or intangible, tacit, uncodified competencies embedded within the resources of the firm (Bell and Pavitt 1995). In this context there is a need for a better understanding of the transfer of knowledge within firms—that is, the ways in which different business processes can be developed to enhance technical capabilities. This needs to be linked to strategies for managing interfirm project-processes—the location of production activities where design and systems integration takes place. One contribution to this will come from the new supply-chain management approaches developed in the field of industrial dynamics and logistics (Towill 1997; Evans et al. 1995).

Figure 14.1
Managing Technology in Project-Based Firms

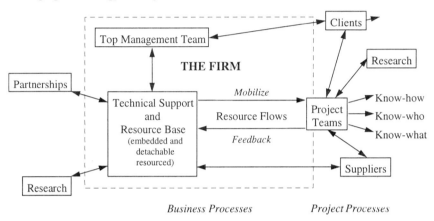

SUMMARY

This chapter has articulated some of the issues for technology policy. particularly within supply firms producing and adapting major capital projects. It has illustrated the need to manage nonlinear dynamic environments in which technical, organizational, economic, and political complexity create uncertainty over long project life cycles. Technology itself, in the shape of new IT-based decision support systems, can help. These systems have proven useful through the involvement of those responsible for technical decision making earlier in project planning and design stages.

Nevertheless, the form of organization of projects is of critical importance for successful outcomes in the deployment of these systems. Approaches to organizing development and production processes vary, with differences in the forms of organization found in Northern Europe, the Anglo-Saxon countries (North America, the United Kingdom, and Australia), and Japan. Each form illustrates a different approach to the management of technology. These will be discussed in further papers, together with the lessons for improving the delivery of large capital projects.

REFERENCES

Barlow, J. *Partnering in the Construction Industry.* Bristol, U.K.: The Policy Press (forthcoming).

Bell, M., and K. Pavitt. "The Development of Technological Capabilities," in *Trade, Technology and International Competitiveness* (I. Haque, ed.). Washington, DC: World Bank, 1995.

Brady, T. "Tools, Management of Innovation and Complex Product Systems." Working Paper prepared for CENTRIM/SPRU Project on Complex Product Systems, June 1995.

Davies, A. "Innovation in Large Technical Systems: The Case of Telecommunications," *Industrial and Corporate Change* 5, 4 (1996).

Evans, G. N., D. R. Towill, and M. M. Naim. "Business Process Re-engineering the Supply Chain," *Production Planning & Control* 6, 3 (1995) 227–237.

Gann, D., K. Hansen, D. Bloomfield, D. Blundell, R. Crotty, S. Groák, and N. Jarrett. *Information Technology Decision Support in the Construction Industry: Current Developments and Use in the United States.* Department of Trade and Industry Overseas Science and Technology Expert Mission Visit Report, 1996.

Groák, S., D. Gann, and K. Hansen. *Process Representation: An International State-of-the-Art Review of Development Tools.* Final Report to EPSRC, SPRU, University of Sussex, U.K., 1997.

Hobday, M. "Complex System vs. Mass Production Industries: A New Innovation Research Agenda." Paper prepared for CENTRIM/SPRU Project on Complex Product Systems EPSRC Technology Management Initiative, 1996.

Hobday, M. "Product Complexity, Innovation and Industrial Organization," submitted to *Research Policy* (February 1997).

Hobday, M., and T. Brady. "Software Processes and Practices in Complex Product Systems: An Exploration of Flight Simulation Domain," submitted to *IEEE Transactions on Engineering Management* (September 1996).

Hobday, M., with R. Miller, T. Leroux-Demers, and X. Olleros. "Innovation Complex Systems Industries: The Case of Flight Simulation," *Industrial and Corporate Change* 4, 2 (1995) 363–400.

Tavistock Institute. *Interdependence and Uncertainty.* London, 1966.

Towill, D. "Successful Business Systems Engineering," *IEEE Engineering Management Journal* Part 1: 7, 1 (February 1997) 55–64; Part 2: 7, 2 (April 1997) 89–96.

Zuboff, S. *In the Age of the Smart Machine: The Future of Work and Power.* New York: Heinemann, 1988.

15

Emerging Patterns of Globalization of Corporate R&D and Implications for Innovation Capability in Host Countries

Prasada Reddy

INTRODUCTION

Transnational corporations (TNCs) have been one of the crucial channels of technology transfer to developing countries. However, TNCs have often been criticized for transferring ready-made and inappropriate technologies. In recent years the debates are focusing on the "triadization" of technology development and marginalization of nontriad countries. Paradoxically, in such a situation, a new development is taking place. Since the mid-1980s, as an off-shoot of the globalization of corporate R&D, TNCs have started performing some of their strategic research and development in some developing countries.[1] Such R&D relates not only to product development for the local markets, but also to developing products for the regional/global markets and generic technologies for long-term corporate use. TNCs involved in this new trend seem to be mostly those dealing with new technologies. This strategic move by TNCs is facilitated by the availability of large pools of scientifically and technically trained manpower, at substantially lower wages compared to their counterparts in industrialized countries, and an adequate infrastructure. This trend shows signs of emerging as a phenomenon similar to that of the establishment of off-shore production facilities in low-cost countries.

This chapter, which is organized into five sections, analyzes the driving forces behind the new trends and its implications for innovation capability in developing host countries. The chapter reviews past studies on globalization of corporate R&D and presents the analytical framework of the present study; discusses three in-depth case studies of TNC R&D activities in India, a developing host country; analyses the implications for innovation capability in the host countries; and draws policy conclusions.

REVIEW OF PAST STUDIES AND ANALYTICAL FRAMEWORK

TNCs, traditionally, tended to confine technology development activities to their home countries. The companies were viewed as deriving competitive advantage, especially technological knowledge, from their distinctive domestic environment, and in turn exploiting this competitive advantage in overseas markets (Hymer 1976; Vernon 1966). Technology development activities are kept close to home-base, not only for security reasons, but also because of economies of scale, and better communication and coordination (Terpstra 1977).

However, the U.S. Tariff Commission (1973) showed that even in the 1960s some American companies carried out technology development activities abroad. At the time, while confining R&D to their home countries, when the necessity arose TNCs performed adaptation or in a few cases product development for the local market types of R&D abroad. Even these limited R&D activities were mostly confined to industrialized countries and a few large developing countries with unique characteristics, such as India and Brazil. Such R&D was considered an additional and inevitable cost of technology transfer (Ronstadt 1977; Behrman and Fischer 1980).

Ronstadt's (1977) survey distinguished between the following four types of foreign R&D units of U.S.-based TNCs: (1) Technology Transfer Units (TTUs)—to facilitate the transfer of parent's technology to subsidiary, and to provide local technical services; (2) Indigenous Technology Units (ITUs)—to develop new products for the local market, drawing on local technology; (3) Global Technology Units (GTUs)—to develop new products and processes for major world markets; (4) Corporate Technology Units (CTUs)—to generate basic technology of a long term or exploratory nature for use by the corporate parent. In recent years, while the markets worldwide are integrating in terms of standards and technologies, some regional clusters are also emerging. National markets in these regional clusters share some common features and needs for specialized products—for example, special types of food, drugs for regional diseases, etc. So, TNCs have started establishing: (5) Regional Technology Units (RTUs)—to develop products and processes for the regional markets (Reddy and Sigurdson 1994).

Socioeconomic developments over the last couple of decades have led to the homogenization of the world's needs and preferences, and such commonalty of preferences is leading to the standardization of products (Levitt 1983). In many industries, radical technological changes allowed companies to develop products on a global basis. Even in industries that were not affected by external forces of change, companies started attempting to achieve global economies, through rationalization of their operations (Bartlett and Ghoshal 1991), by assigning "World Product Mandates," where an affiliate will have exclusive responsibility for the whole range of activities from R&D to marketing worldwide for a particular product (Poynter and Rugman 1982; Bonin and Perron 1986) and/or "Rationalized Product Subsidiary," where the affiliate specializes in the development and manufacture of specific components or parts for final products or executing a specific stage in a vertically integrated production process (Pearce 1989).

Moreover, as analyzed by Chesnais (1988), the basic science is playing an increasingly vital role in major technological advance, and many recent innovations have occurred through cross-fertilization of different scientific disciplines. These ongoing paradigmatic changes in science and technology (S&T) are increasing the pressure on firms, especially in microelectronics, biotechnology, and new materials, due to the requirement of greater breadth of knowledge and skills base for new technology-generation activities than the existing knowledge base of the firms. These pressures can be met partly by increasing in-house R&D, both nationally and internationally, or by the establishment of joint-venture R&D firms or through the external acquisition of knowledge from universities or firms (Chesnais 1988: 509–510).

The convergence of consumer preferences worldwide and the rapid diffusion of technologies have significantly influenced both the pace and the locus of innovation. Companies can no longer assume that their domestic environment provides them with the most sophisticated consumers and the most advanced technological capabilities, and thus the most innovative environment in the world. Today, a consumer trend or market need can emerge in any country, and the latest technologies may be located in another (Bartlett and Ghoshal 1991: 12).

Prior to the mid-1970s, TNCs did not internationalize strategic R&D, mainly because of the difficulties involved with supervision and control (Mansfield 1974). But the introduction of telematics and improvement of information and communication technologies (ICT) has significantly increased the scope for using the new methods of control and in international sourcing of S&T resources (OECD 1992).

The key driving force for globalization of R&D in the 1990s has been the growing demand for skilled scientists and engineers. The mismatch between the outputs of higher education and the needs of the industry is giving rise to shortages of research personnel throughout the industrialized world, especially in electronics and biotechnology. The patterns of cooperation among countries and firms indicate the emergence of an international market for investments in research, and scientific and engineering personnel. The existence of such a market and the necessity of scientific and technological knowledge for competitiveness are leading the firms to direct their investments to those geographical areas which can best meet their research and manpower needs. Uncommon countries are now becoming locations for international corporate R&D, as the TNCs recognize the talent available in countries like Israel, Brazil, and India (OECD 1988: 35–36). TNCs are also sensitive to variations in the cost of R&D inputs from country to country (Mansfield et al. 1979).

Since the mid-1980s, TNCs have started locating some of their strategic R&D in some developing countries. In some of them S&T talents have been lying dormant due to underutilization by the indigenous industry. In most of these countries either the industrialization did not take place at a corresponding level, or the existing industry, because of its low emphasis on R&D, has not been able to fully utilize the available trained manpower (A.S.P. Reddy 1993). TNCs locate technology development activities in countries that offer better comparative advantages. Increasingly, the best location for R&D is not necessarily the United States or European countries. Moreover, the ideal location may differ for subtasks within the R&D function and for different products in the range (Porter 1986: 51).

A few studies have been done on the impact of TNC R&D activities on the host country. According to Behrman and Fischer (1980) the benefits are "nonspecific" and "tied mostly to whatever upgrading of local institutions the firm feels is necessary in order to improve the efficiency of its production and marketing operations." There appeared to be little diffusion of trained manpower into the local communities. However, the authors feel that considerably more data are needed before any conclusions can be drawn. Their studies included only TTU and some ITU type of R&D. When analyzing the implications for the host countries, it is important to consider the type of R&D being performed. Depending on the type of R&D, the impact on the host country varies. Each type of R&D unit displays distinctive linkages with the local affiliate, the corporate headquarters, and with the local S&T system. The local ties are virtually nonexistent for a TTU, whose main technology links are with the parent; somewhat strong for an ITU, which may (but not always) to some extent draw on the local science and technology system to develop products particularly designed for the local market. In this type of R&D unit, its linkages with the local marketing function assume greater importance than linkages with the local S&T system; stronger for a GTU and strongest for a CTU. In these two types of R&D units, the primary motive being that of exploiting local sources of S&T that can not be accessed easily from outside the country, strong local linkages are established (Westney 1988).

This study builds the analytical framework for globalization of R&D in terms of waves. Each wave represents a set of distinctive characteristic features, but reveals the continuation from one wave to the other. Figure 15.1 conceptualizes the evolutionary process of the globalization of corporate R&D.

Most of the technology development activities performed abroad in the 1960s were these TTUs. The main driving force for internationalization of R&D during this "first wave" was to gain entry into a market abroad. This needed adaptation of products and processes to the local conditions and support of technical services. The industries involved were mostly mechanical, electrical, and engineering, including automobile industries.

Figure 15.1
The Evolutionary Process of Globalization of Corporate R&D

	First Wave up to 1960	Second Wave 1970s	Third Wave 1980s	Fourth Wave 1990s
Major Driving Forces	Entry into the local market abroad	Build-up market share in the local market abroad; national government policies	Need for worldwide learning; new technology inputs; and rationalization of TNC operations	Access to scarce R&D personnel; increasing R&D costs; and rationalization of TNC operations
Type of R&D Activity	Adaptation; technology transfer unit (TTU)	Product development for the local market; indigenous technology unit (ITU)	Products and processes development for global markets and basic research; global and corporate technology units (GTU & CTU)	Products and processes development for global and regional markets and basic research (GTU, RTU, & CTU)
Forms of R&D	Own R&D with manufacturing affiliate	Acquisition or green-field investments in own R&D and production facilities	Own R&D affiliates; joint venture R&D; interfirm cooperation; sponsor university research; subcontract R&D	Own R&D affiliates; joint venture R&D; interfirm cooperation; sponsor university research; subcontract R&D
Category of Industries	Mechanical, electrical, and engineering (including automobiles)	Branded packaged consumer goods, food products, chemicals, etc.	Microelectronics (including ICT, biotechnology, pharmaceutical, new materials, civilian aircraft	Microelectronics (including ICT, biotechnology, pharmaceutical, new materials
Facilitating Factors	Large market and proximity to production facilities	Large and protected markets, proximity to customers and production facilities	Communication technologies; flexibility to delink R&D and production; specialization of affiliates and host countries	Divisibility of R&D into core and noncore functions, availability of personnel, changes in IPR and other policies

Source: Adapted from P. Reddy (1997)

By the 1970s, companies started performing R&D abroad in a significant way. The main driving force was to increase the local market share abroad. This required increased sensitivity to local market differences to enhance competitiveness. Moreover, the host country governments started pressurizing the TNCs for more technology transfer. These driving forces triggered what can be considered the "second wave," with a characteristic difference from the earlier wave. ITU laboratories are set up to develop products for the local markets. This type of activity was predominant in branded packaged consumer goods, chemicals and allied products, and so on.

Since the 1980s, increasingly higher-order R&D has been located abroad in what can be considered as the "third wave." Such R&D abroad is carried out as a part of long-term corporate strategy and is often carried out through interorganizational collaboration. The main driving forces for this phenomenon had been: (1) the convergence of consumer preferences worldwide, created a need for worldwide learning; and (2) the increasing science-base of new technologies, necessitated multisourcing of technologies; and (3) the rationalization of TNC operations where the affiliates are assigned regional/global product mandates. These trends are visible mainly in microelectronics, biotechnology, and new materials. The improvement of communication technologies and the flexibility of new science-based technologies that allows delinking of R&D and manufacturing facilitated this process.

The key driving forces for globalization of R&D in the 1990s have been the increasing demand for skilled scientists, and the rising R&D costs. These forces are triggering the "fourth wave" of globalization of R&D, encompassing non-OECD countries also. These pressures need to be addressed without compromising the innovativeness of the company. One way of achieving these twin objectives is to carry out R&D, at least some parts of it, in low-cost locations that have the required S&T capacity. The changing policy and economic environments and the availability of R&D personnel in developing host countries are facilitating their integration into global technology networks.

INTERNATIONAL CORPORATE R&D IN INDIA: CASE STUDIES

Three detailed case studies of TNC R&D activities in India have been carried out. In selecting the case studies, the following criteria were followed: (1) the cases represent new types of R&D-related investments; (2) the cases cover the different forms of TNC R&D investments in the host country; and (3) the R&D units of TNCs have had sufficient time to establish local and global linkages for their activities. When these criteria were applied, all the cases turned out to be those of TNCs dealing with new technologies.

Astra Research Center India (ARCI)

Astra AB is a highly research-oriented Swedish pharmaceutical company.

Astra Research Center India (ARCI)[2] in Bangalore was established by Astra in 1985[3]. As per the recommendation of the Government of India at the time, the research center was registered as a nonprofit organization. The companies, including Astra, utilizing any invention will pay royalties to the center. Astra gets the first right of refusal to commercialize the innovations generated by the center. The objectives of ARCI are: (1) the pursuit of scientific research leading to the discovery of new diagnostic procedures, novel therapeutic products, and targets for rational drug design, for diseases afflicting large populations in both developing and developed countries; (2) to achieve this objective, by employing the powerful tools of molecular biology, immunology, cell biology, molecular dynamics, and molecular graphics in highly focused and time-targeted projects; and (3) to closely interact with the research groups of Astra, Indian Institute of Science (IISc), and universities in India and abroad, through collaborative efforts, extra mural support to projects of mutual interest, and sponsoring scientific meetings on relevant topics (Ramachandran 1991).

Driving Forces. The primary driving force behind the establishment of ARCI is to gain access to scientific personnel. In Astra's opinion scientists educated in India have made significant contributions both within and outside the country to the advancement of chemical and biological sciences. Hence, it was considered that a center for creative applied research in India would be able to attract and retain Indian scientists. Market motives did not play any role in the establishment of ARCI. Several of the staff are scientists who have returned from abroad.

Research Projects. R&D being carried out at ARCI relates to the products to be marketed worldwide by Astra. In addition, ARCI also carries out substantial mission-oriented basic research. The initial task at ARCI was to build the R&D standard up to the international level in terms of laboratory equipment and personnel. Since drug R&D is a long-term process, to begin with it was decided to develop products with a short-term time frame, such as enzymes of use in biotechnology and thereafter diagnostics. During the initial years ARCI successfully developed the know-how for the preparation of several crucial reagents used in molecular biology research. The design and development of diagnostic procedures, using DNA probes and immunological methods, for the diagnosis of some infectious diseases were also successfully carried out. In addition, basic research on molecular biology, protein structure, gene coding, and the like. was also pursued.

ARCI's Global and Local R&D Network. Figure 15.2 shows ARCI's local and global technology network. There are several joint R&D projects in which Astra's laboratories in Sweden and ARCI are collaborating. Astra's R&D structure contributes expertise in pharmacology and toxicology, whereas ARCI provides them with expertise in molecular biology. ARCI has also participated in a number of Astra's projects as experts in biotechnology. Now, with ARCI getting the mandate for drug research and development, the proportion of joint R&D activities with the other laboratories of Astra will decrease.

There are several projects at ARCI which are being conducted in cooperation with research institutes abroad. There is a "tripartite" research effort to develop a drug for malaria, between ARCI, University of California at San Francisco, and Mahidol University, Thailand. ARCI is also involved in a long-term research project on thioredoxin with the Karolinska Institute in Sweden. In another project, ARCI is collaborating with the Massachusetts Institute of Technology (MIT). The know-how is being developed at ARCI, while the downstream processing will be done at MIT.

Astra has endowed two chairs of "Astra Professorship" at the IISc. In its research projects, ARCI has so far received the cooperation of scientists at the IISc, the National Institute of Mental Health and Neurosciences (NIMHANS), and St. John's Medical College, all located in Bangalore. Through extramural grants, ARCI supports projects of mutual interest in all these institutes.

Although ARCI is not an academic institution, it offers training, on specific experimental procedures, to researchers and students from local universities. ARCI also has a postdoctoral program that allows fellows to join an integrated team to work on an on-going project. For the established scientists ARCI offers facilities to spend a sabbatical year or two at the center. ARCI regularly organizes seminars and symposia on relevant topics (Sampath 1990). ARCI, through its R&D activities, is contributing to the diffusion of application knowledge among the scientific community in India.

Figure 15.2
ARCI's Global and Local R&D Network

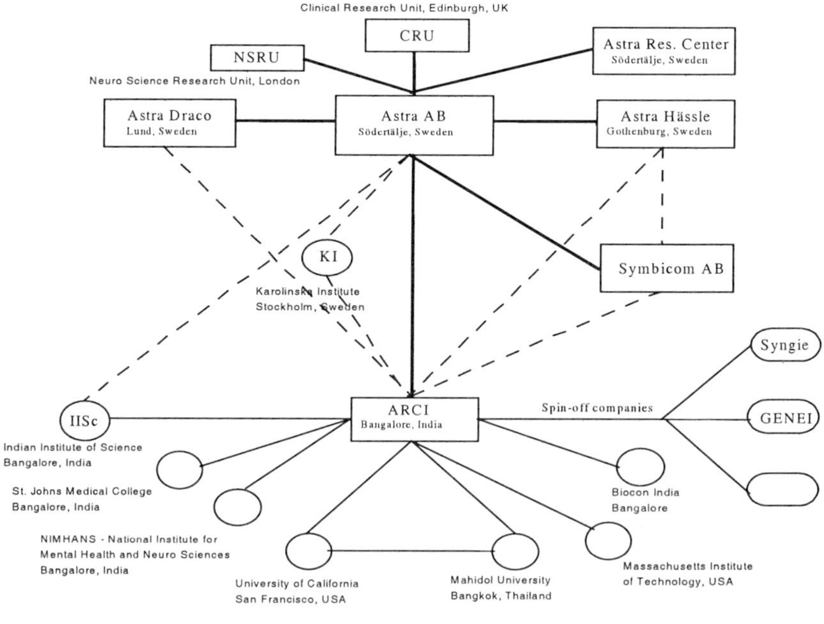

Source: Adapted from Reddy and Sigurdson (1997)

Technology Transfer to Local Firms and Emergence of New Class of Entrepreneurs. ARCI has a policy of transferring the know-how for by-products generated in its R&D to the local firms in the host country. So far, its research results have led to the establishment of three spin-off companies locally. The products being in the high-tech area, the licensing is granted to the type of people who understand the complexities of these technologies well. Such people usually tend to be scientists, who are interested in commercializing them, leading to the emergence of a new class of entrepreneurs. For instance, the know-how for producing the basic tools of DNA recombinant technology has been transferred to a new local company called GENEI (Gene India), which was formed by two Indian scientists. Prior to ARCI's technology these products were being imported. GENEI now exports some of these products to the United States. ARCI's know-how for manufacture of certain biochemical inputs used for drug development has led to another spin-off company called Syngie. The deal is now finalized with another local company for transferring the know-how for manufacturing some of the diagnostic procedures.

Conclusions. Being located in a developing country, ARCI has had to face a few unique problems. First of all the basic tools of DNA recombinant technology were unavailable in India. Hence, the first two years of ARCI's time had to be spent on developing the know-how for producing these tools. Second, ARCI finds it difficult to locate expertise among the local recipients to develop marketable products on a mass scale. Last is the lack of patent protection in India. In DNA recombinant technologies, the novelty is the product. The process of discovery is complicated, but once the product is obtained, its propagation can be achieved by many ways. In pharmaceuticals, the Indian patent laws protect only the process, but not the product.

The ARCI case study shows that, given the opportunity and facilities, scientists in developing countries can also contribute to global technology developments. Joint projects facilitate formal and informal integration of different R&D units, complementarity of research, and internal diffusion of knowledge. Through its transfer of application knowledge and introduction of time- and cost-orientation to research, ARCI is inculcating "commercial culture" among the scientific community in India, which would go a long way to bringing the benefits of science to the society. It also introduced corporate R&D culture in a developing country like India and has become a model for others. This is evident from the number of domestic companies as well as TNCs that have been visiting ARCI to study its operations and set up R&D units themselves.

Texas Instruments (India) Pvt. Ltd.

Texas Instruments Inc. (TI) was founded in 1930 with its headquarters in Dallas. TI is in the business of providing the digital signal processing solutions, including semiconductors, materials, and control systems. Texas Instruments (India) (TI-India) was established in 1986 in Bangalore, with an investment of $15 million and 300 engineers recruited in India. The company also announced additional investment of $10 million over the four years ending 1998 (*DataQuest* 1994: 91).

Driving Forces. The primary driving forces behind location of R&D in India are twofold. First, to gain access to R&D personnel; second, to have a long-term strategic presence in the Asia/Pacific region. TI already has manufacturing sites in several countries of the region and an R&D presence would enable it to identify market requirements and cater to the customers' needs. TI-India, however, is treated as a global resource center and its activities are not confined to the region. India was selected as the location because of its strong educational system in theoretical sciences and engineering, which supplies a large pool of trained manpower. Other reasons include the favorable climate for foreign investments and the common use of English in education, government, and industry. Although not officially stated, the low wages of R&D personnel is one of the important criteria for locating R&D in India.

R&D at TI-India. The initial activity of TI-India was the development and support of proprietary Computer Aided Design (CAD) software systems used for integrated circuit (IC) design by TI's semiconductor design centers worldwide. This activity includes development of applications for simulation, testing, layout, and verification of ICs. In mid-1988, its activities were expanded to include design of ICs. The company focused on differentiated products and developed customized ICs and software, which included digital signal processors (DSP), memories, and mixed-signal ICs. In recent years, TI-India is paying increasing attention to the area of methodology development. Some of its staff are selected to chair worldwide teams in methodology development for DSPs and mixed-signal ICs (*DataQuest* 1994: 91). R&D in TI-India is at present organized into five research groups pursuing different technological directions: (1) digital signal processor design center; (2) memory products design center; (3) mixed signal products design center; (4) software design center for digital printers (RIP); and (5) center for design flows and methodologies for enhancing productivity of TI's semiconductor design centers worldwide. TI-India has a local area network (LAN), which in turn is connected to TI's worldwide data communication network, on a real-time basis through a dedicated 256KB link.

TI-India's Global and Local Linkages. Figure 15.3 shows the global and local linkages of TI-India, which is one of four R&D centers that TI has established to pursue high-tech R&D of this nature. The other centers are located in Dallas, Houston, and Tokyo. TI-India also has linkages with TI's manufacturing sites in Asia, Europe, and the United States. Through these linkages the R&D personnel are drawn into learning and solving the difficulties of larger scale manufacturing of a newly developed product. Several of the R&D projects are joint projects with other R&D units of TI. There is continuous exchange of personnel for short periods between different units of TI. The business groups in TI-India, for their operations, report directly to their respective corporate unit heads in the United States, while TI-India as a unit has administrative linkages with TI's Asia/Pacific regional headquarters in Taipei, forming a matrix-form of command structure.

So far, TI-India's significant contribution is present in 80 international and national patents and its researchers have presented more than 70 papers at international conferences. At TI-India, training of personnel is a major component. It recruits fresh graduates who are only theoretically trained in the universities. First, therefore, they need to be trained in the methods of applying this knowledge to developing new products. Second, these fresh recruits need to be exposed to the state-of-the-art technologies in the area and trained to absorb and use them.

In the initial four to five years of TI-India's operations, the linkages with the local technical universities were limited to the recruitment of personnel. After these initial years, the linkages are extended to include research collaboration with the IISc Bangalore and the IIT Kharagpur. At IIT Delhi, a consortium of TNCs, including TI-India, have sponsored an M.Tech program in VLSI design. In addition, TI-India is embarking on a program of university-buildup, by establishing DSP labs in IITs in Madras and Bombay. Students are also provided with DSP design kits to enable them to design applications. At the IISc, an object-oriented database for analysis of interconnecting data has been established by TI-India. In IIT Madras two fellowships are also instituted for research in mixed signal processors.

Figure 15.3
The Global and Local Linkages of TI-India

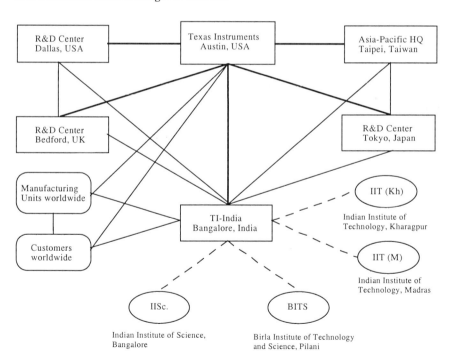

However, there are no instances of TI-India subcontracting research to other companies in India. Because of the type of activities and the high-technology tools with which TI-India performs these activities, there has been very limited need for linkages with local industries, whether in terms of supplies or logistic support. TI-India's R&D so far have not resulted in any spin-off companies in the host country.

Conclusions. TI-India's activities are well integrated into TI's global operations. This is giving opportunities for Indian R&D personnel to acquire expertise and gain access to the latest knowledge in the field. However, linkages with the national systems of innovation in the host country have become significant only in recent years. Hence, the diffusion of technologies to local S&T institutions has so far been limited. The nature of operations at TI-India, being one of designing, the reliance on academic institutions which have expertise in basic sciences is limited. However, with increasing focus on developing methodologies at TI-India, linkages with local innovation systems are expected to strengthen. With a 15% turnover of R&D personnel at TI-India, some of them starting their own software enterprises, wider diffusion of technology and knowledge into the host economy has already begun.

Biocon India Private Limited

Biocon India Pvt. Ltd. was formed in November 1978 as a joint venture between a woman scientist and a biotechnology company from Ireland called Biocon Biochemicals Limited. The Irish company contributed 30% of the equity and the Indian partner 70%. In 1989 Biocon Biochemical Ltd. and its subsidiaries were acquired by Quest International of the Netherlands, a wholly-owned subsidiary of Unilever p.l.c., a leading manufacturer of flavors, fragrances, and food ingredients. As a result of this acquisition, Biocon India has become an associate company of the Unilever conglomerate. Biocon India is mainly in the business of developing and manufacturing industrial enzymes. The company now employs 115 people. Prior to the joint venture, the Irish company had been importing raw materials from India. At the time, Biocon Biochemicals was mainly interested in two products: a plant enzyme called "Papain" and a hydrocolloid called "Isinglass." Biocon Biochemicals felt that by processing them in India, it could gain a cost advantage over its competitors.

Driving Forces. The primary driving forces for the establishment of Biocon India had been the availability of raw materials and the low processing costs in India. After being in the field for some years, Biocon India built up expertise in the field of fermentation technology. As a result, gaining access to R&D personnel and expertise became the primary driving force for Quest International to carry out joint R&D with Biocon India. The low cost of R&D in India has also been one of the important motives. Although some of these products are marketed in India by Biocon India, from Quest International's perspective local market-related motives did not play a role in locating R&D in India.

Research and Development Activities. Biocon India has built a strong R&D base with an initial focus on solid substrate fermentation. This in-house R&D

base is now broadened to submerged fermentation, recombinant DNA technology, and bioreactor design. R&D facilities at Biocon India in the area of microbial enzymes was established in 1984. A substantial amount of basic R&D work has been carried out to develop suitable technologies to produce enzymes through fermentation root. Biocon India has been very successful in its research projects, in terms of building up a track record of commercializing at least two new products every year. Biocon India's strength has been in creating niche markets through innovative products and product applications. A number of novel enzyme applications introduced by Biocon India have qualified for patenting.

Global and Local R&D Network Linkages. Although there was no formal agreement between them, the R&D was a sort of collaborative effort between Biocon India and Biocon Biochemicals, Ireland. R&D investments were made in India, because of the relatively better experience of Biocon India in this area of technology. In joint research projects the majority of the work was done in India. Most of the work carried out in Ireland related to that of testing of the new enzymes, for their efficacy, suitability for plant scale production, performance, and such. By 1989, Biocon India had built up substantial expertise in this field. Biocon India also became strong in production technologies of certain enzymes and has developed certain unique strains and process technologies through its in-house R&D. So Quest International, its new partner entered into a formal agreement with Biocon India to manufacture certain enzymes exclusively to be marketed by them worldwide. In addition, Quest International also requested Biocon India to develop some new products exclusively for them through R&D.

There are several R&D projects that Biocon India is carrying out for Quest International, including the whole range of activities from the laboratory stage through pilot plant to scale up. Under the contractual conditions, Biocon India gets the exclusive rights to apply these new technologies or products in the Indian market, whereas Quest International gets the rights to worldwide markets. Some of the products developed through contractual R&D projects are also manufactured in India exclusively for Quest International. In the case of products where the process facilities at Biocon are not suitable for taking up large-scale manufacturing, Biocon India transfers the know-how to the manufacturing sites abroad. Since the complete expertise and responsibility lie with Biocon India, there has been no need for exchange of R&D personnel between Quest International and Biocon India.

The links with Quest International are also helping Biocon India in several other ways. For instance, Biocon India's knowledge of the complexity of patenting and its procedures was very limited, but now with the help of Quest International it has been building up its knowledge on these issues. For the contractual R&D projects, both national and international patents are jointly held by Biocon India and Quest. Biocon is further entering into a separate agreement with Unilever's corporate laboratory in Fladigan, the Netherlands, to carry out some collaborative and contract research for them. The association with the Unilever group is also helping Biocon India to acquire global market knowledge for Biocon's own products.

Figure 15.4 shows the global and local R&D network of Biocon India, which has built up strong linkages with the national systems of innovation. Biocon India has sponsored research projects in Astra Research Center India, Bangalore; National Chemical Laboratories, Pune; and Central Institute for Food Technologies, Mysore. These research projects are of basic science nature for the results of which Biocon India works closely with these research institutes.

Conclusions. Biocon India feels that international collaborative or contractual R&D activities are beneficial for the host country. First, such activities bring into the country an R&D culture that is essential for survival and growth of the society and which has been sorely lacking in India. Second, TNCs are very experienced in R&D, so collaboration with them helps in upgrading the quality of research in the country and in gaining the application knowledge. Third, such activities give access to global networks, which are very important in terms of widening the knowledge base as well as in learning to compete in international environment. Last, taking up R&D as a business proposition makes the R&D exercise more meaningful, as the culture of time and cost consciousness spreads and scientists become more aware of issues such as intellectual property rights.

Figure 15.4
The Global and Local R&D Network of Biocon India

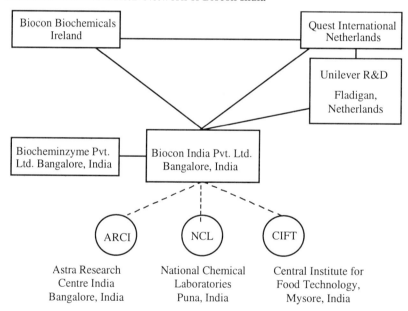

IMPLICATIONS FOR INNOVATION CAPABILITY IN HOST COUNTRIES

While analyzing the implications for the host countries, it is important to consider the type of R&D being performed and its effects. The strength and breadth of the ties with the local systems of innovation vary across the five types. The ties are limited in the case of TTUs, because they only involve adaptation of the parent's technology to local conditions and are better done within the manufacturing unit. ITUs and RTUs were in general supposed to have stronger ties with the local S&T system because of their product development activities, even though they basically redo the designs supplied by
the parent. GTUs and CTUs have stronger ties both to the local innovation system as well as the global research networks. Therefore, the scope for the diffusion of new knowledge to the local S&T system are higher through GTUs and CTUs, whereas TTUs, ITUs, and RTUs mainly utilize the knowledge already available within the local innovation system. However, this does not imply that TTUs, ITUs, and RTUs have no important implications for the host economy. Since the conversion of research results into manufacturing products takes place in the same place, it may lead to other benefits such as the development of supplier networks and technology transfer to domestic small and medium enterprises. On the other hand, GTUs and CTUs being delinked from the operations of production and marketing, their innovations are less likely to lead to production-related benefits in the host country.

The case studies suggest that the scope of R&D activities carried out in some developing countries has broadened in recent years. The primary driving force for locating R&D in India is technology-related—that is, to gain access to R&D personnel. This factor is considered to be the most important for location of higher-order R&D activities, especially for GTU and CTU types.[4] The case studies also suggest that some developing countries have achieved an international competitiveness in certain S&T areas: for example, India in molecular biology, biotechnology, and software technologies. TNCs are attempting to gain access to such knowledge and skills by making R&D investments in such countries, irrespective of whether they are industrialized or developing countries.

Another driving force has been the cost-related factor. In the case studies the cost factor, however, did not assume the same importance as gaining access to personnel as a primary driving force. This is because sometimes, in developing countries, the cost advantages of lower wages of personnel may be offset by the higher material costs. For instance, some of the inputs may not be available locally, as in the case of ARCI, and may need to be imported under special conditions and this adds to the total costs and time. Similarly lack of infrastructural facilities may require TNCs to invest in captive facilities, adding to the total costs—for example, power shortages may require investing in backup facilities or poor communication lines may require investing in a communication network, as in the case of TI-India.

The TNCs that are carrying out higher-order R&D in India are mainly those dealing with new technologies, such as microelectronics and biotechnology. However, this does not imply that the TNCs dealing with conventional technologies are not at all involved in GTU or CTU types of R&D activities in India. There are several cases of such R&D conducted in India by TNCs in collaboration with national research institutes, mostly through nonequity arrangements.[5] The proximity of new technologies to basic science permits personnel with theoretical knowledge, but with little industrial experience, to be employed directly in R&D tasks.[6] It also allows TNCs to utilize the talents in academic establishments in developing countries for their R&D requirements, either by sponsoring research or subcontracting R&D or through research collaboration.

In conventional technologies, in the initial phases of innovation, there is need for continuous interaction between design, engineering, production, marketing, and finance functions. Therefore, R&D tended to be located in the home countries of TNCs, where all these functions are also located. On the other hand, R&D in new technologies can be geographically delinked from manufacturing and marketing.[7] What this implies is that R&D in new technologies can be located in places where R&D personnel of required quality are available and there are academic centers of excellence, irrespective of the presence of TNC manufacturing or marketing operations in that country.

Moreover, in the past, especially in conventional technologies, the scale economies required for R&D were considered to be one of the main reasons for retaining substantial R&D in the home countries of TNCs. But, in the new technoeconomic paradigm the advantages of economies of scope have overcome the barriers of economies of scale. Critical mass can be conceived quite differently today, notably in terms of the size of the "system" needed to acquire the knowledge rather than the size of the firm itself (Mytelka 1993: 702).

The case studies show that the TNC R&D activities are establishing strong knowledge linkages with the local systems of innovation. Such linkages are mainly with local universities and research institutes. The linkage with domestic firms is to a large extent limited. However, in the case of ARCI, R&D activities have led to the emergence of spin-off companies in the host country. It indicates that there is diffusion of technology into the host country, leading to the introduction of new products and processes into the local economy. The case studies also show that these R&D activities are well integrated into TNC global R&D networks. The joint projects involve carrying out specific tasks in a large R&D project, based on the expertise of personnel in the Indian unit, as well as utilizing the expertise available in other R&D units of TNCs. Such a complementary exchange of knowledge helps in building up core competencies in each of the R&D units in a TNC network, including the Indian unit.

ARCI, by licensing its technologies for by-products to local scientists, is contributing to the emergence of a new class of entrepreneurs: technocratic entrepreneurs. Since they were innovation-based products, scientists could gain access to venture capital funds that enabled them to become entrepreneurs. The success of Biocon India in its R&D alliance with Quest International is also an indication of the emerging entrepreneurial zeal among scientists in India. In discussions with researchers involved with TNC R&D activities, it became apparent that more and more scientists and engineers are showing interest in becoming entrepreneurs themselves. In recent years, technocrats have established a number of small firms in India, especially in software design and engineering. The researchers felt that TNC R&D and their need for special talents are giving them an opportunity to take up challenging tasks based on their knowledge and at the same time to try their potential as entrepreneurs. Such opportunities were not available in India in the past. Several Indian scientists and engineers who are educated and residing in the industrialized countries are returning to India not only to work in TNC R&D units, but also to establish their own high-tech industries and R&D centers.

Another characteristic feature of new technologies is that their R&D is divisible into subactivities, which can later be integrated to result in final innovation. For instance, in the initial phases of establishing ARCI, the molecular biology portion of R&D was carried out at ARCI in India and the pharmacology and toxicology were carried out by Astra's R&D units in Sweden, with the final integration for product development taking place in Sweden. This divisibility of R&D functions in new technologies is what enables joint R&D projects and technology alliances, where each partner firm contributes the knowledge in which it has expertise. By implication, this also means that R&D in new technologies can be divided into "core" and "noncore or supplementary" activities. TNCs can save on costs by locating some of the noncore activities in low-cost countries and at the same time release resources in home countries for concentration on core activities. Many of the higher-order R&D units in developing countries seem to initially concentrate on joint projects and as the confidence of TNCs on the capabilities of affiliates grows, they are being entrusted with higher responsibilities or even complete R&D.

As the case studies revealed, one of the most important positive spillovers has been that the international corporate R&D activities are infusing the scientific community in India with commercial culture. The sponsorship of research or subcontracting of R&D to the academic system also contributes to diffusion of such a culture. In other words, international corporate R&D activities are fine-tuning the innovation system in developing host countries to be competitive in generating knowledge. TNCs are also encouraging the scientists in host countries to venture beyond just proving the principles and develop tangible products as a contribution to the benefit of the society. TNCs, as indicated by the ARCI and TI-India case studies, are also establishing chairs and fellowships in local universities as well as adding new research equipment in university laboratories.[8]

In the past, in a developing country like India, one of the main reasons for not reaping the benefits of its scientific capacity has been the lack of application to convert its knowledge into products. TNCs through their R&D activities are contributing to the diffusion of application skills to the researchers. In discussions with the researchers it became clear that TNCs impart training in application techniques to researchers for a few weeks before assigning them to

R&D work. This aspect is also revealed in the case study of Texas Instruments. With the mobility of researchers from one company to another, such skills get diffused through out the economy.

Some TNCs are also collaborating in establishing technology institutions for imparting education. For instance, in India, Motorola is collaborating with the Pune Institute of Advanced Technologies (PIAT) in offering a postgraduate degree in advanced telecommunication engineering with software focus. The faculty will consist of both the staff at PIAT and the experts from Motorola. While such institutes make it easy for the TNCs to recruit graduates of required specialization, they also help in introducing such a specialized subject in the host country. Similarly, TI-India's program of "university buildup" will go a long way in enhancing the capabilities of educational institutions and the pupils trained in such institutions.

One negative spillover of international corporate R&D activities in the host country has been that TNCs are able to recruit and retain the cream of the available talent, due to the higher salaries, advanced training, and other career growth opportunities offered by them. The domestic firms, on the other hand, cannot match the TNCs in these aspects, so they have to do with the rest of the talents. This, in turn, may affect the enhancement of technological capabilities in the domestic firms.

Other effects on the innovation capability of the host country include the diffusion of knowledge related to patents and other intellectual property rights (IPR). As the Biocon India case study reveals, Indian firms and scientists are realizing the importance of patenting and are acquiring the knowledge related to it with the collaboration of TNCs. With the TNC R&D investments in the country increasing, even the academic institutions realize the importance of teaching the IPR issues to the students. The Indian Institute of Science has started a course on IPR in its curriculum.

Although not quantified, the R&D activities of TNCs and liberalization of the economy also led to increase in the R&D by domestic companies.[9] Domestic companies in India have also increased their dependence on the national research institutes. For example, Bharat Electronics and the Indian Institute of Science have started a joint venture R&D to develop high-quality compound semiconductor films for device applications. This project aims to develop gallium wafers grown by a metal organic chemical vapor deposition process, and this has application potential in defense, space, and information industries.[10] Similarly, IISc and Metur Chemicals collaborated in making India self-sufficient in silicon manufacture.[11]

POLICY IMPLICATIONS

At the macro level, one of the key factors that facilitated R&D investments by TNCs in developing countries has been the liberalization of the government policies related to trade, foreign direct investments, and, particularly, IPR. This liberalized environment enabled intrafirm and intraindustry trade and opened up opportunities for these countries to be integrated into TNC global R&D, production, and marketing networks.

However, in the case of India, most of the R&D investments were made prior to its commitment to the World Trade Organization's (WTO) agreement. The reasons for this seem to be twofold: (1) The entire R&D for a product or process innovation did not take place in one location. Each location had only the portion of knowledge related to it; such information without input and collaboration from units in other locations is not of much use for imitation. (2) In some new technologies, the ability to copy and produce a product is not sufficient to be successful in business. The global marketing and distribution networks and the brand names, apart from technological edge, have become equally important for entry into global markets. And these advantages rest with the TNCs, giving the global markets an oligopolistic characteristic. The R&D operations of TNCs, such as TI-India, are very knowledge- and capital-intensive and are globally integrated. It is almost impossible for smaller companies, whether domestic or foreign, to copy the technology and develop products on their own. However, several pharmaceutical TNCs stated that they were waiting for the Indian government to amend the IPR legislation before making higher-order R&D investments in India.[12] Until such time that the Indian Parliament approves the new IPR legislation, TNCs may be hesitant to entrust to their Indian R&D affiliates the complete responsibility for the development of new products, especially in the pharmaceutical sector.

Apart from the IPR-related legislative reform, a national infrastructure for implementing the IPR regime is necessary. Such infrastructure, "according to patent law experts, include a comprehensive information system to support patenting activities, a heightened professionalism among patent examiners, modernization of India's main patent office, and increased awareness of the value of obtaining a patent" (*Science* 1996: 443). Effective enforcement of the legislation is also necessary to create confidence among domestic as well as foreign innovators about the protection of their innovations in the country.

Another policy change in developing countries that has direct bearing on location of TNC R&D activities has been the opening up of access to their national research institutes to foreign firms. Following a reduction in the government financial support, the research institutes are now motivated to earn a portion of their budgets from enterprises, including the foreign-owned. National research institutes should be given the freedom and motivation to collaborate with foreign firms and to undertake subcontract jobs from them. All over the world, academic institutions are forging closer links with industry. In some countries, academic institutions are even launching commercial ventures of their own or in collaboration with the corporate sector.

Dispersed R&D activities of TNCs require excellent communication and other infrastructural facilities. The countries seeking to attract R&D investments should establish such infrastructural facilities on a priority basis.

The location of such multinational firm's plant will depend, heavily however, on the particular regional environment. Whereas the locational choice will often depend on the availability of local skills, infrastructure and access to knowledge, the firm *itself* will also contribute to the long-term growth in the region, the availability of human resources, the access to knowledge, the local suppliers' know-how and networks. It is worth observing that these often scarce and geographically "fixed" factors constitute precisely the "externalities," increasing-return growth features of long-term development" (OECD 1991: 124).

The case studies and a mapping of the location of TNC R&D units in India show a trend of clustering in one location, Bangalore. One of the important reasons for this is the presence of an internationally reputed knowledge center, the Indian Institute of Science, in Bangalore. A study on why such a clustering of high-tech activities is taking place in Bangalore city will throw light on some important aspects that could be exploited in attracting more high-tech investments elsewhere in the country.

In order to enhance the innovation capability, industry-university collaboration, and to derive the externalities of agglomeration of high-tech activities, the establishment of science or technology parks may assume importance. Such parks may attract both foreign and domestic firms to locate R&D, if the parks are situated in proximity to reputed academic establishments. Such parks may also lead to some locational advantages such as exchange of information between the firms located in the park, interfirm technology collaborations and technology transfer, especially for by-products. To strengthen university-industry linkages, the senior managers from both domestic and foreign firms may be appointed to the management boards of universities.

The most important reason for locating international corporate R&D activities in developing countries has been the availability of trained personnel. The host countries should ensure that this supply line is not dried up. For instance, in India, already there are reports of competition for recruitment of trained personnel between the domestic and foreign firms. With the technological specializations increasing, there may be demand from enterprises to develop new specialties in the universities.

The emerging phenomenon seems to offer the developing countries some fresh opportunities. Just as international production activities benefited the newly industrialized economies, international R&D can be expected to benefit the developing host countries. Most important of all, international R&D would be an impetus to the performance of R&D by indigenous industry. If, by creating a proper investment climate, the developing host countries could persuade the TNCs to commercialize the research results in the country, the benefits would be even larger and quicker.

NOTES

1. This does not imply that substantial proportion of corporate R&D is now being conducted in developing countries. It needs to be mentioned here that about 95% of the world's industrial R&D is still being carried out within the industrialized world (Fusfeld 1995: 264). During 1978–1986, of the U.S. patents granted to the largest firms, 55.5% were attributable to R&D in the United States itself. Of those granted for R&D outside the United States, 44.8% were from Japan, 25.2% from Germany and 9.1% from the

United Kingdom, with Europe as a group accounting for 52.1% of patents granted for R&D outside the United States (Cantwell 1992: 83).

2. The organization and management of ARCI considerably changed in 1996.

3. See also Reddy and Sigurdson (1977) for an earlier version of the case study.

4. The organization and management of ARCI considerably changed in 1996.

5. For example, the Netherlands-based AKZO Chemicals subcontracted R&D related to the development of a key ingredient for refining petroleum to the National Chemical Laboratories (NCL) in India. This product, "Zeolite," developed by NCL is being used by AKZO in its operations worldwide (UNDP 1991).

6. According to Ernst and O'Connor (1989), the production systems in conventional technologies primarily require skills based on extensive "learning by doing," whereas the skills required for new technologies can be more easily acquired through formal education and training systems.

7. Wortmann (1990) also reported that R&D activities of a new technology company in Germany, are mainly related to global products and that the product manufacturing need not take place in the same country where R&D took place.

8. For example, Microsoft donated 25 PCs for a software training facility at Jiaotong College, Shanghai University, China, and Matsushita Electric provided machinery required for designing robots (*Business Week* 1994: 50).

9. See *Science* (1996).

10. See *Business Line* (1996).

11. See *Science* (1995).

12. The Indian Patent Act of 1970 recognizes only the process and not the product in the pharmaceutical sector. Pending the amendment of the Act by Parliament, the Indian government is giving a pipeline protection to the new drugs since 1995, through an ordinance.

REFERENCES

Bartlett, C. A., and S. Ghoshal. *Managing Across Borders: The Transnational Solution.* Cambridge: Harvard Business School Press, 1991.

Behrman, J. N., and W. A. Fischer. *Overseas R&D Activities of Transnational Companies.* Cambridge: Oelgeschlager, Gunn and Hain, 1980.

Bonin, B., and B. Perron. "World Product Mandates and Firms Operating in Quebec," in *Managing the Multinational Subsidiary* (H. Etemad and L. Séguin Dulude, eds.). London: Croom Helm, 1986, 161–176.

Business Line. "BE, IISc to Develop Silicon Technology" (July 12, 1996) India.

Business Week. "Special Report: Asia's High-Tech Quest: Can the Tigers Compete Worldwide?" (November 30, 1992) 65–77.

Business Week. "21st Century Capitalism, Part II: Technology & Manufacturing, The New Global Workforce, The Emerging Middle Class, and Where Are We Going?" (December 19, 1994) 20–68.

Cantwell, J. A. "The Internationalization of Technological Activity and its Implications for Competitiveness," in *Technology Management and International Business: Internationalization of R&D and Technology* (O. Granstrand, L. Håkanson and S. Sjölander, eds.). Sussex, England, U.K.: John Wiley & Sons Ltd., 1992, 75–95.

Chesnais, F. "Multinational Enterprises and the International Diffusion of Technology," in *Technical Change and Economic Theory* (G. Dosi, C. Freeman, R. Nelson, G. Silverberg, and L. Soete, eds.). London: Pinter Publishers, 1988, 496–527.

DataQuest. "The DQ Top 20," 2 (August 16–31, 1994) 41–204.

De Meyer, A., and A. Mizushima. "Global R&D Management," *R&D Management* 19, 2 (April 1989).

Ernst, D., and D. O'Connor. *Technology and Global Competition: The Challenge for Newly Industrializing Economies.* Development Center Studies, Organization for Economic Cooperation and Development (OECD), Paris, 1989.

Fusfeld, H. I. "New Global Sources of Industrial Research," *Technology In Society* 17, 3 (1995) 263–277.

Hymer, S. *International Operations of National Firms: A Study of Direct Foreign Investment.* Cambridge: MIT Press, 1976.

Levitt, T. "The Globalization of Markets," *Harvard Business Review* (May/June 1983) 92–102.

Mansfield, E. "Technology and Technical Changes," in *Economic Analysis and the Multinational Enterprise* (J. H. Dunning, ed.). London: George Allen and Unwin, 1974, 147–183.

Mansfield, E., D. Teece, and A. Romeo. "Overseas Research and Development by U.S.-Based Firms," *Economica* 46 (1979) 187–196.

Mytelka, L. K. "Rethinking Development: A Role for Innovation in the Other "Two Thirds," *Futures* 25, 6 (1993) 694–712.

OECD. *Science and Technology Policy Outlook 1988.* Paris: OECD, 1988.

OECD. *Technology in a Changing World.* Paris: OECD, 1991.

OECD. *Technology and the Economy: The Key Relationships.* Paris: OECD, 1992.

Pearce, R. D. *The Internationalization of Research and Development by Multinational Enterprises.* London: Macmillan, 1989.

Porter, M. E. "Introduction and Summary," and "Competition in Global Industries: A Conceptual Framework," in *Competition in Global Industries* (M. E. Porter, ed.). Boston: Harvard Business School Press, 1986, 1–11 and 15–60.

Poynter, T. A., and A. M. Rugman. "World Product Mandates: How Will Multinationals Respond?" *Business Quarterly* 47, 3 (1982) 54–61.

Ramachandran, J. "Strongly Goal-Oriented Biomedical Research: Astra Research Center India," *Current Science* 60, 9/10 (May 25, 1991) 533–536.

Reddy, A.S.P. "Emerging Patterns of Internationalization of Corporate R&D: Opportunities for Developing Countries?" in *New Technologies and Global Restructuring: The Third World at a Crossroads* (C. Brundenius and B. Göransson, eds.). London: Taylor Graham, 1993, 78–101.

Reddy, A.S.P., and J. Sigurdson. "Emerging Patterns of Globalization of Corporate R&D and Scope for Innovation Capability Building in Developing Countries," *Science and Public Policy* 21, 5 (October 1994) 283–294.

Reddy, P. "New Trends in Globalization of Corporate R&D and Implications for Innovation Capability in Host Countries: A Survey from India," *World Development* 25, 11 (1997) 1821–1837.

Reddy, P., and J. Sigurdson. "Strategic Location of R&D and Emerging Patterns of Globalization: The Case of Astra Research Center India," *International Journal of Technology Management* 14, 2/3/4 (1997) 344–361.

Ronstadt, R. *Research and Development Abroad by U.S. Multinationals.* New York: Praeger Publishers, 1977.

Sampath, K. "Astra Research Center India: A Unique Experiment in Strategic Collaboration," *BioSpectra* (November/December 1990) 13–16.

Science. "Joint Research: India Cracks Whip to End Addiction to State Funds," 267 (March 10, 1995) 1419–1420.

Science. "India: Industrial R&D Gets Boost Despite Lack of Patent Reform," 271 (January 26, 1996) 442–443.

Terpstra, V. "International Product Policy: The Role of Foreign R&D," *Columbia Journal of World Business* 12 (1977) 24–32.

UNDP. "Looking South for Answers: The Developing World Is Offering Solutions to Many Global Problems," *UNDP World Development* 4, 5 (September 1991).

U.S. Tariff Commission. *Implications of Multinational Firms for World Trade and Investment and for U.S. Trade and Labor.* Washington, DC: Government Printing Office, 1973.

Vernon, R. "International Investment and International Trade in the Product Cycle," *Quarterly Journal of Economics* 88 (1966) 190–207.

Westney, D. E. "International and External Linkages in the MNC: The Case of R&D Subsidiaries in Japan." Working Paper Y # 1973-88, Sloan School of Management, MIT, 1988.

Wortmann, M. "Multinationals and the Internationalization of R&D: New Developments in German Companies," *Research Policy* 19 (1990).

16

Expenditure, Outcomes, and the Nature of Innovation in Italy

Daniele Archibugi, Rinaldo Evangelista,
Giulio Perani, and Fabio Rapiti

INTRODUCTION

Although technological change is one of the main determinants of long-term economic development, our knowledge of some of its most crucial aspects is still incomplete, and an exhaustive quantification of all key dimensions of innovation activities is still lacking. The following questions are among those which have not yet been fully answered:

1. How many firms innovate? We can assume that, in a competitive economic system, all firms are forced to innovate or to perish in the long run. However, the proportion of firms introducing innovations in each time period, and how this share changes over time, have not yet been quantified. Some firms might be consistent innovators, especially in industries characterized by high technological opportunities, while in other industries the frequency of innovation might be much lower (see Malerba and Orsenigo 1995; Geroski et al. 1996).
2. What is the amount of resources devoted to innovation? It is now widely acknowledged that firms innovate through a variety of sources and that these are industry-specific (Pavitt 1984; von Hippel 1988; Archibugi et al. 1991; Evangelista 1996). However, a quantification of the different inputs devoted by firms to nurture their innovative projects is still needed.

3.   What is the amount and significance of innovated production? Not even the most innovating companies will entirely replace their old products and processes with new ones. It is therefore necessary to quantify the share of new products and processes and to assess their degree of novelty.

Since the earliest of Schumpeter's suggestive analyses (1934, 1942), these issues have been addressed in a great deal of research. The debate is, nonetheless, still open, due to the lack of suitable measuring instruments. The empirical evidence we present in this chapter is a contribution to the debate on how widespread innovation is, on the resources which it absorbs, and on its economic significance. It is based on a new data source, namely a survey carried out in 1993 by the Italian National Institute for Statistics (ISTAT) in collaboration with the Institute for Studies on Research and Scientific Documentation (ISRDS) of the National Research Council (CNR) (hereafter referred to as the Italian Survey) and promoted and coordinated by the European Commission and Eurostat under the Community Innovation Survey venture.

The chapter is organized as follows. The next section examines the methodological aspects of the measurement of innovative activity, and compares two different methods of gathering direct information on industrial innovation. Next, data are presented on the number of firms which have introduced innovations. Only a third of the firms declared that they had introduced innovations in the three-year period 1990–1992, even though 62% of employees and 71% of the turnover of the Italian manufacturing industry have been involved in innovation. Then the breakdown of firms' expenditure on innovation is exported. This is a newly available quantitative indicator of the inputs devoted to innovation. The breakdown of expenditure on innovation by firm size and industrial sector is then analyzed. Whereas small innovating firms manage to keep in step with their larger competitors, the survey has revealed the existence of a vast and often ignored number of noninnovating small firms. Large differences between industrial sectors *vis à vis* the sources of innovation used by firms are also confirmed. Next, the output of innovation is examined by looking at the quantity and quality of new and improved products. Less than half of the turnover of innovating firms was innovated in the three-year period considered. Finally, an outline of policies for industrial innovation is drawn.[1]

MEASURING INNOVATIVE ACTIVITIES

Satisfactory analyses are possible only with the help of appropriate methodologies and measuring instruments. Compared to other economic variables (such as production, value added, investment, exports, and employment), innovation variables are much harder to measure. The difficulties have to do first of all with the very nature of the phenomenon of innovation, which is characterized by a high degree of heterogeneity. In particular, four aspects hamper the measurement of technology and innovative activities:

1. Technological knowledge may be formal or tacit. Only a portion of such knowledge may be written down in books, manuals, patents, and designs, while another part remains tacit.
2. Sources of innovative activity may be internal or external to firms. In the majority of cases, the innovations introduced by firms are based upon both types of sources.
3. While some innovative activities may be easily identifiable in economic terms, through prices and costs, other technological activities occur outside the sphere of market transactions.
4. Technological change consists both of identifiable tangible activities—for example, new machinery and equipment—and intangible activities, which include the generation of new ideas, inventions, and innovations.

In addressing these problems, economists, sociologists, and statisticians have tried to produce indicators capable of describing and predicting reality, but none of them is totally satisfactory. Nonetheless, if they are used properly, they may provide helpful indications for both analysis and economic policy choices.

The intellectual framework of the new measurement tools developed over the last decade is defined by the notion that "the linear model of innovation is dead" (Rosenberg 1994: 139). The onus is now placed on the fact that innovative activity is an interactive process in which the different phases and sources of technological change are interdependent and not hierarchically structured. Thus, whereas in the past a great deal of attention was attached to R&D activities, regarded as the main source of innovations, recently the focus has shifted to the role played by other complementary sources. Hence the development of surveys to measure innovative activity directly. Public research centers, statistical offices, international organizations—including the Organization for Economic Cooperation and Development and the European Commission—and numerous university centers have attempted to supplement statistical information already available with a new indicator based on direct surveys of "the innovative phenomenon." Albeit performed with different methodologies, surveys have followed two main approaches (see Archibugi 1988; Hansen 1992):

- collecting information on the innovations introduced, hence concentrating on the *objects* of innovative activity;
- questioning firms about input, output, and the nature of the innovative process, hence focusing analysis on the *subjects* of innovative activity.

The first group comprises surveys conducted by the University of Sussex Science Policy Research Unit (SPRU) in Great Britain on a set of innovations (Townsend et al. 1981; Pavitt 1984; Pavitt et al. 1987), by the Small Business Administration in the United States (Acs and Audretsch 1990), and those on new products advertised in specialist magazines and publications (Kleinknecht and Bain 1993; Santarelli and Piergiovanni 1996; Coombs et al. 1996). The second group encompasses the surveys carried out by IFO in Germany (Scholz

1992), certain Dutch surveys (Kleinknecht and Reijnen 1991), and the IST AT/ISRDS-CNR surveys (Archibugi et al. 1991; Cesaratto et al. 1991).

Figure 16.1 outlines the distinctive features of the two approaches.[2] Given the relative novelty of this measuring instrument, it is not surprising that the methodologies adopted are still heterogeneous. The diversity of the various surveys has enhanced our knowledge of the advantages and limits of the various approaches. It still has not been possible, however, to obtain temporally and internationally comparative statistical data. It has so far proved difficult to establish how far the differences which have emerged are due to specific methodological factors and how far they are due to the diversity of countries, sectors, and periods covered.

The explanatory value of statistical data increases when comparisons between different populations are possible. This is why in recent years great efforts have been made to harmonize surveys on innovation at the international level. The OECD, for example, supplemented its family of manuals on technological indicators[3] with the Oslo Manual (OECD 1992a) on the methodologies and contents of direct surveys on the innovative activity of firms. Later in 1992, through the joint action of EUROSTAT and Sprint, and in collaboration with national authorities in 13 countries, the European Commission launched the Community Innovation Survey. This was the first-ever survey on innovation to be carried out simultaneously in so many countries on the basis of a common questionnaire. Unfortunately the results for each country are only partially comparable, as a result of modifications to the text of the questionnaire introduced by some of the national contracting parties and inadequate harmonization of the statistical methodologies adopted (cf. Archibugi et al. 1995; Evangelista et al. 1996).

It is nonetheless encouraging to note that various countries outside the European Union—including the United States, Canada, Australia, Hungary, and China—have also undertaken similar surveys. The OECD and the European Commission have hosted international conferences where the methodology, results, policy implications, and perspectives of these new innovation indicators have been explored.[4]

Italy has played a leading role in promoting firm-level survey methodologies that are now enjoying international success. Together with others, the first innovation survey conducted in 1986 provided some guidelines to define a harmonized survey in most industrialized countries. Now, more than ten years later, a great deal of progress has been made. The main priority now is to capitalize on the lessons being learned from the experience gained (and mistakes made) to arrive at long last at a permanent standardization of the methodology for direct surveying of innovative activity.

Figure 16.1
Indicators of Technological Activity

Development	• Internationally comparable	• Underestimates innovation in services • Monetary adjustments required for international comparability
Patenting	• Regular data collection • Detailed breakdown for technological fields • Internationally comparable • Availability of film-level data	• Not all inventions are patented • Not all inventions are patentable • Does not inform on services • Differences in the propensity to patent across sectors • National systems are biased by domestic inventors
Balance of Payments for Technology	• Regular data collection • Detailed breakdown for technological fields • Internationally comparable	• Does not inform on nontransferred technology • Measures only a small part of technological activities • Data biased by financial transaction
Trade of High-Tech Products	• Regular data collection • Direct measure of economic performance • Internationally comparable	• Does not consider innovation in traditional sectors • Does not inform on domestic innovations • Problems of selecting pertinent products
Bibliometrics	• Detailed breakdown for scientific disciplines • Internationally comparable • Direct measure of scientific output	• Databases include a subset of world publications • Differences in the propensity to publish across disciplines • Language barriers
Innovation Surveys	• Direct measure of innovativeness • Potentially includes all activities related to innovation • Applicable to manufacturing and services	• Problems in comparability over time and across countries • Lack of periodicity in data collection • Problems of sample definition • Data often biased by subjective judgements

Source: Archibugi et al. 1995

The following sections of this chapter will assess some of the results emerging from the survey on innovation in the Italian manufacturing industry conducted within the framework of the Community Innovation Survey. The Italian Survey has involved a larger number of firms than in any other European country and accounts for as many as 40% of the total returned questionnaires (cf. Archibugi et al. 1995a; Evangelista et al. 1996).

HOW MANY INNOVATING FIRMS ARE THERE?

One of the first aims of surveys on innovation is to establish how widespread the innovative phenomenon is within the industrial structure. In a dynamic and long-term perspective, all firms are bound to innovate. If they fail to do so, they drop out of the market. Only a very few craft-based firms manage to survive and preserve their original characteristics; however, they are the exception which confirm the rule. Although it is reasonable to expect all firms to innovate, it is still necessary to establish how frequently and intensely they do so.

Table 16.1
Innovating and Non-Innovating Firms by Size

Classes of Employees	Total firms	% Innovating firms of total firms	% Employees of innovating firms of total firms	% Sales of innovating firms of total firms
20-49	15,109	25.9	27.5	29.1
50-99	4,142	40.8	41.6	43.0
100-199	2,012	48.0	48.7	47.8
200-499	1,041	58.5	59.8	67.3
500-999	292	74.0	74.5	79.1
1,000 and over	191	84.3	91.5	95.9
Total	22,787	33.1	61.5	70.7

Source: Elaboration on the Survey on Innovation (1990–92) carried out by the Italian National Statistical Institute and the Institute for Studies on Scientific Research and Documentation. See ISTAT 1995.

Table 16.1 shows that only one-third of the firms involved in the survey introduced innovation during the period 1990–1992.[5] The innovative phenomenon does involve a much larger portion of the Italian industrial structure, with 62% of employees and 71% of turnover in the manufacturing industry covered by the survey being concentrated in innovating firms. There are, however, very significant differences in the percentage of innovating firms across different size classes. Only a low percentage of small firms innovated (26%), while almost 84.3% of firms with over 1,000 employees have introduced innovations. Table 16.2 presents the same information broken down by industries. Data are reported according to the percentage of innovating firms as well as for their economic significance (measured in terms of employees and turnover). In some cases, the ranking of industrial sectors (shown in parentheses) remains unchanged. Aerospace, office machinery, radio, television, and telecommunications, for example, emerge as the most innovative industries on the basis of all three indicators. But in other industries with high industrial concentration, such as motor vehicles, the number of innovating firms is substantially lower (44.7%), despite the fact that 92% of the turnover of the sector is concentrated within them.

Table 16.2
Innovating and Noninnovating Firms by Industry

Industrial Sectors	Total firms	% Innovating firms of total firms		% Employees of innovating firms of total firms		% Sales of innovating firms of total firms	
Office machinery	48	64.6	(2)	94.4	(2)	97.6	(2)
Electrical machinery	989	38.7	(12)	67.7	(11)	72.5	(11)
Radio, TV, telecom.	249	59.8	(3)	91.9	(3)	93.4	(4)
Aerospace	31	67.7	(1)	99.0	(1)	99.5	(1)
Chemical (excl. pharmac.)	561	45.6	(7)	78.8	(7)	80.9	(6)
Pharmaceuticals	198	56.1	(4)	78.6	(8)	80.6	(8)
Synthetic fibers	31	41.9	(9)	82.9	(5)	80.7	(7)
Mechanical machinery	2,713	48.9	(6)	70.5	(10)	75.4	(10)
Precision instruments	435	50.3	(5)	65.8	(12)	67.7	(12)
Motor vehicles	445	44.7	(8)	91.5	(4)	92.0	(5)
Other transport	272	32.7	(17)	75.1	(9)	75.5	(9)
Rubber and plastics	866	41.8	(10)	63.6	(13)	65.1	(14)
Metals	643	37.9	(15)	60.8	(14)	65.2	(13)
Printing &publishing	732	38.3	(13)	54.7	(16)	53.7	(16)
Paper	496	38.3	(14)	58.5	(15)	63.5	(15)
Food, drink, tobacco	1,501	31.2	(18)	53.0	(17)	51.5	(17)
Textiles	2,008	28.1	(21)	38.0	(20)	41.9	(20)
Clothing	1,991	11.3	(24)	17.8	(24)	18.0	(24)
Leather and footwear	1,486	18.8	(23)	24.6	(23)	27.9	(22)
Wood	622	28.8	(20)	36.6	(22)	26.1	(23)
Metal products	2,874	33.4	(16)	42.4	(19)	45.6	(19)
Mineral & nonmineral prod.	1,486	29.7	(19)	47.1	(18)	49.1	(18)
Other manufacturing	1,679	26.0	(22)	36.7	(21)	40.2	(21)
Oil	89	39.3	(11)	81.2	(6)	97.5	(3)
TOTAL	22,445	33.3		60.0		69.3	

Note: Ranking in parentheses.
Source: Elaboration on the Survey on Innovation (1990–1992) carried out by the Italian National Statistical Institute and the Institute for Studies on Scientific Research and Documentation. See ISTAT 1995.

SOURCES OF INNOVATION

The multiform nature of innovative activities and their sectoral specificity have been underlined in a vast amount of literature (Pavitt 1984; Kline and Rosenberg 1986; von Hippel 1988; Archibugi et al. 1991), which has shown the existence of a multiplicity of interdependent sources of innovation. Besides activities generating new technological knowledge, special attention has also been attached to processes of technology adoption and diffusion (both embodied and disembodied), an acknowledged *sine qua non* for technology to express its economic effects to the fullest (OECD 1992b and 1995; Antonelli 1994; Evangelista 1996).

The relative importance of the various sources should also be measured using a common yardstick. The Community Innovation Survey attempted to do so according to an input measure such as innovation expenditure—that is, the breakdown of innovation-related expenditure by items. Table 16.3 shows the breakdown of expenditure incurred to introduce innovations by firm size. The picture which emerges from the table is very clear-cut. Innovative industrial processes consist, first and foremost, of the purchase and use of embodied technologies (innovative machinery and plants), which accounts for 47.2% of total expenditure on innovation, and second, of efforts to generate and develop new knowledge inside firms, as measured by the percentage of innovative spending on R&D activities (35.8%). Other components play a relatively minor role: expenses incurred for design and trial production each account for 7.4% of total expenditure on innovation, while just 1.2% and 1.5% of the latter are allocated, respectively, to the purchase of patents and licenses and to marketing activities related to the introduction of technological innovations.

Table 16.3
Breakdown of Innovation Expenditure by Firm Size (% Values)

Classes of Employees	R&D	Patents and licenses	Design	Tooling-up & trial production	Market-ing	Invest-ment
20-49	14.9	1.5	9.4	7.7	1.9	64.6
50-99	16.3	1.3	8.4	8.5	1.7	63.8
100-199	19.8	1.7	12.8	9.0	2.2	54.5
200-499	27.6	2.2	9.1	9.6	2.2	49.3
500-999	26.0	1.6	13.4	8.1	1.3	49.6
1,000 & over	46.7	0.8	4.8	5.7	1.2	40.8
TOTAL	35.8	1.2	7.4	6.9	1.5	47.2

Note: Rows add up to 100%.
Source: Elaboration on the Survey on Innovation (1990–1992) carried out by the Italian National Statistical Institute and the Institute for Studies on Scientific Research and Documentation. See ISTAT 1995.

The distribution of innovation expenditure shown in Table 16.3 not only reflects the distinctive profile of the Italian manufacturing industry—with its accentuated specialization in medium- and low-technology sectors (Archibugi and Pianta 1992)—but also points to some more general conceptual and methodological issues:

1. R&D activities emerge as a central component of the technological activities of firms and as the most important intangible innovation expenditure. Nonetheless, they account for just over a third of the total expenditure of innovating firms.
2. The largest part of firms' innovation financial efforts is linked to the adoption and diffusion of technologies embodied in capital goods.
3. Expenditure wise, the acquisition of disembodied technology through patents and licenses emerges as a secondary innovation component when compared to other technological sources.
4. Other innovative activities, such as design expenses, play a secondary role with respect to the total expenditure sustained by manufacturing firms for introducing technological innovations.

The importance of the different sources of innovation in business strategies is, however, strongly influenced by firm size, especially as far as R&D expenditure and investment are concerned. Small firms have a high propensity to innovate by acquiring machinery and plants, against the greater propensity of large firms to generate new technologies internally. For firms with fewer than 50 employees, R&D activities account for 15% of total innovation expenditure against a percentage close to 47% in the case of firms with more than 1,000 employees. Data on investment show an opposite pattern. Innovative investments of firms with less than 200 employees account for 50% or more of total innovation expenditure. The other components of innovation also do not appear systematically correlated to firm size. All that emerges is the greater importance of design activities in intermediate size categories, with percentages over 10% of total innovation expenditure in firms with 100–200 and 500–1,000 employees (see Table 16.3).

THE DETERMINANTS OF INNOVATION

The Role of Small and Large Firms in Innovation

The relationship between innovation and firm size has been dealt with over the last two decades by a vast amount of empirical literature. Two models of industrial and technological development have often been contrasted: on the one hand, the model based on large firms, characterized by radical innovative processes centered on R&D activities; on the other, the model of industrial organization based on small firms, characterized by informal innovative activities but technologically "creative" nonetheless (for an overview, cf. Cohen and Levin 1989; Cohen 1995).

On empirical grounds, the analysis of the role small and medium-sized firms play in technological change may be approached in three different ways:

1. by comparing the innovation intensity of large and small firms, considering innovating firms only;
2. by considering the relative contribution of large and small firms to the overall innovation performances of a given economic system; or
3. by considering the innovation intensity of large and small firms, including both innovating and noninnovating firms.

Cohen stressed that most analyses had followed the first or second methodological approach and neglected noninnovative firms:

The most notable feature of this considerable body of empirical research on the relationship between firm size and innovation is its inconclusiveness.... First, most of the samples used in regressions studies are non-random, and with few exceptions...no attempt has been made to study the presence or the effects of sample selection bias. Many of the early firm-level studies confined their attention to the 500–1,000 largest firms in the manufacturing sector, and firms that reported no R&D were typically excluded from the sample (Cohen 1995: 187).

This implies that the relationship between innovation and firm size was mainly studied using biased samples, and samples where small firms were generally atypical. The Italian Survey allows us instead to analyze the relationship between innovation and firm size on the basis of all three methodologies outlined above. The results are shown in Table 16.4. The first two columns show data on firms' expenditure per employee, taking into account total innovation and R&D expenditure, respectively, for the sample of innovating firms only. Column 2 confirms that large firms are much more R&D-intensive than small ones (see, for example, Soete 1979); this is hardly surprising since R&D is an innovative source which requires a minimum threshold, and it does not "capture" the innovative effort typical of small firms. However, when a much more comprehensive indicator such as total innovation expenditure is considered (column 1 of Table 16.4), it emerges that innovative small firms are not substantially disadvantaged compared to their larger competitors. In fact, the data show a U-shaped curve: the innovation intensity of firms with fewer than 100 employees is higher than that of firms in the intermediate size groups, despite the fact that it is still lower than that of firms with more than 1,000 employees.[6] This result is totally consistent with the analyses of Pavitt et al. (1987) and Acs and Audretsch (1990), which take into account the totality of innovating firms without considering firms which do not introduce innovations.

Table 16.4
Intensity and Concentration of Innovative Activities by Firm Size

Classes of Employees	Innovating Firms		Total Firms		Concentration of technological activities and sales (innovating firms-% values)		
	Innovation expend. per employee	R&D expend. per employee	Innovation expend. per employee	R&D expend. per employee	Innovation expend.	R&D expend.	Sales
20-49	14.7	2.2	4.0	0.6	8.1	3.4	15.0
50-99	12.3	2.0	5.1	0.8	6.8	3.1	9.6
100-199	11.7	2.3	5.7	1.1	7.3	4.0	10.3
200-499	11.8	3.3	7.1	1.9	10.2	7.9	12.9
500-999	16.4	4.3	12.2	3.2	11.2	8.1	8.2
1,000 & over	18.3	8.5	16.7	7.8	56.4	73.6	43.9
TOTAL	15.7	5.6	9.7	3.5	100.0	100.0	100.0

Note: Data are expressed in 1992 millions of Italian lire (see Appendix).
Source: Elaboration on the Survey on Innovation (1990–1992) carried out by the Italian National Statistical Institute and the Institute for Studies on Scientific Research and Documentation. See ISTAT 1995

However, the indicators shown in the first two columns of Table 16.4 fail to take into account that the number of innovating firms differs considerably between small and large firms. This aspect is crucial when the overall relationship between innovation and firm size is to be assessed. As shown in Table 16.1, almost 84.3% of larger firms are innovative, whereas the percentage drops to 26% for smaller firms.

Columns 3 and 4 of Table 16.4 present, respectively, the average values per employee of innovation expenditure and R&D expenditure in all firms participating in the survey.[7] The positive relationship between innovation intensity (in a broad sense) and firm size strongly reemerges. The average innovation expenditure per employee of firms with over 1,000 employees is 16.7 million lire, while for firms with 20–49 employees it is only 4 million lire. This difference has to do with the fact that, although small innovating firms are no less innovative than large firms, they are not representative of the overall productivity of small firms.

The last three columns of Table 16.4 enable us to assess the effective economic and technological weight of large and small enterprises in the Italian manufacturing industry. The picture which emerges is one of a high level of concentration of technological activities. The 161 firms with over 1,000 employees account for 56% of total innovation expenditure and 74% of R&D expenditure. The role of the 5,602 innovating firms with fewer than 100 employees appears rather limited. These firms account for only 15% of total innovation expenditure and just 6.5% of R&D expenditure. The technological weight of small firms is thus much lower than their economic weight in terms of turnover (25%).

Intersectoral Differences in Innovation Intensity

Table 16.5 shows the average innovation expenditure per employee in the main industrial sectors, together with a breakdown of total innovation expenditure according to the various sources of technology. Sectors are ranked according to their average innovation intensity. The top industries are office machinery; oil products; motor vehicles; pharmaceuticals; radio, television and telecommunications; metals; and aerospace. The industries which dedicated fewest resources per employee to innovative activities are clothing; food, drink, tobacco; and rubber and plastics.

The distribution of innovation expenditure shows how innovation patterns differ not only across firm size classes but also across industries. The table allows us to identify industries traditionally defined as science-based, in which activities aimed at generating new technological knowledge play a fundamental role. Among the industries which allocate over 50% of total innovation expenditure to R&D are office machinery and computers (64.8%), radio, television, and telecommunications (66.1%), pharmaceuticals (66.7%) and precision instruments (54%). Compared to other science-based sectors, aerospace shows a conspicuously lower percentage of R&D expenditure (39.3%), although it does allocate sizable portions of its total innovation expenditure to trial production (22.7%) and design (12%). On the other

Table 16.5

Innovative Intensity and Breakdown of Innovation Expenditure by Industry

Industrial Sectors	Innovation expend. per employee*	Breakdown of innovation expenditure (% values)					
		R&D	Patents & licenses	Design	Tooling-up & trial prod.	Marketing	Investment
Office machinery	36.8	64.8	0.1	4.5	17.2	1.0	12.4
Oil	30.4	6.8	1.0	8.5	4.3	0.2	79.2
Motor vehicles	30.2	36.7	0.2	1.9	3.2	0.2	57.8
Pharmaceuticals	28.9	66.7	4.5	4.3	4.4	2.1	18.0
Radio, TV, telecom.	28.0	66.1	0.8	12.9	5.3	1.2	13.7
Metals	27.5	8.1	0.4	6.7	4.6	0.3	79.9
Aerospace	23.9	39.3	5.0	12.1	22.7	0.7	20.2
Precision instruments	14.5	54.0	1.4	12.4	8.9	2.0	21.3
Chemicals (excl. pharmac.)	14.4	42.9	1.1	3.6	3.9	4.5	43.9
Paper	12.4	7.2	0.4	4.7	4.6	1.4	81.7
Printing and publishing	11.6	7.9	3.7	6.2	4.0	1.0	77.2
Nonmineral products	11.4	12.8	1.4	8.0	8.6	1.9	67.3
Electrical Machinery	10.7	30.4	1.1	14.6	8.8	2.1	43.0
Other transport	10.3	21.8	0.4	20.7	7.3	5.7	44.2
Wood	10.0	9.7	0.7	3.7	6.5	1.3	78.1
Leather and footwear	10.0	15.5	1.3	8.4	8.6	2.8	63.4
Mechanical machinery	9.9	36.0	1.9	15.0	11.8	2.2	33.1
Other manufacturing	9.9	20.8	0.9	6.5	7.9	2.8	61.1
Metal products	9.8	12.3	1.4	8.8	8.1	1.9	67.5
Synthetic fibers	9.3	30.6	2.4	2.0	7.1	2.3	55.6
Textiles	9.2	12.2	0.5	8.9	8.8	1.9	67.7
Rubber and plastics	8.8	19.8	1.4	9.4	10.1	2.0	57.3
Food, drink, tobacco	8.7	17.5	0.8	6.4	5.8	2.8	66.7
Clothing	7.9	16.5	1.3	25.1	11.1	18.9	27.1
TOTAL	16.5	35.8	1.2	7.4	6.9	1.5	47.2

Source: Elaboration on the Survey on Innovation (1990–1992) carried out by the Italian National Statistical Institute and the Institute for Studies on Scientific Research and Documentation. See ISTAT 1995; *1992 millions of Italian lire (see Appendix).

hand, the acquisition of new machinery and plants is an important source of technology for most traditional consumer good sectors, such as wood (78.1%), textiles (67.7%), leather and footwear (63.4%), food (66.7%), metal products (67.5%), and capital-intensive sectors such as printing and publishing (77.2%), paper (81.7%), metals (79.9%), rubber and plastics (57.3%), and motor vehicles (57.8%).

The importance of design is much higher than the manufacturing average not only in aerospace, but also in the sectors which produce specialized machinery, such as electrical and mechanical machinery and precision instruments. A noteworthy portion of innovation expenditure is allocated to design in clothing.

It is important to note that the ranking of industries shown on the basis of R&D/sales or value added ratios (not reported in the table) does not substantially change if it is measured using a more comprehensive innovation indicator such as total innovation expenditure per employee. Science-based industries, along with a limited group of capital-intensive sectors such as motor vehicles, are leading technological advances. Also some specialized sectors in machinery and electronic components reveal an average innovative intensity. All the more traditional consumer goods sectors concentrate their technological activities into the acquisition of innovations embodied in capital goods, though showing a low innovative intensity.

In synthesis, despite the multiplicity of innovative patterns, a few R&D-intensive and scale-intensive industries emerge as the driving force of technological change. Figure 16.2 shows that R&D-intensive industries and those with high economies of scale account for 71% of the total innovation expenditure of the Italian manufacturing industry. The distribution of innovative activities by industrial sector and firm size thus reveals that even in a country like Italy in which traditional sectors and small firms play an important role, innovation is highly concentrated within a few industries and a restricted club of large firms.

Industrial Sectors and Firm Size as Joint Determinants of Innovation

The Italian Survey confirms that industrial sector and firm size prove to be important factors in determining both the *nature* of the innovative activities performed by firms and their *intensity*. However, the analysis carried out in the previous section did not allow us to isolate the relevance of the two effects (firm size and industrial sectors)—that is, using the usual *ceteris paribus* formula. In other words, a descriptive analysis showing the presence of wide innovation differences across industries and firm size classes is not able to establish whether firm size continues to be a key determinant of the resources devoted to innovation irrespective of the industrial sector considered or, conversely, whether the latter retains its importance irrespective of the size factor.[8]

To check whether any sectoral and size "composition effects" exist, we have estimated a dichotomous qualitative-type model in which the mere presence or absence of the innovative phenomenon is considered as an independent variable. The use of a qualitative model allows us to include in the regression not only innovative but also noninnovative firms.

Figure 16.2
Distribution of Total Innovation Costs Across Pavitt's Sectoral Groups

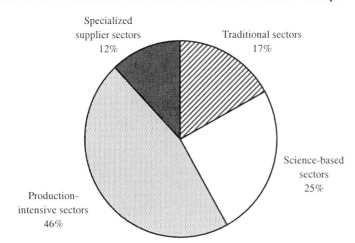

As regressors, we used firm size (measured by the logarithm of the number of employees) and the industrial sector firms belonged to (expressed by 24 sectoral dummy variables). Another two variables were included which are believed to affect the presence or absence of innovative activities (irrespective of sectoral and size determinants). These are the geographical location of the firm (identified by five regional dummies referring to Northwest, Northeast, Center, South, and Islands) and whether the firm belongs to an industrial group.

The estimated model is:

$$\overset{(+)\,(+)}{\text{Pr(inn.)} = \text{f (size, sector, area, group)}}$$

$$0 < \text{Pr(inn.)} < 1$$

where Pr (inn.) is the probability of a firm's being innovative. The expected signs of the coefficients are shown in parentheses. For industrial sectors, we expect an order of probability to innovate consistent with the ranking of sectors observed in Table 16.2 (where an innovative intensity indicator was used). For geographical areas, we expect the probability to innovate to decrease from North to South. We also expect belonging to an industrial group to be positively correlated to innovative activity.

The model has been tested using a normal function (probit) and a logistic function (logit), obtaining virtually identical results in both cases.[9] The results of the regressions are shown in Table 16.6. Estimates refer to the total of the firms included in the survey. The dependent variable [Pr(inn.)] is equal to 1, if the firm has performed some form of innovative activity, or to 0 if it has not.

Table 16.6
Probability of Carrying Out Innovation Activities (Logit and Probit estimates)

	Logit model	Probit model
Number of observations	22.428	22.428
Concordant	70.0%	70.0%
Discordant	29.7%	29.7%
-2LogL	2.774	2.776
Score	2.641	2.641
Intercept	-3,972	-2,398
Belonging to an industrial group	0.166	0.104
Not belonging to an industrial group	reference	reference
Log of number of employees	0.578	0.351
Northwest	0.771	0.447
Northeast	0.454	0.264
Center	0.208	0.119
South	0.067*	0.041*
Islands	reference	reference
Office machinery	1.486	0.886
Aerospace	1.276	0.796
Radio, TV, telecom.	1.212	0.745
Precision instruments	0.962	0.586
Mechanical machinery	0.875	0.535
Pharmaceuticals	0.71	0.437
Rubber and plastics	0.648	0.392
Chemicals (excl. pharmac.)	0.584	0.356
Motor vehicles	0.579	0.353
Printing	0.558	0.338
Paper	0.467	0.282
Electrical machinery	0.435	0.26
Metal products	0.411	0.247
Oil	0.324**	0.193**
Metal	0.296	0.18
Wood	0.203*	0.123
Other transport	0.170**	0.098**
Food, drink, tobacco	0.160*	0.101*
Mineral and nonmineral prods.	0.116**	0.069**
Clothing	-0.913	-0.503
Leather and footwear	-0.319	-0.183
Synthetic fibers	-0.060**	-0.020**
Textiles	-0.015**	0.009**
Other manufacturing	reference	reference

* Significant only at 90% level; ** Not significant at 90% level
All others significant at least at 95% level

Tests confirm that the model has an acceptable capacity to interpret the phenomenon. In all specifications, about 70% of firm behavior is predicted by the model.[10] The intercept represents a transformation of the probability[11] of the firm's being innovative. The value of the intercept refers to the firm with the reference characteristics: namely, nonmembership of a group, location in the Islands, and belonging to the "other manufacturing industries" sector. As the dummy coefficients increase, so the probability of the firm's introducing innovations also increases.[12]

In synthesis, from the regression estimates it emerges that the probability of a firm's being innovative:

- increases monotonically with firm size;
- increases considerably for some sectors with high technological opportunity;
- increases if the firm is a member of an industrial group;
- is much higher among firms in the Northwest and lower in other areas, with a minimum level in the South and Islands.

Although the model takes as a dependent variable a qualitative-type indicator of innovation (limited to the presence/absence of the phenomenon), results seem to confirm that the different factors identified in the previous sections as important for explaining firms' innovative activities remain significant, irrespective of one another.

A considerable variability emerges for the sector coefficients. The ranking of manufacturing sectors according to their probability of being innovative is very similar to the ranking that emerges on the basis of intensity indicators presented in Table 16.4. Office machinery ranks first, followed a long way behind by aerospace; radio, television, and telecommunications; precision instruments; mechanical machinery; and pharmaceuticals. At the bottom end of the table, the ranking does not change: we find clothing, leather and footwear, synthetic fibers; textiles; and other manufacturing industries. In conclusion, it is possible to say that the size factor retains its significance even within the various sectors.

OUTPUT GENERATED BY INNOVATION

We have so far considered the inputs devoted by firms to innovation. This is only one way to measure technological change in industry (the consistency of a variety of innovation indicators is explored in Hollenstein 1996). The intensity of innovation can also be measured according to output indicators. The Community Innovation Survey offers a significant indicator of innovation output, namely the part of total firm sales due to innovation. This does not measure the economic impact of innovation (such as the productivity-based indices reviewed in Griliches 1995), but rather provides direct information on how a firm, an industry, or even an economic system as a whole has changed its production output in relation to innovation.[13]

Table 16.7 shows the distribution of sales according to the nature of the innovations introduced, broken down by industries. It should be borne in mind that here we are considering the output of innovating firms only, since by definition, noninnovating firms will have no innovating output. Within this group of firms, as much as 45.9% of sales have not been affected at all by innovation. If we exclude the office machinery sector, where noninnovated turnover accounts for only 17.1%, even in high-tech industries there is a remarkable share of noninnovating sales. The most striking result is represented by chemicals and pharmaceuticals, with a share of noninnovating sales greater than 60%.

Table 16.7
Distribution of Innovating Firms' Total Sales According to the Type of Innovation Introduced (% Values)

Industrial Sectors	% of Sales*			
	Not innovated	Innovated by process	Innovated by incremental product innovations	Innovated by significantly new prods.
Office machinery	17.1	8.1	16.8	58.0
Electrical machinery	32.3	18.8	28.0	20.9
Radio, TV, telecom.	39.2	21.8	24.6	14.4
Aerospace	37.4	24.8	12.7	25.1
Chemical (excl. pharmac.)	61.9	15.9	14.1	8.1
Pharmaceuticals	62.9	10.6	10.8	15.7
Synthetic fibers	48.3	16.7	24.0	11.0
Mechanical machinery	42.6	15.8	25.2	16.4
Precision instruments	37.8	13.9	27.5	20.8
Motor vehicles	49.3	8.0	26.8	15.9
Other transport	22.2	25.8	30.7	21.3
Rubber and plastics	44.8	23.1	20.2	11.9
Metals	28.3	58.2	9.2	4.3
Printing and publishing	31.2	52.6	6.7	9.5
Paper	52.3	23.9	15.3	8.5
Food, drink, tobacco	57.8	25.3	11.2	5.7
Textiles	40.1	31.4	15.9	12.6
Clothing	24.9	48.1	14.3	12.7
Leather and footwear	36.1	25.4	17.1	21.4
Wood	40.3	35.1	11.6	13.0
Metal products	40.8	32.0	15.1	12.1
Mineral & nonmineral prod.	53.6	23.4	14.3	8.7
Other manufacturing	35.0	29.1	20.0	15.9
Oil	62.5	18.8	10.5	8.2
TOTAL	45.9	26.2	15.5	12.4

* Rows add up to 100%.
Source: Elaboration on the Survey on Innovation (1990–1992) carried out by the Italian National Statistical Institute and the Institute for Studies on Scientific Research and Documentation. See ISTAT 1995.

The data also allow us to break down innovative sales between process innovations, incremental product innovations, and significantly new products. Table 16.7 highlights, first and foremost, the gradual and incremental nature of firms' innovative activities. 26.2% of turnover was innovated by introducing process innovations and 15.5% through the introduction of incremental improvements to preexisting products. Only the remaining 12.4% of turnover referred to totally new products.

A further qualification of the quality of innovations has been made by asking firms to rank the degree of novelty of their products. The results are given in Table 16.8. The innovating sales of this group of firms have been broken down between (1) new for the firm, (2) new for the Italian market, and (3) new in absolute terms. While 42.1% were considered new for the firm only, another 43.7% were considered new for the Italian market. Only 14.2% were considered as new in absolute terms. Interindustry differences again emerge as very important. In some industries firms are strongly characterized by the adoption of innovation developed elsewhere. Rather small shares of sales related to products new in absolute terms are found not only in traditional industries, but also in some science-based industries such as chemicals, radio, television, and tele-communications. Equally, rather low shares are found in specialized industries such as precision instruments.

The last column of Table 16.8 presents the share of new products in absolute terms of the total sales of Italian manufacturing (including both innovating and noninnovating firms). This shows that only 2.7% of the total sales are due to new products. This might reflect the relative backwardness of the Italian production system, which relies heavily on imitation and adoption from other countries. In more general terms, it shows that the pace of change in an economic system is probably slower than generally believed. Interindustry differences show that the share of radically new products is linked, along with the amount of resources put into the innovative process by the firms, to the length of product life cycle characterizing the different industries.

CONCLUSIONS

The analysis of the results of the Italian Survey has allowed us to address empirically some of the main issues of industrial innovation. In some cases, it has allowed us to confirm known findings on the basis of more robust empirical evidence; in others, it has allowed us to quantify phenomena which were not previously measured, and on other occasions to rectify some generally held views. All in all, the evidence presented in this chapter shows that the measurement of innovation by firm-based surveys provides relevant information for both researchers and policy makers. Much more information can be squeezed out by the Community Innovation Survey, and this chapter is meant as just a preliminary contribution to the exploitation of this new data source. So far, growing attention has been given to organizing and standardizing firm-based innovation surveys, but we are confident that in the next few years their use will increase in the economic and technology studies literature.

Table 16.8
Sales According to the Degree of Novelty of the Product Innovations Introduced (% Values)

Industrial Sectors	Distribution of sales of firms Introducing product innovation (a)			% of sales innovated introducing products new in absolute terms (b)
	New for the firm	New for the Italian market	New in absolute terms	
Office machinery	30.3	42.4	27.3	19.9
Electrical machinery	36.1	56.0	7.9	2.8
Radio, TV, telecom.	64.8	20.8	14.4	5.2
Aerospace	9.2	62.9	27.9	10.5
Chemical (excl. pharmac.)	62.0	34.8	3.2	0.6
Pharmaceuticals	38.7	40.1	21.2	4.5
Synthetic fibers	56.5	40.8	2.7	0.8
Mechanical machinery	50.1	23.2	26.7	8.4
Precision instruments	61.0	21.4	17.6	5.8
Motor vehicles	10.4	63.9	25.7	10.1
Other transport	37.7	47.4	14.9	5.8
Rubber and plastics	48.2	28.1	23.7	5.0
Metals	34.9	57.0	8.1	0.7
Printing and publishing	56.0	34.5	9.5	0.8
Paper	77.1	18.9	4.0	0.6
Food, drink, tobacco	55.1	42.5	2.4	0.2
Textiles	49.4	26.9	23.7	2.8
Clothing	56.8	22.8	20.4	1.0
Leather and footwear	39.2	23.0	37.8	4.1
Wood	53.6	40.1	6.3	0.4
Metal products	56.5	24.5	19.0	2.4
Mineral & nonmineral prod.	61.3	21.7	17.0	1.9
Other manufacturing	58.8	26.0	15.2	2.2
Oil	51.9	38.9	9.2	1.7
TOTAL	42.1	43.7	14.2	2.7

(a) Columns add up to 100%.
(b) Industry average values computed including noninnovating as well as innovating firms.
Source: Elaboration on the Survey on Innovation (1990–1992) carried out by the Italian National Statistical Institute and the Institute for Studies on Scientific Research and Documentation. See ISTAT 1995.

The most interesting evidence presented in this article can be summarized as follows:

Firms rely on a varied range of innovation sources. R&D represents a crucial source for the generation of innovations and is the single most important intangible source of innovation. However, R&D absorbs just above one-third of total innovation expenditure. It is thus necessary to establish the quantitative and qualitative importance of other sources of innovation. While this has been shown on the basis of qualitative indicators (see Pavitt, 1984; Levin et al., 1987; von Hippel, 1988; Archibugi et al., 1991; Evangelista, 1996), the Italian Survey confirms it using a larger and statistically representative sample and improved quantitative and qualitative indicators. The evidence presented confirms that innovation patterns vary significantly in relation to firm size and sector. This finding has important implications for industrial innovation policy. It confirms that industries are nurtured by a mix of different technological sources. This represents an important piece of evidence that is necessary to tailor sector-specific policy measures.

Only a fraction of small firms innovate. By encompassing both innovating and noninnovating firms, the survey enables us to broaden our understanding of the role of large and small firms in technological change. We have confirmed that, among R&D-performing firms, small firms are substantially disadvantaged compared to large ones. However, in terms of a more comprehensive indicator such as total innovation expenditure, it emerged that small firms which introduce innovations are not substantially less innovative than their larger competitors. It has also been shown, however, that innovating small firms are only a minority. Once both innovating and noninnovating firms are considered, it emerges that the small ones are much less innovative than the large ones and that they are even more disadvantaged in terms of R&D expenditure. We have shown that this is true irrespective of (albeit important) sectoral specificities and the geographical location of the firms. This also entails significant implications for industrial policy. The main problem appears to be not so much that of stimulating the innovation intensity of small firms as of increasing the number of small innovating firms (or upgrading them in size). In other words, the evidence presented suggests that, to foster the economic performance of small firms, it is more important to *broaden* rather than *intensify* the innovating industrial base.

The bulk of resources devoted to innovation are highly concentrated in a few industries and firms. In recent years the idea that the most technologically advanced industries could be identified according to R&D-based indicators alone has been challenged. It has been argued that more comprehensive indicators need to be used to select industries with the most significant technological performance. In this chapter we have used very comprehensive technological indicators such as total innovation expenditure, and these have shown that the most innovation-intensive sectors remain substantially the same as those identified on the basis of the R&D/turnover ratio.

Table 16.9
Sales According to the Degree of Novelty of the Product Innovations Introduced
(% Values)

Classes of Employees	Distribution of sales of firms Introducing product innovations*		
	New for the firm	*New for the Italian market*	*New in absolute terms*
20-49	58.5	27.4	14.1
50-99	57.7	26.9	15.4
100-199	55.9	28.2	15.9
200-499	51.5	29.3	19.2
500-999	44.9	35.8	19.3
1,000 and over	36.4	51.2	12.4
TOTAL	42.1	43.7	14.2

*Columns add up to 100%.
Source: Elaboration on the Survey on Innovation (1990–1992) carried out by the Italian National Statistical Institute and the Institute for Studies on Scientific Research and Documentation. See ISTAT 1995.

While the Italian Survey has confirmed that there is much scope to enlarge the understanding and measurement of technological change from a narrow R&D concept to a wider innovation concept, it has also shown that the *ranking* of sectors does not change substantially.

Furthermore, the Italian Survey has also shown that the bulk of inputs for innovation are heavily concentrated in a small number of sectors and firms. To increase the innovative capacity of the Italian system it will thus be necessary both to strengthen its technological core and to create the conditions for the technological potential of the most innovative sectors to be transferred to the whole economic system.

Innovating firms and innovated output. The survey has also provided an estimate of the output of innovation according to the share of innovated turnover and to its degree of novelty. Over a three-year period, innovating firms translate only a small portion of their innovative inputs into turnover. For all the innovating firms, we have shown that as much as 46% of their sales is not at all affected by innovation. The innovated output is moreover represented mainly by incremental changes, while only 2.7% of the total output of the Italian manufacturing industry is entirely new. We have also quantified the propensity of each industry to introduce significant and less significant innovations. These results confirm the highly cumulative nature of technological progress (Rosenberg 1976; Pavitt 1984), but provide further evidence of the essentially imitative nature of technological change in Italy.

APPENDIX

The Italian Innovation Survey covers innovative activities undertaken in the Italian manufacturing industry in the period 1990–1992. Firms were asked whether they had introduced innovations during this three-year period. With reference to the same period, firms were asked another set of qualitative questions on objectives of and obstacles to innovation, and sources of information used. Some more quantitative data on firms' innovation inputs and outputs have been collected on a one-year basis. In particular, firms were asked to provide data on their innovation expenditure and innovative sales for the year 1992 only. Accordingly, the figures on innovation expenditure presented in Tables 16.3, 16.4, 16.5, and Figure 16.1 refer to 1992 only. The data on firms' innovated sales presented in Tables 16.7, 16.8, and 16.9 also refer to 1992 although the definition of innovative sales includes product and process innovations introduced during the period 1990–1992.

NOTES

1. The research for this chapter was carried out within the IDEA project of the Targeted Socio-Economic Research of the European Commission. We wish to thank Giorgio Sirilli, Keith Smith, and Mario Pianta for their comments at various stages of the research. Aldo del Santo has also provided fundamental help in performing the statistical survey. The usual disclaimers apply.

2. This is not the place to discuss the methodological aspects of these surveys in detail. See Smith 1992a and 1992b; Hansen 1992; Kaminski 1993; Archibugi and Pianta 1996.

3. The OECD has already compiled manuals for collecting data on R&D expenditure, patents, human resources for science and technology, the technology balance of payments and surveys on innovation.

4. Among these, we would like to recall the Innovation, Patents and Firms' Technological Strategies Conference, Paris, OECD, December 8–9, 1994; and the Innovation Measurement and Policies Conference, Luxembourg, European Commission-Eurostat, May 20–21, 1996.

5. We take into account here firms which took part in the survey by compiling the questionnaire (66.9% of the total of Italian manufacturing firms with 20 or more employees). To ascertain whether the firms taking part were statistically representative, ISTAT made a survey on a subsample of nonrespondent firms. Results show that there are no significant differences between the two groups of firms. This is why we concentrate here only on the firms that responded.

6. In considering average values per size class, we obviously ignore the undoubtedly significant specificities of sectors. For an analysis of the relationship between innovation intensity and firm size at the level of the main industrial sectors, see Archibugi, Evangelista, and Simonetti (1995b).

7. Since it is our intention here to measure the total innovation intensity of each group of firms (by size or industrial sector), the index considers the employees of both innovating firms and non-innovating firms as the denominator, whereas the innovation expenditure of innovating firms alone is obviously shown as the numerator.

8. The most recent literature has argued that precisely because of the presence of marked sectoral specificities in levels of technological opportunity and appropriability, the relationship between firm size and innovation intensity studied at the level of the entire manufacturing industry may furnish spurious indications of the effective importance of the size factor (cf. Cohen and Levin 1989).

9. In general, the econometric literature (see, in particular, Amemiya 1981) suggests that to compare the results of the two models it is necessary to multiply the logit estimate

coefficients by 1/1.6 = 0.625. By effectively multiplying all logit coefficients by this value the results in both cases are very similar.

10. The value of the Concordant test is similar to that of R^2 in a standard regression.

11. For example, in the case of the logit model Pr(innovate) = exp(-4.0648)/[1 + (-4.0648)] according to a logistic function.

12. In econometric terms this means that, for example, in the logit model (see Table 16.6), the intercept for the firms in the office machinery sector (i.e., the general probability of the firms being innovative due to the fact that firms belong to this sector) must be increased by 1.5355 with respect to the firms of the reference sector (other manufacturing).

13. For a comparison at the industry level of inputs and outputs from innovation see Sterlacchini 1996.

REFERENCES

Acs, Z., and D. Audretsch. *Innovation and Small Firms.* Cambridge: MIT Press, 1990.

Amemiya, T. "Qualitative Responses Model: A Survey," *Journal of Economic Literature* 18, 4 (1981) 1488.

Antonelli, C. *The Economics of Localized Technological Change and Industrial Change.* Boston: Kluwer Academic Publishers, 1994.

Archibugi, D. "In Search of a Useful Measure of Technological Innovation," *Technological Forecasting and Social Change* 34 (1988).

Archibugi, D., S. Cesaratto and G. Sirilli. "Sources of Innovative Activities and Industrial Organization in Italy," *Research Policy* 20 (1991) 299–313.

Archibugi, D., P. Cohendet, A. Kristensen, and K. A. Schäffer. *Evaluation of the Community Innovation Survey*, Sprint/EIMS Report. Aalborg: IKE Group, 1995a.

Archibugi, D., R. Evangelista, and R. Simonetti. "Concentration, Firm Size and Innovation: Evidence from Innovation Costs," Technovation 15, 3 (1995b) 153–163.

Archibugi, D., and M. Pianta. *The Technological Specialization of Advanced Countries.* Dordrecht: Kluwer Academic Publishers, 1992.

Archibugi, D., and M. Pianta. "Innovation Surveys and Patents as Technological Indicators," in Innovation, Patents and Technological Strategies. Paris: OECD, 1996.

Cesaratto, S., S. Mangano, and G. Sirilli. "The Innovation Behavior of Italian Firms: A Survey on Technological Innovation and R&D," Scientometrics 21, 1 (1991) 115–141.

Cohen, W. M. "Empirical Studies of Innovative Activities," in *Handbook of the Economics of Innovation and Technical Change* (P. Stoneman, ed.). Oxford: Blackwell, 1995.

Cohen, W. M., and R. C. Levin. "Empirical Studies of Innovation and Market Structure," in *Handbook of Industrial Organization* (R. Schmalensee and R. D. Willig, eds.). Amsterdam: Elsevier, 1989.

Coombs, R., P. Narandren, and A. Richards. "A Literature-Based Innovation Output Indicator," *Research Policy* 3, 25 (1996) 403–13.

Evangelista, R. "Embodied and Disembodied Innovative Activities: Evidence from the Italian Innovation Survey," in *Innovation, Patents and Technological Strategies.* Paris: OECD, 1996.

Evangelista, R., T. Sandven, G. Sirilli, and K. Smith. "*Measuring the Cost of Innovation in European Industry.*" Paper presented to the International Conference on Innovation Measurement and Policies, Luxembourg, May 20–21, 1996.

Geroski, P. A., J. Van Reenen, and C. F. Walters. *How Persistently Do Firms Innovate?* Mimeo 1996

Griliches, Z. "R&D and Productivity: Econometric Results and Measurement Issues," in *Handbook of the Economics of Innovation and Technical Change* (P. Stoneman, ed.). Oxford: Blackwell, 1995.

Hansen, J. A. *New Indicators of Industrial Innovation in Six Countries: A Comparative Analysis.* Final Report to the U.S. National Science Foundation, June 1992.

Hollenstein, H. "A Composite Indicator of a Firm's Innovativeness: An Empirical Analysis Based on Survey Data for Swiss Manufacturing," *Research Policy* 25 (1996) 633–645.

ISTAT. *Indagine sull'innovazione tecnologica, anni 1990–92.* Rome: ISTAT, 1995.

Kaminski, P. *Comparison of Innovation Survey Findings.* Paris: OECD, 1993.

Kleinknecht, A., and D. Bain (eds.). *New Concepts in Innovation Output Measurement.* Houndmill, U.K.: Macmillan, 1993.

Kleinknecht, A., and J. O. N. Reijnen. "More Evidence on the Undercounting of Small Firm R&D," *Research Policy* 20 (1991).

Kline, S. J., and N. Rosenberg. "An Overview on Innovation," in *The Positive Sum Strategy* (R. Landau and N. Rosenberg, eds.). Washington, D.C: National Academy Press, 1986.

Levin, R., A. Klevorick, R. Nelson, and S. Winter. "Appropriating the Returns from Industrial Research and Development," *Brooking Papers on Economic Activity* 3 (1987) 783–831.

Lundvall, B. A., and B. Johnson. "The Learning Economy," *Journal of Industry Studies* 1, 2 (1994).

Malerba, F., and L. Orsenigo. "Schumpeterian Patterns of Innovation," *Cambridge Journal of Economics* 19, 1 (1995) 47–65.

OECD. *Proposed Guidelines for Collecting and Interpreting Technological Innovation Data. Oslo Manual.* Paris: OECD, 1992a.

OECD. *Technology and the Economy: The Key Relationships.* Paris: OECD, 1992b.

OECD. *Industry and Technology. Scoreboard of Indicators 1995.* Paris: OECD, 1995.

OECD. *Innovation, Patents and Technological Strategies.* Paris: OECD, 1996.

Pavitt, K. "Sectoral Patterns of Technological Change: Toward a Taxonomy and a Theory," *Research Policy* 13 (1984) 343–373.

Pavitt, K., M. Robson, and J. Townsend. "The Size Distribution of Innovating Firms in the U.K.: 1945–1983," *The Journal of Industrial Economics* 35, 3 (1987).

Rosenberg, N. *Perspectives on Technology.* Cambridge: Cambridge University Press, 1976.

Rosenberg, N. *Exploring the Black Box.* Cambridge: Cambridge University Press, 1994.

Santarelli, E., and R. Piergiovanni. "Analyzing Literature-Based Innovation Output Indicators: The Italian Experience," *Research Policy* 25, 5 (1996) 689–711.

Schmalensee, R., and R. D. Willig (eds.). *Handbook of Industrial Organization*, Vol. 2. Amsterdam: Elsevier, 1989.

Scholz, L. "Innovation Surveys and the Changing Structure of Investment in Different Industries in Germany," *STI Review* 11 (1992).

Schumpeter, J. A. *The Theory of Economic Development.* Cambridge: Harvard University Press, 1934.

Schumpeter, J. A. *Capitalism, Socialism and Democracy.* New York: Harper, 1942.

Smith, K. "Technological Innovation Indicators: Experience and Prospects," *Science and Public Policy* 19 (1992a) 383–392.

Smith, K. *Quantitative Innovation Studies in Europe with Existing Datasets: Possibilities and Problems*. Oslo: Royal Norwegian Council for Scientific and Industrial Research, 1992b.

Soete, L. G. "Firm Size and Innovative Activity," *European Economic Review* 12 (1979).

Sterlacchini, A. "Inputs and Outputs of Innovative Activities in Italian Manufacturing," *Quaderni di Ricerca*, Università degli Studi di Ancona, Dipartimento di Economia, 78 (1996).

Stoneman, P. (ed.). *Handbook of the Economics of Innovation and Technical Change*. Oxford: Blackwell, 1995.

Townsend, J., F. Henwood, G. Thomas, K. Pavitt, and S. Wyatt. "Innovations in Britain since 1945." Occasional Paper No. 16, Falmer, U.K.: SPRU, University of Sussex, 1981.

von Hippel, E. *The Sources of Innovation*. New York: Oxford University Press, 1988.

17

Fostering an Innovative Society in a Fellow Traveler Country

Annamária Inzelt

INTRODUCTION

Postsocialist economies have been slow to respond to recent global challenges. Their previous attempts to adjust their economies to the new technological and economic environment proved to be abortive. Despite remarkable scientific results in advanced technologies (e.g., lasers, biotechnology, information technology) they were not able to diffuse knowledge. The weak knowledge distribution capability of these economies became one of the heaviest burdens affecting their development. The collapse of the socialist system removed many constraints to innovation, but in many respects these societies are prisoners of their own past. These countries and their economic agents are still outsiders to international networks. Although political changes and economic transition are helping them to join public and private networks, it is a long process to obtain full-fledged membership, even if they gain an "entrance card." It is usually difficult for laggard countries to become members of the key players' clubs of world technological development and competition. It is much more difficult for postsocialist economies that are not only laggard economies but have been isolated to a great extent from existing networks and alliances, and restricted in scientific mobility and freedom of technology transfer. They have to adjust to a global economy, which is also in transition. At the turn of the century, the innovation environment favors transition economies less than newly emerging Asian economies.

This chapter investigates the adjustment process of transition economies. It concentrates on diffusion in the field of biotechnology and information technology. Hungary has significant knowledge assets in both these high-tech industries. In spite of this its diffusion has been weak. Although considerable changes are taking place as part of the transition period, empirical findings concerning Hungarian biotechnology and information technology clearly show that the country's knowledge distribution capability is still weak. The chapter highlights the factors hindering innovation. Its conclusion attempts to answer the difficult question of how this region can improve its competitiveness and innovativeness to build on international experience and its own traditions.

HERITAGE OF THE PAST

The predicament facing socialist economies lies in relatively well-developed science together with inefficient, ad hoc business enterprises. Situated in a socialist system, this type of latecomer, peripheral industry was cut off from major international sources of technology. They operated in isolation from the world centers of science and innovation and lagged behind technologically. The available technological and industrial infrastructure was underdeveloped. Because science was a matter of prestige, these countries spent a relatively greater proportion of GDP on R&D than others at the same economic level. With their historically good scientific capabilities and an enviable level of funding under socialism, good scientific value resulted and helped to accumulate scientific assets. This is in contrast to developing countries.

In socialist countries, science was well-developed and offered a strong knowledge base even if the average local university and other educational and technical institutes were poorly equipped. The socialist system was hardly innovative, and economic actors had little interest in commercializing R&D results. The knowledge distribution capability of these economies was very weak. Under this repressive regime, travel and the exchange of ideas were restricted and the goals and results of scientific research were distorted to suit a political agenda. In the bipolar world economy, international knowledge and technology transfer were limited. Barriers to cooperation preserved the disparities between countries.

INNOVATION MODEL

Using the terminology of the innovation model, we can speak of a *deformed version of the first generation.*[1] In order to transform the R&D resources of former socialist countries into economic success, these countries have to redefine their institutional and behavioral operations. The socialist system was very weak in diffusing knowledge and in commercialization. The scheme of the *chain-link, technology-push, one-way linear model* of planned economies was much longer than this model in market economies (see Figure 17.1).

Figure 17.1
First-Generation Innovation Model by Economic Systems

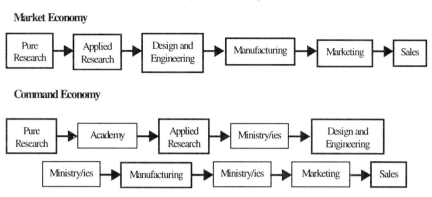

In this model all development issues, not only strategic ones, were decided by the bureaucracy. All links between the actors were indirect and cooperation went through the hierarchy. Integration was vertical with some horizontal elements. Interorganizational relationships were much more important than intraorganizational ones, because most business functions were separated from enterprises. As a reformed socialist country, Hungary had modified the strict Soviet model in several waves since 1956. In the resulting framework, the authorities had allowed limited direct links between universities and enterprises and between enterprises. Enterprises also regained some business functions (Inzelt 1988). All supporting characters were absent from the economic arena, and interest in cooperation was limited by commands.

In 1968 the Hungarian Economic Reform introduced some market instruments and took some steps toward a second-generation, market-pull innovation model, in an attempt to overcome the main feature of the socialist system: the lack of innovativeness resulting from the fact that economic actors had little interest in commercializing R&D results. In the field of S&T policy formulation, the most important efforts were to reestablish links between R&D and business activities, and to shorten the path from inventions to the practical application of new technologies in the production process.

This rethinking process resulted in the large-scale development programs that were launched during the 1970s. They represented a modified bureaucratic approach to research programming, and were able to encourage a shift away from the interrupted innovation model to the one-way linear innovation model. But this transformation was not enough to approach the feedback loop model: the chain of innovation still had gaps. The mixture of top-down programs and enduring elements of the system played a most important role in knowledge distribution. However, economic performance and comparison with the fast-growing, newly emerging nonsocialist world showed very clearly that one-way linear models—as macroeconomic models—could not strengthen the position of socialist economies in the face of global competition. During the late 1970s some attempts were made to introduce a *torso of the third generation*, the interactive or coupling model of innovation (see Figure 17.2).

In the 1970s Hungary launched "complex" large-scale development programs in the field of high-tech industries, specifically in information technology and biotechnology. ("Complex" meant that the program covered all areas of the innovation process in a joint framework: one governmental organization was responsible for the whole program and coordinated the actions of others.) These programs included joint research, education, production, and market orientation agendas. Different actors in the innovation process were encouraged to work together. For example, biotechnology was one of the areas targeted for the large-scale governmental S&T programs. In this particular area it is extremely important to follow an approach which is oriented toward the whole innovation process.

The reformed socialist model was able to treat the different stages of innovation in a dynamic context rather than separately. In principle, these programs stimulated knowledge and foreign technology transfer, but these activities depended heavily on the short-term political situation. They also moved from a policy of self-sufficiency toward one of international collaboration, but participation in international S&T networks was restricted. Nevertheless, this reform helped to establish direct horizontal links and direct cooperation between institutes and universities and business, slowly replacing vertically organized bureaucratic connections. Business enterprises and production areas were less isolated from the domestic R&D sector, and the gap between the level of technology used and state-of-the-art technology diminished somewhat. But incentives for commercialization and innovation remained weak; hence development and testing activities were inadequate.

Because the fundamentals of the system remained untouched, there were gaps in the network of communication paths. Feedback loops occurred in a few segments of the economy. In Hungary, the internal organization of firms was transformed, enabling them to react to some extent to external challenges. These transformations allowed them to introduce modified innovations, but they were unable to adapt to incremental changes.

As is well known from the literature, the above-mentioned generation of the innovation model (Kodoma 1992; Rothwell 1994; Nelson 1984) does not suit high-tech industries of the post-Second World War. The growing awareness of the strategic importance of emerging generic technologies, with increased strategic emphasis on technology and manufacturing strategy, has led to the *fourth- and fifth-generation, system integration and networking model of innovation*. This model describes how front-running corporations work all over the world.

Figure 17.2
Third-Generation Coupling Model

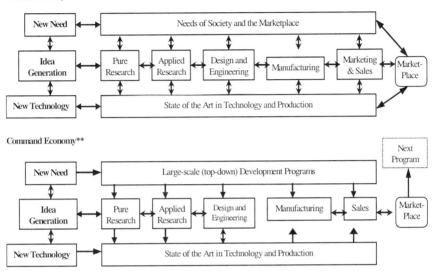

Source: *adapted from Rothwell 1993; **compiled by author

In the 1980s, when fourth- and fifth-generation models were spreading in market economies, socialist countries were able to produce good results in knowledge creation in new generic technologies and create a few archipelagos to produce these technologies. But generic technologies penetrated into the socialist economy only slowly. In the era of these high-tech industries, command economies were clearly at a technological dead end. To emerge from this, they need to create the proper economic environment for commercialization and innovation. This is one of the greatest challenges for transformation of the system.

ADJUSTMENT PROCESS

One of the key questions for transformation is how countries of the socialist system can move from the one-way linear model of innovation to feedback loops, system integration, and the networking model. Establishing the preconditions for developing this new model is the main task of the transition period. The accumulated knowledge of postsocialist countries was locked in a poor structure. Changes can help these countries realize their intellectual potential and use it for the advancement of society.

Recent OECD studies describe the transformation of the S&T institutional system, legal system, and financing structure in transition economies. These transformations are institutional preconditions for moving toward a market economy and strengthening the knowledge creation and distribution capability of these countries. A new type of organization has been set up so that they can strengthen the capacity for knowledge distribution and create closer academy/industry relations. Industry has been transformed, too, with enterprises becoming firms, subsidiaries, and the like. Their transformation, privatization, foreign direct investment, and so on have been accompanied by downsizing, acquisitions, mergers, bankruptcies, and closures.

As a result of these changes, some links have broken down between traditional partners, while at the same time new actors have appeared. In Hungary, for example, Fraunhofer-style R&D institutes, liaison offices, an Innovation Relay Center, and several data banks have been set up to help technology transfer. Their presence shows that the country is likely to adopt a market economy model in which these organizations have a growing role in the dynamic process of innovation. They are working to fill the gaps of interaction between firms and universities, and between firms and contract research organizations. It appears that they can bring industry closer to the work going on at universities, and give them a window on new ideas and access to emerging scientific and technological developments. They can help postsocialist countries use their intellectual assets much more efficiently. Institute/university/industry partnerships pool their expertise to develop and commercialize new products and discover new and better ways of doing things.

The key S&T indicators show a declining trend. The collapse of the command economies has accelerated the deterioration of research and development systems. Table 17.1 illustrates losses in R&D personnel, closures, and decreased R&D capacities, while Figure 17.3 shows the diminishing number of patent applications. These drops illustrate the combined results of reevaluation of the S&T system and the cost of transition. However, recent statistical data enable us to suppose that the cost of transformation will not increase further.

Research and development are essential components of innovation. Hungary devotes 0.8% of GDP to R&D, which is around the level of less developed small OECD countries. There are no clear priorities if we analyze the declared aims from the point of view of distribution of state funds as the financial source. Industrial research carried out and financed by businesses is on a small scale. In-house expenditure on R&D is very limited, although spending on in-house research is important, even if firms do not set out to produce new scientific results. General experience shows that, if a firm has no R&D capability, it is difficult for it to acquire or adopt new technologies.

Expenditure on innovation and the breakdown of costs (R&D, acquisition of patents and licenses, product design, etc.) depend largely on the strategy of the firms involved. Resources devoted to R&D, patents, and other outputs are only elements, albeit important ones, in the innovation process. Innovation requires the integration of a range of diverse activities, such as research, design, prototype development, market research, training and skill development, and so on, as Figures 17.1 and 17.2 highlighted. Since diffusion of new technologies (e.g., information technology and biotechnology) happens primarily through innovating firms, firm-level studies are of vital importance.

Table 17.1
Changes in Key R&D Figures

	R&D Personnel in	R&D Expenditure by Sector, in Million
Sector	1995/91	1995/91
Business	52%	161%
Government	81%	161%
Higher education	75%	188%
Total	66%	154%

Source: Scientific Research and Experimental Development 1995.

Figure 17.3
Number of Patent Applications in Hungary

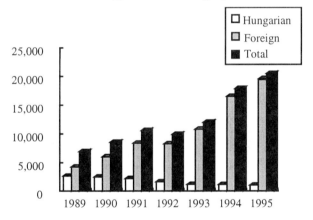

Source: Hungarian Patent Office.

The environment for innovation can help or hamper the diffusion of knowledge, cooperation, penetration into the market, and so on. Here we touch on some elements of the macroeconomic system that are preconditions for improved dissemination of knowledge, innovation, and competitiveness.

1. The financial system avoids innovation. This complicated issue is a limiting factor for the whole economy. Public finance is limited and uncertain, the tax environment is unfavorable, and the willingness for self-financing in the early stages of innovation or for risky innovation is weak. For companies, many potential partners (commercial banks, long-term savings banks, pension funds, venture capital firms, business angels, etc.) hardly exist as financers of innovation.

2.   The hiatus in policy formulation.
3.   Inadequacy of the legal and regulatory environment. This has a transitional character in many areas and some specific regulations are still lacking.
4.   The existing physical infrastructure in Hungary causes bottlenecks in the innovation process. For example, telephone lines and access to e-mail are limiting factors in certain regions. However, it should be said that compared to the recent past, the situation has changed significantly in a positive direction. There are far more telephone lines, and digital exchanges have significantly improved the quality of lines. All communication infrastructures are available but their relative cost is higher than in advanced market economies. Not only are the technical conditions much better for being well-informed in time, but so is the political environment.
5.   Intermediary institutions, associations, chambers of commerce, and the like are nonexistent or hardly functioning.
6.   Research is not generally oriented toward innovation.
7.   Cooperation between the domestic scientific community and business is weak. Hungarian firms are hardly involved in international cooperation.

EMPIRICAL FINDINGS ON HUNGARIAN BIOTECHNOLOGY
(Biotechnology Audit of Hungary 1996)

This new technology emerged in 1971 when the first Hungarian research center devoted to biotechnology was established. Modern biotechnology is a research-intensive pervasive technology with a strong scientific component. Its world significance is well known. Biotechnology is an extremely interesting and important example of how technological change proceeds in modern industrial economies. Advanced industrialized countries consider biotechnology as a key technology for the 21st century.

At the firm level it requires the availability of firms' own competitive research capabilities, which provide a suitable interface for communicating with research sites like universities where the necessary basic knowledge is generated. Recent research concentrates on this issue, on how it can grow and diffuse in a transition economy.

Biotechnology activities cover a broad spectrum in Hungary, with many different modern technologies being used in practice. If we rank them we can observe that *analytical techniques* are used most frequently by firms and institutes. Then firms use *biochemicals* and *computer/automation* techniques to support biotech processes. Institutes are different, being deeply involved in *downstream processing* and then *cell and tissue culture*. Among the 12 groups of biotechnology methods and processes, *genetic engineering* ranks 5th at institutes, and 7th at firms, for frequency of use. This means that genetic engineering is not yet of major importance in the pharmaceutical industry, in strong contrast to the advanced countries. *Proteins* are much less used by institutes, and the firms investigated do not use them (see Figure 17.4).

The diffusion of biotechnology depends on many factors, from highly qualified, properly funded research to commercial potential. If we analyze key sectors of biotechnology (the pharmaceutical industry, food and beverages, agriculture, chemicals, and diagnostics) according to the elements of innovation models, we find that pharmaceutical firms in the sample do not have basic research devoted to biotech. This means that strongly science-linked biotechnology depends on external knowledge. Food and beverage firms are not involved in basic and applied research. This observation fits quite well with the general tendency of this sector, where only a few multinationals carry out research. How efficiently a firm can adapt to new technologies is a question of long-term ability to adapt if it devotes no sources to research. Improving firms' capabilities in this respect is also a key factor of innovation.

Figure 17.4
Biotechnology Methods and Processes

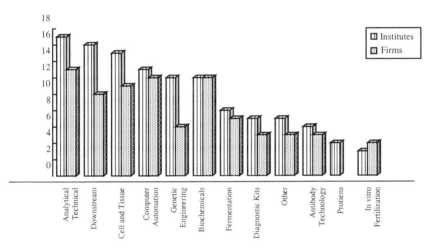

University-industry links are generally very weak in Hungary. All biotechnology-oriented institutes investigated emphasized that their R&D programs are determined by science, not by industry or by policy, and feedback loops are still lacking. Only one institute (Vegetable Crops Research Institute Corp.) said that its R&D programs are determined predominantly by the market.

If we try to evaluate the size of inputs in biotechnology innovation, we have to emphasize that the critical mass for economic success is lacking at every stage.[2] It is well known from the literature that the time of entry into the market and the moment of introduction of a new product are becoming crucial factors in competition. If the capabilities for these are lacking, business has less chance for success. In the last decade biotechnology research has become much more cost-intensive than it was in the 1970s. If basic research organizations cannot recover their research costs in time, far fewer first-rate science results will be generated in Hungary.

Key Sources of Information for Innovation

Among the inputs, the sources of information are very important. If we look only at the pattern of responses by sources, that in itself indicates a problem. Conferences and journals were of key importance for more than half of the responding firms. Second were educational and research institutes, and third in-house R&D. Of course these are relevant sources but they belong more to supply push and less to the market-driven model. The flourishing idea of market economies—network as information source—has much less importance in this sample. They have not yet reached the building-up period. They would like to obtain information about possibilities for partnership and cooperation from outside sources. Acquisition of embodied and disembodied technologies, licenses, clients, and suppliers is of low importance. The sources of information do not include acquisition of innovative firms and investment in innovative technologies.

This pattern of information sources is empirical evidence that there is hardly any feedback loop in the innovation process.

Missing Players for Innovations

With the exception of a few sectors (e.g., horticulture), there is a lack of small biotechnology companies. The general financial conditions mean that there are very few spin-off companies. Scientists from the research community or companies cannot spin off to become entrepreneurs because suitable credit is lacking. There are no business angels, venture capital, and so on. The few spin-off companies that do exist are typical products of the transition economy.

Business network: By its nature, innovation is a collective process which needs the gradual commitment of an increasing number of partners. In the context of the technology audit, we sought to investigate the relations between business and its major partners: clients/customers (companies or industries), suppliers, different actors in the public sector, financiers, alliances, R&D partners, and associations. The above sections on sources of information, inputs, and the like provide some information on this issue. In the following we will go into more detail.

Not only the ranking order of partners, but also the frequency of the relations between them, differ between firms and institutes (see Figure 17.5). R&D partners are not ranked very highly, which is not surprising since they were not among the key sources of innovation. Only four of the firms investigated have permanent contacts, three mentioned occasional contacts, while the remainder did not mention any relations in this area. This empirical evidence is

Figure 17.5
Business Network of Firms and Institutes

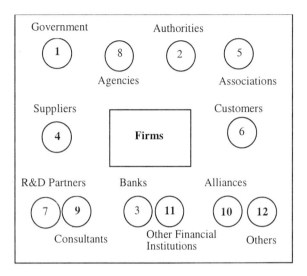

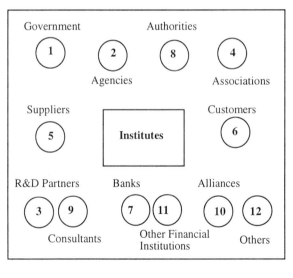

confirmed by the experience of institutes, which are looking for firms as partners to finance their research projects and experimental development.[3] If firms want to remain players on the biotechnology stage, they should consider R&D partners more seriously.

Most of the institutes have *R&D partners*. Partnership usually means exchange of information, joint research, and in some cases cooperation in training. If we examine the lists of their partners we can observe two important features: (1) Many of them have some cooperation with domestic partners, which is not very common in Hungary. Domestic partners are other departments of the same university/institute or other universities/institutes. (2) The other remarkable thing is that they have many foreign cooperation partners.[4] Not only is the number of foreign partners impressive but many of them belong to top-level institutes in world biotechnology, for example DFG, FhG, University of Göttingen, University of Cologne, and the Max Planck Aquaculture Institute in Germany; Texas State University and Columbia University; the Museum of Natural Science in France; and Hampshire College in Great Britain.

Institutes and universities have some cooperation with firms in two types of contract research: joint research with firms and R&D contracted out by firms.

Few firms have R&D cooperation or contract-out research. Besides many formal relationships they have extensive informal relations with Hungarian universities and institutes. Some firms also have scientific and experimental development cooperation with multinational corporations—for example, Bayer, Eli Lilly, Ciba-Geigy, Du Pont, Pharma, Merck, and others.[5] As a new phenomenon worth mentioning, private Hungarian laboratories are among the partners of some agricultural firms.

Creation and diffusion of new technologies involve dynamic interaction between firms and their environments. These environments, called "innovation infrastructures," consist of networks of suppliers and customers, support services, and financial institutions, as well as the publicly supported science and technology infrastructure, such as universities and publicly funded R&D programs.

CONCLUSIONS

Returning to the challenge posed in the introduction, in terms of comparative advantage, biotechnology is a sector which should be promoted in Hungary as an "industry of the future." However, according to our experience, Hungary can expect a very good position in *agrofood biotechnology* and a less good position in world competition in other biotech-related sectors. While the agrofood biotech sector could be a strategic one for Hungary, it is very doubtful whether biotechnology in the pharmaceutical industry or diagnostics will have strategic importance in Hungary. This does not mean we should neglect these sectors. Government policy can concentrate on strengthening knowledge distribution, diffusion, and adaptation capabilities. It has to choose the groups to be targeted for public support on the basis of the *Audit* results and other methods of evaluation. Without such a selection procedure, it may not be able to concentrate on R&D or measurement of the critical innovation mass necessary for success. The *Audit* provides a new tool for policy formulation and government can launch new actions to improve the macroeconomic decision-making process.

The *Biotechnology Audit* highlighted certain weaknesses in the National System of Innovation:

- weak knowledge distribution capability
- slow commercialization of R&D results
- linear model of innovation has been strengthened but feedback loop model hardly exists
- undeveloped financial system for R&D and innovation
- shortage of financial resources
- need to extend the legal framework and enforce existing laws and regulations
- slow and costly authorization procedure (pharmaceutical and food industries)

Government is only one player in this arena. We share the opinion of many of our respondents that *biotechnology could be a strategic issue for Hungary*. However, this does not mean that everything related to biotechnology has strategic importance or that all R&D costs have to be financed from public sources. Government has a key role in creating a suitable environment for the development of biotechnology. It has to extend the legal framework and enforce existing laws and regulations. It can diminish the negative impact of shortage of financial resources with more appropriate policy measures.

The macroeconomic environment for innovation can help or hinder diffusion of knowledge, cooperation, market penetration, and the like. The unstable Hungarian market and changeable economic conditions are not favorable to strategy formulation, and strategic analyses are usually not well-founded.

NOTES

1. The basic deformation was the enterprise itself. The main corporate functions and their organization were external (Inzet 1988). The market was only a simulated one even in reformed socialist economies.

2. This is a problem not only of firms but basic research laboratories too. As one of the young researchers stressed, they need two years for laboratory tests instead of three months because they do not have enough money to repeat tests even if they have the equipment. Many researchers are willing to work day and night to speed up laboratory test phases.

3. Agricultural institutes are usually ready to meet firms' demands, being client-oriented research organizations, but they also have to face the problem of a lack of demand. Some basic research organizations (e.g., university departments) declared themselves ready to continue applied research or experimental development on a contract basis to earn money to develop their laboratories, and to finance appropriate salary levels.

4. Hungarian economic reform (in 1968) had some positive impact on the international relations of the business and scientific communities. Biotechnology was not included in any large Council for Mutual Economic Assistance programs, (the last S&T cooperation program simply declared it a target without any enforcement). In this field cooperation started with advanced market economies and developing countries. Before the transition period there was some R&D cooperation, visiting fellowships, and a few partnerships on the basis of Hungarian basic research results with multinational corporations, Pharmaceutical Research Institute Ltd.). Business cooperation was also established on the basis of individual contracts permitted by government. (Japanese TAKEDA-Chinoin [pharmaceutical firm], Nadudvar [state-owned farm], and a Japanese firm [now a joint venture: Agroferm Hungarian-Japanese Fermentation Corp.]).

5. It is not easy to obtain accurate information on the content of this cooperation. The firms mentioned are involved in tableting, storing, and marketing drugs on the Hungarian market; managing registration; analyzing medicines; and R&D.

REFERENCES

Biotechnology Audit of Hungary 1996. Commissioned by the German Ministry for Education, Science, Research, and Technology (BMBF). Fraunhofer Institute for Systems and Innovation Research (FhG ISI) and Innovation Research Center (IKU).

Inzelt, A. *Rendellenességek az ipar szervezetében (Abnormalities in Industrial Structure).* Budapest: Közgazdasági és Jogi Könyvkiadó, 1988.

Kodoma, F. "Technology Fusion and the New R&D," *Harvard Business Review* (July–August 1992) 70–78.

Nelson, R. *High-Technology Policies: A Five-Nation Comparison.* Washington, DC: American Enterprise Institute Studies in Economic Policy, 1984.

Rothwell, R. "Systems Integration and Networking: Towards the Fifth Generation Innovation Process," Brightonm U.K.: SPRU, University of Sussex, 1993.

Rothwell, R. "Industrial Innovation: Success, Strategy, Trends," in *The Handbook of Industrial Organization* (M. Dodgson, R. Rothwell, and E. Elgar, eds.). Cheltenham, U.K.: Blackwell, 1994.

Scientific Research and Experimental Development. *Tudományos kutatás és kísérleti fejlesztés.* Budapest, 1995.

18

Perspectives on the Autoparts Industry in Late-Developing Countries: A Case Study

Francisco Veloso and José Rui Felizardo

INTRODUCTION: WHY AN AUTO INDUSTRY?

The automotive industry is a massive generator of economic wealth and employment. In Western Europe, Japan, and the United States, it accounts for as much as 13% of GNP, and one in every seven people is employed through the industry, either directly or indirectly (e.g., insurance). Moreover, sectors like oil, rubber, and steel are highly dependent on the 50 million cars produced every year. Another important characteristic of the motor industry is that it does not operate in a classical economic environment of competitive markets, where each of the many firms is too small to influence the future of the market decisively. It is fundamentally oligopolistic in nature and structures. The sophisticated technology embodied in a car, the huge investment required to build a plant, and the complex network of firms and people needed to commercialize an automobile, have created significant entry barriers that have led to the world players lasting as much as an entire century. The demand side is also crucial to understand the role of this industry. Buying a car is usually the second largest investment objective of a family, right after a house, but, unlike the latter, it has a limited life and has to be replaced after a few years. This behavior, which is similar around the world, creates a predictable need for the availability of automobiles in every country. Nevertheless, the size of the investment makes the buying decision highly dependent on the income level of the household. From a country perspective, a relation has been observed between the volume of car sales and its level of development, with an important threshold when nations reach US$5,000 per capita GDP (Maxton and Wormald 1994: 120).

Because of these characteristics, the auto industry has been extremely important to national economies and a source of concern for governments, particularly since the 1950s. In fact, although most of the above-mentioned aspects have been present since Ford started mass-producing the Model T at the beginning of the century, they acquired a different dimension after the Second World War, with the world boom in demand for consumer and industrial goods, and the expansion of road systems. The countries which were late in industrializing started their catch-up process during this period; for them, these were significant concerns. As demand started to grow, imports of cars and parts from global oligopolistic producers started to create significant trade imbalances that affected these countries' ability to acquire much-needed capital goods for their industrialization process.

In the search for solutions to this important issue, most governments realized that this predictable demand for cars could also be seen as a major industrial development opportunity. Besides the employment[1] and trade issues, the industry created a significant demand for intermediate inputs, creating pressure to develop other sectors of the economy. Governments reasoned that it could provide a hub for an integrated industrial structure by triggering the domestic production and technological advance of industries such as steel, machine tools, and components. The problem was how to establish national industrial capability in a context of oligopolistic producing companies. The solution was the adoption of strong trade protection mechanisms (quotas, tariffs), forcing the assemblers to locate their plants in the countries if they wanted to access local demand. Simultaneously, the enactment of policies to stimulate foreign direct investment (FDI) and local content requirements[2] would foster the desired linkages within the national economy. These policies evolved over time and the initial schemes were later complemented with measures devoted to export promotion, finance, quality, R&D, and the like.

Nowadays, governments in most latecomers consider that the auto industry established in the country has fulfilled their expectations, although with varying levels of success. While very few have been able to develop their own brands (Korea is the most prominent exception), they all now have an important indigenous autoparts industry. In Mexico, Brazil, Spain, Taiwan, Korea, and Thailand, the automotive industry is considered a key pillar of their industrial bases and a determining factor for future development. Portugal is no exception in the landscape of latecomer countries that have developed local automotive production capabilities. Although with no national brands, the Portuguese auto industry now exports 70% of its production, representing 13% of total exports.

Firms producing autoparts account for a significant share of the industry. In addition, some of the most important technological and managerial capabilities are located in these firms. They therefore play a major role in the dissemination of manufacturing best practices and in establishing links with other sectors of the economy.

The past success of these countries was based on a protected national environment that is now changing due to the GATT/WTO agreements. In addition, short product life cycles, lean manufacturing practices, and environmental concerns are demanding additional capabilities of assemblers and suppliers. While the main oligopolistic characteristics of the large assemblers still hold, barely impacted by the entrance of a very small number of new competitors, what is also being observed is a growing trend toward concentration of suppliers throughout the globe. This new market environment is changing the competitive conditions for the industry, and new government policies and company strategies are needed to continue the growth path of the past. Portugal, once more, is no exception.

In this chapter the history of the development of the Portuguese auto industry is presented[3], the current situation is analyzed, and the key challenges and future perspectives are discussed. Finally, recommendations for a renewed industrial policy framework are outlined.

THE DEVELOPMENT OF THE PORTUGUESE AUTO INDUSTRY

During the 1950s and 1960s, a large majority of the countries that aimed at industrializing established trade-related policies directly intended to develop a local automotive industry.[4] Portugal started this process in 1963 with a decree that forbade the import of CBU (completely built-up) units, while requiring 25% of national added value to be included in cars to be assembled locally. The major assembler companies responded to this closure of borders through the establishment of plants in Portugal. Five of the big international companies (Ford, GM, Renault, Citroen, and BMC) established producing subsidiaries in Portugal. Besides these, a number of other lines were also started through licensing contracts, particularly for commercial vehicles. As a result, in 1973, 30 assembly lines were producing autos (passenger and commercial) for a national market as low as 50,000 new vehicles per year.

The small scale of the assembly plants made them highly inefficient,[5] a situation that was equally true for their suppliers.[6] The small quantities produced[7] and frequent model changes prevented the autoparts companies from enjoying any profitable operation by producing only for the assemblers. As a result, with the exception of a few companies that aimed at the export market, suppliers were mainly carrying out simple operations of metalwork transformation. This situation became critical toward the end of the decade because of the downward trend in sales in the years following the 1974 Portuguese revolution, which aggravated the scale problems. Despite a number of problems, by 1979 a total of 25,000 jobs had been created in the industry, including both assembly operations and component manufacturing.

In 1980 a new regulatory framework for the automotive industry emerged. On one hand, the European Free Trade Association and European Community trade agreements signed by Portugal required the removal of the severe restrictions on the imports of CBUs. On the other hand, the government realized that the problems outlined above were hampering the development of the industry, and were a result of the existing policy framework. The new policy included (1) quantitative restrictions on the imports of CBUs and CKDs (completely knocked down) and (2) the possibility of exceeding the quota of CBUs through the export of parts and components produced in Portugal. This trade policy was complemented with incentives for FDI. The assumption underlying this policy was that the quality of the Portuguese labor force and its low cost, as well as the country's geographical characteristics and climate, would prove to be complementary ingredients for the development of a strong and modern automotive industry.

In the following years, several important changes occurred in the industry. Inefficient assembly and autoparts producers closed or were reconverted, while new firms with a scale adjusted to the European market were established. The most important was, undoubtedly, the Renault investment project. It included three production lines: one for passenger cars; one for engines, gearboxes, and water pumps; and one foundry. This project and the new legislation led to a significant influx of FDI. From 1980 to 1983, roughly 10 billion Portuguese escudos were invested in the autoparts sector, creating 4,000 new jobs. Simultaneously, with the aim of adapting local suppliers to market demands, 51 technology transfer agreements were signed, mainly with European firms. As a result, as can be observed in Table 18.1, the industry went through a period of rationalization, during which the number of firms diminished, but output increased. Rationalization resulted also from the adverse market conditions that existed in Portugal and throughout Europe in the early 1980s. In fact, as Figure 18.1 shows, the investments and reconversions of national firms only started to have direct results in terms of exports in 1984, while the anticipated boom in the national market started to take place in 1985.

Because of the recession at the beginning of the 1980s, the Portuguese government managed to negotiate a rescheduling of the opening of its internal market. In 1988, two years after the date initially agreed, Portugal completely opened its market to the imports of EC products, ushering in a new phase of development for its auto industry. Despite the significant growth in imports, the initial reaction of the industry was very positive, and exports kept their ascending path, both in terms of vehicles assembled (see Figure 18.1) and particularly in the autoparts sector (see Figure 18.2).

1988 also saw the start of the first specific program for the development of Portuguese industry. PEDIP was a European cofunded industrial policy initiative aimed at speeding the catch-up process that Portugal was undertaking to join its developed new partners. It included a number of measures ranging from R&D (for the first time in a program within an industrial context) to export promotion and financial support to existing firms, as well as strong incentives for the establishment of foreign firms. Although it did not have sectorial priorities, the Portuguese government considered the establishment of a major original equipment manufacturing (OEM) assembler as a major opportunity for the development of this industry and, because of the effects on the economy outlined in the previous section, of overall industrial capabilities.

Figure 18.1
The Portuguese Automotive Market

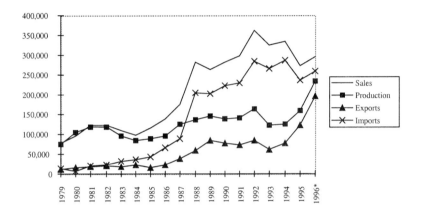

Source: AFIA 1997. *Estimate

The AutoEuropa Ford/Volkswagen joint venture is considered a major milestone in the Portuguese industrialization process. The Greenfield plant is producing multi-purpose vehicles (MPVs) and has a capacity of up to 180,000 units per year. It is the single most important FDI in Portugal, and when it reaches full production will have created 5,000 direct jobs and almost 7,000 indirectly. The investment has attracted 22 new foreign investments established near its plant, to allow just-in-time deliveries, 12 of them joint ventures with Portuguese firms. The direct impact of the plant on the national auto industry figures can be seen in Figures 18.1 and 18.2, where 1995 corresponds to the year when production started. Besides the direct immediate impact, significant indirect results are also expected due to the local content agreement between the government and the consortium, aiming to push up the level of the autoparts companies. By mid-1995, 44 national firms had achieved the highest quality certification level (Q1) from Ford, corresponding to roughly 45% of value added. The agreement covers an increase in the number of firms and value added, to reach 65% by the year 2000.

Figure 18.2
Sales and Exports of the Autoparts Sector

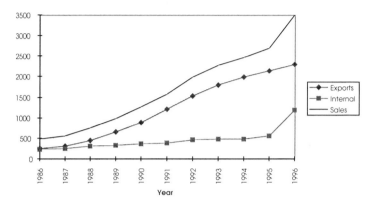

Source: AFIA 1997.

THE CURRENT SITUATION

While history is important to understand the present, it is analysis of the current situation that enables an understanding of the sources of competitiveness of the Portuguese auto industry, as it may reveal its weaknesses. Since the assemblers are essentially foreign, the emphasis will be on autoparts. Three main aspects will be considered: the market to which Portugal is selling, the parts that it is selling, and the levels of the value chain at which there is intervention. In addition, a comparison of some key indicators with foreign countries will be established.

Portuguese auto market production is clearly decoupled from national auto sales. As can be observed in Figure 18.1, the growth in demand experienced since 1988 has been filled through imports. In 1996, almost 90% of the vehicles sold in Portugal were imported. During the first half of the 1990s, production levels were quite steady, until the AutoEuropa plant started its operation, in 1995. This decoupling of sales and production was associated with the opening of the market to imports and the necessary rationalization of plants that did not have the capacity to be competitive with international market production costs. The result was a dramatic fall in the number of assembly lines in Portugal (see Table 18.1).

At present, besides AutoEuropa (120,000 units in 1996), passenger cars are produced only by Renault (12,000 units) and Citroen (14,000 units). The rest of the lines assemble commercial vehicles (a total of 88,000 units), in which the Opel/GM plant reconverted in 1993 plays the most important role (57,000 units). An identical decoupling is seen in the autoparts sector. As Figure 18.2 demonstrates, the national supply of parts and components has been almost constant since the beginning of the decade, a situation that was only changed by AutoEuropa. Exports have therefore been the main growth drive for this sector.

Table 18.1
Plants and Employment in the Auto Industry

Year	Assembly Lines		Autoparts Firms	
	Number	Employment	Number	Employment
1979	31	10,000	212	14,800
1983	24	N/A	160	16,600
1991	17	5,000	130	18,000
1995	13	7,000	150	23,500

Source: AFIA 1997.
N/A = not available

Figure 18.3 illustrates this decoupling phenomenon. Most of the fluxes in the industry are across national borders, rather than internally. This situation is likely to continue, with the sole exception of the link between supply of components and assemblers, which will increase in the coming years with the supplies to AutoEuropa. The achievement in terms of exports is quite significant, since this industry is now the leader in terms of contributions to the national trade balance, with 13% of all exports. Nevertheless, as can be observed, the specific trade balance within the sector is still negative, a situation the government is predicting will change in the next couple of years.

Figure 18.3
Fluxes in the Portuguese Automotive Industry in 1995

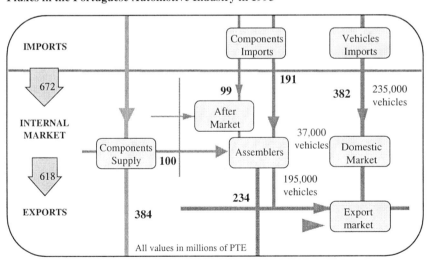

Source: AFIA 1997; some figures are estimates

There are two other important features that Figure 18.3 demonstrates. In first place is the market's dependence on the international context, in particular Europe. The destination of the great majority of the national production of both vehicles and components is the European Union (94% of sales), in which Germany (28%), France (24%), Spain (22%), and the United Kingdom (9%) are the most important countries (according to AFIA sources). Nevertheless, because of the global characteristics of this industry, understanding how other major world players are evolving should not be neglected. The European opening of its market to the Japanese auto industry, to occur in 1999, will certainly have an impact on the Portuguese market. The second important aspect is the fact that the primary market of autopart firms is OEMs and not the aftermarket. This is certainly the case within national boundaries, but it is also true for exports, where 60% of components are destined for OEMs. This shows the commitment of national firms toward actual customers instead of undifferentiated sales, endorsing the view that firms have acquired a significant level of quality, even according to international standards.

While export capacity demonstrates a significant level of market maturity in Portuguese firms, the structure of the parts supplied reflects the lower level of Portuguese technological capabilities and the importance of the low cost of national labor. The autoparts market structure is shown in Figure 18.4, where it can be observed that production is dominated by segment A (engines, transmissions, and brakes), segment C (interiors), and segment D (electrical components). The segments acquire significance as we analyze the categories they account for. Within segment A, the production of the Renault plants established in the early 1980s, manufacturing engines and gearboxes, represents almost a third of exports. In interiors, the production of seats and fenders is the most important activity. Segment D, accounting for the largest block of national exports, is electrical components. Cabling and auto-radio assemblies, two highly labor-intensive activities, represent more than 90% of exports in this segment, which is clearly dominated by foreign firms. Overall, eight categories of production account for almost 80% of the country's exports. These include, besides the six categories referred above,[8] tires and batteries. Within these, only in seats, fenders, and batteries do Portuguese firms play a major role.

The dependence on foreign firms and the secondary role of Portuguese-owned firms is reaffirmed by an analysis of the tier structure among major suppliers. In AutoEuropa only 9 (and these among the least complex) out of 40 main components and functions are coordinated by Portuguese firms. Most suppliers are therefore working at the second and third levels. A similar situation is found in the renovated Opel/GM plant. Out of 17 main suppliers, only five are Portuguese firms (*Competir* 1995). These numbers are expected to grow in coming years, but the fragility of the industry under national ownership is evident. The firms themselves are aware of their shortcomings and the most dynamic units are trying to reverse this situation.

Figure 18.4
Autoparts Market Structure (% of sales)

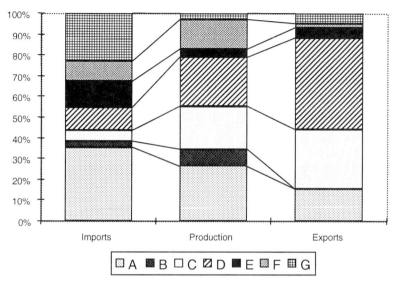

(A = Engines, transmissions, and brakes; B = Chassis and suspensions;
C = Interiors; D = Electrical Components; E = Tires; F = Body; G = Other)

Source: AFIA 1997.

Two major initiatives give a positive indication for the coming years. In 1994, the firm Autosil, a manufacturer of batteries, acquired the French company STECO, becoming the sixth company in the world ranking for the production of this component. It is currently studying the establishment of a plant in India, to take advantage of low production costs and the high-level skill of Indian labor. The second initiative is the ACECIA consortium. Four major companies producing parts and components for car interiors (textiles, injection-molded plastics, seat integration, and stamped metal parts) joined forces with two R&D institutes to be able to present car assemblers with a complete design and production capability for the car interior. The consortium was established this year and no results have so far appeared. Nevertheless, it is an important action that represents a first move of the industry in the right direction.

Clearly, the national autoparts industry is trying to acquire a different status in the international arena, aiming to position itself upward in the supply chain and to establish an international presence. Nevertheless, the poor traditions of the industry in intangible activities such as product design, R&D, and international marketing are still a burden that has to be overcome to reach a sustainable position.

Table 18.2 presents a comparison of key characteristics of the auto industry across latecomer countries that are aiming to play an important role in this industry. In all of them, the promotion of national capabilities has been at the core of industrial policy decisions in recent decades. As can be observed, Portugal has the smallest production base of all the countries considered, both in assembly of vehicles (customer industry for parts) and in parts and components production. Nevertheless, Portuguese autoparts companies export proportionally more than any of the other countries considered, providing a positive sign in terms of the capabilities of the few existing firms. This situation can perhaps be understood, once again, through the country's dependence on foreign firms. Indeed, Portugal is relying heavily on FDI, particularly in terms of fully foreign-owned companies, which are usually highly export-oriented. The challenge for the future is twofold: on the one hand, to be able to keep these firms interested in producing in Portugal, even as wage levels go up; and on the other hand, to create spillovers of knowledge from the foreign companies to national ones, promoting the development of capabilities within the sector and across the manufacturing industry.

CHALLENGES AND PERSPECTIVES FOR THE FUTURE

Because of the international characteristics of the industry, particularly in the case of Portugal, the major challenges for the autoparts industry emerge out of the behavior of foreign players both inside and outside national borders. There are a number of important trends in the industry that are known to manifest themselves independently of geographical location, at the level of both assemblers and suppliers, and that will necessarily affect national firms.

The first aspect is the wave of further expansion outward from its core developed country markets, as OEMs set up assembly and supply capabilities in an effort to gain access to emerging markets in Asia and, more recently, in South America (Auto&Truck International 1997). At the same time, in more developed regions such as Europe, where car ownership has stabilized at two-three persons per vehicle, the same companies are struggling to reduce capacity and rationalize production.[9] It is therefore unlikely that another major OEM plant such as AutoEuropa will be set up in Portugal. There is excess production capacity in Europe and severe competition from other regions to attract OEMs, making it very difficult to base a policy for expansion of the auto sector on the possibility of another important investment. In addition, with the end of European structural development funds in 1999, it is highly unlikely that the Portuguese government will be able to provide a level of incentives anywhere near those provided to AutoEuropa (a third of the total investment of ECU 2,250 million). Without disregarding such a possibility, leveraging the existing base to promote the capacity of the industry may prove to be the optimal strategy. This requires a change in overall policy and the industry's growth patterns, emphasizing the need for careful planning, at the level of both companies and government.

Table 18.2
International Comparison of Autoparts Industries

Country Auto Sales		Portugal (1996)	Mexico (1995)	Taiwan (1995)	Thailand 1995)	Spain (1995)
	Volume (Units)	296,000	580,000	548,000	572,000	2,900,000
	Number	13	14	10	12	11
Assemblers	Production (Units)	233,132	1,200,000*	423,318	526,000	374,998
	Number of Firms	160	600	2,500	400	700*
	Ownership N	49%	79%	N/A	48%	N/A
	F	17%	10%	N/A	3%	N/A
Autoparts	JV	34%	11%	N/A	48%	N/A
	Sales (US$ billions)	3.5	N/A	N/A	5.33	N/A
	Exports (US$ billions)	2.3	5.2	2	1.5	N/A

N/A = not available; N = Nationally Owned; JV = Joint Ventures; F = Foreign-Owned; * Estimate
Sources: Japan International Cooperation Agency; Auto&Truck 1997; U.S. Chamber of Commerce; Taiwan Chamber of Commerce; AFIA 1997.

The second aspect is related to the nature of the competition between regions and nations to provide locational advantages for major producers and suppliers, which is no longer simply on the basis of cost, as more intangible and knowledge-intensive characteristics exert as great an influence as cheap labor (Maxton and Wormald 1994: 132). Only activities that are labor-intensive continue to move according to wage levels, and there are fewer of these activities every day. The heavy dependence of Portuguese exports on foreign-owned activities based on cheap labor is certainly not a desirable situation, since it does not demonstrate other more knowledge-intensive production capabilities and creates a strong dose of uncertainty regarding the future. As East European countries with lower wages join the European market, or establish stronger trade relations, production of labor-intensive components may easily move there. Changing the structure of the industry in favor of components that embody higher levels of technology is essential to affirm the national industry and to keep the multinationals interested in Portugal. Design, material selection, and logistics are key areas where this new breed of suppliers has to be able to respond.

The third aspect is related to the implications of the routinization of just-in-time (JIT) and lean production principles, which change the characteristics of the globalization process (Womak and Jones 1996). OEMs are increasingly demanding that their first-tier suppliers have a global presence together with them, and be able to supply on a JIT basis, with demand intervals as low as two hours—for example, the new Chrysler plant in Argentina (Bradsher 1997). In addition, they are reducing their first-tier supplier base and establishing world supply agreements with the same firms (Taylor 1994). For Portuguese firms to be able to reach the first tier, they must develop the capability to expand production outside of Portugal. The development of investment capabilities that Portuguese firms will have to undertake represents an entire spectrum of issues never before addressed. Domestic resource constraints due to the limited size of the market and the firms themselves make these moves quite risky, and successful integration of firm networks with government policies becomes essential to economize on scarce entrepreneurial and managerial talent, thereby increasing the probability of success.

The need to internationalize production and acquire capacity is also emphasized by the merger and acquisition trend occurring in autoparts (*Wall Street Journal* 1997). In particular, a raid on European companies is being undertaken by Americans, traditionally with a greater presence in the old continent. They are striving to gain the necessary capacity to dilute the growing costs of design and R&D, while offering global supply possibilities for manufacturers. This scenario is bound to be even more important in the coming years due to market deregulation. The trade protection environment, where national capabilities could be nurtured and protected from aggressive international competitors, is now coming to an end. While this had already been true among certain economic blocs (e.g., the European Union), the establishment of the GATT/WTO agreement on free trade is creating a virtually global market for autos and components. The end of restrictions in terms of FDI investment, local content requirements, and tariffs is exposing national competitors to international demand and supply. To prepare for these new market conditions, all governments in industrializing countries are aggressively adjusting their policies. Instruments such as subsidization of learning and acquisition of technological capabilities and proactive internationalization of national firms, among others, are part of the new way in which governments are intervening in the economy.

Growth in this scenario is not an easy task and demands a convergence of national interests. The key question that this chapter aims to address is the role of the state and its industrial policy, which will be debated in the next section.

WHAT ROLE FOR INDUSTRIAL POLICY?

The problems in the Portuguese autoparts industry can be generally defined as those of industrial backwardness. It lacks the resources, capabilities and organization necessary to add enough value to raw labor power. In an economic framework, it is said that domestic production costs[10] (focusing on marginal costs)—MD—exceed international prices—PI (MD > PI). Traditional neoclassical economics tells us that for a national industry to be competitive in the world market it should be exposed to market forces. These will push labor costs down until they offset differences in productivity and regain MD = PI. Because this offsetting is naturally easier in more labor-intensive industries, latecomer countries are destined to specialize in these activities in the earlier stages of their industrialization. Over time, as they learn and increase productivity, they are able to move into higher capital-intensive industries. According to this framework, industrial development is a natural process that evolves over time, and no role exists for the state except, perhaps, adjustment of the exchange rate to assure the competitiveness of national exports.

History has demonstrated that the real market does not work through such linear relationships. In fact, in all late industrializing countries, the state has disciplined labor, driving wages as low as politically possible. Nevertheless, as the case of the competition between the South Korean and Japanese textile industries illustrates, even after massive devaluation, low-productivity late-developing countries (LDCs) have been unable to compete with high-productivity ones. In fact, as we observe the development path of LDCs, we conclude that what accounts for differences in rate of growth of industrial output and productivity among them is the degree to which they were able to discipline capital rather than labor (Amsden 1992).

Information asymmetries and learning patterns are the key reasons that have been explored within nonorthodox economics to explain this fact (see Stiglitz 1989). According to this perspective, and contrary to neoclassical assumptions, technical information does circulate freely among countries. More important, it has a cumulative character that make it impossible for an LDC to jump to the production possibility frontier through the acquisition of technology from the most advanced nations. The development process is therefore accomplished through learning steps in which new technologies are progressively adopted and endogenized in firms. The initial disadvantage of these countries can lead them to suboptimal patterns of development by specializing in technologies with lower learning potentials, resulting in what Stiglitz (1989) calls a low-level equilibrium trap: "a country may not only find itself in a steady-state equilibrium with a low level of capital, using a technology with a low rate of technological change and a low ability to learn, but it may find it optimal (given a high enough discount rate) to remain there."

The suboptimal path possibility is even more acute because of the implications of information asymmetries in capital markets. Problems of adverse selection, moral hazard, and contract enforcement that also exist in developed markets and lead firms to risk-averse behavior are considerably aggravated in LDCs because of the turbulence of the process of change, the smaller scale of the firms, and malfunctions in the institutions that collect, process, and disseminate information. Targeting the right investments, those that are simultaneously profitable and knowledge-augmenting, becomes extremely difficult.

The existence of significant information asymmetries that are not easily solved by market behavior, and the consequent need to plan the investment path, sets the stage for government intervention in the economy. In a developing economy, although the state is facing the same information problems as the firms, it is a single unique actor in terms of availability of resources and the ability to coordinate across various sectors of the economy. The question is what kind of actions it should perform to avoid the low equilibrium possibility and to assure development. More *laissez-faire* approaches limit government intervention to the provision of macroeconomic stability, credit schemes to mitigate risk-aversion of credit institutions toward industrialization projects with higher degrees of uncertainty, and generic institutional development (for example, information centers). At the other extreme there is actual state guidance. This can be enacted through intense widespread industrial targeting and is usually associated with import substitution and export promotion policies (see Shapiro and Taylor 1990).

Both approaches have problems and a middle way may prove to be the best solution. Guidance from the state is usually possible in more closed environments where, using the "infant industry protection argument," a government closes its borders to outside products until the national industry is able to reach either the necessary scale or the productivity that enables a world level price (Itoh 1991). This model, widely used in the Asian new industrializing countries, is of limited application to small open economies such as that of Portugal (Shapiro and Taylor 1990). On the other hand, the *laissez-faire* approach does not address the central problem of suboptimal investment, nor does it consider important aspects such as shortages of national resources. In Portugal, for example, when the government decided to allocate huge investment incentives to the AutoEuropa plant, it naturally created shortages of credit to other projects that were widely felt throughout the economy.

A midlevel, proactive industrial policy is probably the best solution for the Portuguese environment and particularly with regard to the autoparts industry. All the ingredients mentioned above are present. First, firms still lack important technological and managerial capabilities, and they could enter a path of suboptimal investment, remaining as second- and third-level suppliers if they see this as a potential long-term perspective. Moreover, international FDI that has relied mainly on the low costs of Portuguese labor could simply decide to move elsewhere as wage increases hamper the Portuguese development path. Second, if national firms are to enter the first-tier level, they have to invest abroad in order to gain capacity and access to the market. These are risky projects that require the collaboration of several partners—even potential competitors in the national market—a situation unlikely to happen without some mediator, probably the state. Moreover, access to capital for these investments may also be a problem, since banks may see it as being too risky. Another important aspect is related to the choice of parts to bet on. Not all subsectors present the same initial conditions for potentially successful expansion, and the scale requirements associated with scarcity of national resources establish the need for some degree of industrial targeting.

As PEDIP II, the general program for the development of Portuguese industry, comes to an end in 1999, the government has to prepare a new framework for supporting the development of national industry. The auto industry is clearly one of the government's key concerns and could be used as a test bed for evaluating potential initiatives. The following measures for proactive government intervention fall within this evaluation perspective.

1. The government should provide some degree of guidance. This should be done through constant consultation among the state, producers, and export and labor organizations. Flexible but institutionalized channels should be used, enabling a feedback cycle that centralizes information and selectively shares it among firms. Strong national solidarity is necessary within these collaborations to avoid transgressions. An export-marketing cooperative of all autoparts exporters, possibly with compulsory membership to avoid free-riding problems, is a potential example.
2. The state should also provide venture capital for new projects, often at highly favorable interest rates. A development bank, representing forms of lending that traditional banks are not willing to undertake, could be a potential solution. Moreover, the knowledge acquired by an institution that provides this type of credit would prove very important in terms of multiplying effects for other industries. This should not mean that public investment will crowd out private investment, but should rather crowd it in. Several Portuguese autopart firms that are now aiming at expanding to South America are the perfect targets for these credit lines.
3. The state should give support to thrust industries, but specific performance criteria for support should be applied. The obvious measures to be used are exports and technical advances. Additional conditions in terms of scale can also be considered whenever relevant. The present situation, where statistics on R&D spending in the autoparts industry are not widely available and are not part of the annual industry report, demonstrates the poor focus on technical advance that has existed in the past. Criteria that take some of these factors into consideration are being established for foreign firms investing in Portugal, but should be extended to nationals as well.

4. Knowledge spillovers from foreign companies to local firms and universities should be carefully addressed. It is essential that the presence of more advanced companies in the country should create spillovers to the economy to enhance technical advance. The establishment of requirements for R&D funding of national universities and institutes, as well as cooperative research with national companies, ought to be considered.
5. Infrastructure in its widest sense, particularly education in technical skills, has to be enhanced. Portugal is consistently in the rear guard of industrializing nations in terms of technological education statistics.
6. The issue of who bears the cost and who reaps the benefits should be analyzed. As European Union financial aid diminishes, industrial development money and redistributive and labor polices have to go hand in hand (real wages have to rise with productivity growth).

CONCLUSION

Three main conclusions can be reached from the discussion presented here. First, the past record of this industry in terms of overall growth and export capacity is far from assuring a prosperous future. Changing conditions in the competitive environment that value more knowledge-intensive activities and require international production capabilities cannot be tapped by an industry that is limited to national boundaries, still relies on low wages, and is mostly dominated by foreign companies that are not usually willing to disseminate their know-how.

Second, a development path can be charted in a network environment only where national firms pool resources to achieve first-level supply capabilities. This implies the joint development of innovation capabilities, offering advanced design and production services to assemblers. It also requires the creation of investment capabilities, since the establishment of production facilities abroad is a must in the process. In the establishment of these national networks, the creation of a dense multilevel supply structure would ensure that the risk involved in the internationalization process is minimized and that dissemination of best practices and capabilities in Portugal is maximized. The example of the Japanese *keiretsu* is very helpful to understand how the lower levels of a supply chain provide an important part of the value that the upper levels are able to deliver, both in Japan and abroad.

Third, it is important to understand that government plays an important role in a context of resource scarcity, information asymmetries, and risk aversion. Optimization of the development path requires a proactive attitude, through the design of policies that go beyond the resolution of market failures and promote (and enforce) coherent long-term investments. Deciding which national parts and components should be pushed to higher stages may prove to be a necessary step.

ACKNOWLEDGMENTS

The authors acknowledge the research grant of Programa Praxis XXI to Francisco Veloso. They would also like to thank IAPMEI for providing valuable information and fostering insightful discussions.

NOTES

1. The ratio of one in every eight jobs related to the auto industry has held in the United States since 1930 (Shapiro 1993: 39).
2. "Local content requirements" is the expression used to designate the incorporation of parts and components produced in a country.
3. Main sources are IAPMEI (1996), AFIA (1992), Moniz (1995); and ACAP (1997).
4. Brazil started as early as 1949, Thailand in 1960, Taiwan in 1961, Spain in 1964, and Portugal in 1963.
5. It is estimated that 200,000 vehicles per year is the target production level to have the scale economies that allow the costs of production to be competitive worldwide.
6. It is important to understand that efficient production scales for autoparts are of the order of 500,000 per year in stamping, 250,000 in machining, and 200,000 in casting.
7. Even considering the aftermarket
8. Cabling, auto radios, engines, gearboxes, seats, fenders.
9. Renault's desire to shut some of its assembly lines, including the Portuguese one, is public knowledge.
10. The cost dimension includes the cost of the resources necessary to achieve a certain level of quality that is demanded by the buyer.

REFERENCES

AFIA (Associação de Fabricantes para a Indústria Automóvel and AIMA. *O Comércio e a Indústria Automóvel em Portugal: Relatório da Actividade Nacional e Internacional de 1995*. Lisbon, Portugal: AFIA, 1997.

Amsden, A. "A Theory of Government Intervention in Late Industrialization," in *State and Market in Development, Synergy or Rivalry?* (D. Putterman and L. Rueschemeyer, eds.). Boulder: Lynne Reinner Publishers, 1992.

Auto&Truck International. *World Automotive Market Report*. Chicago: Adams/Hunter Publishing, Inc., 1997.

Bradsher, K. "New Factory Resets Sights of Chrysler in South America," *The New York Times* 146 (April 26, 1997).

Competir. "Direcção Geral da Industria," VI, 3–4 (July–December 1995).

IAPMEI. *"The Portuguese Automotive Cluster."* Unpublished report, 1996.

Itoh, M. *Economic Analysis of Industrial Policy*. San Diego: Academic Press, 1991

Maxton, G., and J. Wormald. *Driving Over a Cliff? Business Lessons from the World's Car Industry*. Cambridge: Addison Wesley, 1994.

Moniz, A. *A Indústria Automóvel em Portugal: Tendências de Evolução*. Socius WP 2/95, Instituto Superior de Economia e Gestão, Universidade Técnica de Lisboa, 1995.

Shapiro, H. *Engines of Growth, the State and Transnational Auto Companies in Brazil*. New York: Cambridge University Press, 1993.

Shapiro, H., and L. Taylor. "The State and Industrial Strategy," *World Development*, 18, 6 (1990) 861–878.

Stiglitz, J. "Markets, Market Failures, and Development," *Perspectives on Economic Development*, World Bank Area Papers and Proceedings 79, 2 (1989).

Taylor III, A. "The Auto Industry Meets the New Economy," *Fortune* (September 5, 1994).

Womak, J., and D. Jones. *Lean Thinking*. Boston: Harvard Business School Press, 1996.

19

Technology Policy and Innovation: The Role of Competition Between Firms

Suma S. Athreye

INTRODUCTION

The period since the late 1980s has seen the emergence of industrial policies based predominantly on encouraging more competition between firms in order to obtain industrial efficiency. In countries and sectors where levels of industrial concentration were high, industrial policies that would foster competitive structures were often prescribed as a panacea for the sluggish technological performance of such countries and industrial sectors. Nowhere has this policy of fostering competitive conditions been pursued more vigorously than in the United Kingdom, across a whole range of industrial sectors.

The view that more competition is necessarily better for innovation has historically had many dissenters (Schumpeter 1934, 1942; Galbraith 1952; Scherer 1967). This chapter will focus on the efficacy of relying on competitive policies to promote innovative activity and technological competition between firms. In particular it aims to highlight the role of rivalry and *ex-post* monopoly power, rather than increased numbers of competitors, in the generation of innovations. Supporting evidence for the arguments is presented from a survey of United Kingdom computer firms, in the hardware and software and services segments of the industry.

COMPETITION AND INNOVATION: THE THEORY

There are at least two kinds of theoretical rationales for the belief that more competition promotes innovation and that initial levels of concentration may have an impact on the introduction of new process innovations. First, being a price taker on the product market, the competitive firm has an incentive to introduce a new process innovation because it can pocket the entire difference in marginal revenue and reduced marginal cost resulting from the innovation. In contrast, the monopolist can only sell a larger output by reducing the price of the additional units, as the monopolist faces a downward-sloping demand curve on the product market. Therefore, as marginal costs decrease, so does marginal revenue, which makes the impact on overall profitability less easy to predict. Thus, a firm with an absolute monopoly will have less incentive to innovate. Statically speaking, therefore, the incentive to innovate would seem to be higher under competitive conditions. Since the entire argument rests upon given demand schedules and reduction in unit production costs, it is interpreted as applying to the specific case of process, rather than product, innovations.

In a more dynamic conception where new firms are allowed to enter, the story is obviously rather different. Monopolies earning above-normal profits may attract entrepreneurs to enter the market bearing new process innovations. As Arrow (1971) argued, the new entrant benefits from the higher existing profit levels under the monopoly (relative to competition), as well as from the cost reduction afforded by the new innovation. Indeed the new entrepreneur may have access to the best route for market penetration—that is, undercutting the incumbent. Alternatively, we could have a similar scenario under perfectly competitive conditions. This was the picture painted by Schumpeter in his early work, where the diffusion and settling-down process of the economic system inherent in a competitive situation is destroyed by the temporary monopoly of the "heroic" entrepreneur who innovates in the expectation of monopoly profits. This sets up a new cycle of imitative behavior and diffusion which is broken once again by another heroic innovator. For Schumpeter, the source of economic development lay in this cyclical process of creative destruction.

Dynamically speaking, when new entry is allowed and the number of participants in a market is allowed to change, both competition and monopoly appear to possess different advantages and disadvantages for the innovative entrepreneur. In a monopoly situation, the entrant may benefit from the existing higher prices associated with monopoly, while in the case of competitive markets, entry might be nearly cost-free. The crucial requirement is entrepreneurial heroism and vision, and in this conception innovation happens because of entrepreneurship. The main difference between the two market situations relates to who is more likely to undertake innovation. In incentive terms, the incumbent under monopoly conditions appears to have a less clear incentive to undertake innovation, while the entrant has every incentive. Both statically and dynamically, therefore, more competition may be better for innovation, though the expectation of monopoly profits may be the major incentive that promotes innovation.

This view changes dramatically, however, when innovations˒ are made a function of cost rather than of entrepreneurship alone. If there are predictable costs associated with innovation (such as R&D costs), it seems likely that firms earning above-normal profits will be those more able to bear the costs associated with innovating. Firms facing monopolistic and oligopolistic structures may thus have a greater ability to undertake innovation. This argument was made most forcefully in Galbraith's (1952) work. More recently, Cohen and Klepper (1996) have also argued that a large firm has an advantage in terms of R&D spending vis à vis small firms, as the former can spread the costs of R&D over a larger range of output. Though this argument is also associated with the later work of Schumpeter, the association is a somewhat partial representation of Schumpeter's (1942) view. Schumpeter recognized the advantages that large size may confer in respect of possibilities for undertaking R&D, which was increasingly becoming the activity most associated with the generation of innovation. However, he also believed that monopolistic competition conferred incentives to innovate because of the nature of rivalry in such markets, which tend to concentrate on nonprice factors including technology. In contrast, competitive markets tended to be characterized by price-based competition.

Numerous empirical studies in the Schumpeterian tradition have relied on testing the relationship between concentration and innovation or firm size and innovation. Both concentration and firm size are taken as indicators of imperfect competition and innovation is measured variously by output indicators such as patents or input indicators such as R&D. Reviews of this vast body of literature have generally pointed to the inconclusive nature of the results (Kamien and Schwartz 1982; Cohen and Levin 1989; Cohen 1995). As Kamien and Schwartz (1982: 22–23) point out, the two hypotheses associated with the Schumpeterian hypothesis—that is, a positive relation between innovation and monopoly power with the concomitant above-normal profits, and the proportionately higher innovative activity due to large firms—are independent hypotheses. Possession of monopoly power does not imply large size, nor does large size always imply monopoly power. Sometimes, however, the two may occur together. Unfortunately, most empirical studies have relied on studying large size (a consequence) rather than the role of technological competition in the determination of monopoly power, and hence size (Cohen and Levin 1989: 1078).

It is the point about rivalry and monopolistic structures as facilitating innovation that this chapter would like to concentrate on and explore. The essential point of difference between perfectly competitive markets and imperfectly competitive markets lies in the price-taking or price-making behavior of firms in such structures. This is what is really meant by the phrases "market power" and "monopoly power." Indeed, one measure of monopoly power is the difference between average price and the marginal cost of production. Under competitive conditions this is always zero, implying an infinite elasticity of demand, while under monopolistic conditions this is greater than zero, as long as the price elasticity of demand is greater than one. The degree of market/monopoly power that a firm may exercise is thus related to reducing the elasticity of demand for its product. This point requires further elaboration.

A perfectly elastic demand curve such as a firm would be expected to face under competitive conditions, and which gives it no market power, is what would happen if there were completely homogenous production, so that every customer could buy an identical product from another firm with no loss to its utility.[1] Put another way, every seller competes for the buyer in exactly the same way, so that buyers are indifferent as to which seller to buy from. Contrast this to the situation where customers do in fact have preferences between producers and show distinct preferences for the products of particular firms. There could be many reasons for such preferences: custom, personal acquaintance, proximity. In such a situation firms may produce differentiated goods that are likely to be close substitutes for each other, and spend vast sums of money—for example, on advertising—in order to secure a set of customers for its product. When every firm producing a commodity is in such a position, the general market for the commodity may be subdivided into a series of market segments, within each of which the producing firm may have some market power.[2] Any firm wishing to extend its market to those of its competitors will necessarily incur large marketing costs; but within its own market, protected by the barriers imposed by such marketing costs, the firm enjoys market power similar to that of a monopolist. As competition grows in a firm's particular market segment, so will the elasticity of demand for the product of that particular firm, thus reducing its price-making power.

In the above discussion, which is based largely on the conventional understanding of monopolistic markets, we have assumed product differentiation through advertising costs rather than production. A firm in such a situation maintains its monopoly power but by incurring a cost. Competition and imitation are possible but expensive. Ultimately what sustains the monopoly power of the firm, its price-cost margins, and a lower elasticity of demand is its ability to both appear and be inimitable relative to its rivals. While advertising costs allow the firm to appear inimitable, control of its productive resources, better or different use of these resources in order to increase its own productivity or to improve the quality and characteristics of its product could all secure both better price-cost margins and a far greater and lasting degree of inimitability vis à vis the firm's rivals. Indeed, this is what is usually referred to as technological competition between firms.

Some early work on innovative activity supports this idea and highlights the role of price elasticity of demand in explaining innovative activity. Thus, Kamien and Schwartz (1970) found that price elasticity of demand affected the returns on R&D at the margin. The advantages derived from reducing costs through process innovation were greater the more elastic the demand for the product. However, product innovations, as Spence (1975) demonstrated, benefited more in inelastic markets. The problem with aggregated price elasticity for the sort of argument put forward in this chapter is that the elasticity of demand for a firm under conditions of monopolistic competition is not the same as that for the industry. Indeed the two could vary significantly, as much of the literature on kinked demand curves suggests.

This view of the way in which monopolistic rivalry facilitates technological competition has two implications. First, if the desire for inimitability and the attendant ex-post monopoly power and price-cost margins drive the process of technological competition, then what will sustain such a process is the ability to appropriate the fruits of such investment in technological activities in the form of higher price-cost margins relative to the firms' competitors. A study now exists which supports the hypothesis that the ability of firms to appropriate innovations encourages investment in technological activities (Levin 1988). However, appropriation of innovation is not only a matter of intrinsic profitability of the innovative venture, but also an inverse function of the extent to which the innovative output can be imitated. Notice that the scope for inimitability is in some ways highest when the output of innovation is completely embodied. Where the output of innovation is completely disembodied—the case of patents for example—appropriability is weakest unless supported by institutions such as intellectual property rights.

The most important implication is that where imitability is possible and there are incentives to imitate, this will set off a period of diffusion where more competitive market structures will prevail.[3] Appropriability conditions will thus be defined by the degree of inimitability by other firms, and thus determine whether lasting competitive advantages are conferred by innovation. If appropriability is low and imitation cannot be prevented, then periods of innovation will be accompanied by diffusion and an increase in the number of firms that produce that product. This is the picture that conforms broadly to the outline in Schumpeter's Theory of Economic Development (1934).

If, however, appropriability is high and imitability is low, then firms can achieve lasting competitive advantage as a consequence of innovation. Innovating firms will be characterized by having a different structure of production and organization, and an advantage that will be reflected in the growth of the firm (Nelson and Winter 1982; Geroski and Machin 1993). In this situation, we have the sort of large innovating firm with attendant monopoly that was described in Schumpeter's later work, Capitalism, Socialism and Democracy (1942).

There is a second implication of what has been argued, which might be worth emphasizing. While advertising and marketing help firms to appear inimitable, incurring the costs of technological development may actually make this inimitability more real and controllable by allowing more productivity or allowing firms to endow their product with newer or different sets of characteristics. There is good reason, therefore, to believe that in order to be inimitable, the whole range of strategies may actually go together—that is, innovation costs of R&D will be incurred with marketing costs, with higher product design costs (Evangelista 1996) and the costs of improving the quality of human resources. Several recent studies across different national data sets have highlighted the empirical finding that innovative firms appear to pursue a wider range of strategies conjointly, when compared to noninnovative firms (Napolitano 1991; Arora and Kattuman 1995; Baldwin and Johnson 1996). Flaherty (1983) put forward the same argument for particular industries. All these findings are in line with what we would expect if the desire for inimitability to maintain monopoly power were the driving force behind technological competition and the production of innovative output.

The implications of this view of the importance of monopolistic competition and rivalry in technological competition are less clear for firm size. Consider the first implication. Lack of appropriability because of ease of imitation will tend to make markets more competitive following innovation. Given other factors like similar market sizes, in such markets we may expect to see both smaller-sized innovative firms and technology-bearing entrants. This is because there is very little that can be sustained and retained as a competitive advantage for an innovating firm in such industries.

The second implication—that is, the coordination of a wider range of activities in order to produce innovative output—might imply a larger size. One could argue that coordinating so many specialized functions is only cost-effective with a large output. However, it is also possible that such coordination can take place without an organizational structure with specialized activities such as formal R&D, marketing, and design. As we know from the work of Alfred Chandler, firms set up specialized organizational structures as they expand. An important determinant of the size of a firm is the size of the demand for its product. In large markets a large market share may also mean a large size, but monopoly power depends upon market share in the particular segments in which the firm is involved. If this is large and there is an organizational structure based on specialized activity in each coordinated area of strategy, then adopting a wide range of strategies might also imply a larger size of firm.

The assumption is still that all innovation is embodied in the output of the firm in order to maintain monopoly power. If the innovative output is completely disembodied and intellectual property rights exist to protect such investment, then, as Cohen and Klepper (1996) point out, the R&D and size relationship is likely to break down even further.

To summarize the discussion of this section, two types of innovative dynamics can be associated with monopolistic structures, assuming that successful innovation always confers a temporary monopoly on the innovator. In one, ex-ante (preinnovation) monopoly power and attendant advantages attract innovative entrants. There is a period of creative destruction when innovative outputs eat into the market shares of incumbents. However, the ease of imitability, if it exists, will rapidly promote competitive conditions ex-post (after the innovation has occurred) and prevent innovative entrants from sustaining their competitive advantage by expanding or capitalizing on the innovation for their own growth. In this situation, innovation will be accompanied by competitive conditions and innovators will be both smaller and younger. The same dynamics could characterize an ex-ante competitive market structure as well, but the crucial point is that an ex-post competitive situation makes for diffusion.

The second situation is where ex-ante monopolists are also successful in maintaining ex-post monopoly power. This depends upon the absence of imitability, either because of the inherent costs or because of some form of increasing returns from the innovation. Innovative firms are allowed to build on their competitive advantage and are distinguished by the possession of a wide range of complementary capabilities. One may expect such a situation to be characterized by innovation and monopoly power for innovators, by the larger size of innovating firms, and by older firms appearing to have decisive advantages in innovation.

In the remainder of the chapter, it is argued that the first pattern appears to characterize U.K. hardware firms, while the second appears to characterize U.K. software and services firms.

DATA AND VARIABLES

The evidence we will use is based on the data collected by the Cambridge University Center for Business Research (CBR) survey of 67 randomly selected firms in the computer sector (hardware and software and services).[4] Sectoral breakdown of the data shows that they are largely for software and services firms (only 15 of the 67 firms are hardware firms). This is, however, representative of the total population of firms in this sector.

An output *measure of innovativeness* is used (SALES3) based upon the firms' answers to the following question: What percentage of your current sales is the result of innovations developed by your firm in the last three years? Firms were asked to distinguish between product differentiation and innovative sales. If the answer to this question was greater than 0%, then a firm is classed as innovative, and the variable DINNOV1 takes on value 1. Thus noninnovative firms are those which did not report any sales as attributable to past innovative efforts. This variable has the advantage of being an output measure of innovation but the disadvantage of being a subjective evaluation of a firm's innovative ability. The sales measure (SALES3) is also used as an intensity measure of innovativeness, especially when we try to assess the relationship between innovative sales and competition.

The *environment variables* facing the firms are of two types. The first three are principally based on the response to the questions:

1. How many firms do you regard as serious competitors? This variable (COMPSER1) gives us a direct measure of the scale of competitiveness.
2. How many of these serious competitors are larger than your firm? This variable (COMPSER2) gives a measure of the scale of competition from large firms. We expect it to be higher the smaller the existing firm is relative to its competitors. A more absolute measure of size is derived from a classification based on employment size in 1995 (ESIZ95).
3. How many of your serious competitors are overseas firms? (COMPSER3). This variable measures the extent of foreign competition facing any particular firm.

We use the above three variables to index the present competitive rivalry between firms in two ways. The three variables allow us to capture the extent of competitiveness when looked at cross-sectionally (across firms), and are particularly appropriate in capturing the effect of the extent of rivalry on the extent of innovation. However, for each individual firm, what is being measured in these variables is ex-post competitive rivalry, as the innovative sales we are considering are the result of innovations that were undertaken in the past. This dual nature, in which we can use the same data, and its sector specificity, allow us to say more about the question of rivalry and innovation. Further, in recognition of the fact that software and services are characterized by greater market segmentation than the hardware sector, we also look at the competitiveness of the ith firm in relation to the average amount of competition faced by all firms in each subsector to obtain a better picture of the extent of rivalry.

Another variable we have included under environment is the response of firms to the impact of policy. This is assessed by the score assigned to the question: How would you rank favorable government policies as a factor influencing your competitive success over the last five years?[5]

A set of *strategy variables* is also identified. These are based on the rankings given[6] to various factors suggested as important to firms' competitive success or as barriers to the attainment of firms' growth objectives in the last five years. Table 19.1 gives a full list of these factors, from which strategies and constraints are inferred. We also include a set of *firm characteristics* such as employment size in 1995, turnover in 1995, export share in 1995, and the age of the firm. These form the basis for Table 19.3.

Last, firms that had positive innovative sales were also asked questions about the objectives of their innovative activity. These data are reported in Table 19.4.

The survey data were separated into hardware firms and software and services firms in recognition of the vastly different competitive environments facing the two subsectors.[7] The hardware firm faces competition in a mature and highly standardized product, where leading firms are both large and currently facing declining profit margins. Process innovations are important and, by and large, firms face mass markets. In contrast, the software and services sector has been one where product differentiation and innovation are very important and where niche market strategies are widespread. There has been considerable expansion and high growth for most firms in this sector, and the leading firms here are both large and extremely profitable.

Table 19.1
Strategy Variables Scored on a Scale of 1–5

Factors favorable to competitive success in the last five years

Growth of market demand in the United Kingdom
Growth of market demand in Europe
Growth of market demand globally
Technological leadership and innovation
Product quality or design
Specialized expertise or product
Diversification of products/services
Responsiveness and flexibility to client needs
Development of close/long-term relationships with clients
Marketing and sales expertise
Competitive prices
Established reputation
Research and development expertise
Favorable government policies
Ability to raise finance
Low production costs
For software/services, close relationships with computer hardware manufacturers

Constraints faced in achieving competitive success in the last five years

Availability of finance for expansion
Cost of finance for expansion
Overall growth of market demand in principal product markets
Increasing competition globally
Increasing competition nationally or locally
Availability of highly qualified staff (research, professional, managerial)
Availability of skilled production workers
Management skills
Marketing and sales skills
Difficulties in acquiring or implementing new technology
Availability of appropriate premises or site

COMPETITIVENESS AND INNOVATION: SOFTWARE FIRMS

One of the principal areas of debate, and the most significant for its implications for policy, is the relationship between the scale of competition and innovation. Has more competition been better for innovation? In order to study the relationship between innovativeness and competition, we plotted the variable SALES3 against COMPSER1, both variables as previously described.

In the case of the software and services firms, the relationship between innovation and competition seems unclear. While the pattern is not linear, a non-linear U-shape appears to give as poor a fit, as can be seen from the R-squared value reported in Figure 19.1. This suggests that innovative sales appear to decrease initially as competition increases and then increase again with further competition. It may be worth pointing out that the category of software firms includes some firms who sell all their product as innovative output to one firm. In other words, they are the firms to whom R&D is contracted. This segment of the software market that does purely customized work appears to be populated by a large number of firms.

Figure 19.1
Innovative Sales and Competition: Software and Services

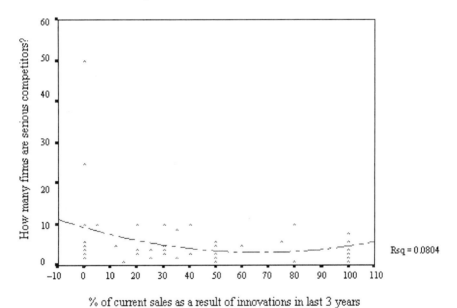

% of current sales as a result of innovations in last 3 years

One way of interpreting the large numbers in the customized segment is suggested by Brady et al. (1992), who studied the development of the U.K. software sector in the late 1980s. In the early 1950s and 1960s, when the seeds of the software industry were being sown, the software firms that existed in the market acted as consultants to military and space agencies and later to hardware manufacturers, who were struggling to develop operating systems for third-generation computers. Brady et al. suggest that many of the emerging, smaller software firms began to offer packages which reflected primarily the converging needs of large numbers of computer users across many sectors. With respect to application software, specialization developed along two lines. Some software firms developed products for general purpose applications such as payroll or accounts where there was a commonality in the requirements of users across many sectors. Other firms evolved to sell to particular large vertical markets such as banking or insurance or the military (1992: 495).

This dynamic of growth in the software sector is still largely true. Firms still start out as consultants and then develop by catering to vertical markets or expanding the range of cross-sector applications. Thus the large number at the customized end of the market may represent new entrants to the industry. If this is true, the correct way to read Figure 19.1 may not be from left to right but from right to left. Starting from competitive conditions, firms slowly carve out their particular niche markets, relying on innovative product sales.

That an increase in innovative sales is associated with a decrease in competitiveness in the case of software firms is also confirmed when we look at Table 19.2,[8] which reports the differences between the environment and strategies of innovative and noninnovative firms in our sample. On average, it appears that innovative firms face less than half the number of competitors compared to noninnovative firms. Further, this difference is statistically significant. The smaller number of competitors generally, for both innovators and noninnovators, confirms the importance of niche markets for growth in this sector. This makes the software sector somewhat closer to the sort of monopolistic competitive situation described earlier in the chapter. However, since the innovative sales reported were a consequence of innovations undertaken three years ago, this also means that innovators face ex-post monopolistic markets.

Might the smaller numbers of competitors simply represent a feature of this sector and not the intensity of competition? In order to answer this, we controlled for the average competition faced by firms in the software sector. By looking at the ratio of the competitors faced by the ith firm to the average competition in the sector as a whole, we obtained a crude measure of the intensity of competition. As can be seen, innovators are still characterized by a lower intensity of competition than noninnovators. So the arguments made above are still valid.

Table 19.2 also provides an illuminating picture of the kinds of strategies that differentiate innovating firms from noninnovating firms. Innovative firms clearly score a different set of strategies as more important compared to noninnovating firms. Thus, R&D expertise, technological innovation, and established reputation are ranked higher by innovating firms.[9] Interestingly, so are marketing and sales skills and difficulties in finding highly qualified staff and appropriate marketing and sales expertise. This suggests that innovating firms adopt strategies that combine effective marketing with technological expertise. They also appear to value trained personnel more.

Table 19.2
Environment and Strategies: Innovators and Noninnovators in the U.K. Software and Services Sector

Variable Type	Mean Score		t-value	P
Description	Innovators	Noninnovators	(d.f.)	(t > T)
Environment				
▪ Number of serious competitors	4.58	10.09	−2.15 (45)	0.04
▪ Number of serious competitors that were larger	3.36	9.70	−2.62 (44)	0.01
▪ Favorable government policies in competitive success	1.36	2.00	−1.80 (46)	0.08
▪ Intensity of competition	0.84	1.85		
Strategies				
Favorable factors				
▪Technical innovation and expertise	4.00	3.25	2.03 (46)	0.05
▪Close relationships with clients	4.37	4.91	−2.19 (47)	0.03
▪Competitive prices	2.69	4.08	−4.23 (46)	0.00
▪Low production costs	1.93	3.20	2.64 (40)	0.01
▪Established reputation	4.08	3.41	1.73 (46)	0.09
▪Marketing and sales expertise	3.22	2.67	1.36 (46)	0.18
▪Growth and market demand in the United Kingdom	2.78	3.58	1.53 (46)	0.13
▪Close relationships with hardware manufacturers	1.67	2.80	−2.24 (31)	0.03
Constraints				
▪Availability of highly qualified staff	2.78	1.75	2.38 (47)	0.02
▪Marketing & sales skills	3.38	2.50	2.01 (47)	0.05

In contrast, noninnovating firms in the software sector adopted a sales strategy based on price competitiveness in the market. Thus, relatively higher scores are assigned by this group to growth of market demand in the United Kingdom competitive prices, low production costs, close relationships with clients (which is extremely important in an industry where 60% or more of current sales consist of orders from previous clients), and close relationships with hardware manufacturers. The last point is interesting because a decade ago relationships with hardware manufacturers were probably characteristic of innovating software manufacturers. Now they appear to be little more than a selling strategy, where software sales get bundled with hardware sales. Interestingly, though selling seems a high priority for noninnovating firms, they do not rate marketing and sales expertise as highly. This is explained by the fact that marketing and sales expertise generally refers to product market development, which might be essential for a firm wishing to commercialize a new technology, but not for a firm wishing to sell on the basis of competitive prices.

Table 19.3 [10] presents some characteristics of innovating and noninnovating firms, notably size and age. It can be seen that innovating software firms are somewhat larger and older than their noninnovating counterparts. They also seem to export larger amounts.

Table 19.4 charts the objectives of the innovating activities of software and hardware firms. The objectives of innovation in the case of software and services firms are dominated by market demand considerations, particularly adapting to changes in hardware technology which would imbue their product with different characteristics. This points to the important role of product innovations. Together with the smaller numbers of competitors and niche marketing, this can probably be interpreted as driven by the need to maintain market/monopoly power.

In short, the software sector exhibits the dynamics of the second kind, where innovation confers *ex post* monopoly and allows firms to build upon it. Thus, innovation is likely to occur in small niche markets protected by reputation and active marketing, both nationally and internationally. Innovative firms are both larger and older and appear to be characterized by a wider range of strategies than noninnovators.

COMPETITIVENESS AND INNOVATION: HARDWARE FIRMS

In the case of hardware, too, the relationship between innovation and competition is not linearly positive but rather more clearly a U-shaped relationship. However, as shown in Figure 19.2, in the case of the hardware firms, we find an inverted U-shaped relationship between competition as measured by the number of serious competitors and innovative sales as a percentage of all sales. This suggests that more competition and higher innovative sales are positively related up to a point, after which additional competition is associated with lower rather than higher innovative sales. In this sense, the pattern for the hardware sector is the opposite of what we argued for the software sector.

Table 19.3
Characteristics of Innovating and Noninnovating Firms: Software and Services and Hardware Firms

Characteristic (average scores)	Software and Services				Hardware Firms			
	Innovator	Non-innovator	t-value (d.f.)	P (t > T)	Innovator	Non-innovator	t-value (d.f.)	P (t > T)
Firms' employment in 1995	44.48	19.00	2.27 (42.82)	0.03	44.00	116.00	-0.89 (6.07)	0.41
Employment size	2.26	1.82	1.84 (25.86)	0.08	2.00	2.83	-1.16 (10.99)	0.27
Firms' turnover in 1995 (£'000)	23,780.92	1,914.25	1.56 (35.38)	0.13	2,527.14	13,311.17	-1.68 (5.47)	0.15
Age of firm in 1997	58.00	11.33	1.01 (42.13)	0.32	11.28	13.33	-0.42 (7.72)	0.69
Exports and a % of all sales	21.19	7.54	1.88 (33.00)	0.07	31.29	19.83	0.63 (8.99)	0.55

Table 19.4
Differences in the Objectives of Innovating Firms: Software and Services and Hardware Firms

Nature of objective	Software and Services (n = 37)	Hardware Firms (n = 7)
Increase markets by:		
Replacing products being phased out	2.92	3.43
Adapting to changes in hardware technology	3.16	2.71
Extending product range	3.22	3.57
Creating new geographical markets	1.83	3.29
Lower production costs by:		
Reducing wage costs	1.51	1.86
Reducing materials consumption	1.30	2.00
Reducing design costs	1.81	1.85
Other objectives:		
Improving labor flexibility	1.41	1.14
Improving product mix	2.05	2.14
Improving product quality	3.21	4.00

Note: None of these differences are statistically significant primarily because of the small numbers in the hardware firms group.

Figure 19.2
Innovative Sales and Competition: Hardware Firms

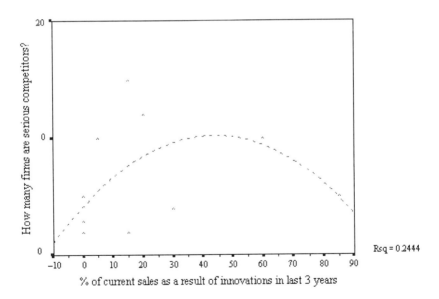

This difference is also confirmed if we look at the average number of serious competitors faced by innovators and noninnovators in the case of hardware firms, as presented in Table 19.5. From this table, it is clear that innovating firms appear, postinnovation, to face more than twice the average number of competitors, more competition from larger firms, and also more competition from overseas firms, relative to noninnovative firms. Furthermore, all these differences are statistically significant, even given the small number of hardware firms. *Ex-post* innovation, the market structure therefore appears to be associated with more competitive situations, compared to the software sector. Again this seems to be true even if we adjust for the average level of competition in this sector. Thus, the intensity of competition is higher for the firms that are innovators.

Table 19.5
Environment and Strategies: Innovators and Noninnovators in the UK Hardware Sector

Variable Type Description	Mean Score		t-value	P
	Innovators	Non-innovators	(d.f.)	(t > T)
Environment				
▪ Number of serious competitors	8.28	3.33	2.37 (11)	0.04
▪ Number of serious competitors that were larger	7.00	1.80	2.85 (10)	0.02
▪ Number of serious competitors that were overseas firms	5.57	0.80	2.05 (10)	0.07
▪ Favorable government policies in competitive success	1.00	2.16	−1.94 (11)	0.08
▪ Intensity of competition (average number for all hardware = 8.5)	0.97	0.39		
Strategies				
Favorable factors				
▪ Technical innovation and expertise	3.57	4.66	−1.94 (11)	0.08
▪ Close relationships with clients	2.14	4.00	−1.96 (11)	0.08

Note: Intensity of competition is defined as number of competitors faced by the ith firm/average number of competitors in the sector.

The results on both competition and innovation suggest that diffusion of new products is fairly rapid in this sector, which is confirmed in Table 19.4, where replacement of products being phased out due to obsolescence is rated very highly by hardware firms as an objective of innovative activity.

The characteristics of innovating and noninnovating firms reported in Table 19.3 are also the opposite of those observed for the software sector. Thus, hardware firms that are innovative tend to be younger and smaller compared to noninnovative hardware firms. This result also reflects the composition of the U.K. hardware sector, where 58% of the industry's output comes from subsidiaries of non-U.K. firms (Athreye and Keeble 1996: 31). Overseas subsidiaries are more likely to substitute for exports through production, and while the parent firms may undertake innovative activity, this would not be reflected in the operations of the U.K. company.

There does not appear to be a great difference between innovative and noninnovative firms in terms of strategy. This could be due to the thinness of the data, which is reflected in high standard errors for most mean values. Of the two significant differences, the first is that innovating firms in the hardware sector appear to be more specialized than noninnovating firms. This might suggest the importance of process innovations and the dominance of a system based on modular production. Innovation in this system is characterized by specialized firms producing different but compatible products that are assembled into a more complex end product. Table 19.4 also shows that hardware firms rate the objective of reducing production costs more highly than software firms. Alternatively, it might just be a sign that innovative activity proceeds in vertically disintegrated segments, such as various kinds of component manufacture. The second difference is that established reputation is ranked higher by noninnovating than innovating firms. This should reflect the importance of entrenched incumbent positions in this market, which must constitute a formidable barrier to entry.

To summarize, the evidence from the rather sparse data available for the hardware sector suggests new entrants bearing new technology in what appears to be a large and concentrated market. Innovative firms, however, do not show many systematic differences in strategy with respect to noninnovating firms, except that they depend less on established reputations for marketing and they seem to base their innovating activities on specialization of production in the product market. This suggests that process innovation may be more important than in the software sector, though some product innovation, in the form of replacing products being phased out, and the creation of new geographical markets are also deemed very important.

SUMMARY AND POLICY IMPLICATIONS

To conclude, we would like to restate the basic propositions being put forward in this chapter. It is argued that the role of *ex-post* monopoly power is of crucial importance in innovation and the dynamics of innovative growth. Both rivalry and expectation of *ex-post* monopoly power can make a difference to the dynamics of the innovation process. Two types of innovative dynamics can be associated with monopolistic structures, assuming that successful innovation always confers a temporary monopoly on the innovator. In one, *ex-ante* (preinnovation) monopoly power and its attendant advantages attract innovative entrants. However, the ease of imitability, if it exists, will rapidly promote competitive conditions *ex-post* (after the innovation has occurred) and prevent innovative entrants from sustaining their competitive advantage by expanding or capitalizing on the innovation for their own growth. In this situation, innovation will be succeeded by competitive conditions and innovators will be both smaller and younger. The same dynamics could characterize an *ex-ante* competitive market structure as well, but the crucial point is that an *ex-post* competitive situation makes for diffusion. We have shown that this dynamic seems to characterize the hardware sector.

The second situation is where *ex-ante* monopolists are also successful in maintaining *ex-post* monopoly power. This depends upon the absence of imitability, either because of the inherent costs or because of some form of increasing returns from the innovation, as in the case of software where buyers are often tied into a particular kind of software. Innovative firms in such a situation are allowed to build on their competitive advantage and are distinguished by the possession of a wide range of complementary capabilities. One may expect such a situation to be characterized by innovation and *ex-post* monopoly power for innovators, by the larger size of innovating firms, and by older firms appearing to have decisive advantages in innovation.

The implication for industrial policy of the argument put forward in this chapter, which links technological competition to monopolistic rivalry, is threefold. First, while monopoly and a single seller are bad for innovation, so is too much competition. What regulation needs to do is to control the degree of competition in national industries and to foster enough competitive rivalry to induce technological competition. This would require specific rather than general industrial policies, which the U.K. government seems unwilling to implement.

Second, it also suggests that there can be no *a priori* leaning of policy toward firms of any particular size, since innovators can be small or large. This last point is rather well illustrated by our data for the computer sector. Hardware firms that innovate appear to be small relative to their competitors, while software firms that innovate appear to be large relative to their rivals.

Third, in the design of industrial policy, the government needs to trade off the objective of diffusion, which consists of inducing firms to improve their productive efficiency by utilizing the best available technology, against a policy of encouraging innovation, which typically involves a shift or increase in such available technologies.

Industrial policy toward the computer industry in the U.K. switched from one of promoting a national champion[11] (under the Labour government in the 1970s) to withdrawal of any specific policy for the computer sector in the early 1980s. At that time, the only policies favorable to U.K. manufacturers in the computer sector were the government's policy toward the development of particular regions, its promotion of small-scale industry, and a general policy that popularized and raised awareness of information technology use. An attempt was made to encourage innovation by the provision of various information inputs—something which is likely to help diffusion more than innovation. The emphasis was on maintaining a suitable climate for competition and diffusion rather than on encouraging competitive rivalry and innovation in particular sectors.

Contrast this to the policy pursued, for example, in Japan where both the above points were clearly recognized. The Ministry of International Trade and Industry there actively encouraged five national companies to engage in the computer sector, with the threat of IBM being allowed to expand its Japanese operations (Anchordoguy 1987). Not only did the threat of IBM force the companies to adopt best-practice technology, but the pressure on the five companies to work in different areas and on joint patents meant that the results of these different policies were quite dramatically different. Despite having a lead in terms of technology, the U.K. has lagged behind in terms of performance, while Japanese firms have both caught up and overcome the competition from American firms in several segments of the hardware market. This result was not, however, achieved by a general policy for all industry but by a specific policy for the computer sector.

Ironically, over the same period the regional development policies of the U.K. government have virtually wiped out the domestic industry in hardware. The entrenched positions of leading multinational firms in this sector have been made stronger and have probably made survival and innovative entry by domestic firms much harder. In the case of software, the expansionary needs of the industry are not well understood, especially with respect to their needs for finance. As a consequence, the larger-sized firms in the software sector are increasingly being acquired by U.S. companies.

Thus, the most telling result regarding the efficacy of current industrial policy toward innovation is the difference between innovating and noninnovating firms in their perception of the favorableness of government policy (see Tables 19.2 and 19.3). While on average firms did not consider government policies favorable to their success (the average score is less than three), there is a statistically significant difference between the value assigned to it by innovating and noninnovating firms. Noninnovating firms found government policies more favorable than innovating firms.[12]

NOTES

1. This is why such a market is also called a buyer's market.
2. The discussion in this paragraph is based on Sraffa (1926: 544-545).
3. Empirical testing of this implication obviously suffers from the difficulties of constructing a measure of appropriability based on the difficulty of inimitability.
4. The CBR survey covered 85 firms in all. However, 22 of the questionnaires were targeted specifically at large firms in the North and South of England. We do not include the data obtained for these firms as they are not representative of the population.
5. Rankings are based on a scale 1 to 5 with 1 being "not important" and 5 being "very important."
6. See Note 5.
7. The U.K. Standard Industrial Classification categories used and their description are contained in the following table:

SIC	Description	Description includes	Description excludes
3002	Manufacture of computers and other information processing equipment (Hardware firms)	Includes the manufacture of automatic data processing machines, including all types of computers, peripheral units, printers, terminals, etc., magnetic or optical readers, and machines for transcribing data onto data media in coded form.	Manufacture of electronic parts found in computing machinery; manufacture of electronic games; repair and maintenance of computer systems.
7220	Software consulting and supply (Software firms)	Analysis of user needs and problems; development, production, and documentation of customized and non-customized software.	Reproduction of noncustomized software; software consulting related to hardware.
7260	Other computer-related activities (Services firms)		Maintenance and repair of office computing machinery, hardware consulting, software consulting, data processing, and database activities.

Technology Policy and Innovation: The Role of Competition Between Firms   321

8. Table 19.2 and 19.3 report t-values assuming equal variances.

9. Though the ranking is higher for innovative firms in our sample, the difference in ranking is not very large. This reflects the importance of rapidly changing technology, in the form of rapid changes in software platform, for example, for product development in the software sector.

10. Table 19.3 reports t-values based on parametric tests, as an assumption of equal variances of the two subpopulations could not be made.

11. The results reported in this section are preliminary since they are incomplete in many respects. As mentioned in the previous section, a large part of the additional data yet to be encoded corresponds to hardware firms.

12. See Henry (1989) for an analysis of policy failures in the early computer industry.

REFERENCES

Anchordoguy, M. *Computers Inc.: Japan's Challenge to IBM.* Cambridge: Harvard University Press, 1989.

Arora, A., and P. Kattuman. "Complementarities and the High and Low Roads to Technological Competence: Evidence from Small Manufacturing Firms in India." Mimeo, Pittsburgh: Carnegie Mellon University, Heinz School of Public Policy, 1995.

Arrow, K. "Economic Welfare and the Allocation of Resources of Invention," reprinted in *The Economics of Technological Change: Selected Readings* (N. Rosenberg, ed.). Harmondsworth U.K.: Penguin Books, 1971.

Athreye, S., and D. Keeble. "Technological Convergence, Globalization and Ownership in the UK Computer Sector." Mimeo, Cambridge University Center for Business Research, 1996.

Baldwin, J. R., and J. Johnson. "Business Strategies in More and Less Innovative Firms in Canada," *Research Policy* 25 (1996) 785–804.

Brady, T., M. Tierney, and R. Williams. "The Commodification of Industry Applications Software," *Industrial and Corporate Change* 1, 3 (1992) 489–513.

Cohen, W. M. "Empirical Studies of Innovative Activity," in *Handbook of the Economics of Innovation and Technological Change* (P. Stoneman, ed.). Oxford: Basil Blackwell, 1995.

Cohen, W. M., and S. Klepper. "A Reprise of Size and R&D." *Economic Journal* 106, 437 (1996) 925–951.

Cohen, W. M., and R. C. Levin. "Empirical Studies of Innovation and Market Structure," in *Handbook of Industrial Organization* (R. Schmalansee and R. D. Willig, eds.), Vol. 2, 1060–1107. Amsterdam: North-Holland, 1989.

Evangelista, R. "Embodied and Disembodied Patterns of Innovation and Industrial Structure." Unpublished Ph.D. dissertation. Science Policy Research Unit, University of Sussex, U.K., 1996.

Flaherty, M. T. "Market Share, Technology Leadership and Competition in International Semiconductor Markets," in *Research in Technological Innovation, Management and Policy* (R. S. Rosenbloom, ed.). Greenwich: JAI Press, 1983.

Galbraith, J. K. *American Capitalism.* Boston: Houghton Mifflin, 1952.

Geroski, P., and S. Machin. "Innovation, Profitability and Growth over the Business Cycle," *Empirica* 20 (1993).

Henry, J. *Innovating for Failure: Government Policy and the Early British Computer Industry.* Cambridge: MIT Press, 1989.

Kamien, M. I., and N. L. Schwartz. "Market Structure, Elasticity of Demand and the Incentive to Invent," *Journal of Law and Economics* 13 (1970) 241–252.

Kamien, M. I., and N. L. Schwartz. *Market Structure and Innovation.* Cambridge: Cambridge University Press, 1982.

Levin, R. C. "Appropriability, R&D Spending and Technological Performance," *American Economic Review Proceedings* 78 (1988) 424–428.

Napolitano, G. "Industrial Research and Sources of Innovation: A Cross Industry Analysis of Italian Manufacturing Firms," *Research Policy* 20 (1991) 171–178.

Nelson, R. R. and Winter, S.G. *An Evolutionary Theory of Economic Change.* Cambridge: The Bellknap Press of Harvard University Press, 1982.

Pavitt, K. *Technical Innovation and British Economic Performance.* London, U.K.: Macmillan, 1980.

Scherer, F. M. "Research and Development Resource Allocation Under Rivalry," *Quarterly Journal of Economics* (August 1967) 359–394.

Schumpeter, J. A. *The Theory of Economic Development.* Cambridge: Harvard Economic Studies, Vol. 46, 1934.

Schumpeter, J. A. *Capitalism, Socialism and Democracy.* New York: Harper Row, 1942.

Spence, A. M. "Monopoly Quality and Regulation," *Bell Journal of Economics* 6 (1975) 417–429.

Sraffa, P. "The Laws of Returns Under Competitive Conditions," *The Economic Journal* 36, 144 (1926) 535–550.

20

Innovation and Quality in the Service Sector: Application to SMEs

Manuela Sarmento-Coelho

INTRODUCTION

Many important transformations and developments, such as the globalization of markets, the formation of economic blocs, and the increasing importance of technological change, are taking place worldwide. The economic and societal model accordingly requires a highly efficient and competitive manufacturing and service industry. This will depend on continuous innovation and systematic quality improvement in enterprises.

The key to competitiveness lies in customer satisfaction, which means that product quality (goods and services) perceived by customers has to equal or exceed their expectations. In order to achieve customer satisfaction, the enterprise needs qualified human resources.

Universities and training/educational organizations have an important role in this. They can also improve the links between research and development, and advanced education and training activities. The objective of *lifelong learning* should be followed by individuals and organizations.

The European Commission's *White Paper on Growth, Competitiveness and Employment* refers to the importance of human resources development to improve Europe's competitiveness and dissemination of R&D results. The importance of strong links between education institutions and enterprises through cooperation is also emphasized.

This chapter presents several factors that should be considered by the enterprise to succeed in the present global and competitive market, and special attention is given to small- and medium-sized enterprises (SMEs) in the service sector.

DEVELOPMENT OF THE SERVICE SECTOR

The service or tertiary sector is very heterogeneous: it is a heavy user of new technologies and provides intangible and immaterial products to society in different sectors such as transportation, distribution, communication, health, tourism, insurance, banking, education, trade, management, consulting, and research, among many others.

The numbers of employees of SMEs in the service sector vary widely. Thus a enterprise with 10 employees will have completely different needs and strategies from an enterprise with 500 employees.

The service sector represents an increasing share of global production activities, with a growing percentage of gross national product and of salaried employees. The increasing importance of the service sector does not correspond to a parallel decrease in manufacturing, but to modifications of both sectors. The manufacturing industry will never disappear, and the service industry will never replace the manufacture of goods. Indeed, products are now incorporating more and more services, and it is often hard to distinguish the two.

CONSEQUENCES OF GLOBALIZATION

Globalization is understood as a process which makes different countries' markets more interdependent in the trade of goods and services. Among the most widely acknowledged factors promoting globalization are recent advances in information technology, manufacturing, and services.

The role of technology in the globalization of knowledge is inducing enterprises to build new forms of cooperation and is becoming vital for achieving economic development. The processes of globalization and liberalization are shaping economic and social changes throughout the world, influencing all sectors of activity and sustaining competitiveness among enterprises, countries, regions, and economic blocs. onsequently the globalization of manufacturing, services, markets, and technological knowledge involves cross-national business ventures. Two of the major issues of globalization for countries and regions are capability development and sustainable development.

GOVERNMENT POLICY IMPLICATIONS

Government and public institutions have an important role through policies and measures that can implemented to provide support for strengthening enterprises. Economic policy should be stable and credible. Public institutions should provide relevant information on: (1) foreign and domestic SMEs, (2) incentives for investments in technological partnership, (3) research data, and (4) aid for reducing risk and building mutual trust between enterprises and institutions.

Effective legal protection of intellectual property rights and protection of research results are also needed, as they give a vital incentive for innovation. Such protection offers innovators the guarantee of a rightful profit from their innovative work.

RESEARCH AND DEVELOPMENT

The economies of developed countries are in a state of continuous innovation. The absence of innovation means stagnation and eventually the death of the enterprise. Scientific knowledge and technological education are prerequisites for the development of society. Thus, progress in societies and enterprises is based on parallel progress in research, education, and training. The R&D of universities and research institutes is the core of this progress.

Cooperative relations between the service industry and universities are important for both parties. The advantages of this cooperation can be: (1) faster industrial application of research findings, (2) improvement of the output of technical qualifications, (3) more effective technology transfer, (4) lower R&D investment by SMEs, as they can exploit know-how to achieve innovation, and (5) additional R&D contracts for universities.

Universities can improve the effectiveness of mechanisms for the development of R&D in conjunction with education and training, allowing companies access to a full spectrum of technology transfer and innovation.

EDUCATION AND TRAINING

Training systems are important for the implementation of innovation and for the progress of society in general. The evolution of technology implies the modification of human resource skills. A better-educated, better-trained, and better-informed workforce helps to strengthen innovation and thereby enterprises.

The lack of skilled and qualified people in some SMEs may, in the near future, hinder the application of new technologies vital for the preservation or enhancement of their competitiveness. Therefore SMEs need contracts with universities for access to R&D results and suitable training and education in order to have competent human resources.

As far as education and training is concerned, SMEs may have a unique learning attitude. SMEs are aware that they can survive only by developing unique core competencies. However, they have neither the means nor the staff to do the same as large enterprises. In many cases they may try to learn as much as possible from their suppliers and clients. They may also be prepared to follow practical courses, organized in the evening, not far from the company's premises.

Human resources have turned into a major strategic asset in terms of company economics, and the predictive management of jobs and skills is a crucial objective for today's enterprise. Beyond technical skills, new qualities are now demanded from employees: (1) good basic knowledge, (2) adaptability, (3) versatility, (4) good communication skills, and (5) ability to grasp new situations, and (6) creativity and ability to innovate. SMEs are aware that in order to achieve their objectives all their employees must have good performance, because the battle to improve quality, increase margins, and retain customer loyalty is now a daily one.

A good human resources policy, which includes innovation and quality cultures, can be achieved only if SMEs implement the following: (1) proper selection and recruitment, (2) high level of motivation and performance, (3) low staff turnover, (4) appropriate training and education procedures, (5) good internal climate among all members, and (6) good entrepreneurial image and culture.

INNOVATION

Innovation is not confined to the manufacturing sector. Innovation and dissemination are playing an increasing role in the service sector. According to the definition proposed by the OECD in its *Frascati Manual* (1993) innovation involves the transformation of an idea into a marketable product or service, a new or improved manufacturing or distribution process, or a new method of social service.

According to the European Commission's Green Paper on Innovation, innovation is defined as the renewal and enlargement of the range of products and services and the associated market; the establishment of new methods of production, supply, and distribution; and the introduction of changes in management, work organization, and the working conditions and skills of the workforce.

Innovation is an essential precondition for growth, to maintain employment and competitiveness, and needs the commitment and motivation of enterprises' employees and partners. The key elements for innovation are: (1) human resources with high-level skills, (2) research and development, (3) the use of new technologies, and (4) continuous education and training.

The creative process requires from human resources: (1) knowledge, (2) awareness of the problem, (3) motivation to reach objectives, and (4) teamwork capabilities. Incentives are the starting point for all innovation processes and can be summarized as: (1) customer needs, (2) market needs, (3) suppliers' influence, (4) intra- and interindustry competition, (5) enterprise strategy, (6) influence of enterprise members, (7) sociopsychological motives, (8) influence of other research results, (9) influence of R&D institutions, and (10) government incentive policies.

To solve an innovation issue, it is necessary to define the problem and then to clarify the objectives. Technology transfer involves the following seven stages: (1) identification of market problems, (2) characterization of the enterprise's external and internal human resources, (3) determination of resources for the problems identified, (4) evaluation of alternatives, (5) selection of the most appropriate alternative, considering service characteristics, costs, market and profit, and (6) development and implementation of a technology research plan.

There are three classes of innovations: (1) radical innovation—a completely new service, good, or process; (2) progressive innovation—a new service, good, or process; and (3) indispensable innovation—a service, good or process modified through successive improvements. Table 20.1 shows the classes and types of innovation. The innovation process and technology transfer for the development of a new product is shown in Figure 20.1, where new product development stages are shown: conception, invention, innovation, and diffusion.

Table 20.1
Classes and Types of Innovations

Classes	Types	Characteristics	Examples
Radical	• Introduction of completely new product	• very infrequent • very large financial investment • result of long-term research • long-term for implementation • very detailed planning	• new vaccines • lasers • printing on paper
Progressive	• Introduction of new product • New processes	• infrequent • large financial investment • result of medium-term research • frequently technological rupture • medium-term implementation • relatively large modifications • detailed planning	• just-in-time • Internet • personal computer
Indispensable	• Improvement of existing product (new features, design, size, raw materials) • Release of new brands of existing products • Modifications of processes • Introduction of new items in product lines • Introduction of new product lines • Modifications of packages (size, design, raw materials)	• common • medium financial investment • result of short-term research • necessary for SMEs' competitiveness • short-term implementation • relatively minor modifications • not detailed planning • influence of economic factors	• cost reduction • enhancement of service • new design of consumer electronic equipment

Figure 20.1
The Innovation Process and Technology Transfer for the Development of a New Product, Good, or Service

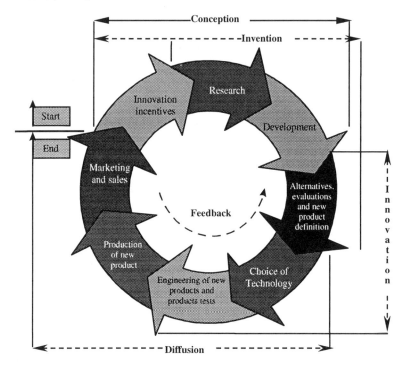

PARTNERSHIP

The enterprise and its environment—clients, suppliers, competitors—are all becoming increasingly complex systems, constantly changing and with multiple interactions. All business is affected by the internationalization of trade. No company is safe from the arrival of a foreign competitor in its domestic market.

The process of acquiring innovation and technological capability is complex and many factors have to be taken into account: time, effort, cost, human resources, risk, and complex interactions between enterprises, and between enterprises and institutions.

Potential technology partners range from research-related institutions (for example, technical institutes, universities, and national laboratories) to consortia, comprising competitors or noncompeting companies. Therefore, in parallel with

Figure 20.2
Access to Technology

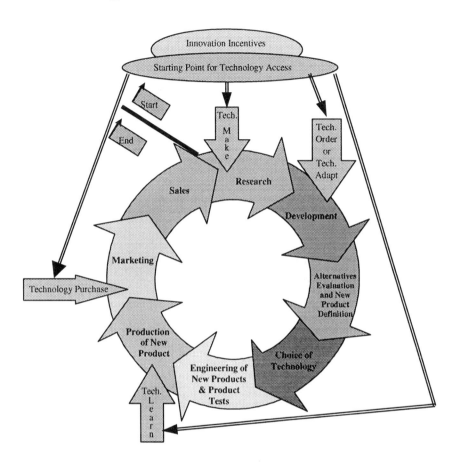

their own internal efforts for strengthening key innovation activities, enterprises have a large range of possibilities for obtaining technology from external sources. To innovate, SMEs can have access to technology in different ways and at different stages, as shown in Figure 20.2.

The ways and stages to access technology can be summarized as follows:

1. Technology Make: the enterprise is responsible for the stages of the innovation process using all its human and technical resources. This is very expensive and involves various risks. Only large enterprises (LEs) like pharmaceutical companies can afford this type of investment.

- SMEs in a horizontal consortium can carry out this process of innovation, having research activity in common. Different enterprises of the same branch of industry will thus share risks, costs and results. The cycle that starts with development is usually independent for each enterprise.
- SMEs are commonly found in a vertical consortium. The research is done by a large enterprise which transfers knowledge to its suppliers, in order that they can develop and produce the necessary components. This kind of partnership is good for both sides. Research costs may be shared, the large enterprise will have components with the required characteristics, and the suppliers—SMEs— will have a loyal customer and new technology or processes.

2. Technology Order: The research is carried out by a university or research institute. This type of partnership is very common with SMEs. The cycle that starts with development, followed by production, marketing, and sales, is the responsibility of the SMEs.
3. Technology Adapt: The research and technology are sold by one enterprise to an SME that adapts the product to its situation. This solution is less expensive and less risky. "Technology push" and "market pull" can also motivate this type of partnership.
4. Technology Learn: the know-how is sold by an enterprise to an SME, which then has to pay royalties to the owner of the research. Nevertheless the receiving enterprise can develop other products based on this R&D. For the owner of the R&D, this is a way to profit from investment. This kind of technology transfer, called licensing, is more common between large enterprises and SMEs. In most cases, LEs can have two objectives: to penetrate closed markets, or to respond to a public contest with an SME, when a national presence is required.
5. Technology Purchase: the most frequently used form of innovation. For instance, an SME purchases software and sells it to a customer without development, or buys new equipment to modernize its production line.

Innovation can be achieved by any SME with the help of intermediary institutions that can collaborate by identifying technological needs, advising on how to innovate, searching for technologies and seeking or supporting private and public financing.

Table 20.2
Intermediary Organizations

Type		Capabilities
Autonomous organizations		• International contacts • Matching demand and supply • Financing demand
R&D institutions	• **Universities** • **R&D institutes** • **R&D laboratories**	• Technological knowledge • Know-how • Access to new technologies
Industry or trade associations		• Identification of sectoral needs

New forms of partnership between enterprises and between enterprises and R&D organizations and universities involving new technology-based companies and government-sponsored incubators have emerged worldwide. Intermediary institutions can be classified in three groups (Table 20.2).

Partnership and the innovation system are important not only for SMEs but also for countries. In developed countries like Germany, the government pays special attention to intermediary institutions. According to the German Ministry of the Economy, there are 1,038 technology transfer organizations, 515 being R&D institutions, 284 autonomous institutions, and 239 industry and trade associations.

CUSTOMER SATISFACTION

Product lifetimes are constantly decreasing, so they must be renewed in order to remain competitive and to respond to customer needs and expectations. However, to render a service or to produce a good of high-quality level means that customer expectations are equaled or surpassed.

For manufactured products, quality level is measured through well-established quality standards for each type of product, but the quality level of a service is measured as a direct function of customer appraisal and satisfaction.

Nevertheless, customer satisfaction depends on (1) quality level of the service, (2) physical conditions of the environment, (3) emotional interaction during provision of the service, (4) demographic characteristics, and (5) psychographic characteristics. Demographic characteristics influence buying standards. Among others, demographic data include age, gender, family details, profession, income, education, nationality, religion, and political affiliation. Psychographic characteristics, such as beliefs, feelings, wishes, needs, values, and emotions, influence the behavior and attitude of customers while the service is being rendered.

Psychographic characteristics are particularly important in the service sector, because the relation between customer and employee is constant while the service is being rendered. It is thus very important to choose the correct employee profile for each service, considering education, training, professional experience, competence, and professionalism, to minimize failures and to maximize customer satisfaction.

COMPETITIVENESS AND QUALITY

Business competitiveness is the core of the modern economy. Competitiveness and innovation keep enterprises alive and fighting for market leadership. To innovate is thus one of the outstanding goals of an enterprise. Competitiveness is the result of total innovation efforts, concerning production and procedures, human resources, finance and administration, and marketing and sales.

A market leader aims at having (1) the best quality, (2) the best customer service, (3) total flexibility, (4) total adaptation, (5) the best response to changing market and customer demands, (6) a continuous process of innovation, (7) a distinctive difference, (8) company internationalization, and (9) ethics in all kinds of negotiation. Today's enterprise has to be aware not only of new technologies, such as information technology and telecommunications, but also of new management techniques, such as commercial distribution and customer satisfaction.

In the service sector, enterprises have to implement a strategy of total quality and innovation management, which involves the following aspects: (1) rapid adaptation to technological changes, (2) cooperative research and development efforts, (3) top management totally involved with the strategic objectives of innovations and quality, (4) a global strategy focused on customer satisfaction, (5) market research on services, goods, and procedures regularly adjusted to suit customer needs and to avoid existing faults, (6) appropriate human resources training for all members, (7) stimulation of motivation, quality, professionalism, and competence among its members and collaborators, to improve workers' performance, (8) responsibility and decision making given to all members, especially to those dealing directly with customers, and (9) quality incorporated in all business stages, from project innovation to delivery to the customer.

CONCLUSIONS

This chapter presented different factors in the success of SMEs in meeting the challenges of a competitive market, the limits of which are international. The following matters were discussed: (1) development of the service sector, (2) consequences of globalization, (3) government policy implications, (4) research and development, (5) education and training, (6) innovation, (7) partnership, (8) customer satisfaction, and (9) competitiveness and quality.

In summary, SME management has to implement innovation and quality and also consider the added value in appropriate investment in human resources, services offered, customer satisfaction, R&D, and the capacity for building partnerships with other institutions or enterprises.

Nevertheless, to respond to the specific skill requirements of SMEs for innovation and quality, a series of four conditions must be satisfied: (1) government support with an adequate level of funding and diffusion of information on European Community and other programs, (2) an appropriate range of research activities, (3) effective mechanisms for transferring the results, and (4) a high level of education and training.

REFERENCES

Albrecht, K., and L. Bradford. *The Service Advantage*. New York: McGraw Hill, 1992.

Amaral, L., and E. Rodrigues. *Service Industry and Enterprise Competitiveness*. Lisbon: Cadernos de Divulgação 39, D.G.I., 1995.

Barsky, J. D. "Building a Program for World Class Service," *Total Quality Management* 37, 1 (1996)

Carneiro, A. *Innovation: Strategy and Competitiveness*. Lisbon: Texto Editora, 1995.

European Commission. *White Paper on Growth*, "Competitiveness and Employment." Luxembourg European Commission Publishing Services, 1994.

Green Paper on Innovation. Luxembourg: European Commission, 1996.

OECD., *Frascati Manual*, "Proposed Standard Practice for Surveys of Research and Experimental Development." Paris: OECD, 1993.

Reynoso, J., and B. Moores. "Towards the Measurement of Internal Service Quality," *International Journal of Service Industry Management* 6, 3 (1995) 64–84.

Ribault, J. M. et al. *Technology Management*. Lisbon: Publicações D. Quixote, 1995.

Sarmento-Coelho, M. "Quality Strategic Groups in Tourism Lodgement Industry," *Técnica*, Lisbon: Instituto Superior Técnico, 1997.

Simões, V. C. *Innovation and Management in SMEs*. Lisbon: GEPE do Ministério da Economia, 1997.

Verma, R., and G. Thompson. "Basing Service Management on Customer Determinants," *Cornell University Review* (March 1996).

Waissbluth, M. et al. *Creación de Pequeñas Empresas Innovadoras*. Santiago, Chile: Alfabeta Editores, 1994.

PART IV:
CHALLENGES FOR NEWLY
INDUSTRIALIZED REGIONS

21

Technological Capability, Policies, and Strategies in Asia

Benjamin J. C. Yuan

INTRODUCTION

Normally, the formation of national industrial technology policy is closely related to the government's administrative structure and the ambitions of its leaders. It is essential to consider the industrial technology policy of relevant nations when designing the contents of national policy. To establish an advantageous industrial technology policy, it is advisable to compare the evolution of industrial technology policy in other nations. Therefore, for the purpose of designing national industrial technology policy, it is important to adapt proved outstanding foreign industrial technology policies in accordance with one's national infrastructure.

In the Group of Seven developed nations, the ratio of research expenditure to GDP and the proportion of research expenditure in the private sector are equally high, followed by the next group of South Korea, Hong Kong, and Singapore. In Singapore, more and less as in Taiwan, the government aims to achieve the goal of allocating 2.2% of GNP as national expenditure on R&D activities, and to make the private sector bear a minimum 50% of total R&D expenses. In addition, the government wants to ensure that people engaged in R&D activities will account for 4/1,000 of the total working population. These countries appear, moreover, to make much of the role private enterprise plays in research activities and regard aggressive research as the key to industrial development.

The role technology plays in economic development has become crucial in fast-growing Asia, as well as in developed countries. Technology policies being adopted by Asian nations are similar in many respects to the policy pursued by Japan from the 1970s to the 1980s. Generally, Asian countries have worked out mid- and long-term technology development plans. These plans reflect their philosophy on what industrial technology policy they should take to ensure greater development of their industries. They have a common notion that technology is essential to industrial development, but basic industrial strength varies depending upon the country.

On the other hand, the so-called NICs (newly industrialized countries), which are highly concerned to promote high-technology and information-intensive development policies, are now shifting from the position of recipients of technology transfer to that of pushing technology transfer to Southeast Asian nations in cooperation with developed countries. China has traditionally shown great interest in the establishment of an industrial technology policy that will greatly help in the applied use of its wide range of fundamental research activities to turn out industrial products.

South Korea is making aggressive efforts to train technologists and promote international cooperation, with technological innovation given top priority in its new five-year economic plans. Under these plans, the government is giving support to large-scale technological development projects in civil aviation, railways, and nuclear energy, in a bid to increase technology development investment to between 3 and 4% of GNP by 1998. In addition, in order to push ahead with technological innovation by the private sector, the government is promoting its Industrial Technological Information Center and building technological databases. It is also assisting the public sector to bolster technological competitiveness in the interests of smaller enterprises, while at the same time working on more aggressive research cooperation with developed and neighboring countries.

Taiwan is actively pursuing high technology as it becomes an example of the global economy. The Industrial Technology Research Institute (ITRI) and the Institute for Information Industry are the main bodies for nurturing technology and they work together with the government to set up and implement technology policy.

The Industrial Technology Research Institute was founded as a private entity in 1973, with a mission to undertake private research with the objective of accelerating the development of industrial technology. There are two ways to achieve this mission: first, by developing innovative technologies for the establishment of new high-technology industries; and second, by integrating relevant technologies in such a way that existing industries can improve their manufacturing efficiency and product quality. The first step in developing a technology strategy is identifying which technologies require investment. These may be traditional, core technologies, or emerging technologies that may have a substantial impact in the future.

Although Taiwan provides a relatively good foundation in the field of industrial technology, R&D activity is insufficient in comparison with advanced countries.

There are major bottlenecks that affect its industrial R&D activity: (1) incomplete legislation related to R&D; (2) insufficient allocation of financial and human resources to R&D; (3) insufficient involvement in R&D by industries, especially small and medium-sized enterprises; (4), lack of technological independence and difficulties in obtaining key technologies; (5), improvements are required in the areas of industrial manufacturing technology, technological capability, and capability for technology development.

Principally, industrial technology should match industrial development so as to support economic progress. Therefore, in accordance with the related objectives of economics, technology, and industrial development and adapting the process of evolution of other nations, the future objective of industrial technology development is basically related to expenditure on R&D, numbers of R&D personnel, technological trade balance, and technological and R&D capabilities. In the short term, Taiwan is still at the stage of technology transfer, and, as an important issue, R&D by itself acts as a complement so as to improve efficiency.

Technology is increasingly recognized as one of the key areas on which the competitiveness of an economy turns. Technology policy means forecasting a country's development and planning its strategy: to start research today to make technology ready for tomorrow and the day after tomorrow. The philosophy of technology policy is based on confidence in its own excellence and the ability to reach an even higher level of development. The aim of technology policy is to help to create a technological advantage in certain sectors of industry, implement it on the world market, and keep this advantage until the next innovative move forward is prepared.

COMPARING THE TECHNOLOGICAL CAPABILITY OF TAIWAN AND MAINLAND CHINA

This research attempts to create a model measuring national technological strength by integrating related theories, and is applied in the cases of Taiwan and Mainland China. We will also provide further research with reference to assessing technological strength. The study has two objectives: first, to discuss the dimensions of measuring national technological strength and the measurement indexes of each dimension; second, to analyze approaches to measuring technological strength.

Basically, technological strength can be divided into three levels: national, industrial, and enterprise. In this chapter we consider measurement at the national level, since the other two levels involve too much to consider. National technological strength includes industrial technology and national defense technology; however, in this study we do not consider the latter.

Seeing that only since the liberation revolution of 1978 has Mainland China started to pay attention to the statistics of technological data, we decided to choose 1988 to 1992 as our research period in both Taiwan and Mainland China.

When comparing technological strength in Taiwan and Mainland China, we center on the factors and indicators that appear most tangible and important, such as R&D expenditure, patent counts, and the like; those abstract factors for which statistics and data are lacking, like the added value of the whole system or the length of R&D time, are not taken into account.

Inevitably, the data obtained do not correspond precisely to the definition of each technology indicator. It can be predicted that the discrepancy in indicators will hamper comparisons of technological strength at the same standard, and the research results may conflict with the actual situation. The reliability of our assessment may accordingly be lower than expected.

In Mainland China, statistical analysis as well as literature on technological achievements are lacking. In 1989, the fourth *White Paper on Technology Policy* appeared; it was not until 1992 that the fifth was published. This is an obstacle to the completeness of our statistics.

The Measurement Model of National Technological Strength

Regarding the composition of technological strength as well as the measurement of technological ability, scholars propose diverse explanations and viewpoints. Some suggest that technology ability comprises, according to its resource attributes, technology, information, personnel, and organization. Others argue that it should also include culture, fundamental facilities, systems, and such. As far as the dimensions of technological strength measurement are concerned, some discuss them from the viewpoint of the system and measure inputs as well as outputs, to evaluate a country's technological strength. On the other hand, others measure it purely in terms of technology—for example, R&D ability of private enterprises, the length of R&D time, the number of key technologies; however, none of these can provide an overall examination of technological strength.

To conclude, there are various factors which affect national technological strength; the interaction between these factors is quite complex, so a complete assessment system is required to grasp the configuration of national technological strength. This study, referring to domestic and foreign literature, seeks to set up a "Measuring Model of National Technological Strength." We believe that, in measuring national technological strength, one should start with seven major factors: culture, talented personnel, technology, organization, system, inputs, and outputs. Figure 21.1 shows the model.

Comparison of the Technological Strength of Taiwan and Mainland China

We shall adopt the same model as that of national technological strength measurement to compare the technological strength of Taiwan and Mainland China. Nevertheless, due to the lack of adequate statistical data from both nations, we

Figure 21.1
The Measuring Model of National Technology Strength

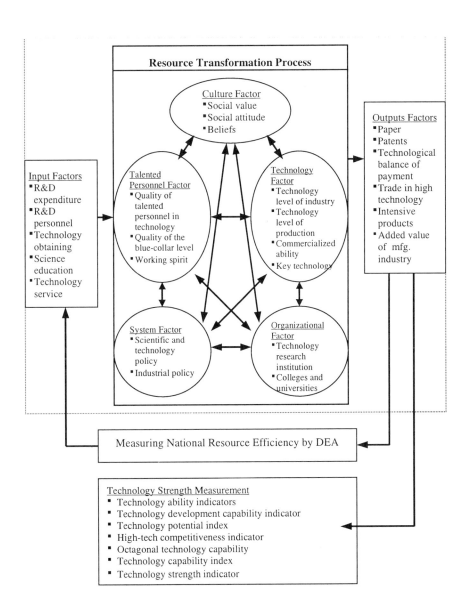

decided not to deal with those measurement dimensions and indicators for which there is insufficient information. In addition, in assessing national technological strength, we reason that the data processing stage is highly complicated, and, what is worse, that it is hard to clearly describe each factor's influence on national technological strength. We regard the data processing stage as, so to speak, a black box. We will therefore not go into details on this aspect. Instead, we will emphasize the measurement of original resource inputs and production outputs to evaluate the national technological strength of Taiwan and Mainland China.

In a comparison of the technological capability indicator in Taiwan and Mainland China, as shown in Table 21.1, in 1988 this indicator for Mainland China was about four times as high as that of Taiwan; in 1989 it decreased to nearly three times, and in 1990 it dropped again to two times. In 1991 it rose to become three times as much as that of Taiwan. In respect to the growth rate of the technological capability indicator, Taiwan maintained a steady growth from 1988 to 1991, and its growth rate increased from 3.28% in 1989 to 5.53% in 1990, then finally to 18.22% in 1991. By contrast, the growth rate in Mainland China was quite unstable: in 1990 it declined dramatically by 26.41%, whereas in 1991 the rate soared to 66.82%.

In a comparison of the technological development capability indicator in Taiwan and Mainland China, this indicator can offer an overall picture of the inputs and outputs for both countries in terms of national technology. From Table 21.2, it can be seen that in 1988, the technology development capability indicator of Mainland China was 5.2 times as high as that of Taiwan; in 1989 it grew to 8.5 times; in 1990, it shrank slightly to 7.3 times; and in 1991 it increased to 7.9 times.

In a comparison of octagonal technological capability indicators in Taiwan and Mainland China, we classify the indicators in two major parts. One is the input indicators of science and technology, including R&D expenditure, R&D personnel, and so on, while the other is the output indicators of science and technology, including technology trade volume (including imports and exports), technology export volume, export value of technology-intensive products, number of patents granted to Chinese, number of overseas patents granted, total value added of manufacturing, and so on. We use these eight input and output indicators in order to draw an octagon, and the bigger its area, the stronger technological capability it shows.

From Figure 21.2 we can note that technological capability in Mainland China is shrinking, in contrast with technological capability of Taiwan. Rising and falling growth of this kind can be explained by relative growth rates. That is, the technological growth rate of Mainland China is lower than that of Taiwan, while by contrast; the national technological strength of Mainland China appears to be shrinking.

Table 21.1
Comparison of Technological Capability Indicator in Taiwan and Mainland China, 1988—1991

	Number of Patents Granted to Chinese [A]	Technological Balance of Payments [B]	Export of Technology-Intensive Products [C]	Total Value Added of Manufacturing [D]	Technology Capability Indicator $\frac{A+B+C+D}{4}$	Growth Rate of Technology Capability (%)
Taiwan	6,586 (59.29)	2.891 (59.90)	117.175 (77.30)	470.324 (87.91)	(71.1)	
1988 Mainland China	11,293 (101.67)	38.343 (794.51)	12.859 (5.61)	1,106.516 (206.83)	(277.16)	
Taiwan	10,397 (93.60)	4.89 (101.33)	196.372 (85.68)	526.492 (98.41)	(94.76)	33.28
1989 Mainland China	15,480 (139.36)	38.192 (791.38)	18.48 (8.06)	985.34 (184.18)	(280.75)	1.30
Taiwan	11,108 (100)	4.826 (100)	229.198 (100)	534.988 (100)	(100)	5.53
1990 Mainland China	1,9304 (173.78)	22.63 (468.92)	26.86 (11.72)	919.985 (171.96)	(206.60)	26.41
Taiwan	13,555 (122.03)	6.072 (125.78)	246.566 (107.58)	628.61 (117.50)	(118.22)	18.22
1991 Mainland China	21,178 (190.66)	47.36 (981.35)	28.768 (12.55)	1,038.148 (194.05)	(344.65)	66.82

Unit: US$100million, number of items, %
Numbers in parentheses indicate the relative indexes based on that of Taiwan in 1990.
Sources: Indicators of Science and Technology, Republic of China (1994), Science and Technology Indicator P.R.C. (1992), Science and Technology Yearbook R.O.C. (1992)

Table 21.2
Technology Development Capability Indicator of Taiwan and Mainland China, 1988—1991

		Technological capability indicator A	Resource inputs of technological development $B = \sqrt{ab}$			Technology development achievement $C = \dfrac{d+e}{2}$			Technology development capability indicator $\dfrac{A+B+C}{3}$
			R&D expenditure (sum of money) a	R&D personnel b	$\sqrt{ab}$	Technology export (sum of money) d	Number of overseas patents granted (points) e	$\dfrac{(d+e)}{2}$	
1988	Taiwan	(71.1)	15.6 (59.11)	63,903 (84.94)	(70.85)	0.1257 (43.49)	458.8 (62.62)	(53.06)	(65.0)
	Mainland China	(277.16)	24 (90.94)	477,237 (634.35)	(240.18)	2.86 (989.62)	49.5 (6.76)	(498.19)	(338.51)
1989	Taiwan	(94.76)	20.9 (79.20)	69,024 (91.75)	(85.24)	0.1313 (45.43)	594.2 (81.10)	(63.27)	(81.09)
	Mainland China	(280.75)	23.8 (90.19)	492,100 (654.10)	(242.89)	8.96 (3,100.35)	52.6 (7.18)	(1,553.77)	(692.47)
1990	Taiwan	(100)	26.39 (100)	75,233 (100)	(100)	0.289 (100)	733 (100)	(100)	(100)
	Mainland China	(206.60)	24.02 (91.02)	617,100 (820.25)	(273.24)	9.889 (3421.8)	50 (6.82)	(1,714.3)	(731.38)
1991	Taiwan	(118.22)	31.75 (120.31)	82436 (109.57)	(114.81)	0.365 (126.3)	906 (123.6)	(124.95)	(119.33)
	Mainland China	(344.65)	26.19 (99.24)	617,100 (820.25)	(285.31)	12.77 (4,418.69)	53 (7.23)	(2,212.96)	(947.64)

Unit: US$100million, personnel, number of items, %

Given that information on the number of Taiwan's and Mainland China's patents which are registered overseas is lacking, we are compelled to substitute points for influential patents in China. Numbers in parentheses indicate the relative indexes based on that of Taiwan in 1990. Given the lack of statistical data concerning Mainland China's R&D personnel in 1991, this study uses 1990 figures instead, on the assumption that there was no growth from 1990 to 1991.

Figure 21.2
Growth and Decline of Technology Capability in Taiwan and Mainland China, 1988—1991

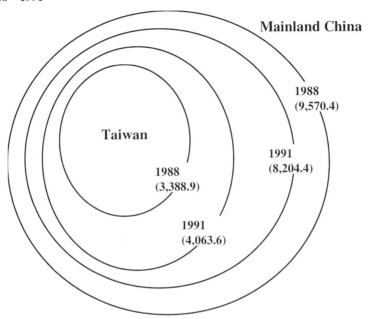

(Numbers in parentheses indicate the relative circle area.)

Comparison of Technology Indices of Taiwan and Mainland China

By means of factor analysis, we shall assess the relative national technological strength of Taiwan and Mainland China. First of all, we will calculate the relative weight of factors which influence technological strength; then a single-multiple indicator centered on these factors (the technology index) will be produced. In this section, we refer mainly to the theory proposed by Blackman (1973).

In assessing the technology indicators of Taiwan and Mainland China, this study adopts the science and technology indicators of the highly reliable Institute for Management Development (IMD) World Competition Report, 1994. This report includes four major categories with a total of 42 individual indexes. However, we exclude ten indicators in this study, because the necessary data of Taiwan and Mainland China for these indicators is lacking, so in total, we have four major categories with only 32 individual indicators. Meanwhile, in order to have sufficient countries for observation during factor analysis, we include Germany and Japan for comparison of technological strength. By doing so, we are able to understand the differences of technological strength between Taiwan and Mainland China and advanced countries. Table 21.3 presents detailed technology indicators and statistical data.

Table 21.3
Comparison of Technological Strength of Taiwan, Mainland China, Japan, and Germany in 1992

Dimension of Assessment	Code	Assessment Indicators	Taiwan	Mainland China	Japan	Germany
	A1	Total expenditure on R&D (US$ millions), 1992	3,049	2,600	109,825	49,103
	A2	Total expenditure on R&D /percentage of GDP (%)	1.74	0.7	2.97	2.53
	A3	Annual real compound percentage growth in total expenditure on R&D, 1988–1992 (%)	18.24	10.05	4.37	3.14
R&D Expenditure	A4	Business expenditure on R&D, 1992 (US$ millions)	1,635	1,344	75,488	33,286
	A5	Business expenditure on R&D /percentage of total R&D expenditure (%)	53.61	51.68	68.73	67.79
	A6	Annual real compound percentage growth in business R&D expenditure (%)	15.36	16.99	4.68	1.47
	A7	Research cooperation between companies and universities is sufficient or insufficient (scores)	4.38	3.6	6.25	5.45
	B1	Total R&D personnel nationwide /full time work equivalent (FTE)	46.2	1969	939.4	478.6
	B2	Total R&D personnel nationwide per 1000 of labor force	6.32	0.33	14.1	14.1
	B3	Annual compound percentage growth of R&D personnel nationwide, 1985–1992	10.92	10.61	3.18	2.66
R&D Personnel	B4	Total R&D personnel in industry /full time work equivalent (FTE)	14.9	886	584	306.7
	B5	Total R&D personnel in industry /percentage of total R&D personnel	32.2	45	64.1	62.2
	B6	Number of scientists and engineers in R&D /university graduate percentage of total R&D personnel	68	44.5	66.3	41.4
	B7	Qualified engineers are lacking or sufficient (scores)	6.65	2.82	6.34	7.87
	B8	Engineering sciences attract talented young people (scores)	6.78	3.06	5.97	6.75

		0	2	4	24
	C1	Nobel prizes/Number of awarded in physics, chemistry, physiology, medicine or economics from 1950 to 1993	5.73	2.98	6.63	7.89
Intellectual Property	C2	Intellectual property in the country is adequately or inadequately protected (scores)	4.28	3.62	6.41	6.58
	C3	Basic research supports or does not currently support long-term economic and technological development (scores)				
Generation	C4	Science and education is or not adequately taught in compulsory schools (scores)	6.83	2.52	6.81	5.74
	C5	*Number of published Science Citation Index articles	4,432	9,225	55,027	50,547
	C6	Total number of patents granted in 1992	21,264	3,966	92,100	46,520
	C7	Number of patents granted in U.S.A. in 1993	1,189	53	22,293	6,890
	C8	Average annual number of patents granted to residents per 100,0000 inhabitants	62.93	0.12	44.09	21.49
	C9	Annual compound percentage change in patents granted to residents, 1986–1992	13.34	72.83	7.47	2.53
	C10	Number of patents secured abroad by country residents per 100,000 inhabitants in 1992	0.61	0.01	57.56	86.96
	D1	Technology strategies of companies is usually or not often well thought out (scores)	5.06	5.08	7.21	6.26
	D2	Sourcing of technology by domestic companies is superior or inferior compared to international competitors (scores)	4.86	1.98	6.35	5.74
	D3	Technological cooperation between companies is common or lacking (scores)	5.39	2.48	6.44	5.6
Technology	D4	R&D in key industries is often ahead or behind foreign competitors (scores)	4.54	2.56	7.14	5.96
Management	D5	R&D spending of the firm is likely to increase or decrease in real terms over the next two years (scores)	6.39	5.76	6.07	6.42
	D6	Production technologies are generally more advanced than those of foreign competitors or not (scores)	5.67	7.86	8.63	6.67
	D7	Lack of sufficient financial resources usually constrains technological development in the firm or not (scores)	5.78	1.6	6.97	7.06

Source: IMD 1994.

Note: * represents the new technological indicator we added to this study.

Regarding the comparison of the national technology index of Taiwan and Mainland China, first of all we convert the four countries' statistical figures, in terms of the 32 indicators, into standardized scores, so as to avoid large differences among variants, which would have a negative influence on the factor analysis. Second, we use factor analysis to obtain the factor loading. At this stage, SAS statistical software is used, and assessment of the variants' communality is performed by Squared Multiple Correlation (SMC). Finally, the method of Principal Axes contributes to the derivation of common factors.

We then key in the standardized statistical figures of Table 21.3, and let the computer proceed with the method of principal axes factor analysis, finding respectively the factor matrix of R&D input index, R&D personnel input index, intellectual property generation index, technology management input, etc.

Finally, we screen the 32 technology indicators and obtain 17 indicators for measurement of the technological strength of Taiwan and Mainland China. Figure 21.3 is a comparison of the four countries' technology index in 1992.

Figure 21.3
Comparison of Technology Strength Indices in Taiwan, Mainland China, and Germany, 1992

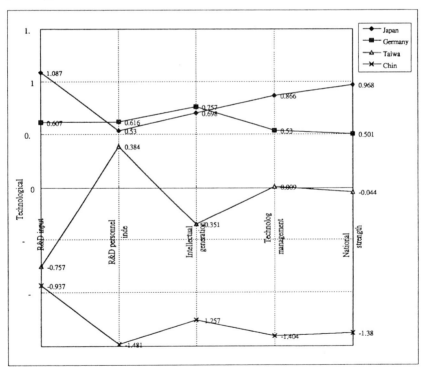

CONCLUDING REMARKS

To sum up, there are a variety of approaches to measure national technological strength, such as technology indicators, questionnaires and experts' evaluations, technology content assessment, factor analysis, data envelopment analysis, and so on. Among them, some are described qualitatively; others are measured quantitatively. Both modes have their strength and weaknesses.

In order to understand a nation's overall technological strength, it is advisable to utilize systematic methods to discuss the interaction between the factors of technological strength. This study offers a survey of the literature, and we propose that, in measuring a country's technological strength, one should start with the measurement of resource conversion efficiency and seven major factors: culture, talented personnel, technology, organization, system, inputs, and outputs.

In 1988 the technology development capability indicator of Mainland China was 5.2 times as high as that of Taiwan. In 1988 it grew to 8.5 times, in 1990 it dropped to 7.3 times, and in 1991 it rose to 7.9 times. According to the octagonal technological capability shown, the relative growth and decline of technological strength in Taiwan and Mainland China from 1988 to 1991 demonstrate that Taiwan's growth rate is much higher than Mainland China's. Consequently, in terms of quantity of technology ability, the discrepancy between Taiwan and Mainland China will be smaller in the near future.

As shown by the empirical factor analysis, in terms of technology competition, Taiwan's technology index is higher than Mainland China's by 1.3 points, whereas it is slightly lower than those of Germany and Japan, by 0.5 and 0.9 points, respectively. It goes without saying that Taiwan has the advantage of greater technological strength. In quality of technological strength, Taiwan is dominant over Mainland China.

As for the R&D input index, Taiwan's is higher than Mainland China's by 0.18 points. R&D expenditure does not include R&D expenditure on national defense. Regarding the technology index of intellectual property generation, Taiwan's is also higher than Mainland China's by 0.91 points. To be more specific, the technology index shows that Taiwan is able to convert input resources into more added value with efficiency—that is, with the same investment. Taiwan's output volume is larger than Mainland China's.

REFERENCES

Abdul Haq, A.K.M. "A Methodology for Measuring Technological Capability of a Country," *Engineering Management International* 3 (1985) 191–198.

Basberg, B. L. "Patents and the Measurement of Technological Change: A Survey of the Literature," *Output Measurement in Science and Technology* (1987) 77–87.

Blackman, A., J. R. Wade, E. J. Seligman, and G. C. Sogliero. "An Innovation Index Based on Factor Analysis," *Technological Forecasting and Social Change* 4 (1973) 301–316.

Chen, C. "Paper in Science and Technology Policy," *Taiwan Institute of Economic Research* (1986).

Dai-yang, L., and Y. Wen-ruei. "A Study on Technology Potential and Industrial Technology Development in R.O.C.," *Proceedings of the 1993 National Conference on Management of Technology*, 1993, 363–380.

350 Science, Technology, and Innovation Policy

350 Science, Technology, and Innovation Policy

Huang, S. "On the Methods and Applications of Data Envelopment Analysis for Measuring the Efficiency of Non-profit Organizations," Ph.D. dissertation, Institute of Management Science, National Chiao Tung University, April 1993.

IMD. *The World Competitiveness Report.* Lausanne, Switzerland: World Economic Forum, 1994.

IMD. *The World Competitiveness Yearbook 1996.* Lausanne, Switzerland: World Economic Forum, 1996.

Jen-yuan, J. "Establishing Input and Output Indicators of Science and Technology Development," *National Science Council Monthly* 15, 10.

Ming-jie, H. "On Technology Ability and Technology Introduction Achievement," Ph.D. dissertation, Graduate School of Business Administration, National Cheng Chi University, June 1992.

Moller, K. "Technology and Growth: Measures and Concepts: A Case Study of Denmark," *Technovation* 11, 8 (1991) 475–481.

Narin, F., M. B. Albert, and V. M. Smith. "Technology Indicators and Corporate Strategy," *Science and Public Policy* 19, 6 (December 1992) 369–381.

Porter, A. L., J. D. Roessner, and H. Xu. "High Tech Competitiveness: Comparing 29 Countries with a Set of Three Indicators," *Technology Management (Proceedings of Portland International Conference on Management of Engineering and Technology).* Portland, Oregon, October 27–31, 1991, 804–807.

Sharif, M. N. "Measurement of Technology for National Development," *Technology Forecasting and Social Change* 29 (1986) 138–171.

Shie, C. "Experience Exchange and Expansion Programs of Technology Management," *Science and Technology Advisory Group, Executive Yuan,* III (June 1992).

Shie, J., and M. Chang. "On Problems of Small and Medium Enterprises' Technology Introduction from Mainland China: A Case Study on Introduction of Infrared Heating Technology from Japan," China Productivity Center, R.O.C., 1992.

Shi-yung, D. *Illustrated Technology Policy.* Taipei, Taiwan: Lan De Publishing Co., 1992a.

Shi-yung, D. *Illustrated Ten Leading Tendencies in Industry and Technology.* Taipei, Taiwan: Lan De Publishing Co., 1992b.

Shi-yung, D. *Methodology of Modern Technology Management.* Taipei, Taiwan: Lan De Publishing Co., 1992c.

Technology Atlas Team, "Assessment of Technology Climate in Two Countries," *Technology Forecasting and Social Change* 32 (1987a) 85–109.

Technology Atlas Team. "Measurement of Technology Content Added," *Technology Forecasting and Social Change* 32 (1987b) 37–47.

Tsai, D., and D. Jou. "On Formation and Development of Technological Capability," in *Proceedings of Symposium on R&D Management in Industrial Technology* (Department of Industrial Technology, MOEA, R.O.C), 1994, 179–193.

22

An Empirical Approach to the Legacy of the Korean Industrial Policy

Junmo Kim

INTRODUCTION

In understanding and analyzing economic growth, the roles of the government and the market have been treated as the major factors that promoted it. Especially in the cases of developing countries, from the discussion of the late developers in the 19th century to newly developing East Asian countries, the question of whether the government or the market has more significance for economic growth leaves room for discussion until today.

With this backdrop, what seems to be more realistic is to accept that the truth lies in between the government and the market. In other words, it is reasonable to attribute economic growth in the developing countries to both of them. After reaching this idea, a sensible way to advance academic discussion is to present an empirical indicator that can show the legacy of contributing factors to industrial and economic growth and thereby approach the issue of the role of the government and the market.

This chapter will present a wage analysis utilizing discriminant functions in order to address the issue raised above with the Korean Heavy and Chemical Industrialization as the prime example and as a metaphor. Thus, the purpose of this chapter is to present a better empirical approach to industrial policy evaluation by utilizing wage as an indicator, and thus resolve the schism between the two fundamentally important constructs of the government and the market.

Wage is selected as the empirical industry performance indicator for two important reasons. The first one is that wage can be a proxy for the competitiveness of an industry, since wage reflects "labor-rent" of workers in the industry. Second, wage data are available in time series format, and this allows a wide possibility to employ different statistical techniques.

In this chapter, Korea's wage performance from 1971 to 1991 is utilized; during this period, the Korean economy changed its status dramatically with a massive industrial policy. We will first present a brief historical overview of the Heavy and Chemical Industrialization. Second, we will review the existing literature on the role of government in industrialization. Third, we will introduce methodology and data employed in this research, followed by a discussion of the results.

A BRIEF OVERVIEW OF HISTORY: THE HEAVY AND CHEMICAL INDUSTRIALIZATION

After successfully implementing its first and second economic development plans, the Korean government launched the more ambitious Heavy and Chemical Industrialization (HCI) in 1973. Despite the official announcement by the president in 1973, foundational sectors for the HCI were already being invested (Enos and Park 1987; Pack and Westphal 1987; Haggard et al. 1994), including the Pohang Iron and Steel Co. (POSCO), construction of which was started in 1968 with its first phase being completed in 1973.

In the HCI drive, six strategic industries were selected for major support: steel, nonferrous metal, shipbuilding, electronics, machinery, and chemical industries (Republic of Korea 1976; Office of the Prime Minister 1978; J..R. Kim 1992). These industries required a huge scale of investments, and government industrial policy was a crucial element in understanding the undertaking of the industrialization (Stern et al. 1995).

To promote the HCI, various policy measures were prepared. While conventionally understood measures such as tax breaks and the provision of industrial parks were included, the essence of industrial policy measures was financial support. To serve this aim, new financial institutions such as the National Industrialization Fund (NIF) were created (J.-H. Kim 1990; Leipziger 1987),[1] while commercial banks virtually under government's control until the early 1990s were the major industrial policy resources.

In this chapter, as a crucial case, I am utilizing Korea's HCI, and will try to present an approach to show how government policy can shape industrial competitiveness using labor-rent based analysis.

REVIEW OF EXISTING LITERATURE

State-Based Theorization and Industrial Policy

As is true in other academic disciplines, the study of industrial policy has evolved through achievements. With the advent of statism in the 1980s, a mainstream was formed in the study of the state including its role in industrialization. State's intervention in stimulating industrialization was not a completely new idea. Historically, mercantilism has served the same goal; Alexander Gerschenkron's historical analysis depicts how German government-backed industrial banks commanded industrialization in the late 19th century (Gerschenkron 1962).

A new element from the 1980s, however, was a more refined theorization of the entity of the state. Succeeding the advent of the dichotomous metaphor of strong state versus weak state (Katzenstein 1978), some authors, depending on their level of analysis, focused on bureaucracy or individual bureaucrats and key policy makers in describing the state (Johnson 1982, 1987). A more refined version utilized a theoretical construct of government-business relations, the essence of which in the statist argument was that specific contexts of government business relations in certain countries make government industrial policy and government leadership over industries meaningful (Samuels 1987).

One example of the government business relationship is suggested in the form of "reciprocity" (Amsden 1989). Amsden's idea was to find "reciprocal" relationships existing between big business and the Korean government since the 1960s. She blends the "relations" with her idea of how technological development has occurred in Korea—namely, combination of mid-tech and shop floor innovation.

While statism was emphasizing the efficacy of government policy, the theory has been confronted with the other theory that focused on "market" side evidence, arguing that Korea's development track can be classified as neoclassical one. The essence of this argument can be summarized as follows: Authors have claimed that the main force to growth has been private, which is coupled with business investment demand, private savings, and a well-educated workforce operating in market-oriented environments (Patrick and Rosovsky 1976; Hughes 1985, 1988; Westphal 1978). In presenting their theories, they argued that high rates of investment were made possible by high savings rates, and together with this, low labor costs helped the economic growth in Japan's case, which is analogously applicable in the Korean case as well. One thing to note here is that these authors are not unaware of the existence of the role of government; however, they have put more "weight" on market factors, because they thought market mechanisms were the decisive forces.

With the two conflicting ideas, research has developed several directions. One is to provide a more in-depth investigation on industrial policy utilizing case study methods. This approach has yielded a comprehensive and clear vision of how actual government supports were provided to firms and how government and firms interacted with each other (Okimoto 1989). One clear example with the Japanese computer industry shows that the government not only provided financial support for a start-up of the industry, but also created demand to make the investment worthwhile (Anchordoguy 1988).

A second stream is to combine state side and market side ideas. Robert Wade, in this line of thought, presented his "Governing the Market" concept (Wade 1990), in which he claimed that government managed quasimarket systems. Taking Wade's synthesis as an example, the other examples in this second stream, provided by both economists and political scientists, made a more refined contribution for understanding industrial policy in Korea in the sense that their presentation of data is more systematic and persuasive than their predecessors in the field of industrial policy (Leipziger 1987).

Another improvement of this group of authors is that they opened their eyes clearly to the Korean context. Earlier statist authors had not treated the fact that Korea's HCI had experienced very serious crises after the two big pushes (1973–1974 and 1978–1979). In comparison, some of the second-tream scholars began incorporating the problems of lagging performance of the HCI sectors. The most explicit attempt on this aspect can be found in Auty (1992, 1994), who described the slow learning of the HCI sectors in Korea as "the maturation of the HCI sectors."

Weaknesses of the Earlier Literature on Industrial Policy

As Korea's HCI case dramatically shows, government industrial policy can be inefficient and regarded as a failure, depending on the position authors take and on the time frame specific authors may take into account. As almost two decades have passed by since the initial pushes, the HCI sectors are the locomotives of Korea's export growth. In the export of year 1995, the HCI products made up almost 30% of all exports, excluding electrical and electronics sectors. Considering that electrical and electronics sectors were also targeted as a part of the HCI promotion, the percentage of exports the HCI products made up are almost 64% of the total exports (*Korea Economic Weekly* 1995). With this reversal of performance, what is possible to argue can be articulated as follows based on the summary discussion of the preceding section.

First, the existing literature, in general, suffers from the problem of insufficient empirical support. When the literature offers evidence in the form of data, the data are usually what the government offered in its policy in "raw" data form, rather than analyzed empirical results of the state's policy.

Second, to specify what I meant by the first point, it is possible to present different degrees of empirical support according to each theoretical approach.

For approaches that generally base their positions on statism, there has been a tendency to overemphasize the role of government. Instead of showing to what extent government produced efficacy, the authors ended up providing raw data and transferred the burden of proof of their arguments to readers who have to read through the data and understand what it has to do with their arguments. For those approaches with more refined use of data and better understanding of the Korean contexts, they tend to offer "supply side evidences," which is what governments have provided. In my view, what is needed is to show variables to evaluate industrial policy.

With the above discussion, we have shown that the existing frameworks, while they have contributed to the understanding and assessment of the industrial policy, still contain limitations in the assessment of industrial policy. Bearing these limitations in mind, in the next section we will present the analysis employed in this chapter. First, we will introduce the theoretical background of the research. Then we will proceed to the data source and methodology section. Finally, we will present the findings yielded from the wage analysis utilizing cluster and discriminant analysis.

METHODOLOGY AND DATA

Theoretical Background

In this chapter, wage is selected as the industry performance indicator. Now it is possible to present on what grounds wage is an appropriate measure in this research. First, in the analysis here, we will employ the notion of industry-specific labor rent, which distinguishes each industry from another and forms a clear indicator for industry competitiveness (Katz and Summers 1989).

Second, in utilizing the notion of labor rent, it is claimed that there is industry rent-sharing between firms in the industry and workers in the industry (Blanchflower et al. 1996). Third, since wage data are available in time series format, and factors exist that affect industry performance change over time, those changes will be reflected in the wage performance of industries. Furthermore, if there is a pattern of change that is identical but differs only in degrees for two industries, over an extended period of time, it becomes possible for us to acknowledge the "forces" behind the wage performance of those two industries, and reject the possibility of it being an accidental occurrence.

The different degrees of response by each industry against the underlying common change factor (force) can also be utilized as an industry classification scheme. When compared to the conventional scheme of industry classification, such as SIC, the new classification scheme shows natural evolution of industry through changing factors.

With these characteristics, it becomes easier to see how each industry's competitiveness has changed due to economic conditions and government policies. In other words, wages, when coupled with appropriate techniques, can work as a litmus test or chromatography of factors that affected the industry performance.

Use of Annual Wage Change

In utilizing wage data, we used the yearly change of wages for each industry. The reason for this is to avoid a problem in using cluster analysis. Since cluster analysis is sensitive to the units and scales of variables, by using annual change rates, we can avoid the problem of changing the implicit weight of each industry, which can lead to a change of the group structure.

Data

In this analysis, the Occupational Wage Survey (OWS) collected and published by the Ministry of Labor in Korea 1971–1991 was used as the data source. The OWS covers approximately 18% (1971) and 29% (1991) of the total economically active, nonagricultural population in Korea. Also, the OWS covers both manufacturing and service industries. The sampling frame used in this survey is all workplaces that hire more than ten workers. Due to the systematic sampling frame this survey takes, the data set reflects Korea's mainstream labor force trends.

Methodology

Cluster Analysis. The eventual goal of this research, in relation to discriminant analysis, is to find underlying economic forces in wage performance of industries, which is yielded by using discriminant functions. To do that, it is necessary to assign group structure to industries. Cluster analysis, utilized for this purpose, is a technique that classifies objects (industries) based on the similarities found in their characteristic variable (Aldenderfer and Blashfield 1984; Lorr 1984). In this research, the characteristic variable is the annual wage change rate for each industry.

In actual calculation, a typical matrix shows a form of $N \times (T - 1)$ matrix A, in which $A(i, t)$ means the annual wage change of industry i in year t. Each row, a profile of an industry, is filled with annual change rates of the industry for the period of research.

In this study, as for the similarity measure, Euclidean distance is selected; this measures the distance between industry profiles, which is the distance of the paths across $(T - 1)$ years of wage change patterns between industry. The next step in cluster analysis is to choose a structural model and an appropriate clustering method. In this chapter clustering is used with Ward's method as the clustering method. Ward's method was chosen for its advantage of maximizing between group variance and minimizing within group variance (Ward 1963). Based on the above description, the following group structure was yielded from the Occupational Wage Survey of the Korean Wage Data.

Discriminant Analysis. The main purpose of performing a discriminant analysis is to find out the factors that yielded the clustering pattern acquired from the cluster analysis.[2]

From the discriminant analysis, we get several canonical discriminant functions, and among them we can select statistically meaningful ones. Each discriminant function, denoted as F, can be understood as a function that expresses a force that underlies the pooled wage variation across industries (Tatsuoka 1988; Klecka 1980). These functions are matched with real-world economic data series in the later stage of this research (Galbraith and Calmon 1990, 1996).

Among the selected roots, different degrees of importance are attached to them (Morrison 1969). The first root takes the responsibility for the greatest percentage of wage variation and has the greatest eigen values. As the number of roots (functions) goes up, its relative importance and eigen value is diminished.

In a functional form, a discriminant function can be written as $F = a_1 \times 1 + a_2 \times 2 + \ldots + ap\ xp$, and in the context of this research it can be expressed as $F = a_1$ (wage change in year 1) + a_2 (wage change in year 2) + ... + at − 1 (wage change in year t − 1).

Deriving the above discriminant functions can be explained in the following way. Since each discriminant function is to distinguish group means in a way to maximize between group variance and minimize within group variance, when coefficients from a1 to ap form a vector, Lambda (λ), the discriminant criterion, can be expressed as follows.

$$\lambda = \frac{a'Ba}{a'Wa}$$

Using calculus, we can get following (Galbraith and Lu 1997).

$$\frac{\partial \lambda}{\partial a} = 0$$

$$(W^{-1}B - \lambda I)a = 0 \ \text{ or } \ W^{-1}Ba = \lambda Ia$$

Here, λ, the discriminant criterion, is the eigen value and a is the eigen vector or eigen root. The eigen vectors are the coefficients of the discriminant functions (Morrison 1969).

Then, the last stage for discriminant analysis is to match eigen vectors with real-world economic data to provide contextual meanings of underlying forces to wage performance (Ferguson and Galbraith 1996).

FINDINGS

In this section, we will present the results we have yielded by analyzing industrial wage data of Korean industries in light of industrial policy assessment to integrate existing discussion with better empirical support.

Extraction of Roots

By performing a discriminant analysis based on the grouping structure with the cluster analysis method, five canonical roots are extracted. Among the five roots, three statistically significant roots are selected. With these three roots, nearly 89% of wage change variation over a 21-year period (1971–1991) in Korea is covered. With this, the next step in analysis was to find "real" world macroeconomic data series that match with these three roots.

The First Root: The Real Interest Rate Linkage

The first root which explains about 48% of the total wage change variation over the 21-year period in Korea is best matched with the annual change rate of real interest rates.

Then, a question arises: what does this linkage imply to the study of industrial policy in Korea? From the existing literature on Korea's industrialization, it is well known that the government rationed capital in the form of policy loans (Stern et al. 1995; Cho and Cole 1992; Amsden 1989; Wade 1990). On this point, different theoretical approaches nearly agree. In a society where capital is a scare resource, access to capital is an advantage. Even clearer is the advantage that comes from lower interest rates bearing policy loans for targeted industries in comparison with conventional loans based on market rates for nontargeted sectors (see Figure 22.1).

Figure 22.1
Evolution of the Real Interest Rate and the First Root

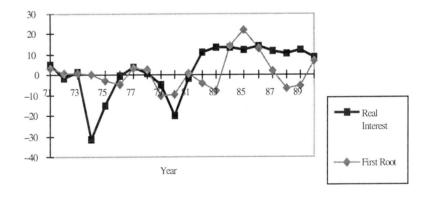

During its massive and ambitious industrial policy implementation, especially in the big push of the HCI in the 1970s, the Korean government offered preferentially lower interest rates to newly starting heavy and chemical industries. This reduced the burden of the HCI sectors during their initial years of investment. Currently available raw data cited by different academic publications present the different interest rates given to the HCI sectors and how much capital has been allocated *vis à vis* the other sectors.

From wage analysis presented in this chapter, it is possible to show the impact of that policy treatment on the HCI sectors. In Figure 22.2, the horizontal axis shows the scores each industry gets on the real interest root; the higher the score, the better that industry's wage performance in relation to the real interest rates. The vertical axis is the cumulative wage change percentage (actual percentage divided by 10).

Figure 22.2
Real Interest Rate and the First Root

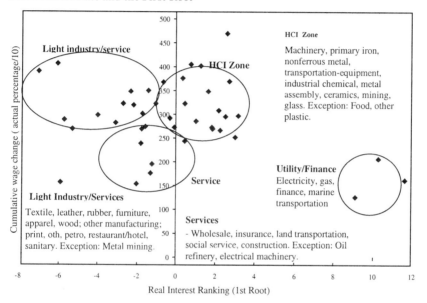

When real interest rates are higher, financial sectors, electricity, gas & utility and marine transportation industries performed well, although their overall wage gains over the 21-year period were not significant.

Insulation of the HCI Sectors. Our prime focal points, the HCI sectors, are well insulated in the middle, and their overall wage increases over the 21-year period were higher than average (between 2,500% and 4,000%). Among the HCI sectors, nonferrous metal and machinery sectors outperformed in terms of cumulative wage growth, followed by the first iron and steel industry, the transportation equipment (auto) industry, and the industrial chemical industry.

On the other side of HCI emphasis was the reflexive underemphasis of light industries and service industries. This has been discussed by earlier studies. This wage analysis clearly shows their relative position. Light industries, such as apparel and textile, and service industries, like wholesale, real estate, and sanitary industries, marked lower on the real interest ranking, which means that these industries were hurt when real interest rates were higher.

The point that real interest rates effect against light manufacturing industries and service industries becomes more understandable when one considers the terms of loans these industries would usually get. While heavy and chemical industries mainly rely on long-term capital for their equipment, light industries and especially service industries are more likely to depend on short-term capital, which aggravates these industries' wage performance portion that has been affected by the real interest rates.

Real Interest Rates: A Window for Receiving International Economic Influence. While real interest rates had an important role in presenting the legacy of industrial policy and the trace of government's targeting that sacrificed other sectors, the real interest rates linkage has another important dimension in understanding the mechanism that received international economic influences.

Conventional wisdom dictates that an international economic fluctuation, such as an oil shock, would influence the economy overall, including the wage performance of an economy. Indeed, international economic variables do affect the economy. It is very well known that foreign exchange rates are one of the key variables that affects the export performance of Korean industries (Jwa 1990).

A deep inspection of the econometrics-based research prompts one to make an inference that these international economic variables affect the "output" side of the Korean economy directly, rather than the wage (distribution) side. What does this imply? On this we can suggest the following argument. Scholars in historical institutionalism and statism have presented the role of domestic institutional structures that buffer international shock, and shape domestic spectrum of an economy (Steinmo et al. 1992; Gourevitch 1986; Katzenstein 1985).

When we analyzed the Korean wage data with discriminant analysis and tried to match canonical roots, it was impossible to find international variables that match well with the roots. As we will discuss later in this chapter, the other two roots, other than the real interest root, are not international variables. They are domestically determined ones.

Reconciling the idea that an economy like Korea's is exposed to international economic influences (and therefore the impacts of international economic changes should be reflected in the wage performance) with the other finding that all the economic series data that match with canonical roots are domestically determined ones, suggests to me that there is a connection between the two in the following way.

There must be a mechanism through which the foreign impacts are relayed to domestic areas; the answer comes from the interpretation of the real interest rates. One especially interesting point is that real interest rates act as a receiving point or a window for receiving international economic impacts. More striking is that not only do real interest rates transmit foreign impacts, but also the real interest rate itself is a reflection of domestic institutions.

Thus it is plausible to argue that the real interest rate is a mechanism where foreign impacts are received and mediated in the wage performance. In other words, a domestic institutional structure that consists of financial systems including the real interest rates buffered wage performance. This is empirical support for the historical institutionalists' argument on the role of domestic institutions (Gourevitch 1986; Hall 1986).

Some positive evidence already exists to support the role of real interest rates in Korea as well as in other East Asian countries (Edwards and Khan 1985). In Reuben's research (Glick and Hutchison 1990), the U.S. real interest rates are affecting other countries' real interest rates. This is clearly applicable to the Korean case as well (Jwa 1993; T.- J. Kim 1993).

Overall, with real interest rates, it is possible to show the industrial policy legacy, and real interest's role as a window to receive international economic impacts. Real interest rates, together with the other two roots, were the essence of the domestic institutions that made the conduct of industrial policy work.

The Second Root: Investment Root

The second root which covers about 22.6% of the total wage change variation over the 21-year period from 1971 to 1991 is best matched with the four-year moving average of investment volume.

Figure 22.3
Investment and the Second Root

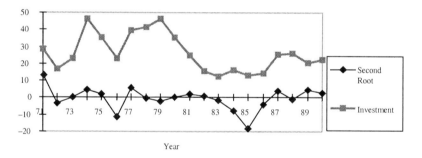

Year

In Figure 22.3, the second root and the annual change rate of the four-year moving-averaged investment behave in the same direction. The importance of this root is that investment is an actualization of industrial policy.

To see how different industries performed on the investment root, we can present the following discussion. In Figure 22.4, the horizontal axis is the canonical scores on the investment root. The higher score an industry has on this axis, the higher the industry's wage performance when investments are strong. The vertical axis shows the cumulative wage change over the 21-year period.

Figure 22.4
Moving Averaged Investment and the Second Root

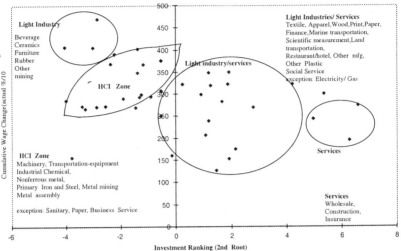

Interpreting Figure 22.4 provides some ubiquitous wage performance patterns (Galbraith and Calmon 1994) and some nonconventional findings. As would be similar in other countries, construction and wholesale industries were in the top group in terms of wage performance on the investment root, although wholesale industry's overall wage gain was smaller than that of the construction industry.

The middle group includes light manufacturing industries, such as textile, apparel, and printing industries, and service industries, like land transportation, marine transportation, social service, and finance industries. The electrical machinery industry was also a member of the middle group.

The unconventional finding that was contrary to my prior expectation was the location of the core HCI sectors. Machinery, first iron and steel, industrial chemical, and transportation equipment industries were all located in the minus zone on the investment root. Interpreting this can produce at least two possibilities. The first one would be that despite heavy influx of capital, the HCI sectors' wage performance portion influenced by investment linkage was not substantial. The second interpretation is that the HCI sectors may have suppressed wages to compensate for heavy investment.

The Third Root: Credit Availability

The third root covers approximately 18.6% of the total wage change variation over the 21-year period in Korea. This root is best suited with change rates of credit availability. Domestic credit availability is a measure that expresses the extent government policy can be exercised. In the earlier part of this chapter, we have shown the portion of policy loans among total domestic credit available (Stern et al. 1995). Overall around 50% of the total available domestic credit was allocated by government-backed banks as policy loans. Thus, credit availability is an indicator to show the degree to which industrial policy can assist target industries (see Figure 22.5).

Location of the HCI Sectors. In Figure 22.6, the most outstanding observation is the location of the HCI sectors. In terms of location on the credit availability root, starting from nonferrous metal, first iron and steel, industrial chemical, and machinery industries are all located in the frontal area, which indicates that these industries did perform well in terms of wage performance when credit availability was increased.

Supply-side Industrial Policy versus Demand Creation Industrial Policy. There is strong evidence to link the findings with the existing description and historically based studies on industrial policy in Korea. From the early start, there have been different financial mechanisms that were designed to help the HCI sectors. In our observation, these financial mechanisms or funds can be divided into two groups: one is investment related (supply side) and the other is demand-side related.

Figure 22.5
Evolution of Credit Availability and the Third Root

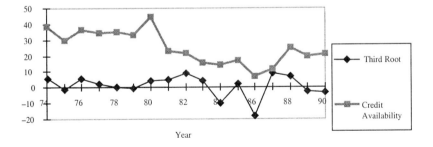

Figure 22.6
Credit Availability and the Third Root

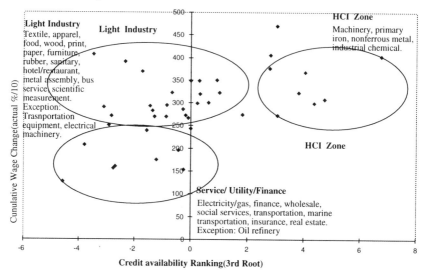

The supply-side mechanisms that helped the HCI sectors include provisions for preferentially low interest rates that these sectors could get and the provision for the funds. Considering the amounts of investments required for the HCI sectors, the lower interest rates could have been helpful *vis à vis* non-HCI sectors.

The demand side is related to financial mechanism and provisions that have created and stimulated the demand for domestically produced HCI products (Anchordouguy 1988).[3] The planned shipbuilding fund was to increase merchant fleets owned by private firms, but in doing so shipbuilding industries could get orders for the ships. The machinery localization fund was distributed to firms that would desire to purchase domestically produced machineries (Haggard et al. 1994).

The demand-side provisions did not end in the domestic arena, but can also be found in export deals. In general, machinery exports accompany long-term financing for those products. This has also been the case for Korean HCI sectors. To serve this end, Korea Export and Import (EXIM) Bank has provided a fund to assist deferred payment loans to industries. Thus, sales of the HCI sectors products were, to a great degree, a function of how much credit is available.

This element of industrial policy is not a unique recipe confined to the Korean case. The Japanese Ministry of International Trade and Industry (MITI) also has conducted a demand-stimulating mechanism in promoting the Japanese computer industry in the 1960s. In my view, demand creation is the second stage of an industrial policy to ensure that the effects of the first-stage industrial policy, which would be the influx of capital and subsidy, are fully materialized.

It is not surprising to find the existence of a similar policy instrument in Korea and Japan. Also interesting is that the legacy of the demand-creation policy is found by the wage analysis utilizing discriminant analysis.

Other Industries on the Third Root. The transportation equipment industry, which includes the auto industry, is not located in the HCI concentration zone in the analysis of the third root. This is different from the "behavioral pattern" of the transportation equipment industry in the analysis of the first and second roots. On this point, there is a convincing argument. While most of the heavy industrial sectors depend on "industrial orders," which in turn depend on credit availability, demand for automobiles is more consumer driven. Even if we take into account the point that consumers also use auto loans, demand for autos is still a more individual-based decision.

A big circular concentration includes, in Figure 22.6, light industries, most of the service industries, and the transportation equipment industry. The construction industry is after the HCI group on the credit availability root. This is understandable because most construction projects need financing.

CONCLUSION

In this chapter, I have attempted to present the Korean industrial policy legacy from an angle not explored by previous scholarship by pursuing discriminant-analysis-based wage analysis. One advantage over the existing industrial policy literature was that we were able to offer a better empirical rediscovery of policy impacts. In this sense, we can claim that we have moved the assessment of industrial policy one step forward.

The fact that real interest rates and credit availability were two of the three prime eigen roots that captured wage-change variation in Korea suggests that there are complex arrays of domestic institutional arrangements, which include government policy measures. Thus, instead of arguing that government policies are a panacea or the major force, an implication we can glean from this research is that government policy does work in a network of domestic institutions.

Another interesting finding was the role of domestic settings, which worked as an insulation to keep the HCI sectors from economic fluctuations. An implication from this point is that the insulation buffers the HCI sectors while they mature in their competitiveness.

NOTES

1. According to Leipziger (1987), about two-thirds of the NIF portfolio were invested in the HCI projects.

2. In discriminant analysis, as in cluster analysis, the following concepts are used. Industries are objects, an attribute of characteristic variable is wage change rate, and group membership from cluster analysis is inherited.

3. Similar examples of demand creating industrial policy can be found in the Japanese government's computer industry promotion.

REFERENCES

Aldenderfer, M. S., and R. K Blashfield. "Cluster Analysis," *Sage Quantitative Applications in the Social Sciences* 44. Beverly Hills: Sage Publications, 1984.

Amsden, A. *Asia's Next Giant: South Korea and Late Industrialization.* London: Oxford University Press, 1989.

Anchordoguy, M. "Mastering the Market: Japanese Government Targeting of the Computer Industry," *International Organization* 42, 3 (Summer 1988).

Auty, R. M. "The Macro Impacts of Korea's Heavy Industry Drive Re-evaluated," *The Journal of Development Studies* 29, 1 (October 1992).

Auty, R. M. *Economic Development and Industrial Policy.* New York: Mansel Publishing Limited, 1994.

Blanchflower, D. G., A. J. Oswald, and P. Sanfey. "Wages, Profits, and Rent-Sharing," *Quarterly Journal of Economics* 1 (February 1996) 227–251.

Cho, Y.-J., and D. Cole. "The Role of the Financial Sector in Financial Adjustment," in *Structural Adjustment in a Newly Industrializing Country: The Korean Experience* (V. Corbo and S-M. Suh, eds.). Washington, DC: World Bank, 1992.

Edwards, S., and M. S. Khan. "Interest Rate Determination in Developing Countries: A Conceptual Framework," *International Monetary Fund Staff Paper* 32 (September 1985) 377–403.

Enos J. L., and W. H. Park. *The Adoption and Diffusion of Imported Technology: The Case of Korea.* London: Croom Helm, 1987.

Ferguson, T., and J. K. Galbraith. "The Wage Structure 1920–1947," Paper presented at the Harvard Economic History Seminar, March 1996.

Galbraith, J. K., and P.D.P. Calmon. "Relative Wages and International Competitiveness in U.S. Industry," Working Paper No. 56, LBJ School of Public Affairs, University of Texas at Austin, 1990

Galbraith, J. K., and P.D.P. Calmon. "Industries, Trade and Wages," in *Understanding American Economic Decline* (M. Bernstein and D. Adler, eds.). New York: Cambridge University Press, 1994, 161–198.

Galbraith, J. K., and P.D.P. Calmon. "Wage Change and Trade Performance in U.S. Manufacturing Industries," *Cambridge Journal of Economics* 20, 4 (July 1996) 433–450.

Galbraith , J. K., and J. Lu. "Linear Decomposition of Multiple Time Series." Manuscript 1997.

Gerschenkron, A. *Economic Backwardness in Historical Perspective.* Cambridge: Harvard University Press, 1962.

Glick, R., and M. Hutchison. "Financial Liberalization in the Pacific Basin: Implications for Real Interest Rate Linkage," *Journal of the Japanese and International Economies* 4 (1990) 36–48.

Gourevitch, P. *Politics in Hard Times: Comparative Responses to International Economic Crises.* Ithaca: Cornell University Press, 1986.

Haggard, S., R. N. Cooper, S. Collins, and C. Kim, eds. *Macroeconomic Policy and Adjustment in Korea 1970–1990.* Boston: Harvard University Press, 1994.

Hall, P. *Governing the Economy: The Politics of State Intervention in Britain and France.* Oxford: Oxford University Press, 1986.

Hugh, H. "Policy Lessons of the Development Experience," Occasional Paper 16, Group of Thirty. New York, 1985.

Hughes, H., ed. *Achieving Industrialization in East Asia.* New York: Cambridge University Press, 1988.

Johnson, C. *MITI and the Japanese Miracle.* Stanford: Stanford University Press, 1982.

Johnson, C. "Political Institutions and Economic Performance: The Government-Business Relationship in Japan, South Korea, and Taiwan," in *The Political Economy of the New Asian Industrialization* (F. Deyo, ed.). Ithaca: Cornell University Press, 1987, 136–164.

Jwa, S.-H. "Korea's Export Competition against the Asian NICS and Japan in the U.S.: An Emphasis on the Exchange Rate Effect," *Korea Development Review* 12, 2 (1990).

Jwa, S.-H. "Korea's Interest Rate and Capital Controls Deregulation: Implications for Monetary Policy and Financial Structure," *Joint Korea-U.S. Academic Symposium* 3 (1993) 65–84.

Katz, L. F., and L. H. Summers. "Industry Rents: Evidence and Implications," *Brookings Paper on Microeconomics* (1989) 209–290.

Katzenstein, P. *Between Power and Plenty.* Madison: University of Wisconsin Press, 1978.

Katzenstein, P. *Small States in the World Market: Industrial Policy in Europe.* Ithaca: Cornell University Press, 1985.

Kim, J.-H. "Korean Industrial Policy in the 1970s: Heavy and Chemical Industrialization Drive," Korea Development Institute Working Paper, July 1990.

Kim, J.-R. *Thirty Years of Economic Policy in Korea.* Seoul: The Joong Ang Ilbo, 1992

Kim, T.-J. "Perspectives on Korea's Financial Liberalization," *Joint Korea-U.S. Academic Symposium* 3 (1993) 29–39.

Klecka, C. O. "Discriminant Analysis," *Sage Quantitative Applications* (1980).

Korea Economic Weekly. (November 3, 1995).

Leipziger, D. M. *Korea: Managing the Industrial Transition: A World Bank Country Study. Vol. I: The Conduct of Industrial Policy.* Washington, DC: The World Bank, 1987.

Lorr, M. *Cluster Analysis for Social Scientists.* San Francisco: Jossey-Bass, 1984.

Morrison, D. G. "On the Interpretation of Discriminant Analysis," *Journal of Marketing Research* 6 (1969) 156–163.

Office of the Prime Minister. *Evaluation Report of the First Year Program: The Fourth Year Economic Development Plan.* Republic of Korea: Office of the Prime Minister, 1978.

Okimoto, D. *Between MITI and Market: Japanese Industrial Policy for High Technology.* Stanford: Stanford University Press, 1989.

Pack, H., and L. E. Westphal. "Industrial Strategy and Technological Change," *Journal of Development Economics* 27 (October 1987) 87–128.

Patrick, H., and H. Rosovsky. *Asia's New Giant.* Washington, DC: Brookings Institution, 1976.

Republic of Korea. *The Fourth Fifth Year Economic Development Plan 1977–1981.* Seoul: Government of the Republic of Korea, 1976.

Samuels, R. *The Business of the Japanese State: Energy Markets in Comparative and Historical Perspective.* Ithaca: Cornell University Press, 1987.

Steinmo, S., K. Thelen, and F. Longstreth. *Structuring Politics: Historical Institutionalism in Comparative Analysis.* New York: Cambridge University Press, 1992.

Stern, J., J.-H. Kim, and D. H. Perkins. *Industrialization and the State: The Korean Heavy and Chemical Industry Drive.* Cambridge: Harvard University Press, 1995.

Tatsuoka, M. *Multivariate Analysis.* New York: Macmillan, 1988.

Wade, R. *Governing the Market: Economic Theory and the Role of Government in East Asian Industrialization*. Princeton: Princeton University Press, 1990.

Ward, J. H. "Hierarchical Grouping to Optimize and Objective Function," *Journal of the American Statistical Association* 58 (1963).

Westphal, L. E. "The Republic of Korea's Experience with Export Led Industrial Development," *World Development* 693 (1978) 347–382.

23

Technopolis Development: Korean Experiences

Tae Kyung Sung

INTRODUCTION

A new type of public and private alliance, captured in the term "technopolis," is gaining popularity and attracting attention from central and local governments, the private sector, universities, and research institutes in Korea. Technopolises are having more far-reaching consequences on technology commercialization and economic growth than just the terms "techno" and "polis" might suggest. The modern technopolis links cutting-edge research with technological development and commercialization in order to spur economic development and to promote technology specialization and diversification (Smilor et al. 1988).

An alliance is a cooperative relationship between two or more independent organizational entities. It is an agreement to work together to achieve specific, predefined objectives. The guiding principle is that participants in an alliance agreement should each specialize in performing individual tasks related to core competencies, thereby eliminating duplication as much as possible as well as gaining as much synergy as possible. Alliances typically extend over a considerable period of time and involve procedures to share benefits and risks and exchange information. By its very nature, an alliance such as a technopolis is complex and demands significant commitment. Thus, technopolis development needs careful planning, management, and coordination.

In this chapter, a comparative analysis of the development strategy of Korean technopolises is attempted. Which sector initiates and operates the technopolis? What development strategies has each technopolis devised? What kind of configuration is each technopolis taking? What are its major functions? The objective of this chapter is to answer these questions. While American technopolises develop spontaneously as the result of the uncoordinated actions of private business, universities, and governments, Korean technopolises are planned, being created by the strategic efforts of central government (Luger 1996). Thus, a comparative study of the development strategies of Korean technopolises will shed new light on the technopolis literature.

LITERATURE REVIEW

Technopolis: History

The first archetype of the technopolis can be traced back to research parks in the United States in the 1960s. Science parks were beginning to be established in England in the 1970s and technology innovation centers were attempted in Germany in the 1980s. More recently, technopolises in Japan and science-based industrial parks in Taiwan have emerged as more functions were added to cope with fiercer competition (MITI 1997).

Central (federal) and local governments, universities, research institutes, and corporations are beginning to realize that it is impossible to react to a rapidly changing world economy and fierce international competition without integrating or networking research and development, technology innovation, commercialization of new technology, technology transfer, policy, and financial support functions.

As a result of these integration efforts, a new type of alliance called a technopolis has been developed. The technopolis, which is based on the traditional industrial park, is augmented by functions such as technology innovation, technology transfer and commercialization, and business incubation. There are several distinct types of technopolises throughout the world, since each technopolis reflects the economic, political, cultural, societal, and administrative factors of its own environment.

Concept of the Technopolis

In a fiercely competitive, mass customized, and rapidly changing international environment, every nation is desperately seeking to enhance its competitive advantage (Porter 1990). One of the most promising and attractive alternatives is to build concentrated industrial and/or research parks which aim at close linkage between governments, universities, research institutes, and corporations to innovate new technology and products, transfer and commercialize such technology and products, and begin and incubate small businesses. (MITI 1997).

According to Smilor, Kozmetsky, and Gibson (1988), the modern technopolis is one that interactively links technology development with the public and private sectors to spur economic development and promote technology diversification. In developing technopolises, the following three factors are especially important and provide a way to measure their dynamics:

1. The achievement of scientific preeminence
2. The development and maintenance of new technologies for emerging industries
3. The attraction of major technology companies and the creation of home-grown technology companies

In Korea, "technopark" is a much more popular term than "technopolis." The purpose of a technopark is defined as "to effectively achieve technology innovation and high-tech development in a particular region; functions such as research and development, venture and incubation, education and training, assistance and services, and sample production are integrated into one site through close cooperative interaction among universities, research institutes, and corporations" (MITI 1997).

Functions of the Technopolis

Depending on the participating sectors, leading sector, objectives, scope, and formation, each technopolis provides a variety of functions:

1. Research and Development
2. Education and Training
3. Technological Innovation
4. Technology Transfer and Commercialization
5. Venture and Incubation
6. Production
7. Providing a high quality of life
8. Promoting Start-ups and Spin-offs

FRAMEWORK FOR ANALYSIS

To comparatively analyze technopolises in Korea, a framework for analysis needs to be developed. Based on the study performed by MITI, the following four analyzing factors are identified: initiating and operating sector, development strategy, configuration, and critical success factors (CFSs).

Initiating and Operating Sector

According to the initiation and operating sector, there are five types of technopolises initiated by the central government, the local government, the university, the private sector, and the third sector.[1] In most cases, the initiating sector assumes operating authority and responsibility. The balance sheet for each type of technopolis is summarized in Table 23.1.

Table 23.1
Comparison Based on Initiating and Operating Sector

Type	Strengths	Weaknesses
Public		
Central Government	▪ Strong planning, administrative, legal and financial capacities ▪ Stability ▪ Credibility	▪ Insensitive to environmental changes ▪ Budgetary constraints ▪ Inflexibility
Local Government	▪ Fit to structure and characteristics of local environment	▪ Dependence on local government ▪ Budgetary constraints ▪ Requires a certain level of industry infrastructure
Private		
University	▪ Low cost for technopolis site ▪ Strong R&D ▪ Utilization of existing facilities ▪ Flexibility	▪ High dependence on university ▪ Possibility of losing the identity of the university
Private firm	▪ Profitability ▪ Flexibility ▪ Sensitive to environmental changes	▪ Unclear authority and responsibility ▪ Possibility of easy breakdown
Third sector	▪ *Strong planning* ▪ *Flexibility* ▪ *Division of initiating and operating sectors* ▪ *Sensitive to environmental changes*	▪ *Unclear about public vs. private nature* ▪ *Manpower shortage* ▪ *Restrictions on operation*

Source: MITI (1997).

Development Strategy

The development strategies of technopolises can be divided into two major types: new establishment, and augmentation of the functions of an established industrial park. To establish a new technopolis, the strategies can be further categorized as globalization through technology specialization and vitalization of the local economy. Depending on what kind of established industrial park the functions are added to, three types of strategy can be utilized (see Table 23.2).

Table 23.2
Comparison Based on Development Strategies

Type	Strengths	Weaknesses
Public		
Industrial Park	▪Close linkages to production ▪Easy management and control	▪Weak R&D ▪Difficulty in leveraging with universities due to geographical remoteness
Research Park	▪Active R&D ▪Good start-up and incubation ▪Easy technology outsourcing	▪Difficulty in commercialization
Industrialized Area	▪Synergy effect ▪Integration of related industries	▪Difficulty in locating initiating sector ▪Difficulty in locating proper site
Private		
Globalization through technology specialization	▪Attracting corporations ▪Technology specialization	▪Concentration in capital city area ▪Difficulty in implementation
Vitalization of local economy	▪Boosting local economy ▪Vitalizing local industry	▪Difficulty in financing and commercialization ▪Manpower shortage

Source: MITI (1997).

Technopolises can be built based on industrial parks by augmenting R&D, venture and incubation, assistance and services, and simulation functions, and may be initiated by universities, local government, or the third sector. Research parks can be developed into technopolises through augmenting venture and incubation, assistance and services, technology transfer, and sample production by initiatives of universities. Established industrial parks also can be transformed into technopolises by augmenting venture and incubation, assistance and services, and technology transfer functions, and may be initiated by local government or the third sector. New technopolises aiming at globalizing through technology specialization and/or vitalizing the local economy can both be initiated by universities (networked), local governments, and the third sectors.

Configuration

Depending upon location, objectives, and functions, the configuration of technopolises can be classified as either centralized (park-type), distributed (network-type), or mixed. The characteristics of each type are summarized in Table 23.3.

Table 23.3
Comparison Based on Technopolis Configuration

Configuration	Strengths	Weaknesses
Centralized	Synergy effect Cost efficiency	Low participation from remote sector Too much standardization
Distributed	Meet the needs of each business Flexibility in site location	Difficulty in implementation Hard to provide one-stop services Low synergy effect Cost inefficiency
Mixed	Inherits both strengths and weaknesses of both types	

Source: MITI (1997).

Critical Success Factors

To successfully develop a technopolis, the following CSFs should be taken into consideration (Bopp 1988; Gibson and Rogers 1994; Gibson and Sung 1995; Goto 1993; IC² Institute 1990; Sedaitis 1997; Smilor, Kotzmetsky, and Gibson 1988):

1. Strong leadership of the initiating sector
2. Balance of power and coordination of participating sectors
3. Information sharing among participating sectors
4. Financial assistance and taxation incentives
5. Low rental cost and utility costs
6. Existence of research institutes
7. Existence of good universities
8. Well-established industry infrastructure
9. Specialized and unique technology or industry
10. Systematic implementation of plan
11. Benchmarking
12. Attracting highly skilled technicians and experts
13. Providing high quality of life
14. Strong investment from participating sectors
15. Drawing interest and support from local residents
16. Attracting visible and prominent corporations
17. Providing one-stop administrative services

KOREAN TECHNOPOLISES

Pohang Technopark International (PTI)

According to C. Lee (1996), the plan for the Pohang Technopark International (PTI) was initially conceived in 1992 by POSTECH (Pohang University of Science and Technology). But there was no progress due to the lack of commitment and interest of Pohang city until 1996. Currently, POSTECH is devising a master plan to build PTI by the early 2000s with strong sponsorship and cooperation from the city. PTI, with an area of about 220 hectares (ha) (about 500 acres), will be located north of the POSTECH campus in a hilly area of Jamyung-dong.

Participating Sectors. POSCO (The Pohang Iron and Steel Co., Ltd.) was established in the late 1960s as a strategic industry and is currently rated as the world's second largest steel producer after Japan's Nippon Steel. POSCO serves as the main financing source for PTI. POSCO has also contributed significantly to R&D by establishing RIST (Research Institute of Industrial Science and Technology) and POSTECH, which are the two most important participating sectors in PTI.

POSTECH was founded in 1986 as a research-oriented university by POSCO to meet POSCO's own R&D needs as well as to contribute to the advancement of science and technology for the whole nation. There are a total of 15 research centers and laboratories at POSTECH. The most active research centers of excellence are AFERC (Advanced Fluids Engineering Research Center), ARC (Automation Research Center), RCCT (Research Center for Catalytic Technology), CBM (Center for Biofunctional Molecules), and CAAM (Center for Advanced Aerospace Materials), all of which are funded by the Korea Science and Engineering Foundation (KOSEF).

POSTECH also has a Graduate School of Information Technology (GSIT) and a Graduate School of Iron and Steel Technology (GSIST), which were established in 1991 and 1995, respectively. GSIST was established to fulfill the demand for specialists in the iron and steel industry and to strive to enhance the industry's international competitiveness.

The Pohang Accelerator Laboratory (PAL) completed the third-generation "2-GeV synchrotron radiation accelerator" project in December 1994 after six years of project planning and construction. It was the largest project ever in a scientific field in Korea. The total investment was about US$200 million, of which POSCO funded 60% and the government invested the remaining 40%. Currently, about 200 scientists, engineers, and technicians are actively involved in the operation of the laboratory. The government is committed to support PAL in the coming years.

To facilitate technology innovation, transfer, and commercialization, POSCO funded RIST in 1987 to provide scientists and engineers with a highly sophisticated research environment and to offer ample opportunities for them to link with related industries. There are about 600 researchers and technicians working at RIST at present. The value of research contracts reaches $60 million per annum.

Infrastructure

1. Harbor: 56 ha (approximately 125acres) in size, with a total investment of $1.7 billion over the next 16 years
2. Transportation: Super-speed train which will be completed by the early 2000s will pass Kyongju, located within 30 minutes' drive from Pohang
3. Airport: will be upgraded to international airport
4. Utilities: Ample and available
5. Quality of life: Good

Looking Ahead. With a strong research university (POSTECH) and research institutes (RIST and PAL), prominent corporations (POSCO), and the determination of Pohang city, coupled with outstanding industrial infrastructure, PTI will be a dominant figure in the 2000s.

Taedok Science Town (TST)

Taedok Science Town (TST) has been the subject of many national and regional development efforts and policies for the last 20 years. The plan itself represents a concerted attempt by the central government to create a technopolis outside the capital region (Oh 1996). Currently, TST is spread over an area of about 278 ha (about 700 acres) with a population of approximately 50,000 inhabitants. There are two national universities in TST as well as the Atomic Research Institute and 32 government and private research institutes.

Roles of Universities. Chungnam National University is acting as the intermediary between research centers in TST and a number of potential industries in the Taejon region because of its location, excellent research capabilities, and outstanding students.

Korea Advanced Institute of Science and Technology (KAIST) focuses on research activities and has closer links with research centers in TST as well as in other areas throughout Korea.

At present both universities are building TBIs (Technology Business Incubators) to accelerate technology transfer and commercialization between research institutions (universities and research institutes) and private firms.

Development Policy. The development of TST sends an important message to all the major cities throughout Korea, since most development policies and practices have been concentrated in the Seoul area. To be successful in developing a technopolis, a strong bond between the R&D of universities and research institutes and industry is indispensable to promote regional development. TST development policy is paying primary attention to this linkage.

Incubation Policy. The purpose of an incubation center is to support and facilitate start-ups and spin-offs, which will inevitably vitalize local businesses and industries. The problem with TST is that resident labor forces in Taejon, the mother city of TST, are disproportionately concentrated in commercial and service occupations and employed mostly in semiskilled and low-skilled jobs (Oh 1996). To overcome this structural constraint, innovative strategies are needed to attract entrepreneurs from universities, research institutes, and R&D centers. One of the best strategies is to provide incubation services to start-ups and spin-offs, and TST is putting tremendous efforts into developing incubation centers.

Kwangju Advanced Science and Industrial Complex

While Korea has experienced unprecedented economic growth during the last three decades, economic development has not been equal among different regions of the country (Kim 1996). This unequal regional development results from deficiencies in social overhead capital and manpower for science and technology as well as from uneven commitment from central government. The Kwangju Advanced Science and Industrial Complex is designed to reverse this trend and to vitalize the Kwangju area's economy and industries. It is planning a two-phase development, which is summarized in Table 23.4.

Table 23.4
Kwangju Advanced Science and Industrial Complex: A Quick Look

	Size (ha)	Development Period	Investment (billions of $)	Specialized Industries
Phase 1	98.5	1991–1995	0.81	Nonmetallic mineral products
Phase 2	95.2	1996–2001	0.74	Transportation equipment

Source: Kim (1996).

Effects on Regional Economy. Both the industrial mix effect and the locational share effect are positive on manufactures of nonmetallic mineral products and transportation equipment in the Kwangju region. A summary of the economic impact is presented in Table 23.5.

Table 23.5
Economic Impact of Kwangju Technopolis

Industries	Production (M Won*)	Number of Jobs
Agriculture, Forestry & Fishery	482	1
Manufacturing	55,820	814
Construction, Utilities	281,845	3,046
Others	54,578	2,708
Total	392,975	6,569

*Won = Korean currency
Source: Kim (1996).

The Next Step. To fully develop the Kwangju Advanced Science and Industrial Complex for vitalizing the local economy, the following problems need to be solved. First, there is a lack of appropriate government development policies. Second, locational strategies for small firms need to be revised to diffuse and commercialize technology more effectively. Third, the social infrastructure is not well established (Kim 1996).

Taegu-Kyungbuk Technopark (TKT)

TKT is a university-initiated, company-participated, local government-supported, central government-promoted technopark, located at the center of the Taegu-Kyungbuk region. The main center is located on the campus of Yeungnam University; it is about 3.3 ha (approximately 7.35 acres) in area and has centers on the campuses of ten other participating universities.

Overview. The basic objective of TKT is to form an industrial technology center in order to prepare for globalization and localization as well as to revitalize the regional economy (S. Lee 1996). To fulfill this objective, TKT is putting efforts into connecting regional universities with small businesses. The TKT project started in 1997 and will be carefully implemented stage-by-stage by the TKT Promotion Conference, which consists of leaders from all participating sectors.

Basic Directions

1. Coping with the era of globalization and localization
2. Serving the needs of the Taegu-Kyungbuk regional economy
3. Directing the future development of Taegu-Kyungbuk
4. Utilizing the linkage of prominent regional universities, companies, and local government
5. Securing economies of efficiency
6. Maximizing the positive effects of technopark development

Strategies. According to S. Lee (1996), a model of integrated industrial technology strategy which combines internal and external technology renovation strategies is desirable. Basically through the potentials of universities with neighboring industrial complexes, strategies need to be developed which facilitate close cooperation between the research capabilities of universities and the demands of industries.

Financing

1. Total investment: 230,000,000,000 Won ($255,000,000)
2. From universities: 115,000,000,000 Won ($127,500,000)
3. From corporations: 50,000,000,000 Won ($55,500,000)
4. Central government: 50,000,000,000 Won ($55,500,000)
5. Local government: 15,000,000,000 Won ($16,500,000)

Key Functions

1. Technology development
2. Technology support
3. Incubations
4. External cooperation

COMPARATIVE ANALYSIS

A comparative analysis is attempted based on the above described four technopolises by factors such as development strategy, configuration, initiating sector, operating sector, and the like. The comparison of the four technopolises is summarized in Table 23.6.

Table 23.6
Comparison of Four Korean Technopolises

Region	Pohang	Taejon	Kwangju	Taegu
Name	Pohang Technopark International (PTI)	Taedok Science Town (TST)	Kwangju Advanced Science and Industrial Complex	Taegu-Kyungbuk Technopark (TKT)
Development period	1992–early 2000s	1992–2000s	1991–2001	1997–
Development strategy	Globalization through technology specialization	Research park	Vitalize local economy	Vitalize local economy
Configuration	Centralized	Centralized	Centralized	Distributed (Mixed)
Initiating sector	Government (POSCO)	Government	Government	Network of universities
Operating sector	POSTECH	Government	Government	Third sector
Size	220 ha	278 ha	197 ha	3.3 ha (main center only)
Major universities	POSTECH	KAIST Chungnam National University	Chunman University	11 regional universities
Number of firms	More than 200		1,600 SMEs	
Investment			$15,555M	$255M

Development Strategy

While PTI is aiming at globalization through technology specialization, Kwangju and TKT are building technopolises to vitalize local business and industries. By contrast, TST is more concerned with research-oriented activities.

Configuration

Except for TKT, the centralized technopolis is the typical configuration in Korea. Since most Korean technopolises are planned and initiated by central and/or local governments, the centralized technopolis is more common. In the case of TKT, a network of 11 universities in the TKT region acts as the initiator and this fact is reflected in deciding on a mixed (to some extent, distributed) configuration.

Initiating Sector and Operating Sector

TST and Kwangju technopolises are initiated and operated by government and most of their funds are provided by government. But PTI demonstrates a totally different approach. PTI is heavily funded by POSCO and is operated by POSTECH. POSTECH is very well-known for its international perspective. The development strategy for PTI is accordingly to globalize through technology specialization.

TKT was initiated by a network of 11 regional universities and each university is focused on specialized technology for technology innovation and commercialization. To coordinate 11 universities as well as other participating sectors, a promotion conference has been organized to play the role of operating sector.

SUMMARY AND CONCLUSIONS

In a fiercely competitive, mass customized, and rapidly changing international environment, every nation is desperately seeking to enhance its competitive advantage. One of the most promising and attractive alternatives is to build technopolises, which link cutting-edge research with technology development and commercialization in order to spur economic development and to promote technology specialization and diversification.

While Korea has experienced unprecedented economic growth during the last three decades, economic development has not been equal among different regions of the country. This unequal regional development results from deficiencies in social overhead capital and manpower for science and technology as well as from uneven commitment from central government. The development of technopolises has reflected exactly the same pattern.

But technopolis development in Korea experienced a major directional change in the early 1990s, when local governments were endowed with the autonomy to plan, manage, and control their own territories. Since then, many city and local governments have made strenuous efforts to become competitive in industries where they have advantages over others.

Comparative analysis of the four technopolises shows that each one has a different development strategy, configuration, initiating sector, and operating sector to reflect the economic, political, cultural, societal, and administrative factors of its own environment. But there seems to be strong central government influence in the area of financial assistance as well as in initiating and operating roles. As the autonomy of local governments grows, technopolis development patterns will change from "planned" to "spontaneous" as in most cases in the United States.

This comparative study is only the beginning of a long journey. This journey will focus on more detailed analyses on a much larger sample of technopolises throughout the world to understand the technopolis phenomenon and to provide useful guidelines to successful implementation of technopolis development. The journey will require a painstaking longitudinal study.

NOTE

1. The "third sector" refers to entities which are neither public nor private sector. Third-sector entities can include collaborations between academia, government, and industry, consortia or committees comprised of various members from across sectors, or specific professional groups. All of these third-sector entities are formed to maximize the synergistic effect and the balance of power through the combination of various sectors.

REFERENCES

Bopp, G. R. *Federal Lab Technology Transfer: Issues and Policies.* New York: Praeger, 1988.

Gibson, D. V., and E. M. Rogers. *R&D Collaboration on Trial: The Microelectronics and Computer Technology Corporation.* Boston: Harvard Business School Press, 1994.

Gibson, D. V., and T. K. Sung. "Technopolis: Cross-Institutional Alliances," *International Business Review* 18 (December 1995) 199–217.

Goto, K. *Science, Technology and Society: A Japanese Perspective.* Austin: IC² Institute, 1993.

IC² Institute. *The Technopolis Phenomenon,* Austin: IC² Institute, 1990.

Kim, I. "Advanced Science and Industrial Complex and Its Regional Development Effect: The Case of Kwangju Advanced Science and Industrial Complex," *Proceedings of the International Symposium on the Technopolis, Its Vision and Future* (November 1996) 133–146.

Lee, C. "The Development Strategy of Pohang Technopark," *Proceedings of the International Symposium on the Technopolis, Its Vision and Future* (November 1996) 333–348.

Lee, S. "Modelling and Strategy of Kyungsan Techno-Research Park As Pole in Taegu-Kyungbuk Region," *Proceedings of the International Symposium on the Technopolis, Its Vision and Future* (November 1996) 147–163.

Luger, M. "Spontaneous Technopolises and Regional Restructuring: The Case of Research Triangle," *Proceedings of the International Symposium on the Technopolis, Its Vision and Future* (November 1996) 3–30.

MITI (Ministry of International Trade and Industry). *A Study on Establishing Legal and Systematic Infrastructure to Facilitate Technopark-type Research Complex.* MITI, 1997.

Oh, D.-S. "Technopolis Development in Korea," *Proceedings of the International Symposium on the Technopolis, Its Vision and Future* (November 1996) 51–73.

Porter, M. E. *The Competitive Advantage of Nations.* New York: The Free Press, 1990.

Sedaitis, J. B. *Commercializing High Technology: East and West.* Lanham: Rowman & Littlefield, 1997.

Smilor, R. W. G. Kozmetsky, and D. V. Gibson. *Creating the Technopolis: Linking Technology Commercialization and Economic Development.* Boston: Ballinger, 1988.

24

An Empirical Study of R&D Effects on High-Tech Firms in Taiwan: The Case of the Hsinchu Science-Based Industrial Park

Pao-Long Chang, Po-Young Chu, and Ming-Yung Ting

INTRODUCTION

Government efforts to foster homegrown innovations are paying off as firms have invested more and more in research and development in Asia. The pace of innovation is speeding up and gradually cutting Asia's dependence on imported technology. Taiwan provides a good example of the process. The government of Taiwan has been aggressively pursuing economic growth by building up strong R&D infrastructures and incubating indigenous companies to specialize in specific forms of technology since the late 1970s. Now, Taiwan has earned a growing high-technology reputation based largely on the successful development of the Hsinchu Science-Based Industrial Park (HSIP).

The success of Taiwan's experience in HSIP could be regarded as a good model for developing countries in South Asia in their current and future economic development. The firms in HSIP, the epitome of high-tech firms in Taiwan, contributed to the success of Taiwan's high-tech industry. The firms based at HSIP had an annual turnover of US$11.3 billion in 1995. The average productivity per employee is about three times that of the aggregated manufacturing of Taiwan. Why are HSIP firms so successful? This chapter will explore the key success factors of these firms.

The chapter focuses on exploring how R&D investment has affected Taiwan high-tech firms' competitiveness, in addition to the critical success factor. We focus on R&D investment and its effects on the firms, and will emphasize the success factors that the Taiwan government has emphasized. First, HSIP will be described. Then we will analyze the R&D investment patterns along a firm's three attributes: capital source, industry category, and age. Third, we attempt to systematically link firms' R&D investment and performance. Finally, the observed common factors related to the success of the high-tech firms at HSIP are described. The conceptual framework of the chapter is summarized in Figure 24.1.

Figure 24.1:
The Conceptual Framework

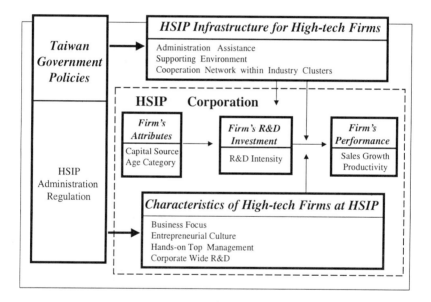

THE HSIP: A BIG CORPORATION

To stimulate the research and innovation of industrial technology, and to promote the development of high technology industries in Taiwan, the ROC government selected Hsinchu for the establishment of a science-based industrial park in 1980. It has provided an ideal fully integrated environment for investment and employment and training of technically skilled personnel. The administration of HSIP has played an important role in the consolidation of subsidiary and support enterprises.

Overview of Park Industries

Since its establishment in 1980, HSIP has been a huge success. Total sales in 1995 were $11.3 billion, up from $75 million in 1983, to $1.7 billion in 1989, and $6 billion in 1993. Per-employee productivity in 1995 averaged $269,260.

The number of companies in HSIP grew to 203 in 1996. HSIP companies are classified into six categories: integrated circuits (IC), computers and peripherals, telecommunications, optoelectronics, precision machinery and materials, and biotechnology. HSIP firms' combined sales grew by an average 2% (see Table 24.1) to reach $11,564 million in 1996. Of the companies in the park, 36 are foreign-owned and 167 are domestically owned. Aggregate investment increased by 72% from 1995 to reach $9,600 million by the end of the year. Domestic sources accounted for 87% of HSIP investment capital, while foreign sources accounted for 13% (see Figures 24.2 and 24.3).

In the area of new investment, 32 new firms entered the park in 1996, four more than in 1995. New investments amounted to $1,127 million, 48% less than in 1995, when the park's semiconductor manufacturers invested heavily in eight-inch wafer production facilities.

Table 24.1
1996 HSIP Industries

Industry	Firms	Employees	Sales (US$M)	Growth (%)
Integrated Circuits	71	29,510	5,709	2
Computer and peripherals	43	14,187	4,407	–4
Telecommunications	33	4,385	700	9
Optoelectronics	28	5,386	637	68
Precision Machinery and Materials	18	1,070	100	6
Biotechnology	10	268	11	38
Total	**203**	**54,806**	**11,564**	**2**

By the end of 1996, 61 firms in the park had received permission to increase their capitalization, one more than in 1995. The total capital raised came to $4,457 million, 77% more than in the previous year. Of these firms, 26 were in the integrated circuits sector and raised a total of $3,690 million in new capital.

In order to raise additional capital and allow all members of society to share the benefits of their success, 20 HSIP firms have gone public and are now listed on the Taiwan stock exchange. The listed companies include ten in the integrated circuits industry, eight makers of computers and peripherals, and two telecommunications firms. More HSIP companies were expected to go public in 1997.

Figure 24.2
Capital Sources of HSIP

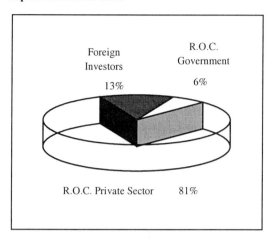

Figure 24.3
HSIP Firms by Region of Ownership

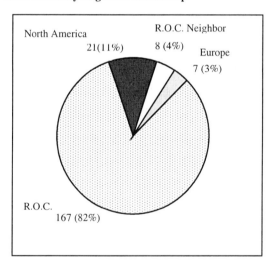

To encourage the flow of capital into high-technology industries, the government established a wide range of financial incentives offered to HSIP-based companies, as well as research grants. These incentives include low-interest loans (up to 49% of the total investment), the right to retain earnings of up to 200% of paid-in capital, a five-year exemption from tax within the first nine years of operation, and no tariffs on imported components for finished products intended for export.

HSIP's optoelectronics, integrated circuits, and computers and peripherals industries enjoyed growth rates of 112%, 76%, and 69%, respectively, in 1995, breaking all previous growth records. The park industries' growth rates, and the percentages of sales spent on R&D, not only far surpass figures for all other ROC industries, but are comparable with those of high-technology industries in Japan and the United States.

Since 1980, the ROC government has appropriated a total of $520 million for land acquisition and development of the park's infrastructure. Currently, further expansion is actively under way, and a new science park has been established in Taiwan to attract more high-tech investment and make the ROC more competitive in international markets.

Land Allocation and Development

HSIP at present includes 480 hectares (1,186 acres) of developed land. Another 100 hectares (247 acres) of land is now under development. Land development at HSIP is primarily geared toward providing industrial, residential, and recreational areas.

The industrial areas contain standard factory buildings and sites for custom-designed factories and laboratories: the residential areas include not only housing but also sports facilities, a lake in a tranquil setting, restaurants, and a bookstore, all adding to the quality of life for those living in and around HSIP. The sculpture "Thai Chi" by renowned artist Chu Ming sits on the lakeshore, testifying that humanistic and cultural values must not be forgotten in the pursuit of technological progress.

As part of its commitment on environmental protection, the Science Park Administration maintains a wastewater treatment plant. Water purified by the plant is piped to an automatic sprinkler system for park greenery. A small-scale incinerator is in operation to dispose of common wastes; a larger incinerator and a sanitary landfill site are under construction and will eventually give the park a fully self-sufficient waste processing system.

Human Resources

By the end of 1996, the number of people working at HSIP was 54,806 (see Figure 24.4). The proportion of employees with at least a junior or technical college education was 59%. HSIP workers' average age was 30 years; men made up 48% of the workforce and women 52%. Returning expatriates have played a vital role in the growth of the park. By the end of 1996, the number of returned expatriates was 2,563. Of the companies in the park, members of this group founded 40%.

Along with three major institutions—Industrial Technology Research Institute (ITRI), the National Chiao Tung University, and the National Tsing Hua University—HSIP creates a conductive intellectual climate for R&D, and fosters cooperative research with its steady source of researchers at hand (Gwynne 1992).

Figure 24.4
Growth of HSIP Employment

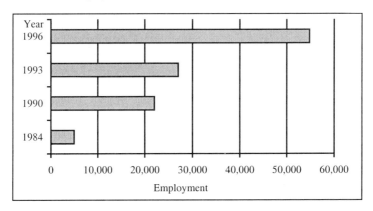

HSIP's administrations stipulate that companies within HSIP spend a minimum proportion of their revenues on R&D and require that a minimum percentage of each firm's employees be scientists and engineers. The average R&D intensity of firms' in HSIP is over 4.5% of sales. About 12% of employees have Ph.D.s and master's degrees; workers with bachelor's degrees or postsecondary technical education make up over 40% of the work force.

R&D by HSIP Firms

According to a survey conducted in 1996 by the Ministry of Economic Affairs' Productivity Center, HSIP firms spent $503 million, or 4.4% of their total sales revenues, on R&D in 1995. The IC industry spent the largest amount, devoting $302 million to R&D, while biotechnology companies allocated the largest percentage of their income for R&D, earmarking 42.8% of their sales revenues for this purpose. R&D personnel in the park numbered 5,412, or 12.8% of the workforce. The IC industry, with 2,414 employees dedicated to research and development, had the largest number of R&D workers (see Table 24.2).

As for individual enterprises, the top 12 companies with R&D expenditures of 5% or more of sales revenues were Winbond, UMC, Macronix, SIS, Acer Laboratories, Vanguard, Microelectronics Technology, Holtek, Umax, Realtek, Accon, and Elan.

Integrated as an industrial cluster, HSIP functions like a big corporation not only with economics of scale but also with enormous flexibility. This big corporation operates with well-built infrastructures to incubate many efficient private modules. As a result, private corporations and public officials work together to contribute to the success of this prosperous park.

Table 24.2
1995 R&D Expenditures and Personnel

Industry	R&D Outlays	Percentage of Sales Revenues	R&D Personnel	Percentage of Employed Personnel
Integrated Circuits	302	5.4	2,414	10.7
Computer and peripherals	108	2.3	1,737	15.6
Telecommunications	52	8.1	698	17.2
Optoelectronics	30	7.8	369	11.3
Precision Machinery and Materials	8	9.0	137	13.2
Biotechnology	3	42.8	57	24.7
Total	503	4.4	5,412	12.8

Because firms in science-based industrial parks play an increasingly important role in job creation and technological innovation, it is important to understand how government-related policies affect the behavior of those firms, and what the ultimate effect of government efforts is on firms' innovation and success.

The R&D Investment Patterns of HSIP Firms

There are many factors that contribute to the future prospects of a company. R&D is one of the most vital factors, especially for high-technology industries. But R&D expenditures directly reduce current-year profits (Gilman 1978; Rosenau 1980). Prudence requires top management to refrain from spending too much on R&D. Thus we need to answer the question: How much should we spend on R&D? This section will provide some additional evidence from HSIP which can assist governments who will set regulations for science-based industrial parks in their countries.

Although many studies have attempted to determine the optimum level for R&D spending based on performance indices (Gilman 1978; Reynard 1979; Ellis 1980), no previous study has been carried out considering a firm's age and capital sources.

In examining the influence of capital source on R&D investment decisions, we employed Wilcoxon rank sum to test the difference between domestic-owned and foreign-owned firms at HSIP. The results showed, at significance level • = 0.0001, there were significant differences in R&D intensity between these two types of firms. The R&D intensity of ROC firms is significantly higher than that of their foreign counterparts.

We also examined the difference in R&D intensity among the six various industries by the Kruskal-Wallis test. At significance level • = 0.0001, we could say that firms from these six industries differ from one another in R&D intensity. Domestic-owned firms have different R&D intensity among various industries at significance • = 0.0001. We could not find a significantly different R&D intensity among the foreign-owned firms across various industries.

Do firms with various ages have different R&D intensities? To answer this question we examined the patterns of R&D intensity among firms through various ages by the Kruskal-Wallis test. At significance level • = 0.05, there were R&D intensity difference among firms. We further grouped the firms by different captal sources. We found domestic-owned HSIP firms with various ages had significant differences in R&D intensity at significance level • = 0.05. But such differences become insignificant among foreign-owned firms with various ages.

According to these analyzed results, we suggest that governments that want to regulate the R&D investment of firms in science-based industrial parks, should consider a firm's age, its industry category, and its capital source. Governments need to force foreign-owned firms to invest at the same level on R&D intensity as domestic-owned firms, so that they can have the same contribution level to the domestic technology development. Domestic-owned firms with different ages, should set different ratio rules on R&D investment, but need to do so for the foreign-owned firms.

THE EFFECTIVENESS OF HSIP FIRMS' R&D INVESTMENT

In this section we will investigate the effectiveness of firms' R&D investment to check whether it is reasonable to assume that government regulations have caused firms in science-based research parks to invest in R&D. We also want to know whether a firm investing more in R&D will get poor current-year performance.

It is a widely held belief that R&D investments are essential to the long-term health of a company, and we have empirical studies to support this belief. Many studies show that higher growth companies invest in R&D much more consistently than companies with low growth rates, and tend to have higher R&D intensities (Hambrick, MacMillan, and Barbosa 1983; Guerard, Bean, and Andrews 1987; Brenner and Rushton 1989). Companies that spent more on R&D during a recession performed better than those that held back (Morbey and Dugal 1992).

We know that R&D spending can lead to higher rates of new product introductions and higher quality products, and that gains in market share lead to improvements in gross margin and higher returns on investment (Collien, Minz, and Collin 1984). Probably due to the high impact of R&D on new product sales, R&D intensity is more often correlated with sales growth (Schoeffler, Buzzell, and Heany 1984; Hambrick and MacMillan 1985; Morbey 1988; Brenner and Rushton 1989). Some research has also found a strong relationship between R&D and the companies' subsequent productivity (Griliches 1987; Morbey and Reithner 1990).

Although many studies have concluded the impact of R&D on a company's performance, we have seldom obtained evidence coming from developing countries. Here, we continue to seek such empirical evidence based on the data of high-tech firms at HSIP in Taiwan. The results are shown in Table 24.3, Table 24.4, and Table 24.5. The evidence from HSIP has reconfirmed results from Morbey and Reithner (1990) that a firm's R&D intensity is positively correlated with its sales growth. But one of our results shows that R&D intensity is negatively correlated with a firm's current-year performance.

Based on the results shown above, we provide evidence to assert that it is reasonable for governments to force firms in science-based industrial parks to invest more in R&D. We also find that governments should assist firms with heavy R&D investment to solve the problem of the immediately negative impact.

Table 24.3
Correlation Between Firm's R&D Intensity and Sales Growth

	Sales Growth			
	Current Year	1 Year Time Lag	2 Years Time Lag	3 Years Time Lag
Current Year R&D Intensity	–0.07823 + (0.0923)	0.22467 *** (0.0001)	0.23563 *** (0.0001)	0.17317 ** (0.0053)

Spearman correlation coefficients; P-Values in parentheses.
+ Indicates significance at 0.1; ** Indicates significance at 0.01;
*** Indicates significance at 0.001

Table 24.4
Correlation Between Firm's R&D Intensity and Productivity

	Productivity			
	Current Year	1 Year Time Lag	2 Years Time Lag	3 Years Time Lag
Current Year R&D Intensity	–0.35970*** (0.0001)	–0.24120 *** (0.0001)	–0.18389 *** (0.0001)	–0.14589* (0.0129)

Spearman correlation coefficients; P-Values in parentheses.
* Indicates significance at 0.05; *** Indicates significance at 0.001

Table 24.5
Correlation Between Firm's R&D Intensity and Productivity Growth

	Productivity Growth			
	Current Year	1 Year Time Lag	2 Years Time Lag	3 Years Time Lag
Current Year R&D Intensity	−0.03535 (0.4881)	0.24368 *** (0.0001)	0.23724 *** (0.0001)	0.24015 *** (0.0001)

Spearman correlation coefficients; P-Values in parentheses.
*** Indicates significance at 0.001

THE KEY SUCCESS FACTORS OF HSIP FIRMS

Technology and innovation are vital to high-technology firms. The high-technology sector is composed of industries in which both R&D expenditures and the proportion of technology-oriented workers are at or above the all-industry average (c. Chakrabarti 1991). The competitiveness of a high-tech firm depends on the firm's willingness to invest in R&D today to ensure continued opportunities for technological innovation and sales growth in the future.

Many studies have highlighted the essential factors for high-tech firms to sustain success (Burgelman, Maidique, and Wheelwright 1996; Cooper and Kleinschmidt 1996; Rothwell 1992). Some of the essential success factors that the governments could use to assist the high-tech firms to develop have been helping most HSIP firms to prosper. The common factors characterizing successful high-tech firms, which should be supported or affected by government policies, are utilized to explain how the HSIP firms succeeded.

As shown in Figure 24.1, we divided the key success factors of HSIP firms into two categories: the environment infrastructure and the characteristics of HSIP firms, and suggested influence of government policy. The following sections will describe the factors of those two categories.

HSIP INFRASTRUCTURE SUSTAINED HIGH-TECH FIRM'S SUCCESS

The government of Taiwan is aggressively pursuing trade opportunities by building up strong innovation infrastructures and encouraging indigenous firms to specialize in specific forms of technology. HSIP plays a major role in building innovation infrastructure for high-tech firms. It brings together government institutions, universities, and large and small firms in the joint effort to create high-technology products for world markets (Gwynne 1993). HSIP is narrowly focused as it serves the few industrial sectors where Taiwan excels. Since 1980, the government has allocated $520 million to build the park's physical and nonphysical infrastructure for HSIP firms. Accessible HSIP administration assistance, plentiful supporting environment, and good cooperation network have been key infrastructure factors contributing to the success of firms in HSIP.

HSIP Administration Assistance

HSIP administration built clusters of "standard" factories to enable newcomers to start their businesses quickly with a minimum of investment. There are prefabricated factories for them to move into on the first day. They get a five-year tax exemption and generous grants; imported tools are duty free; exports are whisked through an automated customs system. HSIP also built housing for several thousand workers and a school for their children. In addition, the park's administration works with tenants to solve their problems, steering them around difficulties with regulations and cutting through red tape. The park administration also provides such services as training, consulting on investment, market research, labor management, business services, building construction, land administration, computer network services, medical services, warehousing, sanitation, emergency services, and security.

Besides tax incentives, the park offers competitive grants for R&D purposes. To encourage R&D, the Science Park Administration presents annual Innovative Product awards and provides grants for innovative high-tech R&D projects and for projects to develop strategic products and components. Since its inception in 1986, the Science Park Administration's system of grants for innovative high-tech R&D projects has become an important resource for HSIP companies seeking to increase their competitiveness. Grants of this type were approved for 36 projects in 1995; a total of $3.4 million was awarded, making up 26% of the $13.2 million overall costs of the projects. Responding to the need for all-round development of science and technology in the ROC, the Science Park Administration in July 1992 began offering grants for projects to develop strategic products and components. In 1995, 51 such grants were approved, and $57 million was awarded, accounting for 41% of the $140 million overall costs of the projects.

When developing countries wish to upgrade the competitiveness of their industries, how to select industries with potential competitiveness as a target for development and how to build competitive infrastructure to develop these industries are important topics. Taiwan is an island with scarce natural resources and limited domestic market, so it should set up an export-oriented strategy for economic development and should develop high-technology industries so that economic development can be maintained (Chang et al. 1994). The industries selected by HSIP as the target industries to be developed all have the future developing potential with high market opportunities.

Because of the scientific and technological nature of firms at HSIP, they have strong market orientation. Those firms are usually willing to build efficient customer linkages, involving potential users in the development process. And the nature of the market, which the firms faced, also has a higher intensity of market need and market growth rate. As we know, strong market orientation and high market opportunity are important success factors to high-technology firms (Rothwell 1992).

Environment Supports for High-Tech Firms at HSIP

Firms in HSIP have plentiful technological resources. In addition to the three major institutions mentioned earlier, the ROC Executive Yuan's National Science Council sponsors four national research laboratories in the Hsinchu area. Of these, three are located within the park: the Synchrotron Radiation Research Center, the National Space Program Office, and the national Center for High-performance Computing. The fourth, the National Nano Device Laboratory, is located on the campus of neighboring Chiao Tung University. Other research and support agencies in the park include the Precision Instrument Development Center and the Chip Implementation Center.

HSIP creates a conducive intellectual climate for high-technology firms' R&D, and fosters cooperative research with its steady source of researchers at hand, which can expand firms' technological prowess at relatively low cost. Therefore HSIP's firms usually have a willingness to take on external ideas, and can make effective linkages with external sources of scientific and technological know-how.

There are well-qualified personnel in HSIP. Proximity to universities and other institutions of higher education has specific advantages. They provide ready sources of well-trained graduates for the workforce. Firms not only get the technologies, they also get many qualified scientists and engineers from the institutions.

Returning expatriates also have played a vital role in the growth of HSIP. The park has a sophisticated network for recruiting repatriates, particularly in the United States. The architecture and services of the park are also designed to make repatriates feel at home. By the end of 1996, the number of repatriates in HSIP was 2,563. With the knowledge and ideas they have brought back, returned expatriates have done much to raise the levels of technology and R&D in HSIP.

Cooperation Within Industry Cluster

HSIP has made a strong effort to mix small firms with larger companies in the park, and together they make a joint effort to create new products for global markets. Large and small firms play interactive and complementary roles; they separately play an important role in technological innovation.

Small firms derive innovative advantages when technical innovation and skilled labor play important roles, which are mainly behavioral; large firms have innovative advantage with high levels of concentration as well as capital and advertising intensity, which are mainly materials (Chakrabarti 1991; Rothwell and Dodgson 1991). Large firms have the ability to gain scale economies in R&D, production, marketing, and to offer a range of complementary products (Rothwell and Dodgson 1991). Small firms have the ability to react quickly to keep abreast of fast-changing market requirements, they are much more productive with introducing new products than their larger competitors (Chakrabarti 1991; Rothwell and Dodgson 1991). The interaction of large and small firms can assist both partners to integrate their respective material and behavioral innovative advantages (Rothwell and Dodgson 1991).

At HSIP small and large firms interact through manufacturing subcontracting relationships, subcontracting R&D, collaborative development, joint ventures, sponsored spin-outs, producer and customer relationships. Small firms, with ability to react quickly, heavily invest R&D in product innovation, while some large firms concentrate on the production process development. Many small firms supply components and subassemblies or finished products to large firms.

CHARACTERISTICS OF SUCCESS FOR HIGH-TECH FIRMS

It is safe to say that the relationship between R&D investment and a firm's performance is based on the characteristics of a firm. The characteristics that are peculiar to HSIP firms are explained below.

Burgelman, Maidique, and Wheelwright (1996) grouped their findings into six themes of success for high-technology firms: (1) business focus, (2) adaptability, (3) organizational cohesion, (4) entrepreneurial culture, (5) sense of integrity, (6) hands-on top management. They suggested that "no one firm exhibits excellence in every one of these categories at any one time, nor are the less successful firms totally lacking in all. Nonetheless, outstanding high-technology firms tend to score high in most of the six categories, while less successful ones usually score low in several."

Most of HSIP's firms, based on our surveyed data and our experience with those firms, had high scores in such areas as business focus, entrepreneurial culture, and hands-on top management. Besides the above-mentioned characteristics of success, the firms in HSIP usually have the important characteristic of successful innovation: treating innovation as a corporatewide task (Rothwell 1992). Government policy and the Park Administration regulation foster these characteristics.

Business Focus

Obviously, small firms cannot hope to provide a complete line of products. When a small firm grows and establishes a secondary product line, it is always closely related to the first. Therefore most firms in HSIP had only closely related products. Those firms focused R&D on their major area. Moreover, their commitment to R&D is both enduring and consistent.

Entrepreneurial Culture

One of the park's roles is to cultivate entrepreneurship ideas from the government labs and the universities regularly grow into firms at HSIP. The region's cultures encourage risk taking and accepted failure. As a science-based industrial park, HSIP is probably appropriate for any small technology-based firm in the process of starting up or entering a phase of growth.

Hands-on Top Management

Many of the HSIP firms' founders have a technical background. They are usually actively involved in the innovation process to such an extent that they are sometimes accused of meddling. Managers of firms in HSIP not only understand how organizations work, and in particular how engineers work, they understand the fundamentals of their technology and can interact directly with their people about it.

Corporatewide R&D

Firms that wish to set up in HSIP must have a substantial R&D budget, and must have a high proportion of their working force in technical or research work. Therefore, almost all firms in HSIP treat innovations as a corporatewide task, and have a long-term corporate strategy in which innovation plays a key role. They always pursue the efficiency in development work, and the achievement of technical and production synergies.

CONCLUSIONS

Sustainable competitiveness has been one of the hot managerial topics of recent years. And R&D was found to be a necessary condition for competitive ability. In a world filled with rapid change, investing in innovation is the equivalent of holding options for the future. Much more can be done if Asia is to continue advancing its technological abilities. Science-based industrial parks can play a major role in building R&D infrastructure for the developing countries in South Asia. Taiwan has long encouraged the development of high-tech products through its network of science and technology industrial parks. It seems that Taiwan's experiences in establishing HSIP to create an infrastructure could provide a good example for the developing countries in South Asia to enhance their economic development.

What leads a firm's commitment to innovation? A firm's attributes do affect its R&D investment decision, as shown in this study by differentiating firms based on their capital sources and R&D intensity patterns during various stages. Evidence also supported that R&D investment could improve the future performance of a firm, but it might beat the cost of its current productivity. This result could serve as a reference to the developing countries' governments in Asia in fostering their industry policies regarding to R&D issues.

This research also has reconfirmed the results obtained from Morbey and Reithner (1990) that a firm's R&D intensity is positively correlated with its subsequent sales growth. We also found that R&D investment would have a negative impact on a firm's current performance. Therefore, we need to pay attention to how much current productivity could be affected to justify R&D. Relative R&D intensity is an important driving force and thus a predictor of corporate growth. And corporate R&D intensity emerges as a, perhaps, the principal means of gaining more market share under global competition.

One of the key factors to the success of HSIP is that it has effectively brought together government institutions, universities, and large and small companies in the joint effort to create a new high-technology industry infrastructure such that competitive high-tech products are developed and manufactured for world markets. The developing countries in Asia, which want to improve their competitive position in the world market, are highly dependent on their choice today. Government policies to foster nationwide R&D, education, and training and firms' willingness to invest R&D today ensure continued opportunities for technological innovation and economic growth in the future. New product development must be given higher priority. There must be sufficient spending on R&D and closer links between the academic and industry worlds, ensuring the speedy commercial application of business innovation.

REFERENCES

Brenner, M. S., and B. M. Rushton. "Sales Growth and R&D in the Chemical Industry," *Research Technology Management* (March–April 1989) 8–15.

Burgelman, R. A., M. A. Maidique, and S. C. Wheelwright. *Strategic Management of Technology and Innovation*, 2nd ed. Homewood: Irwin, 1996.

Chakrabarti, A. K. "Industry Characteristics Influencing the Technical Output: A Case of Small and Medium Size Firms in the U.S.," *R&D Management* 21, 2 (1991) 139–152.

Chakraborty, C. "Sources of Growth, Input Structure and Technical Progress in American High Technology: A Business Cycle Analysis," *Proceedings of the IEEE 1994 International Engineering Management Conference*, New York, 1994, 415–420.

Chang, P-L., C. Shin, and C.-W. Hsu. "The Formation Process of Taiwan's IC Industry-Method of Technology Transfer," *Technovation* 14, 3 (1994) 161–171.

Collien, D. W., J. Minz, and J. Collin. "How Effective Is Technological Innovation?" *Research Management* (September–October 1984) 11–16.

Cooper, R. G., and E. J. Kleinschmidt. "Winning Businesses in Product Development: The Critical Success Factors," *Research & Technology Management* 19 (July–August 1996) 18–29.

Ellis, L.W. "Optimum Research Spending Reexamined," *Research Management* (May 1980) 22–24.

Gilman, J. J. "Stock Price and Optimum Research Spending," *Research Management* (January 1978) 34–36.

Griliches, Z. "R&D and Productivity: Measurement Issues and Econometric Results," *Science* 237, 3 (1987) 31–35.

Guerard, J. B., Jr., A. S. Bean, and S. Andrews. "R&D Management and Corporate Financial Policy," *Management Science* 33, 11 (November 1987) 1419–1427.

Gwynne, P. "Science Parks: Greenhouses for Ideas," *Asian Business* 28, 1 (1992) 34–35.

Gwynne, P. "Directing Technology in Asia's Dragons," *Research & Technology Management* (March–April 1993) 12–15.

Hambrick, D. C., and I. C. MacMillan. "Efficiency of Product R&D in Business Units: The Role of Strategic Context," *Academy of Management Journal* 28, 3 (1985) 527–547.

Hambrick, D. C, I. C. MacMillan, and R. R. Barbosa. "Business Unit Strategy and Change in the Product R&D Budget," *Management Science* 29, 7 (July 1983) 757–769.

Morbey, G. K. 'R&D: Its Relationship to Company Performance," *Journal of Product Innovation Management* 5, 3 (1988) 191–200.

Morbey, G. K., and S. S. Dugal. "Corporate R&D Spending During a Recession," *Research Technology Management* (July–August 1992) 42–46.

Morbey, G. K., and R. M. Reithner. "How R&D Affects Sales Growth, Productivity and Profitability," *Research Technology Management* (May–June 1990) 11–14.

Reynard, E. L. "A Method for Relating Research Spending to Net Profits," *Research Management* (July 1979) 12–14.

Rosenau, M. D., Jr. "Problem with Optimizing Research Spending," *Research Management* (November 1980) 7.

Rothwell, R. "Successful Industrial Innovation: Critical Factors for the 1990s," *R&D Management* 22, 3 (1992) 221–239.

Rothwell, R., and M. Dodgson. "External Linkages and Innovation in Small and Medium-sized Enterprises," *R&D Management* 21, 2 (1991) 125–137.

Schoeffler, S., R. D. Buzzell, and D. F. Heany. "Impact of Strategic Planning on Profit Performance," *Harvard Business Review* (March–April 1984).

25

Innovation Policies in East Asia

Mark Dodgson

INTRODUCTION

There is a wide diversity in innovation policies in East Asia,[1] which reflects the differing roles and broader policies of government, and the variety of industrial structures and business systems found in East Asian nations. This chapter will illustrate this diversity, and consider some of the major features and challenges of innovation policy in this region. It will describe the diversity in national technological capacities and the reliance that many governments have placed on direct foreign investment (DFI) as a means of encouraging technological development, and it questions the role that DFI has played in developing the innovation capacities of local firms. It argues that innovation policies have to be focused directly on building these in-house capabilities in the future. Important in this process, it is argued, is the development of a variety of intermediary institutions, which encourage links between the users and suppliers of technology, and this provides a major policy challenge for the future. For those countries which have established such intermediaries, such as Korea and Taiwan, the policy challenge involves upgrading the level of technological and managerial support they offer.

WHAT IS INNOVATION POLICY?

Innovation policy is considered here to be different from science policy and technology policy. Although there are considerable overlaps and blurred boundaries, science policy is here understood to involve government policies for promoting science in universities and research laboratories, while technology policy addresses the development of important generic technologies, such as information technology (IT) and biotechnology. By contrast, innovation policy is considered to be those efforts by governments which encourage the accumulation, diffusion, and commercial use of new products, processes, and services by firms. Effective government innovation policies build the capacity of firms to innovate.

Although the target of effective innovation policies is individual firms, these policies are commonly network-orientated. In line with recent innovation theory which analyzes systems of innovation—national, technological, local—their focus is cross-organizational learning. An important aspect of effective innovation policies is that they are adaptable and adept at learning from overseas.[2]

NATIONAL TECHNOLOGICAL CAPACITIES IN EAST ASIA

There are large differences in the technological capacities of East Asian nations, but in aggregate the rapid technological development of the East Asian region has been one of the most impressive and important phenomena of the 20th century. Less than 40 years ago, for example, Korea was one of the poorest countries, on par with the poorest African nations. In this short time period it has become a relatively wealthy and technologically strong nation, and has a number of world-class companies operating at the world technological forefront (Kim 1997).

Asia has a positive balance of trade in sales of high-tech products (largely electronics) with the United States According to one estimate, by 2005 the East Asian economies will in combination spend more on R&D than the United States (Sheehan 1995). The number of patents granted within Asia between 1985 and 1990 increased at double the rate of growth of that in the United States (although, of course, these grew from a much smaller base). As for patenting in the United States, Asian patents, primarily from Taiwan and Korea, increased fourfold during the 1970s and tenfold during the 1980s (NSF 1995).

Table 25.1 shows the commitment to R&D in the East Asian economies, and reveals the wide disparities between Korea and Taiwan, which spend as much as many leading nations, and the relatively underdeveloped nations like Indonesia and Thailand.

In a study of high-tech competitiveness of 28 countries, Porter et al. (1996) develop three indicators: high-tech competitiveness (current high-tech production and export capability); high-tech emphasis (the extent to which a nation's exports depend on high-tech); and the rate of change of high-tech standing (see Table 25.2). Their indicators are derived from UN trade data, the Elsevier Yearbook of electronics production, sales and exports, and expert panel opinion.

Table 25.1
Gross Expenditure (GE) and Business Expenditure (BE) on R&D as a Percentage of
Gross Domestic Product

Country	GERD/GDP	BERD/GDP
Korea (1993)	2.41	1.72
Taiwan (1994)	1.8	1.03
Singapore (1993)	1.2	0.75
China (1994)	0.49	0.11
Malaysia (1992)	0.4	0.17
Indonesia (1993)	0.2	0.04
Thailand (1991)	0.2	0.04
Hong Kong (1995)	0.1	n.a.
OECD Average	1.94	1.19
Japan (1994)	2.64	1.87
United States (1994)	2.53	1.80

Source: OECD.

Table 25.2
High-Tech Standing and Emphasis, 1996

Country	High-Tech Standing	High-Tech Emphasis	Rate of Change
Japan	90.7	67.8	31.1
United States	90.0	53.2	35.7
Germany	60.4	32.3	27.8
UK	49.3	46.7	38.3
France	45.5	35.8	32.4
Italy	31.5	21.0	34.8
Sweden	28.0	31.5	28.5
Canada	24.0	15.1	21.1
Australia	15.6	7.8	30.7
Singapore	35.7	100.0	29.8
South Korea	28.6	59.3	26.6
Taiwan	26.9	58.9	28.5
Hong Kong	22.0	54.7	15.0
Malaysia	24.3	77.4	41.3
China	20.6	24.1	55.5
Thailand	17.1	44.2	47.9
India	13.4	7.4	19.4
Philippines	12.6	33.3	50.5
Indonesia	10.9	6.5	79.9

Source: Porter et al. (1996).

According to these measures, a number of broad characteristics are discernible among Asian NIEs. First, in respect to technological standing, a group of Asian nations—Singapore, Korea, Taiwan—are equivalent in performance to Italy, Sweden, and Canada, and these Asian nations plus Hong Kong, Malaysia, and Thailand have a greater high-tech emphasis than all the European nations cited here (explained primarily by electronics exports). Although it is occurring from a relatively low base, the rate of change of technological standing in the South East Asian nations plus China is larger than those Asian nations with larger technology bases.[3]

There are ambitious development plans among those nations with low technology standing. Malaysia's seventh five-year plan, 1996–2000, aims to increase gross expenditure on R&D as a proportion of GNP to 1% (from 0.4 in 1994), and increase the number of scientists and technicians to 1,000 per million population (compared to 400 in 1992). Its science budget of $1.2 billion targets information and communication technologies (ICT), electronics, biotechnology, and manufacturing. Plans for its "Multimedia Super Corridor" south of Kuala Lumpur are estimated to require an investment of $20 billion (Far Eastern Economic Review 1997). Indonesia has similarly announced the intention to reach the level of 1% gross expenditure on R&D/GDP by 2003 (Nikkei Weekly 1996), and 2% by 2018 (Scott-Kemmis and Rohadian 1995).

The major disadvantages East Asian nations face in respect to science and technology are in the comparatively underdeveloped university and research sectors, and in the paucity of skilled scientists and technologists. While remarkable efforts are being undertaken to develop the university and research capacity of most East Asian nations, none have yet to achieve the level of breadth and depth of research systems found in North America and Europe (with the possible exception of China). Using a variety of indicators, the National Science Foundation (NSF) has calculated that patents from Asia, including Japan, have weaker ties to fundamental sciences than in the United States (NSF 1995). Despite being one of the most advanced nations technologically, 99% of patents in Singapore are registered to nonresidents (NSF 1995).

The relative paucity of science and technology professionals can also be seen in skilled managers of technology and innovation. There are weaknesses in many East Asian firms' ability to manage technological innovation. While leading East Asian companies, like Samsung in Korea, have identified their weaknesses in this area and instigated strategies for overcoming them, shortcomings remain (Dodgson and Kim 1997). Generally, policies for training technology managers in East Asia are variable in quality, commonly limited in scale, and often focus on inappropriate European and American models (Minden and Wong 1996). Much of the technological strength of East Asian companies lies in production and project execution (Amsden and Hikino 1994). These are important but relatively easily replicable skills. They do not provide the basis for longer term, sustainable competitive advantage at a global level. They are transferable by means of blueprints and explicit information. They exclude the tacit, creative, and intuitive skills necessary to be truly innovative.

Another problem East Asian firms face in their science and technology is the way they are excluded from international technology collaborations. Databases on the geographical location of partners in international strategic alliances show that over 90% of them occur within the European/Japanese/U.S. Triad (Dodgson 1993). Many East Asian firms are therefore excluded from major technology projects (although leading companies, like Samsung, are increasingly attracting the partnership of other world-class companies because of their technological prowess).

The weakness of East Asian science and technology is revealed in the comparative lack of R&D expenditure and also in the fact that royalties and fees paid to U.S. firms to license use of their proprietary industrial processes nearly doubled in Asia during the 1987–1991 period; these were, on average, 10 times that paid to Asian firms by U.S. companies (NSF 1995). While this is not always a weakness if firms are using the licenses as a learning opportunity to build innovative capacity, few firms have been capable of doing this.

East Asian nations also suffer comparatively in their capacity to develop, produce, and market technological innovations themselves. One reason for this, it will be argued, is the shortage of intermediary institutions which can assist firms in their technological and managerial development. East Asian nations' innovative capacity is weakened by two other major factors:

1. Reliance upon the IT industry and original equipment manufacturing (OEM) agreements: Technology-based, export-oriented East Asian firms operate primarily within the IT industry, major components of which, such as the semiconductor industry, are notoriously cyclical. Most firms in this sector are relatively small and compete on the basis of cost advantages rather than technological innovation. Furthermore, although it varies between countries and is declining in aggregate, most East Asian nations depend heavily upon Japan for investment (Abegglen 1994). A major element of Korea's $13 billion annual trade deficit with Japan, for example, is advanced capital equipment. Even the most technologically advanced East Asian computer company has difficulty building up its international brand recognition, undertakes incremental rather than world-leading R&D-based innovation, and is to a significant extent controlled by overseas buyers. For example, Acer, perhaps the most market-recognized Taiwanese electronics firm, produces half of its personal computers (PCs) under OEM arrangements for other firms (*Far Eastern Economic Review* 1995). This does not argue that OEM has not played a very important role in the development of some East Asian nations' technology (Hobday 1995). It does, however, question the extent to which these relationships encourage the development of in-house innovative capacity.
2. Dependence upon investment by multinationals (MNCs) and inward licensing: In the South East Asian nations there is a high reliance upon government policies encouraging overseas investment and licensing for technological development. Reliance on MNC's investment is exceptionally high in some countries: for example, three-quarters of Malaysia's total value of exports are estimated to be provided by MNCs (Lall 1995) and it plans to continue to use DFI as a major element in its technology development policies (Jegesthesan et al. 1997). Such reliance brings high levels of dependence, of course, and vulnerability should labor costs increase so much as to make lower wage economies more attractive to the MNCs. The question for effective innovation policies is, have these policies developed the capacity of indigenous firms to be innovative?

Investment by overseas MNCs may *potentially* stimulate and develop innovative capabilities by means of the transfer of technology and management know-how and expanded opportunities for self-learning. Local firms may be exposed to new technology and management practices through their dealings with MNCs, and advanced production skills may be transferred to local businesses.

The literature tells us that the extent of technology transfer between MNCs and local firms depends upon:

- Strategy of the MNC: Japanese firms, for example, are generally believed to desire long-term relationships with suppliers which may facilitate technology transfer (Yamashita 1991). Technology transfer may occur more readily when the MNC has a controlling interest in the local firm.
- Nature and complexity of the technology: "strategic" R&D is believed almost always to be located in the MNC home nation; the more complex the technology the greater the difficulty in transferring it.
- Technological and absorptive capacity of the local firm: without significant technological and managerial capacity, many advanced technologies cannot effectively be integrated into local firms.
- Government policies: the existence of government incentives on the one hand, and fears of sudden localization on the other, affect the propensity to transfer technologies.
- Concern about the comparative lack of protection for intellectual property (IP): while there has been considerable progress in the development of IP protection in most East Asian countries and many countries have been signatories to international intellectual property law reforms, few have the administrative structures to effectively police the agreements (Turpin and Innes 1995).

So what evidence have we of the development of innovation capacity as a result of the technological activities of MNCs?

In *Indonesia*, where the technological infrastructure is among the weakest in the region, subcontracting in electronics, food processing, pharmaceuticals, chemicals, and autos has involved little range or depth in the relationship between local companies and MNCs, with little transfer of technology apart from some assistance with quality control techniques and minor product adaptations (Gultom-Siregar 1995; H. Hill 1988; Thee 1990). However, some firms have developed minor product development capacities; quality improvements and increased design activities are noted in the textile and garment industry; and within the car industry up to 0.5% of firms' expenditure is spent on domestic R&D (S. Hill 1995). The minor product developments described by Hill include: auto firms developing stronger shock absorbers for local roads, electrical appliances being designed to deal with voltage fluctuations, and bottled drinks and canned foods being modified to suit local tastes.

In *Thailand*, another comparatively weak nation technologically, subcontracting by local firms for MNCs has similarly been in relatively simple components (Supapol et al. 1995), and where there are technological applications involved in joint venture arrangements these tend to be in low-tech, low-value-added manufacturing (J. Wong 1995).

In the comparatively technologically stronger nations the situation is slightly different. In *Malaysia*, particularly in the Penang region—which employs over 100,000 electronics workers and where 80 of its 150 electronics factories are foreign-owned (Goh 1996)—the intense competition between MNCs in the electronics industry has led to some transfers of product and process technologies to local subcontractors which have had to possess high-precision and good quality control techniques. However, this transfer has, not surprisingly, only occurred when the local firm is relatively strong technologically (Rasiah 1995). And generally both local and foreign firms perform relatively few high value added and technologically demanding tasks like design and development (Lall 1995).

The situation in *China* is similar. In one of the most comprehensive, case-study-based, studies of technology transfer through direct foreign investment in China, Lan (1996) found that while there is valuable technology transferred through DFI, this occurs in a limited number of cases and it is dominated by transplants of hardware only. Over 70 percent of interviewed foreign investors could not identify technological advantages through their investment. Only a small minority intended to undertake any R&D.

Singapore has possessed the most aggressive policies in Asia for attracting MNC investment. Yet even among Japanese MNCs, there has been little transfer of R&D capacity: product and process innovation remains a strategic task back in Japan. Singaporean firms are expected to concentrate on incremental improvement and modification of products (Tang 1996). Foreign firms investing in R&D in Singapore do so primarily for commercial reasons—such as the desire to develop and adapt products to local markets and be close to lead users and customers—rather than accessing local technological capabilities and manpower resources (Wong, Loh, and Roberts 1994).

In summary, subsidiaries of foreign MNCs still contract out a relatively small fraction of the component requirements to unaffiliated local suppliers, and hence the possibility for technology transfer is rather limited (Supapol 1995). Bell and Pavitt (1993) are correct in arguing the depth of accumulated technological capabilities in industrializing countries is limited when technology is incorporated in new production capacity through turnkey projects and direct foreign investment. From this brief review, it appears that while reliance on policies which encourage investment from overseas MNCs may improve the technological resources of local firms in terms of facilities and equipment, they do not build the capacity for innovation.

There are a variety of innovation policy tools available to governments, ranging from direct grants, taxation, and R&D employment incentives, to the use of demonstration techniques (using particular companies as examples of good practice). Intermediary institutions are an important element of effective innovation policy, and they can play an important role in fostering innovation capacity.[4] Of all the East Asian nations, Korea has the most complete system of intermediary institutions (see Kim 1997 for an analysis of these). This chapter will now present a brief examination of the role of intermediary institutions in Taiwan, and the way in which they are, somewhat belatedly, being rapidly developed in Singapore.

DEVELOPING INTERMEDIARY INSTITUTIONS

The role of intermediary institutions elsewhere in the world is extremely diverse. They range from SEMATECH-type organizations focusing on early-stage R&D to relatively simple industrial extension services provided by, for example, the Japanese Regional Technology Centers (Shapira 1992) and the U.S. Manufacturing Extension Partnership (Kelley and Arora 1996).[5]

There are comparatively few of these intermediary institutions in East Asia outside of Korea and Taiwan, and latterly in Singapore. This institutional deficiency has been recognized by the major development banks, such as the World Bank (Najmabadi and Lall 1995), and the Asian Development Bank (ADB 1995).

Taiwan

Government-supported technological and scientific research institutes have been indispensable in Taiwan's high-technology industrial development (Castells and Hall 1994). In contrast with most East Asian nations, Taiwan possesses an old and large system of R&D institutions. Government has spent extensively on R&D and has had a succession of programs designed to encourage indigenous technological capabilities in strategic technologies. The Industrial Technology Research Institute (ITRI), for example, which was established in Hsinchu in 1973, has played a major role in developing local technological capabilities in firms. Weiss and Mathews (1994) and Wu (1995) describe how semiconductor wafer fabrication technology developed by ITRI has been spun off into some of Taiwan's most successful semiconductor firms. Numbers of commentators point to the strong business orientation of ITRI's leaders (Rush et al. 1996)

ITRI employs over 6,000 and has an annual operating budget of $500 million. Its technology focus ranges from the high-tech integrated circuit (IC) industry to the textile industry, and its work on factory automation and advanced materials has also been applied in traditional industries. It coordinates multipartner consortia: the Taiwan New Personal Computer (TNPC) Alliance formed in 1993, for example, involves 31 partners including IBM, Apple, and Motorola. The aims of the Alliance are to "bring together firms from all aspects of the IT industry with a clear focus on transferring, uptaking and diffusing the new PowerPC technology in a series of products spanning PCs, software, peripherals and applications such as multimedia" (Poon and Mathews 1995). The initiative behind TNPC lay with the Computer and Communications Laboratory (CCL), one part of ITRI. CCL selected the PowerPC as an important "generic" technology and built the consortium, including negotiating with the U.S. partners. Poon and Mathews (1995) make the observation that this consortium is as much about diffusion of existing technology as it is about technology generation. ITRI also provides technical services, such as consulting, training (both technical and managerial), and trouble-shooting.

As another example of its carefully developed role as an intermediary, in 1996 ITRI opened its Open Laboratory Program based in an extensive new R&D complex in Hsinchu. The Open Lab Program mainly provides space and facilities for joint R&D between ITRI researchers and local business, and also has space for business incubation, conference, and training facilities. Business incubatees receive "packaged" business and management consulting, financial and legal assistance, and office and administrative support. Entry to the business incubator requires the formal approval of a business plan. Such consulting activity is, of course, assisted by government policies. Taiwan's technology and management guidance incentives, for example, subsidize 60% of the cost of total consulting exercises of firms.

The co-located Hsinchu Science-Based Industrial Park has operated since 1980. It was established to serve Taiwan's high-technology industries and accelerate their development. It offers a wide range of tax incentives, low-interest loans, R&D and manpower training grants, and duty-free importing of equipment and materials. In 1995 it had 170 resident companies, employing 36,000, with sales in excess of $6.5 billion (in 1994). The park is expected to reach combined sales of $50 billion by 2003, and it has extensive development plans. Its main technology foci are electronics, particularly ICs. It has proven successful at attracting returned expatriates: over 1,000 work at the park. Initially the government directly invested in the small start-ups, but increasingly private venture capital companies are assuming this role. Park companies have proved successful technologically, and many have demonstrated sophisticated management strategies when, for example, it comes to managing international strategic alliances (Yuan and Wang 1995).

As Castells and Hall (1994) note, a common feature of successful technology institutions is their spatial collocation and integration, and they identify the success of ITRI and Hsinchu Science Park as being dependent upon their coexistence and also being collocated alongside two universities. Local small firms, as well as having the opportunity to link with particular local institutions, are part of an "innovative milieu" (Cooke and Morgan 1994).

Singapore

Although organizations such as the Singapore Institute of Standards and Industrial Research (SISIR) have been operating since 1973, much of the institutional development in Singapore is of a relatively recent vintage. Between 1985 and 1995 Singapore established 9 research centers and institutes focusing on IT, electronics and biotechnology. At present it has 13 Institutes which the National Science and Technology Board (NSTB) estimates to have collectively undertaken over 800 joint research projects. According to Tang and Yeo (1995) the mission of these institutes and centers is to: (1) provide specialized training, (2) develop precompetitive technologies, (3) provide services to companies, and (4) transfer technology to industry. The NSTB was established in 1991 from the Singapore Science Council, and the same year a National Technology Plan was announced.

Until recently, the overall thrust of Singapore's innovation policy was rather one-dimensional, and NSTB (and its predecessor) focused on attracting MNCs and then providing incentives to undertake R&D and design products locally.

Additionally, the Singaporean Economic Development Board (EDB) offers incentives for MNCs to upgrade the technological content of their operations, and subsidies are available for MNCs wishing to send Singaporeans to headquarters to be trained to implement the transfer of new product lines to Singapore. The EDB has also established the Local Industry Upgrading Program (LIUP) to facilitate technology transfer. LIUP involves an experienced engineer from a selected MNC working full-time and wholly paid by government to provide technical and managerial upgrading assistance to several local supplier firms (P. Wong 1995).

The government plans to dedicate $2.85 billion to science and technology between 1996 and 2000. As described by P. Wong (1996) Singapore's National Technology Plan, while still attempting to encourage R&D by overseas firms, has as its major target the development of indigenous R&D capability among local universities, public research institutes, and firms, including new policies promoting R&D consortia. This recognizes the past problems with reliance upon MNC R&D, which although it has had some success—for example, according to Singapore's former deputy prime minister, 87 MNCs have significant R&D programs in Singapore (Goh 1996)—it has also had some deficiencies: as Goh admits, "the R&D activities of these companies must be regarded as peripheral."

The largest deficiency lies in the development of indigenous innovation capacity (remembering that local firms account for only 1% of Singaporean patenting). In their survey of the strategic management of technology in Singaporean firms, Wong, Loh, and Roberts (1994) found marked differences between foreign and local firms. Specifically, foreign-owned firms developed and accepted technology strategy more often; allocated more funds to research; monitored technology more extensively (with Japanese firms being particularly involved with university research); and more often considered themselves as technology leaders rather than followers). The survey's authors considered among the implications of the study that technical managers and professionals need to acquire more business skills, and Singaporean companies should make greater use of resources in local universities. This shows a marked lack of innovative capability.

In summary, Singaporean policy has altered in recent times to focus on the active encouragement of R&D and the improvement of local technological capacities. Despite rich resources in production technology, Singapore has not developed the capacity to innovate significantly. As in the past, its small size will prove advantageous when attempting to implement the new strategies.

One of the most astute observers of the Asian technology scene, Wong Poh-Kam (1995), contrasts the success of state intervention in hastening technology diffusion among local firms in Taiwan with the past lack of government concern with public R&D in Singapore. Public research institutes, he argues, "serve initially to assimilate advanced technology from overseas and rapidly diffuse them to local enterprises, but increasingly also to serve as the coordinating nodes to promote indigenous technology creation via R&D consortia and strategic R&D programs as well." So, "While the development of the electronics industry in Singapore has been largely driven by an innovation network model centered on MNCs as the key nodes, indigenous small and medium entrepreneurial firms—in close relationship with public research institutes—have been the driving force in the case of Taiwan" (P. Wong 1995:18).

CONCLUSIONS

Despite the remarkable technological successes enjoyed by East Asian nations, there remain pressing policy challenges for governments concerned with building innovative capacity in local firms. This chapter has argued that in contrast to policies which encourage investment from overseas MNCs, those which focus on the development of intermediary institutions are more likely to develop such innovation capacity. Intermediary institutions, as we have seen in the case of ITRI, can assist firms to develop their R&D and innovation management skills (often through international links), and package the wide range of support mechanisms often necessary to encourage high-tech firm growth. Such institutions have traditionally received little attention from analysts, and while the major development banks have begun to identify their importance, there is little consideration by them of the policies that encourage their development and growth.

There is no simple blueprint for developing intermediary institutions. Each will develop according to its specific policy environment, the industry it is a part of, the technologies it uses, and the structure and strategies of its client firms and technology suppliers. However, some broad characteristics of innovation policies which encourage the development of these intermediary institutions can be identified.

There are numerous examples of good practice which can be adopted and adapted from overseas. Learning from overseas experiences is an important element of effective innovation policy. There are numerous examples of this having occurred. The Korea Technology Development Corporation and the Malaysian Technology Development Corporation, for example, were based upon the British National Research Development Corporation. These institutional replicas, however, are in many ways different from their progenitor and reflect different levels of innovative capability in different industries. It is always important sensitively to recognize and adapt institutions to national differences. Bureaucratic capacity varies enormously in East Asia, so it is highly unlikely that one single model is likely to work.

A question arises of what learning can occur within Asian nations about institutions more appropriate to the specific challenges within the region. The transfer of learning about institutions, and the search for complementarities in national technological strengths, between various European nations described in Dodgson and Bessant (1996) is, as Stephen Hill (1995) points out, also possible and desirable in Asia. The successes of intermediary institutions in Korea and Taiwan hold many lessons for other nations, but it is important to recognize that these institutions themselves may need to adapt and change continually as the clients they serve become increasingly technologically sophisticated.

Within the European Union the European Commission has played a central role in the coordination and transfer of international best practice in intermediary institutions. Such a role could valuably be replicated by Asia Pacific Economic Council within the Asia Pacific region.

Good management is the key to successful intermediation and intermediaries, and is critical to the success of both new initiatives and the adaptation of existing organizations. The success of ITRI has to be ascribed in major part to the quality of its management. The skills needed extend beyond those of matching technological and market opportunities, and include those of building adaptable, "learning" organizations. This will, for example, involve the need for sophisticated human resource policies, such as dual-career ladders—which enable scientists and engineers to be promoted and paid similarly to managerial staff—to enable the development and retention of technological learning.

Dodgson and Bessant (1996) note the importance of a new generation of professional services consultant, the so-called innovation consultant. The development of managers who are both technology literate and business oriented will be a major policy challenge in East Asia, where there is a general shortage of technological consultants. Training programs for these innovation consultants, and for managers of intermediary institutions, should be a priority policy area.

Achieving a balance between public good and commercial interests in intermediary institutions is perhaps the most challenging aspect of this element of effective innovation policy. Ideally, initial government funding should give way to private-sector investment as effective policies which build sustainable innovative capabilities enable the disengagement of the government. In practice, intermediary institutions will continue to play an important public-good role and this has to be supported by the government. The challenge for the intermediary institution is to integrate, through its networking activities, the discipline of the market from commercial sources of finance (as is increasingly occurring in Hsinchu Science Park), and the strategic aims of government policy for the long-term development of technology and managerial capability.

ACKNOWLEDGMENTS

This chapter draws heavily upon research undertaken, and a paper written, for the Korean Science and Technology Policy Institute (STEPI) conference on Innovation and Competition in Asian NIEs, Seoul, May 25–26, 1997. The author wishes to acknowledge the support of STEPI and its Director, Professor Linsu Kim, in the production of this study.

NOTES

1. East Asia is commonly defined as including the Association of South East Asian Nations: China, Japan, Korea, and Taiwan. For the purposes of this chapter, Japan is not included in this definition as its level of technological development and the challenges confronting it are broadly different from other East Asian nations. Reference to matters "East Asian" is at best a simple convenience, as the historical, economic, social, religious, and cultural diversity in the region is so extraordinarily wide.

2. There is huge diversity in the context in which innovation policies are formulated and implemented in East Asia. Governments in the region vary in approach from among the world's most centrally planned to the most *laissez-faire*. Korea and Taiwan have tended to discourage direct foreign investment, most of the South East Asian nations actively encourage it. Korea's industrial structure is dominated by large firms, Taiwan's by small firms. Hong Kong is resolutely noninterventionist; Indonesia has a policy of developing national champions in strategic industries. Japanese models of management differ from Korean, which differ from Chinese. Legal systems vary considerably.

3. The major difficulty with these data (apart from the inherent problems of trying to encapsulate the richness of international technological diversity in relatively simple indicators) is the heavy reliance on electronics. While electronics is the major high-technology industry in East Asia, this indicator leads to some distortions in underestimating technological strengths and weaknesses in other areas.

4. See Dodgson and Bessant (1996) and Rush, et al. for detailed accounts of how this occurs.

5. Numerous case studies of these organizations are presented in Dodgson and Bessant (1996) and Rush, et al. (1996).

REFERENCES

Abegglen, J. *Sea Change*. New York: The Free Press, 1994.

Amsden, A., and T. Hikino. "Project Execution Capability, Organizational Know-how and Conglomerate Corporate Growth in Late Industrialization," *Industrial and Corporate Change* 3,1 (1994) 111–147.

Asian Development Bank. *Technology Transfer and Development: Implications for Developing Asia*. Manila: ADB, 1995.

Bell, M., and K. Pavitt. "Accumulating Technological Capability in Developing Countries," *Proceedings of the World Bank Annual Conference on Development Economics 1992*. Washington, DC: World Bank, 1993.

Castells, M., and P. Hall. *Technopoles of the World: The Making of 21st Century Industrial Complexes*. London: Routledge, 1994.

Cooke, P., and K. Morgan. "The Creative Milieu: A Regional Perspective on Innovation," in *The Handbook of Industrial Innovation* (M. Dodgson and R. Rothwell, eds.). Cheltenham, U.K.: Edward Elgar, 1994.

Dodgson, M. *Technological Collaboration in Industry: Strategy, Policy and Internationalization of Innovation*. London: Routledge, 1993.

Dodgson, M., and J. Bessant. *Effective Innovation Policy: A New Approach*. London: International Thomson Business Press, 1996.

Dodgson, M., and Y. Kim. "Learning to Innovate Korean Style: The Case of Samsung," *International Journal of Innovation Management* 1, 1 (1997) 53–67.

Far Eastern Economic Review. February 27, 1995.

Far Eastern Economic Review. July 25, 1997.

Goh, K. "The Technology Ladder in Development: The Singapore Case," *Asian-Pacific Economic Literature* (1996) 1–12.

Gultom-Siregar, M. "Indonesia," *Transnational Corporation and Backward Linkages in Asian Electronics Industries* (A. Supapol, ed.). New York: UNCTAD, 1995.

Hill, H. *Foreign Investment and Industrialization in Indonesia*. New York: Oxford University Press, 1988.

Hill, H. "Indonesia's Great Leap Forward? Technology Development and Policy Issues," *Bulletin of Indonesian Economic Studies* 31, 2 (1995) 83–123.

Hill, S. "Regional Empowerment in the New Global Science and Technology Order," *Asian Studies Review* 18, 3 (1995) 2–17.

Hobday, M. *Innovation in East Asia*. Cheltenham, U.K.: Edward Elgar, 1995.

Jegathesan, J., A. Gunasekaran, and S. Muthaly. "Technology Development and Transfer: Experiences from Malaysia," *International Journal of Technology Management* 13, 2 (1997) 196–214.

Kelley, M., and A. Arora. "The Role of Institution-Building in U.S. Industrial Modernization Programs," *Research Policy* 25, 2 (1996) 265–279.

Kim, L. *From Imitation to Innovation*. Boston: Harvard Business School Press, 1997.

Lall, S. "Malaysia: Industrial Success and the Role of Government," *Journal of International Development* 7, 5 (1995) 759–773.

Lan, P. *Technology Transfer to China through Foreign Direct Investment*. Aldershot, U.K.: Avebury, 1996.

Minden, K., and P. Wong. *Developing Technology Managers in the Pacific Rim*. New York: M.E. Sharpe, 1996.

Najmabadi, F., and S. Lall. *Developing Industrial Technology* Washington, DC: World Bank, 1995.

National Science Foundation. *Asia's New High Tech Competitors*. National Science Foundation 95-309, 1995.

Nikkei Weekly. June 24, 1996.

Poon, T., and J. Mathews. "Technological Upgrading through Alliance Formation: The Case of Taiwan's New PC Consortium," paper presented at the Business Networks, Business Growth Conference, Sydney, October 19–20, 1995.

Porter, A., D. Roessner, N. Newman, and D. Cauffiel. "Indicators of High Technology Competitiveness of 28 Countries," *International Journal Technology Management* 12, 1 (1996) 1–32.

Rasiah, R. "Malaysia," in *Transnational Corporation and Backward Linkages in Asian Electronics Industries* (A. Supapol, ed.). New York: UNCTAD, 1995.

Rush, H., M. Hobday, J. Bessant, E. Arnold, and R. Murray. *Technology Institutes: Strategies for Best Practice*. London: International Thomson Business Press, 1996.

Scott-Kemmis, D., and R. Rohadian. "Indonesia: Science and Technology Policy and Development for Industrial Development," mimeo. Jakarta: Australian Embassy, 1995.

Shapira, P. "Lessons from Japan: Helping Small Manufacturers," *Issues in Science and Technology* 8, 3 (1992) 66–72.

Sheehan, P. Center for Strategic Economic Studies, Victoria University of Technology, Australia, 1995.

Supapol, A., ed. *Transnational Corporation and Backward Linkages in Asian Electronics Industries*. New York: UNCTAD, 1995.

Supapol, A., S. Suebsubanunt, and P. Arbhabhirama. "Thailand," in *Transnational Corporation and Backward Linkages in Asian Electronics Industries* (A. Supapol, ed.). New York: UNCTAD, 1995.

Tang, H. "Hollowing-out or International Division of Labor? Perspectives from the Consumer Electronics Industry and Singapore," *International Journal of Technology Management* 12, 2 (1996) 231–241.

Tang, H., and K. Yeo. "Technology, Entrepreneurship and National Development: Lessons from Singapore," *International Journal of Technology Management* 10, 7/8 (1995) 797–814.

Thee, K. "Indonesia: Technology Transfer in the Manufacturing Industry," in *Technological Challenge in the Asia-Pacific Economy* (H. Soesastro and M. Pangestu, eds.). Sydney: Allen and Unwin, 1990.

Turpin, T., and J. Innes. "Intellectual Property Law in the Asia-Pacific Region," Center for Research Policy, University of Wollongong, Australia, 1995.

Weiss, L., and J. Mathews, "Innovation Alliances in Taiwan: A Coordinating Approach to Developing and Diffusing Technology," *Journal of Industry Studies* 1, 2 (1994) 91–101.

Wong, J. "Technology Transfer in Thailand: Descriptive Validation of a Technology Transfer Model' *International Journal of Technology Management* 10, 7/8 (1995) 788–796.

Wong, P. "Competing in the Global Electronics Industry: A Comparative Study of the Different Innovation Networks of Singapore and Taiwan," presented at the International Symposium on Innovation Networks: East Meets West, Sydney, August 30–31, 1995.

Wong, P. "From NIE to Developed Economy: Singapore's Industrial Policy to the Year 2000," *Journal of Asian Business* 12, 3 (1996) 65–86.

Wong, P., L. Loh, and E. Roberts. "Global Benchmarking Study on the Strategic Management of Technology: The Case of Singapore," NUS/NSTB/MIT Report, 1994.

Wu, S.-H. "The Dynamic Cooperating Relationship between the Government and Enterprises: The Development of Taiwan's Integrated Circuit Industry in Retrospect," mimeo, National Chengchi University, Taiwan, 1995.

Yamashita, S. *Transfer of Japanese Technology and Management to the ASEAN Countries.* Tokyo: University of Tokyo Press, 1991.

Yuan, B., and M.-Y. Wang. "The Influential Factors for the Effectiveness of International Strategic Alliances of High-Tech Industry in Taiwan," *International Journal of Technology Management* 10, 7/8 (1995) 777–787.

PART V:
OPPORTUNITIES FOR CHINA

26

The Construction of China's Information Infrastructure and International Cooperation

Wu Yingjian

INTRODUCTION

With China's economic growth and opening to the outside world, it becomes an urgent necessity to develop the country's information infrastructure and to develop its information industry. This will provide the world with excellent opportunities in international trade, business, and academic and technological cooperation. Macau and Hong Kong can play an important role in this international cooperation.

CONSTRUCTION OF CHINA'S INFORMATION INFRASTRUCTURE

Information technology not only creates large amounts of public wealth, but also triggers changes in modes of social production and peoples' way of living and thinking. The economic status of a nation is closely related to its information technology level and the state of its information infrastructure. Construction of information infrastructure is recognized by developed as well as developing countries as the key to building up the state's economy and competitive strength in the 21st century.

The key points of China's information infrastructure development are: to establish the nation's information network; to enhance the utilization and exploitation of information resources; to promote the application of information technology; to train personnel working with information; to advance information-related legislation; and to promote the information industry as the country's basic industry.

Construction of the State Information Network

Since the 1980s, telecommunications, as the foundation of the nation's economic development, has enjoyed a faster rate of development than the average rate of the nation's economic development (Table 26.1). The capacity of the country's public telecommunications network has now reached more than 100 million subscriber lines. Long-distance lines have reached more than 1 million. The mobile telephone network covers the entire country and has a capacity of over 410,000 channels. The capacity of data communication networks has reached 320,000 terminals. While in some cities, 22% of the population has telephone access, the average across the whole country is 6.2%.

With the worldwide rise of the Internet, China is also seeing rapid development in Internet-related services. Since the opening of the international channel in 1994, four interconnected domestic networks with international links have been built: CHINANET of the Ministry of Post and Telecommunications; GBNET of the Ministry of Electronics Industry; CERNET of the State Education Commission, and CSTNET of the Chinese Academy of Science. The first two of these networks mainly provide commercial services, while the second two concentrate on scientific research and educational services. There are about 300 Internet Service Providers (ISPs) and more than 100,000 users on this network, according to the Ministry of Electronics Industry (MEI). The number of users will exceed $1 million in the year 2000. From 1994 to 1996, the Information Center of MEI performed some basic tests to ascertain the increasing number of network servers and ISP addresses. These data are not precise, but we can still see the rising tide of Internet development in China (Table 26.2).

Besides telecommunications and the Internet, China's broadcasting and television network has also developed very rapidly.

Table 26.1
Increases in China's Public Telecommunications Network

Year	Million Subscriber lines	Annual Increase	Rate of Increase
1990	19.69	2.66	15.6%
1991	23.38	3.69	18.7%
1992	30.00	6.62	26.6%
1993	42.06	12.06	40.2%
1994	61.62	19.56	46.5%
1995	85.10	23.48	38.1%
1996	109.00	23.90	28.1%

Source: 1997 China Electronic Products Market Forecast (Electronic Industry Publication).

Table 26.2
Development of Domestic Internet

Date	Number of Servers	Rate of Increase	Number of IP Addresses	Rate of Increase
January 1994	0			
July 1994	325			
January 1995	569	75%		
July 1995	1023	80%	95	
January 1996	2146	110%	153	61%
July 1996	11282	426%	475	210%

Source: Information Center of MEI.

Developing Application Systems and Utilizing Information Resources

China launched the "Golden" series of information infrastructure construction projects, based on construction of a national information network and aimed at optimum utilization of information resources. First were the Three Golden Projects: the state economic information network (Golden Bridge Project), the foreign trade information network (Golden Customs Project), and an electronic currency system (Golden Card Project). Implementation of the Three Golden Projects also accelerated development of the information infrastructure for banks, railroads, posts and telecommunications, the oil industry, airlines, and department stores. Table 26.3 gives some examples of progress in the Golden series projects. In addition, many cities have also launched programs to develop their local information infrastructure.

Development of the information infrastructure has increased the number of computers in China and the domestic computer market has grown rapidly, as shown in Table 26.4. Analyzing China's information technology market, we can see that although China's market is growing very rapidly, it is still significantly smaller than those of the developed countries. This indicates strongly that there is still a huge potential demand for information technology in China in the near future.

Table 26.3
Progress of Golden Series Projects

Project name	Progress
▪ Golden Bridge Project	Opened in 24 cities
▪ Golden Customs Project	Includes four application systems: quota and certificate management, export return tax, import & export statistics, export foreign currency management system
▪ Golden Card Project	In 12 experimental cities, six have implemented a bank card network operation, and the application of nonbanking cards is expanding
▪ Golden Tax Project	Computer system put into operation in 50 major cities and 800 county tax departments
▪ Golden Macro Project	Interlink and data sharing between economic information centers of all ministries and provinces
▪ Golden Health Project	Network system in state health departments
▪ Golden Communication Project	Computer network management in state highway and waterway transportation
▪ Golden Aviation Project	Network between all aeronautics institutes and factories

Table 26.4
Development of Domestic Computer Market (in billion yuan RMB)

	1994		1995		1996	
	Sales Volume	Rate of Increase	Sales Volume	Rate of Increase	Sales Volume	Rate of Increase
Hardware	30	54%	47	57%	71.5	52%
Software	4.9	23%	6.8	39%	9.2	35%
Information Services	5.8	18%	7.7	33%	11.3	47%
Total	40.7	40%	61.5	51%	92	50%

Source: Electronics Industry Yearbook.

The Development of China's Information Industry

Construction of China's information infrastructure has expanded China's information technology market, promoted international technology cooperation and technology transfer, and accelerated the commercialization of R&D achievements. A group of domestic information technology enterprises is growing, accompanied by a rapid extension of the information technology market. Domestic-made switches reached over 30% of the domestic market last year, while domestic-made telecommunications transmission equipment accounted for 20%. Chinese-built PCs are also very competitive and had a domestic market share of over 50%, as seen in Table 26.5. Integrated application systems and information services have sustained significant growth in domestic software enterprises. A large number of foreign investment joint ventures and cooperative information technology enterprises have also enjoyed considerable benefits from China's information infrastructure construction.

Table 26.5: Increase of PC Market in China (1000 units)

Year	1994	1995	1996
Sales	720	1,000	2,100
Growth rate	60%	39%	110%

INFORMATION TECHNOLOGY AREA OF HIGH-TECH R&D PROGRAM (863 PROGRAM)

Along with the construction of China's information infrastructure, it is essential for the country to have its own research and development, to commercialize research results, and to raise its own information industry. The Chinese government accordingly lays great emphasis on supporting R&D activities. The relevant national programs include the Basic Research Program, the 863 Program, the Key Technologies R&D Program, and the Torch Program. The Basic Research Program's objective is to develop new concepts and principles. The focus of the 863 Program is high-tech research and development. The aim of the Key Technologies R&D Program is to develop the technologies that are critical to the national economy and social development, while the Torch Program is for high- and new technology industries. Here we focus our discussion on the information technology area of the 863 Program.

The 863 Program

The goal of this high-tech R&D program is to aim for the very peak of world high-tech development and narrow the gap between Chinese and advanced foreign technologies, to train a group of highly creative high-tech talents, to commercialize R&D achievements, and to lay the foundation for economic development and national security in the 21st century. The 863 Program selected eight major areas as key topics: biotechnology, space technology, information technology, laser technology, automation technology, energy technology, new materials, and ocean technology. This program resulted from a decision made by the Chinese government in March 1986, hence its title.

The information technology area of the 863 Program includes four subjects: intelligent computer systems (subject 306), optoelectronic devices and integrated system technology (subject 307), information acquisition and processing techniques (subject 308), and communications (subject 317). The Department of Basic Research and High Technology (SSTCC), is in charge of organizing and implementing this area's programs. Expert groups on the four subjects, organized by SSTCC, are responsible for the program's operational management.

After ten years, the 863 Program has raised a new generation of high-tech R&D teams, achieved many important results, enhanced the state's scientific strength, and played an important role in national economic development. According to statistics, 540 important results have been achieved in the information technology area, 9,500 papers announced, 60 patents registered, and a team of 4,000 R&D specialists has been maintained. Enterprises like Dawning Information Industry Co., Sunshine Electronic Information Industry Group, and Tsinghua Wentong Company make up a high-tech enterprise group.

Some Research Progress

High-Performance Computers and Their Application. Supported by the 863 Program, the National Intelligent Computer Research and Development Center successfully developed the Dawning 1 symmetrical multiprocessor (SMP) server and the Dawning 1000 massive parallel process (MPP) computer. On this basis, work on upgrades, applications, and commercialization is also under way.

The Dawning 1000 parallel computer system has 36 knots, with a peak processing speed of 2.5 GFLOPS. It was completed in 1995 and more than 60 universities and institutions have used the Dawning 1000 in large-scale scientific and engineering calculations. These calculations have greatly improved China's research level in related areas. The modified type Dawning 1000A was improved in scalable and the communication ability between knots. The Dawning 1000A with 16 knots has been exported to Cameroon. This machine has a 4 GFLOPS peak processing speed, 2GB of main memory, and a high performance-to-price ratio. The national Intelligent Computer Center is now working on the Dawning 2000 superserver, which will enhance the ability for both large-scale transaction processes and scientific calculations.

The Dawning SMP server can support different operating systems and can be applied in such areas as finance, commerce, communications, and telecommunications. The National Intelligent Computer Center absorbed investments and in 1995 established the Dawning Information Industry Company based on the industrial property of Dawning computers. After over a year, the Dawning company has sold over 250 units of the Dawning series. Customers find that the Dawning computer has good performance and comparable machine functionality to foreign products, and that the price and software package can satisfy their needs. A Dawning computer has been chosen as the model application machine in the Tianjin urban information infrastructure construction project, which will use 206 Dawning computers in four information systems: the tax collection system, scientific research and education networks, commercial and trade automation, and the foreign trade Electronic Data Interchange service center.

Chinese Information Processing. The 863 Program has achieved a series of results in Chinese speech recognition, printed Chinese character recognition, and handwritten Chinese character recognition. Most of the results have been commercialized. For instance, the THOCR printed Chinese character recognition system developed by the Tsinghua Wentong Company can easily handle the input of text and characters. The Hanwang handwriting recognition system developed by the Automation Institute of the Chinese Academy of Science has gained advantage in the market with 8,000 units of direct distribution and 80,000 units of authorized distribution last year. The CPDA863 pen input personal digital assistant (PDA) developed by Shenzhen Sunshine Company has also achieved good sales results.

Telecommunications Technology. Work on upgrading the HJD04 series stored program control switch by China Great Dragon Telecommunications Company was greatly supported by the 863 Program. HJD04 switching equipment can support all types of modern international telecommunications operations. It fully accords with international standards and regulations set by the Chinese Ministry of Post and Telecommunications. Its core technique has reached the international advanced level of the 1990s. In 1996, the sales volume of the HJD-04 reached 2,920,000 lines. This switching equipment has also received certificates from the Russian Ministry of Telecommunications to enter the Russian telecommunications network. Under the support of the 863 Program, the Wuhan Research Institute of Post and Telecommunications has successfully developed a 2.5GB/s SDH fiber-optic transmission system. The nonrelay transmission distance of this system is over 100 kilometers with major technical parameters meeting ITU-T international standards.

Optoelectronics. Supported by the 863 Program, research institutes such as the National Optoelectronics Research Center have made breakthroughs in the key technology of making quantum well semiconductor lasers. They are now able to produce different wavelength and multipurpose quantum well semiconductor lasers. Erbium-doped fiber amplifiers (EDFA) under the scope of the 863 Program have been used in domestic communication trunk lines. High-power semiconductor lasers and pumped solid laser products are supplied in batches to the market.

Remote Sensing. Research into and application of synthetic aperture radar are making good progress. The synthetic aperture radar airborne real-time image processor played an important role in flood monitoring and disaster evaluation in Guangdong and Guangxi Provinces in 1994 and in the Poyang Lake area of Jiangxi Province in 1995. A 21-element adaptive optics system, which can correct for the influence of atmospheric turbulence, has been installed in the 2.16m telescope of Beijing Astronomical Observatory.

In the next few years, the important information technology research fields in the 863 Program will include: Dawning high-performance computers, Internet application software platforms, Chinese information processing products, the CDMA mobile communication system, a 82.5GB/s WDM fiber optic transmission system, quantum well optoelectronic components for display, fiber optic communications and optic interconnections, and geographic observation technology and applications.

INTERNATIONAL COOPERATION

Competition and cooperation in information technology are global. In a more open and cooperative world, no country can isolate itself to develop its own national information infrastructure. The Chinese government lays great stress on international cooperation and particularly encourages cooperation that is beneficial to the development of the Chinese information technology industry, and to the improvement of R&D levels. We welcome all types of international cooperation such as academic exchanges and collaboration, joint research, establishing of joint research organizations, technology transfer, and joint ventures.

The information technology area has jointly established a Joint Development Lab for Advanced Man-Machine Communication with the Motorola Company. The lab is shared by both sides. Subjects of mutual interest will be jointly researched. MPEG2, MPEG4, speech recognition, and mobile telecommunications research are now in progress.

SSTCC recently cooperated with the European Union to hold a China-Europe Joint Workshop on the Software Industry in China. Some 20 famous European software enterprises and more than 80 Chinese software enterprises were invited to have face-to-face discussions and achieved good results. In addition, we are promoting the China Software Industry Association to establish long-term and solid cooperative relationships with its European counterparts. The Association will gather and announce development information and cooperative intentions from the respective software enterprises on a regular basis, organize exchanges between them, and promote Chinese-European cooperation on the software industry.

In the future, the information technology area of the 863 Program hopes for more international cooperation, especially in the area of high performance computers (Dawning series), stored program control switching equipment (HJD-04), intelligent application software, Chinese network environment and application programs, semiconductor quantum well optoelectronic components, and geographic observation technology applications, among others.

Although there is great potential and demand for international cooperation, there are also difficulties, such as strong market competition, the incomplete market economy system in China, adaptability to the Chinese cultural background, and certain short-term behavior by Chinese or foreign cooperative parties. In the matter of intellectual property protection, there are still some disputes concerning intellectual property, even with strengthened protection by the Chinese government. These problems should be given serious attention when considering cooperation. Satisfactory cooperative results can be achieved only if both sides start with the same target, based on the principle of equality and mutual benefit, take long-term views, and give full consideration and preparation to the difficulties.

IMPORTANT ROLES OF HONG KONG AND MACAU

On July 1, 1997, Hong Kong returned to China and in 1999, Macau will also return to the motherland. All Chinese greatly rejoice at this and we expect Hong Kong and Macau to continue their prosperity and development. According to the "one country, two systems" formula, Hong Kong and Macau will continue to keep their social systems and ways of life. As important financial and trade centers, Hong Kong and Macau have a more direct understanding of the world economy and market trends. They have the advantage of faster acquisition of information on technology and market developments. Enterprises from Hong Kong and Macau have more flexible management ability and experience, and better ability to adopt advanced techniques and to obtain financial support. In recent years, Hong Kong and Macau have absorbed many outstanding scientists and technical experts from both overseas and Mainland China. These newcomers have greatly enhanced the scientific and technical strength of Hong Kong and Macau and their connections with overseas and the mainland. Hong Kong and Macau can function as a bridge or junction between Eastern and Western cultures. All these facts demonstrate that Hong Kong and Macau will certainly play important roles in international academic and business cooperation.

Hong Kong and Macau can become centers of joint research and technical exchanges, bridges for technology transfer, and windows or bases for commercialization of mainland research achievements. SSTCC has close contacts and cooperation with Macau University and the Macau Software Institute of UN University (UNU/IIST). SSTCC supported UNU/IIST in building network servers based on Dawning computers and expects further cooperation in large-scale parallel computing research and applications.

Following an invitation by the Hong Kong Productivity Council (HKPC), SSTCC and the HKPC have jointly started research on "establishing a cooperative environment for the mainland-Hong Kong information industry." Large numbers of cases of cooperation were analyzed. Five new projects—a Chinese speech recognition chip, embedded software, HJD-04 switching equipment, Dawning high performance computers, and an operating system for the Chinese Personal Digital Assistant—were proposed as cooperative projects between Hong Kong and the mainland. Ways of cooperation include product development based on research achievements of the 863 Program, establishing cooperative research centers in Hong Kong, joint product sales, and expansion of production.

The future of Hong Kong and Macau is closely related with the development of the mainland. China's political stability and economic development will ensure Hong Kong and Macau's continued prosperity and development.

ACKNOWLEDGMENTS

Significant contributions were given to the writer by Prof. Zhu Pengju of the Information Center of the Ministry of Electronics Industry, Prof. Li Qiang of the China Software Industry Association, and many colleagues from the Department of Basic Research and High Technology, SSTCC, and the China Software Industry Association. I would like to express my sincere appreciation to them.

27

Technology Systems, Strategy, and Organizational Culture of Chinese Enterprises

Xuelin Chu

INTRODUCTION

Previously, the Chinese research and development system was governed by a typical central planning system. As a result, R&D was separated from technology generation, production, and sales, as illustrated in Figure 27.1. The highly capable research institutes are commonly accepted as a national team for scientific research and technology development. In the past, they conducted most of the country's frontier research and built world reputations in almost all leading research areas. When, about 15 years ago, the country adopted an open and reforming strategy, a process was started to transform the Chinese science and technology system in order to make it fit into a market system. The institutes became the focus of concern for effective technology transfer and a greater contribution to national economic development. The Chinese Academy of Sciences (CAS) initiated new policies and measures designed to improve its research and development system. The key idea is to move away from the traditional central planning system and to build a market system where resources are allocated more efficiently through competition. Toward this end, the CAS has initiated a strategy of "one academy, two operating systems." This system has now been adopted by all research-oriented institutions in China.

The aim of the policy is to force the traditionally research-oriented institutions to expand their activities in technology development, and to improve the effectiveness of their technology transfer so that research and development will make a greater contribution to the country's economic development. A practical measure taken by government is to formulate a policy to maintain a small but capable team for basic research, and to encourage the majority of staff to concentrate on technology development and commercialization. In implementing this policy, the government has changed its funding system by dramatically cutting operating funds for institutes, and forcing them to win more funds through competing for research projects supported by new systems like the National Science Foundation of China and the R&D programs administered by government organs like the State Planning Commission, the State Science and Technology Commission, and the State Economy and Trade Commission. Under the new science and technology system, CAS headquarters encourages the national institutes under it to establish technology development and transfer systems.

Figure 27.1
The Technology Development System under the Central Planning System

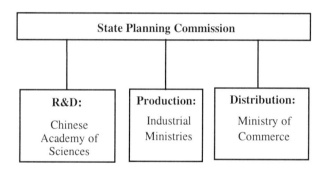

Since the 1980s, over 100 technology development bodies and more than 460 technology companies or enterprises have been established by CAS institutes, of which the Hope Group, the Legen Group, Dayang Graphics, the Chinese Science Group, Daheng, Kejian, Nisaila, and USTC S&T Development are among the most notable. It is clear that Chinese research institutes have accepted technology development (TD) and commercialization as an important aspect of their work, and everybody is aware of, and actively concerned with, the effectiveness and efficiency of technology transfer (TT) and commercialization.

What is the situation now? Is the new system effective in developing technology? What are the key problems in successful technology transfer? Will the institutes be able to maintain their research? An analysis of a trial group of institutes was conducted by a task force consisting of a group of management experts and consultants.

OBSERVATIONS

The trial group consisted of eight major research institutes, all of which are technology-oriented in different fields. Each was studied during a two-week investigation. All relevant documents, financial reports, sample products, and related materials requested by the experts were presented for their examination. They were allowed to visit work sites and to meet people representing different opinion groups to receive personal views from members. Close observations in the field allowed the experts to see that considerable changes have been made to these research institutes. In addition to still-powerful research teams for basic and applied research, technology development systems have been established, which have capable leaders, wide-ranging development projects, and a few technological companies. Direct government financial support is now only one-third or one-fourth of their operating funding. The sales generated by each have been quite impressive.

Though the changes are remarkable compared with the past, they are still some way from what had been hoped. The weakness is most striking in that transformation of the management system seems to be blocked or diverted by organizational changes. Factors affecting technology transfer have been observed in areas such as management philosophy, policy guidelines, structural change, promotion policy, and personnel training.

KEY FACTS

Following site visits, the experts decided to group their findings into the following major points, reflecting the management status of the institutions, as shown in Table 27.1.

MANAGERIAL ANALYSIS

From these findings, it can be concluded that technology development and technology transfer in these institutions do not meet with the expectations of either government or staff. The relevant management system should be held responsible for all problems. The observed situation gives pointers to several management problems.

Management Philosophy

When visiting a research body or technology company, it is reasonable to expect an independent or indeed unique development strategy. It was felt that the institutes under study were strongly dependent on higher authority. There are documents to describe academic or scientific development strategy, but most institutes have no clear picture of a technology development strategy. Staff are rather passive in the technology development area, waiting for their superior to push or give directions instead of making independent decisions. Technology development is treated as a sideline rather than an important part of their duty.

Table 27.1
Findings in Brief

built development department	100%
organized technology companies	100%
top leader in charge of development	80%
allocated funding for development	100%
technology generates sales	100%
have developed products	100%
technology development leader empowered	80%
other leaders very supportive	45%
difficulties in coordination	80%
other leaders indifferent	50%
criticism by members from below	80%
wanted to quit	70%
strategy for technology development	25%
modified promotion policy for TD	40%
have separate planning system for TD	40%
plan for management training	10%
training marketing personnel	10%
research culture dominant	70%
attention paid to development culture	25%
people feel TT effective	0%

Staff Resistance

When the institute management tried to formulate a relevant policy to encourage technological development, it was often challenged by members of its staff when new policies were announced. People appeared unprepared for such technology development, or for the ideological changes leading to a market system. A strong inertia could be felt, a desire to keep the old or traditional system to which people had been accustomed. When such a situation of change occurred, the leading group often responded with a more cautious or conservative attitude, or simply disrupted or even withdrew from management reform.

Human Resource Policy

When technology development is accepted as a part of the responsibility of an institute and new systems are adopted, the development staff are often irritated or demotivated by human resource policies in promotion, reward, performance, and compensation regulations. A lack of suitable personnel policy in these areas was found almost everywhere. As a result, demotivation caused high staff turnover and the loss through resignation of key technology inventors and creators.

Organizational Atmosphere

In the institutes, the dominant atmosphere is research-oriented. This can be seen in their value systems: organizational structure, human relations, planning and budgeting, and the reward system are all designed for research. Typical is the planning and material supply system, which is tailored to the needs of research rather than of technology development and production. This naturally generates frequent conflicts and other problems. Technology development cannot be carried out smoothly under these circumstances, let alone production.

Marketing Management

Marketing management was barely organized. Almost everywhere, an incomplete marketing system could be identified, represented in the structure and in the product design, production, promotion, and distribution systems. People said that they badly needed marketing personnel in all these areas. Market research, product design, promotion, distribution, and international business were still new concepts to them. This situation has been in existence for years, but there is as yet no sign of change.

When a marketing system was introduced into the country, the general environment of the R&D institutions changed. Such change dictated a change of strategy in the institutions. Then, according to the organizational design concepts presented by Anderson (1988: 648), the tasks, structure, system, and staff of the organization should be changed in coordination with its strategy. But the fact is that the institute management has no power to bring about such organizational change, and this inability causes serious management problems. Consequently, the only possible result is the interruption of reform that we saw. Naturally, the technology development and technology transfer systems could not function properly, even though they had been restructured. As stated by Tosi, et al. (1986: 67), management philosophy, values, organizational structure, the reward system, training, promotion policy, and marketing concepts are all parts of organizational culture.

We can conclude that the organizational culture must be changed before a healthy technology development and transfer system can be built and run effectively. The situation revealed in these institutes is representative of that prevailing in R&D institutions throughout the country.

WAYS TO IMPROVE

1. Effective technology transfer needs full support from the management system. We may therefore suggest that changing the management system is the only way to bring in a more effective technology transfer system. Such change should be introduced into all relevant parts of the organization, namely the structure, organization, reward and promotion system, performance appraisal, compensation, training, and so forth. A sample structure is illustrated in Figure 27.2.

Figure 27.2: Technology System of an Institute

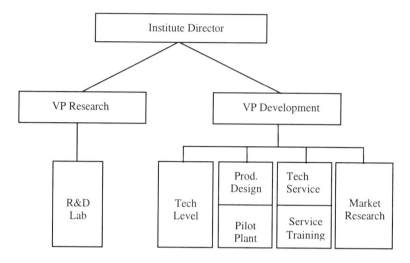

2. Personnel with proper training in both technology and management are essential to management improvement. From our observations, it was found that people need training and indeed requested it. But the existing administration did not put such training on its agenda, nor did it support training with effective measures. This is owing first to the administration's attitude, and second to financial constraints. People in administration fear changes; they are afraid of losing power or even their positions. But capable personnel are critical to the success of technology development and technology transfer. As the Chinese saying goes, "if we have people, we will have everything."

3. Redesign of structure, strategy, and systems is needed to change the situation. This means organizational development (OD). But OD concepts are still new in China, and not many people have the know-how to carry them out. In addition, OD usually requires considerable time and energy. But it is inarguable that OD is needed to build a truly functional technology development system in an institute with a strong research culture.

4. Organizational culture can only be changed systematically, not overnight. There must accordingly be a strategy to carry out such change step by step in order to reach the final goal.

REFERENCES

Anderson, C. R. *Management.* Boston: Allyn and Bacon, Inc., 1988.

Tosi, H. L. et al. *Managing Organizational Behavior.* Cambridge: Ballinger Publishing Co., 1986.

28

University Technologies and Their Commercialization in China

Bing Wang, Zhu Qin, and Zhicheng Guan

INTRODUCTION

The Chinese universities have strong manpower in R&D, not only full-time researchers and part-time professors but also a great number of graduates who are creatively and continually renewed, more advanced equipment and instruments in their laboratories, and better conditions for information exchange at home and abroad. Therefore the universities in China are an important channel of invention-creation. For instance, Tsinghua University has about 1,800 full-time professors and about 700 part-time professors conducting research and development, and nearly 5,000 master's degrees and Ph.D. candidates doing research work under their supervisors. There are about 1,600 research projects being conducted and more than 200 items of research achievements obtained annually.

THE OWNERSHIP OF THE TECHNOLOGIES MADE IN A VARIETY OF COOPERATIONS BETWEEN UNIVERSITIES AND OTHER ENTITIES

For invention-creation or other technologies (hereafter, I-Cs) made by a university in conducting a project that is supported financially by government or foundation, the university can own or hold all rights to them and handle them independently. Of course, there should be no offense against the relevant rules or regulations of the government or foundation. If the sponsoring organization requires the university to implement the technology in designated industrial sections or enterprises, the university shall do it according to the requirements and conditions.

For I-Cs finished in cooperation between universities and institutes, the ownership or application rights will be distributed among them according to actual creative contribution given by each side. In many cases the I-Cs are shared by both sides, since they jointly conduct research work and make the technologies.

For the I-Cs made in a cooperative project between a university and an enterprise, the ownership or application right of the I-C also often will be distributed according to actual creative contribution given by each side. In many cases the university can own the I-Cs, but the enterprise can use them under reasonable conditions. When the enterprise makes creative contributions to the I-Cs or provides materials or research fees, the ownership can be shared by both sides.

It is very complex and difficult to deal with the ownership or application right of the I-Cs finished a in commission project given by an enterprise to a university. For the commissioned project the enterprise proposes the topic to be done and provides research funds, and the investigators of the university conduct research work using the funds and the university's laboratory space, equipment, instruments, and information. Since the university actually makes the I-Cs, it often needs to have ownership according to the relevant article of the Patent Law of the People's Republic of China. On the other hand, since the enterprise provides research funds and often selects the topic to be done, the enterprise also requires to have ownership according to the rule, "I invest in it, you do it for me and, your research result should belong to me naturally." Therefore both sides should negotiate before making an agreement on the ownership of an I-Cs. There are several ways to share the ownership, which can be determined by both sides through discussion. If the research funds given by the enterprise do not cover all costs of R&D and give at least a part of the profit coming from the result of R&D of the project to the university, the ownership of the I-Cs will be shared by both sides. If the research funds given by the enterprise cover all costs of R&D and the enterprise also give at least part of the profit coming from the result of the R&D to the university, the university can usually agree that the ownership of the I-Cs belongs to the enterprise.

Some universities, which have good material conditions, can conduct some research or development projects independently and cover all costs of the projects by themselves. Then the ownership of the I-Cs belongs to the university.

UNIVERSITY TECHNOLOGY COMMERCIALIZATION

State Programs

The Chinese government attaches great importance to strengthening relationships between academic institutions and industry in order to promote economic growth and the development of science and technology. The central government has established a variety of programs to support applied R&D such as the State Key Program, High-Tech Program, Technology Propagating Program, Torch Program, State Spark Program, Eighth Five-Year Propagating Program, New Product Prototype Program, New Product Trial Manufacture Program, and so on. As a key comprehensive university with engineering as its main focus, Tsinghua University has been involved in all these state programs from the very beginning.

The State Key Program is organized by the State Planning Commission for developing important technologies for the Chinese economy. Tsinghua University has some projects of this program, such as Nuclear Heating Reactor at Low Temperature, Air Cooling Nuclear Reactor at High Temperature, Clean and Efficient Combustion of Coal, Computer Simulation of Power Generation Plant Operation, and others.

The High-Tech Program was designed by the State Science and Technology Commission (SSTC) in 1986 to develop some high technologies available for Chinese industry. Some of important programs such as Computer Integrated Manufacturing System (CIMS), Artificial Intelligence Technology, and Customs Large Container Inspection are conducted by Tsinghua University.

The Technology Propagating Program, designed for promoting the propagation of key technologies, was carried out in 1989. The SSTC provides technology fees to the university, whose technology is selected to be propagated as they key technology. The enterprise where the key technology is applied can get interest-free loans for the technology and the priority of exemption from sales taxation or tax discount. Some of these key technologies come from Tsinghua University. For example, Betheng structure steel, a new kind of steel developed by Tsinghua University, has been produced by several Chinese iron and steel plants through the Technology Propagating Program.

The Torch Program was designed for high-technology industrialization and internationalization. The technology in a Torch Project should be advanced, perfected, and useful in production and should have a big potential market to earn high profits. This program provides loans with reduced interest and exemption from taxation or tax-discount. Tsinghua University has provided several high technologies such as computer simulation of power generation plant operation.

The State Spark Program was implemented in 1986. It aims at propagating technologies to the vast countryside and guiding a great number of peasants to use the technologies for agriculture development. One of the best technologies provided by Tsinghua University is a highly reinforced cast iron containing several rare-earth elements which made a township enterprise become a large one.

In the past years, the State Economy and Trade Commission has designed the Eighth Five-Year Propagating Program, New Product Prototype Program, and New Product Trial Manufacture Program. These programs were carried out in 1991. Tsinghua University has 26 projects from them. From all programs mentioned above, Tsinghua University has received a great amount of funds to support its applied R&D activities and has made great contributions to the Chinese economic growth through the implementation of the projects.

Commission Projects Supported by Industry

Tsinghua University also has a lot of applied R&D projects funded by industry through contracts or agreements. In this case, the enterprise asks the university to solve some technical problems and provides funds for the projects. The university conducts R&D and provides research reports or technical documents to the enterprise. Some of the commission projects are applied basic research, which can meet the long-term or middle-term needs of enterprises. For example, the Sino Petrochemical Engineering Company has given some applied research projects on extract separation and distillation to Tsinghua University in the past ten years. Most of the commission projects are development projects such as "continuous milling production line of train wheel ring" developed by the university for Maanshan Iron and Steel Company, one of the largest in China. There are also some engineering projects, some of which are turnkey projects. In this case, the university is responsible for not only research, development, design, and equipment purchase, but also installation, adjustment, and trial production of qualified products.

Joint Research and Development Institutes with Industry

Some large enterprises in China are willing to establish a long-term cooperation with universities in order to use the resources of faculty and technologies of universities. They have set up jointly with universities some R&D organizations. For these units, generally the universities offer manpower resources, equipment, and laboratories and the enterprises provide funds to support the institute's operation. In this case, the institutes are usually located on campus. Sometimes the institutes are placed in enterprises and their equipment, laboratories, junior researchers, and funds come from the enterprises, but senior researchers are professors of the universities. Tsinghua University established a Chemical Engineering and Applied Chemistry Institute (North) jointly with Sino Petrochemical Engineering Company in 1984. Since then, the company has given more than 30 million yuan RMB to support the R&D activities and daily operation of the institute, already has seven advanced technologies from the institute, and has recruited more than one hundred quality graduates from Tsinghua University. It has greatly improved the technical level of the company and increased the company's profit.

A technology propagating center is another kind of long-term joint organization formed by enterprise and university. Supported by the SSTC, some universities have established technology centers which aim at propagating specific research achievements. The university provides specific technologies and asks some related enterprises to joint the center. The SSTC provides financial support to the center and offers some favorable conditions such as government loans and exemption from taxation for the enterprises. For example, the New Air Cooling Betheng Steel Spreading Center and the New Ceramic Cutting Tools Spreading Center have been established at Tsinghua University, since the university provided the related technologies.

Engineering Research Center on the Campus

Up to now nearly 70% of the applied R&D achievements of Tsinghua University are applied to different extents in Chinese industry, but only about 15% of the achievements are well used by the industry. Although there has been a lot of input from the industry to the university, the output from the university is yet insufficient. Major barriers to the application of scientific achievements in industry include lack of recognition of the importance, arduousness, and complexity of the engineering research phase in the whole chain from laboratory R&D to industry. In fact, there is a long distance to go from the prototype in a laboratory to the product manufactured in a factory. The Chinese enterprises still have a weak capability of applying technologies. There is no perfect mechanism to coordinate the relationship between research, development, and economic growth.

In order to put up a bridge between scientific achievements and final products, some engineering research centers have been established with the support of the State Planning Commission and the SSTC. There are several Engineering Research Centers on the campus of Tsinghua University, including CIMS, The Driving Technology of Optical Disk, The Coal Combustion of Industrial and Civil Boilers, Nuclear Technology, and Computer CAD. Although the engineering research centers are located on the campus, they are independent legal entities and operate according to the company's mechanism instead of the university's mechanism—that is, they are industry-supported, government-aided, plan-guided, market-oriented, and financially balanced through transferring their technologies and selling some products.

Technology Licensing

The universities in China have conducted a lot of applied R&D as mentioned above; therefore they have become one of the important resources of technologies which the Chinese industry needs. Many of the technologies can be directly applied in production and create high profits for industry. The relevant rules and regulations allow universities to own industrial property such as patents. For this reason Chinese universities have their own patents and know-how and can commercialize the technologies through technology licensing. Tsinghua University files about 100 patent applications and obtains nearly 90 patents every year. The university has set up the

Tsinghua University Patent Affairs Office, Technology Transfer Office, and Tsinghua University Technology Service Company, which are responsible for its patent affairs and technology licensing at home and abroad. Through technology licensing Tsinghua University transfers its technologies not only to large and middle-sized enterprises but also to smaller town-owned enterprises and to both domestic and foreign enterprises. For example, Tsinghua University has transferred its technology of computer recognition of printed Chinese characters to Motorola Company. The university earns more than US$1 million in royalties from foreign companies and 20 million yuan RMB from domestic enterprises annually.

University Technology Enterprise

The universities in China have operated their factories since 1949 to allow students to get industrial practice in the factories. But in recent years, because of the development of a market economy and the strengthening of R&D activities on the campuses, the factories have become the universities' technology enterprises. Some of the enterprises having advanced technologies are important units to commercialize university technologies. Generally speaking, these university enterprises have kept close relationships with university researchers, and some of the presidents of university enterprises are also researchers with knowledge of important technologies. They can select the newest high-level technologies of their universities to be developed commercially for large potential markets and high profits. Tsinghua University has five such high-tech companies. The profits are used for running the companies and some of the profits should be submitted to the university to support teaching and R&D. For instance, Tsinghua University Special Materials Company produces a lot of crystal display materials and gets a 70% share of its domestic market and a good profit. It is a model to commercialize the university's technologies.

In the past years, under the policy of "the development of science and technology should suit to the economy, and the economy should rely on the development of science and technology," although we have made great progress in technology commercialization, we still have a lot of difficulties to overcome and problems to solve. We will continue our work and try our best to make more contributions to the growth of our national economy.

29

The Role of Technology in Industrial and Economic Development in Hong Kong

Otto C. C. Lin

NEW CHALLENGE

The year 1997 started the most important chapter in the history of Hong Kong: the return to Chinese sovereignty. As Hong Kong looked forward to its new political environment, it also drew near a crossroads in its economic development. What does the future hold for the Hong Kong economy? What course of action should it take to maintain the remarkable growth of the last two decades? In this chapter, we suggest that science and technology will be the engine of economic growth for the future of Hong Kong. We further suggest that institutions of higher education will play an important role in the social and economic development of Hong Kong.

Accomplishments

With barren land and no natural resources, the economic accomplishment of Hong Kong is truly miraculous. Over the past three decades, as seen in Figure 29.1, its GDP has increased at an average annual growth rate of 8%. Per capita GDP rose from US$410 in 1960 to $23,210 in 1995 (Figures 29.2 and 29.3), a level enjoyed by the developed countries of the world. As a reference, during the same period, the per capita GDP of the United States rose from $2,840 to $27,550.

Figure 29.1
Hong Kong's Gross Domestic Product, 1975—1996 (year-by-year growth rate in real time)

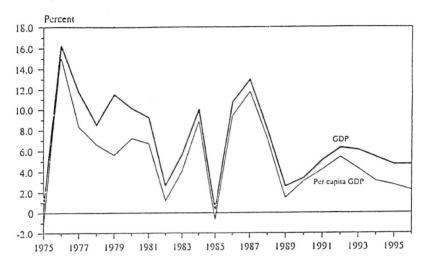

Note: Over the past two decades, the Hong Kong economy has been expanding rapidly, with GDP growing by 7% per annum and per capita GDP by 5% per annum.

Figure 29.2
GDP per Capita, Hong Kong and the United States, 1994 (in thousands of US$)

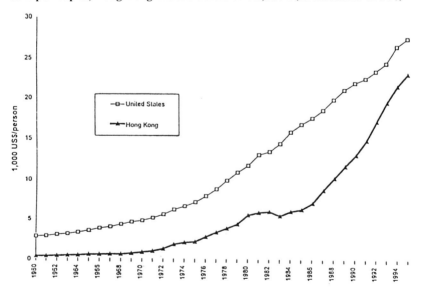

Figure 29.3
GDP Per Capita, 1994

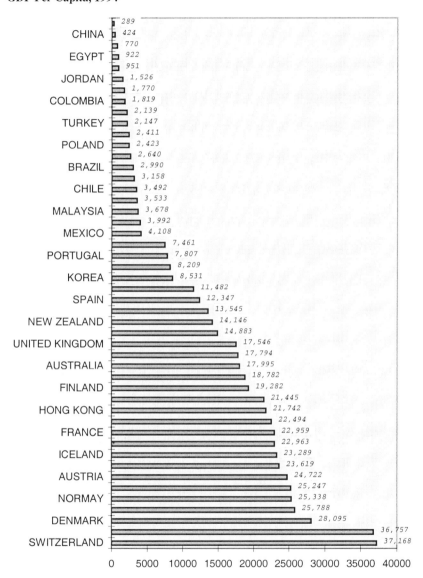

In 1996, Hong Kong was the eighth largest trading entity in the world. It was the world's first container seaport in terms of throughputs, and ran the fourth busiest airport in terms of passengers. It was also the fifth largest banking center in the volume of international banking transactions. According to the 1995 Institute for Management Development/World Economic Forum (IMD/WEF) studies, Hong Kong was the third most competitive economy in the world.

Service-Dominant Economy

A recent survey (see Table 29.1) showed that manufacturing companies headquartered in Hong Kong employ about 10 to 12 times more workers in their Chinese facilities. Thus it can be estimated that about 3.5 to 4 million jobs in inland China can be attributed to business based in Hong Kong.

Table 29.1
Distribution of Employment Between Hong Kong and China, 1996

	Hong Kong	China
ALC	350	0
Albatronics	250	3,000
ALCO	1,400	14,000
ASM	800	1,500
Avantec	25	300
China Aerospace	50	15,000
Elec and Eltek	670	2,100
Epson	540	2,000
Gold Peak	1,300	10,000
Johnson Electric	560	12,530
Lafe	292	5,000
Mabuchi Industry	400	30,000
Meadville	700	1,040
Primatronix	75	1,500
QPL	800	0
Silicon Electronics	30	150
Team Concepts	250	2,000
Valence Semiconductor	27	0
Varitronix	350	500
Vitelic(HK)	210	0
V Tech	1,000	16,000
Wing Sang Bakelite	52	1,000
Wong's Circuit	1,130	610
Mean	460	5,800

Note: This is just a partial sampling of electronics firms.
Source: Company data.

As it stands (see Tables 29.2 and 29.3) Hong Kong's sales and trade now account for about 27.4% of the economy, and has become the largest sector. It is followed by financial and business services at 24.9%, community and personal services at 17.1%, transportation and communications at 9.8%, manufacturing at 8.8%, construction at 4.9%, electricity, gas, and water at 2.3%. This composition is typical of a service-oriented economy such as Singapore. Actually, in Singapore, manufacturing remains the most significant sector, at nearly 25% of the economy.

Table 29.2
Gross Domestic Product at Current Prices by Economic Activity, 1993—1995

Economic Activity	1993 $ *	1993 % **	1994 $ *	1994 % **	1995 $ *	1995 % **
▪ Agriculture and Fishing	1,612	0.2	1,596	0.2	1,453	0.1
▪ Industry	153,459	18.5	156.103	16.4	163,302	16.0
▪ Mining and quarrying	197	+	249	+	268	+
▪ Manufacturing	92,582	11.2	87,354	9.2	89,719	8.8
▪ Electricity, gas & water	17,591	2.1	22,175	2.3	23,562	2.3
▪ Construction	43,089	5.2	46,325	4.9	49,753	4.9
▪ Services	675,098	81.3	792,472	83.4	853,648	83.8
▪ Wholesale, retail & import/export trade, restaurants and hotels	224,462	27.0	249,167	26.2	247,581	27.4
▪ Transport, storage & communication	78,993	9.5	92,109	9.7	100,129.	9.8
▪ Financing, insurance, real estate & business services	214,550	25.8	254,346	26.8	253,492	24.9
▪ Community, social & personal services	130,408	15.7	151,293	15.9	174,448	17.1
▪ Ownership of premises	89,862	10.8	115,659	12.2	128,864	12.7
▪ Adjustment for charges of financial intermediation services indirectly measured	−63,177	−7.6	−70,101	−7.4	−81,866	−8.0
▪ Gross Domestic Product at factor cost (production–based estimates)	830,169	100.0	950,172	100.0	1,018,403	100.0
▪ Taxes on production & imports	53,278		56,286		52,971	
▪ Gross Domestic Product at market prices (production–based estimates)	883,447		1,006,458		1,071,374	
▪ Gross Domestic Product at market prices (expenditure–based estimates)	897,463		1,010,885		1,084,570	
▪ Statistical discrepancy	−1.6%		−0.4%		−1.2%	

Notes: preliminary estimates; * = in millions; ** = distribution; + = less than 0.05.
Source: Census and Statistics Department of Hong Kong.

Table 29.3
Gross National Product of Hong Kong, 1993 and 1994

	$ Million	
	1993	**1994**
GNP at current market prices	907,808	1,018,225
Per capita GNP at current market prices ($)	153,840	168,709
GNP at constant (1990) market prices	698,432	732,943
Per capita GNP at constant (1990) market prices ($)	118,358	121,441

Note: GNP is derived from GDP by (a) adding total income earned by Hong Kong residents from outside Hong Kong and, (b) subtracting total income earned by non-Hong Kong residents from within Hong Kong.
Source: Census and Statistics Department of Hong Kong.

It should be noted, however, that due to rapid changes in technology, the definitions of manufacturing and services are gradually broadening and beginning to overlap. The boundary between manufacturing and service sectors is increasingly becoming obscured. The conventional segmentation of the economy based on these sectors is losing its traditional meaning. Taking this into consideration, the "manufacturing sector" in Hong Kong would be at a significantly higher level.

Sustainability of Growth

A recent analysis conducted by the Massachusetts Institute of Technology focusing on this issue has clearly shown that the current system cannot be sustained for long. There are many reasons for this. First, the cost of doing business in Hong Kong has been rising. Labor and real estate costs have climbed steadily. Second, as has become evident, Hong Kong's prosperity in the last 30 years was largely due to its unique position of being the only, or most dominant, entry port for China. This position is now being challenged. Many seaports and centers of transportation have rapidly developed along the Chinese coast. Thus the growth of re-export business in Hong Kong has steadily decreased in recent years, as seen in Figure 29.4. Third, traditional external rivals in the international marketplace such as Singapore, Taiwan, Japan, and Korea have shown growing competitive strength, especially in high-technology fields. Finally, to continually move manufacturing and facilities deeper inland has stretched the management ability and resources of Hong Kong-based business. There are yet other factors.

Figure 29.4
Growth in Hong Kong's Visible Trade, 1985—1996 (year-by-year growth in real time)

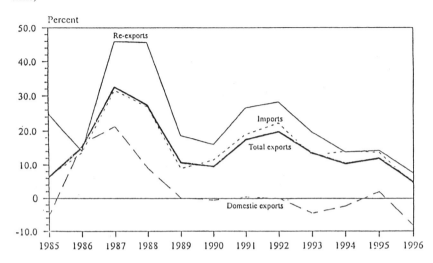

Technology has been shown to be a prime mover of modern society. The extent and level of sophistication in technology constitute the comparative advantage of a product or service in modern industry. Likewise, the overall competitiveness of a nation is determined more by its ability to develop and exploit technology than by the wealth of its natural resources. Thus developed countries like the United States, Japan, Netherlands, Sweden, and Switzerland, and the three other dragons in the Asian Pacific, have all placed priority on developing science and technology to increase the value of their products and services. This experience should be enlightening to Hong Kong as well as other developing economies.

Hong Kong's Weakness

The weakness of Hong Kong is clearly shown by the 1995 IMD competitiveness study. Although Hong Kong occupies the third place in overall world competitiveness (Table 29.4), it ranks 23rd in science and technology and 19th in people. A U.S. National Science Foundation study also showed (Figure 29.5) that Hong Kong had the poorest technology infrastructure among Asian countries. Thus, for Hong Kong to remain a competitive economy in the future, aggressive action in education and R&D must be taken.

Table 29.4
World Competitiveness, 1995

Ranking	Country	Domestic Economic Strength	Internationalization	Government	Finance	Infrastructure	Management	Science & Technology	People
1	United States	1	1	6	2	2	1	1	10
2	Singapore	2	2	1	1	12	5	10	1
3	Hong Kong	3	3	2	4	17	8	23	19
4	Japan	4	9	27	6	28	4	2	6
5	Switzerland	10	18	7	3	10	3	4	4
6	Germany	8	5	13	8	11	14	3	9
7	Netherlands	14	4	26	5	13	11	13	11
8	New Zealand	22	23	3	10	6	6	22	12
9	Denmark	16	11	20	7	9	7	11	2
10	Norway	19	30	22	17	1	13	17	3
11	Taiwan	7	14	5	12	29	15	8	18
12	Canada	24	15	19	11	3	16	18	8
13	Austria	17	13	18	13	14	12	9	13
14	Australia	23	31	9	16	4	19	20	16
15	Sweden	31	17	33	22	2	2	6	14

Figure 29.5
Comparison of Asian Technological Infrastructures

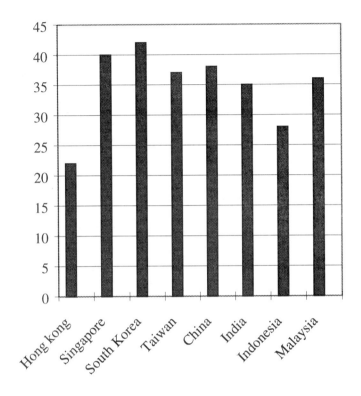

Source: National Science Foundation (1995).

Traditionally, investments in both education and R&D in Hong Kong have been far below those of developed countries, or even the newly industrialized economies (NIE) in the Asian Pacific. For example, up until the mid-1990s, public expenditure on education per capita in Hong Kong was much lower than that in Taiwan, Singapore, New Zealand, Australia, and the United Kingdom. This reflects on both secondary and tertiary education. For a long period, a majority of young adults in Hong Kong were deprived of higher education, the space for which was so limited that it was largely accessible only to the elite or the well-to-do.

As of 1995, total R&D expenditure as a percentage of GDP for Hong Kong was the lowest of all the NIEs. While Taiwan, Singapore, New Zealand, and Australia have shown an R&D investment of about 1–1.5% of GDP, Japan and the United States have consistently invested 2–3% for decades. The R&D expenditure data for Hong Kong, as seen in Table 29.5, was not available for 1993, but for 1994 was estimated to be at about 0.105% at best. It may be of interest to note that the United

Kingdom, for the same time frame, spent twice as much as Hong Kong on education for its people and over 2% of GDP on R&D for its society and industry.

Table 29.5
Total Spending on Research and Development as Percentage of GDP, 1994

Japan	2.88
United States	2.44
South Korea	2.29
Taiwan	1.80
Singapore	1.18
China	0.50
Hong Kong	(0.10)*

Source: IMD 1996: 524.
*No data provided in IMD report. Figure in parentheses is author's estimate based on data provided by government and industry sources.

HKUST AS A MODEL FOR TECHNOLOGICAL DEVELOPMENT

Thus in Hong Kong, the role of tertiary institutions in strengthening education and in R&D is of primary importance. University curricula and research should aim at providing human resources to lead industry and the technology to fuel industry. Under pressure the government has indeed been making efforts in recent years to support both education and research in the universities. Hopefully the Special Administration Region (SAR) administration will further strengthen measures for developing education, science, technology, and industry to attain competitiveness for the future Hong Kong.

Higher education in Hong Kong is also undergoing a transformation to become more research-oriented, starting from the early 1990s. As a major effort, the Hong Kong University of Science and Technology (HKUST) was established in 1991. It has proclaimed as its mission to excel in teaching and research and to assist the social and economic development of Hong Kong. A conscientious effort is being undertaken in teaching and technology transfer. In addition to the traditional academic departments, many multidisciplinary institutes and research facilities have been established, focusing on cutting-edge technologies. Information science and technology and related fields are important parts of the designated high impact areas. The university, with an R&D branch in addition to the academic affairs branch, has striven to facilitate the development and transfer of relevant technologies for the Hong Kong region, as shown in Figures 29.6 to 29.9. Although in existence for less than a decade, HKUST is recognized as a major institution in Asia. The university has taken measures to recruit accomplished expatriate professionals to its staff. These experienced scientists and engineers from North America and the rest of the world have facilitated the progress of the university.

Figure 29.6
The Organization of the Hong Kong University of Science and Technology

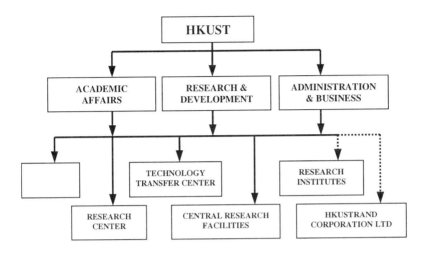

Figure 29.7
From Basic Research to Product Commercialization

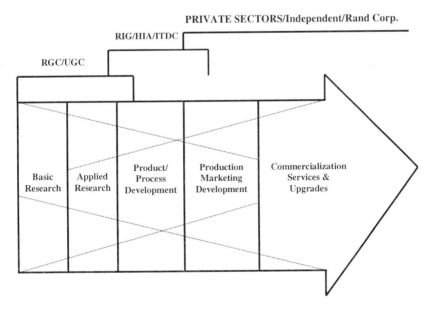

Figure 29.8
The R&D Infrastructure of Hong Kong University of Science and Technology

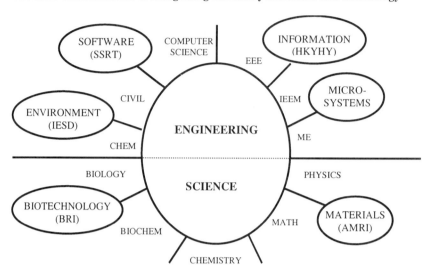

<u>Central Facility</u>

Materials Characterization and Preparation	CAD/ CAM	Micro- Fabrication	Advanced Engineering Materials	Geotech Centrifuge	Wind/Wave Tunnel
Electrical/ Mechanical	Plant Growth	Animal Care	Liquid He	Glass- Blowing	

CONCLUSION

Considering the current environment and its impact on industrial technology policy, in order to achieve the objective of industrial technology development in the near future, there are some effective strategies that are important for all levels of decision makers to consider. These strategies are: to reinforce design and direction in the area of industrial technology; to lay the foundation for industrial technological R&D so as to promote the upgrading of industrial technology; to provide incentives for the public to become involved in research and development, and promote the establishment of R&D units by private and public enterprises; to reinforce the organizational function of R&D in nonprofit organizations; to accelerate the build-up of self-contained and self-sufficient technological capabilities; to promote the creation of R&D infrastructures so as to attract enterprises into research and development; and to modify and readjust the future mode and scale of operation for science and technology projects.

Figure 29.9
The Role of R&D Branch: Facilitating the Research, Development, and Transfer of HKUST Technology

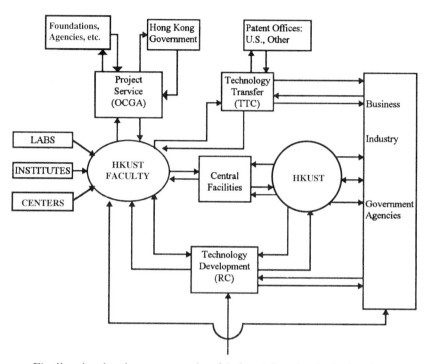

Finally, the development trends of industrial technological policies of all governments have much in common as they face the coming of the 21st century, of which a driving ambition to upgrade national competitiveness and maintain economic development is one example. Industrial technology development has become a critical concern for governments, and its importance has gradually reached the same level as economics, politics, and culture. A focused industrial technological strategy is adopted for selecting types of technology and allocating resources to more value-added and effective technologies. Major government-subsidized research projects are becoming one of the most important industrial technological strategies for all nations. Last, research collaboration is another effective development trend and is also becoming a way to leverage research and development for industrial technological development.

REFERENCES

IMD. *The World Competitiveness Report 1995*. Lausanne/Geneva, Switzerland: IMD and the World Economic Forum, 1995.

National Science Foundation. *Asia's New High-Tech Competitors*. NSF 95-309. Washington, DC: National Science Foundation, 1995.

30

Science and Technology Policies for Macau Toward the 21st Century

Rui Martins and Vai Pan Iu

INTRODUCTION

The success of science and technology development in Macau is definitely hinged on the cooperative efforts of higher education institutions, research and development organizations, and local industrial and commercial sectors in Macau. This chapter compiles an overview of the situation in Macau regarding application of advanced technology, human resources, and R&D activities in science and technology. With consideration of availability of human resources in Macau, the experience and competence gained in R&D, and the diversification of local industry for improvement and transformation of local economy, an integrated approach focused on particular directions should be adopted in making the science and technology policies.

HUMAN RESOURCES IN SCIENCE AND TECHNOLOGY

Higher education in science and technology started in 1989 in the Faculty of Science and Technology (FST) at the University of Macau (UM), which now offers B.S., M.S., and Ph.D. studies in civil engineering, electrical and electronics engineering, electromechanical engineering, software engineering, and mathematics. Bachelor degree programs in business information systems and computer science and a certificate in computer studies are also offered by the Faculty of Business

Administration of UM, Asia International Open University (AIOU), and Macau Polytechnic Institute (MPI). Obviously, FST/UM is the most important entity to produce engineers in science and technology for Macau, comprising currently around 30 advanced engineering laboratories in many diversified areas. Since 1993 FST has produced around 300 B.S. graduates and 20 M.S. graduates in engineering and is expected to have 100 B.S. graduates per year. FST graduates are starting to dominate the engineering profession in Macau, and most of the researchers in research institutes in Macau are FST graduates. Because of the strong ties with Portuguese research institutes and universities, many FST graduates are sent to Portugal for training and have contributed to form a group of young researchers for Macau. However, the leading figures in science and technology research are the professors of FST. In summary, FST plays a key role in educating young people in science and technology and leading the research and development in their specializations in Macau.

APPLICATION OF ADVANCED TECHNOLOGY

Since the mid-1980s, Macau has witnessed important industrial development and applications of advanced technology for want of efficiency, raising competitiveness of production, protection of environment and public health, and improvement of quality of life. Some of the more important and significant developments and introduction of advanced technology are described below.

SPIC Concordia Industrial Park was created in 1993 by the Macau government and other investors in response to the need for high-tech and high-value-added industry. SPIC has attracted six new high-technology factories in 1996 with sites covering 58% of the park and investment of US$96 million. These new factories will have higher labor production per capita and higher percentage of technical personnel. One of the new high-technology factories is manufacturing fiberglass for electronics and aerospace industries, which is not produced in Hong Kong or Singapore.

There are recent successful examples of technological development of modernization of traditional industry with advanced technology. Three large textile and garment factories have adopted Unit Production Systems with production flow controlled by computers. These systems boost the production rate by 30 to 50%, and also improve adaptability and quality of products. CAD/CAM technology is also adopted in the prototyping of toys in the manufacturing sector.

The Macau government has also played a leading role in the application of advanced technology to protect the environment and public health in waste and sewage disposal. Due to geographical constraints and considering the refuse problem in Hong Kong, Macau brought in the latest refuse incineration technology to solve the problem of industrial and domestic solid waste disposal. A refuse incineration plant was designed and built in 1992 with capacity to burn all solid waste produced and with stringent compliance of environmental protection standards regarding emission and noise. The plant also generates electrical power for consumption and is regarded as the most advanced incinerator in the Far East. Regarding sewage disposal, the Macau government decided to build three wastewater treatment plants to have secondary treatment of the sewage before discharge to the sea, whereas in Hong Kong only primary treatment of sewage is conducted. Again due to geographical constraints, the location of the Macau plan is close to residential areas and hence the government imposed zero environmental impact on the design of the plants, which demands introduction of various state-of-the-art technologies for treatment of sewage, gases and odors, noise reduction, and neutralizing of ash from burning the sludge. These sewage treatment plants have set models of excellence in Asia.

The Macau Electricity Company also acquired advanced technology in power generation with installation of the world's largest generators powered by diesel oil, which is less polluting.

The Macau Water Supply Company deployed a fully automatic filtration system for treatment of drinking water and has been experimenting with the use of the latest membrane filtration technology for possible replacement of the conventional filtration technology.

The population of Macau is also enjoying the convenience and efficiency of digital communications and the application of information technology in the form of mobile phone communications, telebanking, multimedia computing, and Internet services. In summary, the application of advanced technology is the fundamental contributor to the improvement of industrial development and quality of life.

DEVELOPMENT OF SCIENCE AND TECHNOLOGY

Fostering research and development for science and technology requires incentives generated from local industry, expertise, desirable environment and conditions, and funding. The Macau Foundation has been one of the main sources for financial support of research in Macau. Considering its small population by Asian standards, Macau has a good number of nonprofit research institutes with strong groups of technical staff and good laboratory facilities. The following gives a brief introduction to these institutes and their activities.

Center of Scientific and Technical Research (CSTR)

CSTR was established in 1993 in the Faculty of Science and Technology, University of Macau. Its main objective is the promotion of R&D projects in cooperation with public and private institutions and companies. CSTR has been attracting funds from the Research Committee of UM, the Macau Foundation (MF), the Chinese Natural Science Foundation, and other sources for various projects. Good cooperative relations with universities and research institutions in China and Portugal were also established. The research fields covered are in the areas of civil engineering, electrical and electronics engineering, electromechanical engineering, and software engineering. In 1997, 22 projects were initiated and granted financial support. Some examples of research projects within the Macau context are modeling vehicular exhaustion emission dispersion in the heavy-traffic urban core, pile dynamics and application in Macau, safety and stability of electrical supply in Macau, optimal technique of power distribution systems, microelectronics chip design, signal prediction of cellular telephone network, knowledge-based systems for aviation control support, laser stereolithography for 3D toy prototypes, laser marking and scribing technologies, Portuguese-Chinese-English multilingual computer aided translation, electrical energy saving, ecological assessment of high turbidity coastal waters in the Pearl River Delta, telemedicine, and remote sensing techniques for environmental assessment of West Pearl River estuary.

Institute of Computers and Systems Engineering of Macau (INESC-Macau)

INESC-Macau is a nonprofit, private, scientific and technical research institute created in 1996 with partnership of University of Macau, the Macau Foundation, Macau Polytechnic Institute, INESC (Portugal), Macau Electricity Company (CEM), Macau Telecommunications Company (CTM), Macau Airport Management Company (ADA), and Post Office of Macau (CTT). Its objectives are applied research and development in high technology, technology transfer, support to incubation activities in high technology, and research cooperation with FST/UM for M.S. and Ph.D. theses. With research personnel mainly from FST/UM, INESC-Macau acts as an interface for cooperation of UM with local private and public sectors. The current R&D areas include Internet, WWW, intelligent systems, database and information systems, computer networks, control and automation, and microelectronics. Some of the current projects are a database system for the Macau government, remote detection and user notification system of post office boxes for CTT, Web application and a virtual library for Macau Productivity and Technology Transfer Center (CPTTM), an EDI study for CPTTM, and a remote detection alarm for the elderly for the Social Work Department.

Institute for Development and Quality of Macau (IDQ-Macau)

IDQ-Macau is a private, nonprofit, scientific and technical research institute created in 1997 in partnership with the University of Macau, Macau Polytechnic Institute, Macau Civil Engineering Laboratory, CPTTM, and Leal Senado de Macau. The main activities are research and development in quality engineering, maintenance engineering, and safety engineering for local private and public companies.

United Nations University/International Institute for Software Technology (UNU/IIST)

UNU/IIST is a research and training center of UNU created in 1991, funded by UNU, and the governments of Macau, China, and Portugal, with the objective of assisting developing countries to attain self-reliance in software technology through activities in development of their own and exportable software, university education curriculum development, and participation in international research. The current research focuses on formal methods for reactive systems, real-time systems, and hybrid systems with high global standing. UNU/IIST has also been cooperating with FST/UM in the M.S. program in software engineering.

Macau Civil Engineering Laboratory (LECM)

LECM is a semiprivate, research and development consulting institution, providing experimental and technical support for quality control including monitoring and testing for government and private civil engineering projects, and is appointed by the Macau government for drafting Macau codes, standards, and technical specifications to be practiced by local civil engineers. LECM also conducts research related to the local civil engineering conditions and cooperates with FST/UM on various civil engineering research projects at undergraduate and postgraduate levels. This cooperation has been making contributions to the provision of the desirable balance of academic and practical aspects in research projects as a means for training students of UM and staff development for LECM.

Macau Productivity and Technology Transfer Center (CPTTM)

CPTTM is a nonprofit organization established by the Macau government and the private sector aimed at the support and development of productivity and competitiveness of local industry. Its main activities are in-company consulting, vocational and technical training, dissemination of information on new technology, adoption of International Standards Organization certification, and support for application of new technology. In the area of science and technology, CPTTM plays the role of promoter and facilitator for joint research and development projects.

Inter-University Institute of Macau (IIUM) has planned to establish expertise in food processing science and it is anticipated in the near future this will be accomplished, thus adding another important research potential for science and technology in Macau.

Some of the laboratories in government departments and public companies are of high standard in terms of technical staff and facilities and can complement the overall research establishment in Macau. Such laboratories are those in the Macau Weather and Meteorological Center, Macau Health Department, and Macau Water Supply Company.

The aforementioned research organizations or units have been providing platforms for training people to do research and to manage research activities and are advantageous for Macau for development in science and technology.

CONCLUSIONS

Considering the availability of human resources and sufficient competence gained in past research activities, Macau has built a foundation for research and development in science and technology in certain areas which are useful for the modernization of traditional industry in Macau and also for the diversification of Macau's economy. One possibility is to upgrade the existing industry with advanced technology—namely, computer technology and automation. Recently, a Science, Technology, and Innovation Committee was established as a consultative body for assisting the government in defining policy for modernization and science and technology development in Macau with the objectives of elevating the value of Macau and elaborating on policies for the improvement of Macau's economy. To accomplish the intended objectives, the committee is chaired by the governor with membership composed of undersecretaries for higher education, economics and public works; rectors of UM, MPI, AIOU; presidents or directors of MF, FST/UM, CPTTM, LECM, IDQ-Macau, UNU/IIST, INESC-Macau; and an elite group of appointed members from the science and technology community and industrial sectors. As the composition of the committee includes the key players for science and technology in Macau, it will be easier to reach an integrated policy for the development of science and technology which will benefit Macau to a greater extent.

31

Regional Integration, Networked Production, and Technological Competition: The Greater China Economic Circle Through and Beyond 1997

Xiangming Chen

INTRODUCTION

When economic historians at the dawn of the 21st century look back at the last two decades of the 20th century, they are most likely to focus on two crucial years—1979 and 1997—which bracket the emergence and evolution of the "Greater China" economic circle (hereafter the GCEC) of China (particularly southeastern China), Hong Kong, Macao, and Taiwan into an integrated transborder regional economy with truly global impact and implications. The combined exports of the GCEC as a share of the world's total almost doubled from 4.8% in 1985 to 8.5% in 1996, ranking third behind the U.S.'s 11.9% and Germany's 9.9%, and ahead of Japan's 7.9% (Mainland Affairs Council 1997: 54). The GCEC's rapid economic and export growth has led to an optimistic projection that the combined gross domestic product of the GCEC might surpass that of the United States in the early part of the 21st century. Even if this extreme scenario does not materialize, the GCEC will still be growing sufficiently to become a legitimate fourth pole in the tripolar world economy anchored on the United States or North American Free Trade Agreement (NAFTA), the European Union (EU), and Japan. Under a different scenario, the GCEC may gain enough economic strength to challenge Japan for the third pole of the tripolar world economy.

THE GCEC AS A NEW FORM OF REGIONAL INTEGRATION

Perspectives on Intra- and Cross-National Regional Integration

Distinctive economic regions form both within and across nations. Much of the theory and research on the question of regions deals with regional economic differentiation and development within countries. Rigorous theorizing on regions goes back to Christaller's (1972) original formulation of central place theory (CPT) based on the regional hierarchies of towns in southern Germany in the 1930s. The French economist Perroux pioneered the notion of growth poles which are foci or centers "from which centrifugal forces emanate and to which centripetal forces are attracted. Each center, being a center of attraction and repulsion, has its proper fields, which is set in the fields of other centers" (1950: 50). Growth poles may produce either "backwash" (negative) or "spread" (positive) effects on the regional hinterlands (Myrdal 1957), which Hirschman (1958) differentiated as "polarization" versus "trickle-down." Critics of spatially based regional theories (e.g., Simon 1990) argue that space must not be treated in isolation, but as an integral element of politicoeconomic systems.

Theorizing on cross-national regions has focused almost exclusively on regional economic and political cooperation and integration as reflected in the European Union and its various antecedents. The level of integration was the most common dependent variable to be analyzed, while typical independent variables included national goals, size of regional groupings (number of member countries), perceived costs and benefits, and extraregional factors such as superpower influence (Axline 1994). Cross-national regional integration involves an evolutionary process that starts with trade liberalization, moves toward wider economic cooperation, and may eventually reach complete economic union. The neoclassical perspective suggests that economic integration tends to produce trade diversion (among member countries) rather than trade creation (more trade with external countries) (Axline 1977). While the coexistence of the normative (constitutional) and practical (operational) attributes of sovereignty of the national state serves as the premise for cross-national regional integration (Drake 1994), deep and comprehensive integration across nation-states tends to weaken both the institutional and territorial aspects of their national sovereignty, making the integrated larger politicoeconomic entity the most important and powerful organization in the international system.

These theoretical perspectives would adequately account for the process and outcomes of regional integration cases like the European Union and NAFTA. The EU, for example, can be traced back to the European Coal and Steel Community

(ECSC), formed in 1950 with the signing of Treaty of Paris. The ECSC subsequently evolved, with modified goals and through new treaties, into the European Economic Community (EEC), the single European Community (EC), and eventually into the European Union today. The EU typifies *formal* cross-national regional integration, which is characterized by an institutionalized governance structure based on jointly negotiated and signed agreements on issues of common concern such as trade among member countries. In contrast, the GCEC represents *informal* cross-national regional integration. While formal cross-national regional integration generally takes a long time to materialize, informal cross-national regional integration may occur more rapidly. The former focuses on supplying the markets of member countries, while the latter may be more oriented toward external markets (Oman 1994; Tang and Thant 1994).

The Evolution of the GCEC

Prior to 1970, the GCEC was defined by a pair of asymmetrical bilateral ties between Hong Kong and China and between Hong Kong and Taiwan, without any linkage between China and Taiwan for obvious political reasons. China and Taiwan enjoyed a dependent yet beneficial relationship with Hong Kong by maintaining a trade surplus with the latter. Hong Kong occupied an embryonic pivotal position by being the only node with ties to the other two parties. During 1971–1978, limited trade between China and Taiwan through Hong Kong occurred. China's exports to Taiwan via Hong Kong totaled US$20 million, whereas Taiwan's exports to China were less than $100,000 (Kan 1994). The 1979–1986 period saw the rapid growth of trade between China and Hong Kong, Taiwan, and Macau, and thus the formation of the GCEC (see Table 31.1), though the various bilateral ties remained asymmetrical. Hong Kong's growing pivotal position allowed it to serve as an important intermediary, which, according to Townsend's (1987) trade model, reduces the transaction costs and inefficiency of bilateral links in a multiple exchange structure.

The period 1987–1990 was marked by the strengthening of Taiwan's largely one-way beneficial relationship with China through Hong Kong as a result of the Taiwanese government loosening foreign exchange controls, and permitting Taiwan residents to visit the mainland in 1987. By the end of 1990, Taiwan's investment in China amounted to $2 billion, topping the previous cumulative total (Kan 1994). Taiwan's trade surplus with China rose fourfold from 1987 to 1990 (see [5] and [6] in Table 31.1). Taiwan's exports to China, however, continued to soar, resulting in a much larger trade deficit with, and greater trade dependency on, China (see [9] in Table 31.1). Total trade between China and Hong Kong in 1995 ballooned to $126.5 billion, far larger than the $21.1 billion in Hong Kong-Taiwan trade (HKTDC April 1996: 8). By the end of 1995, Taiwan's investment in China through Hong Kong topped $20 billion (X. Chen 1996), 10 times as much as the end of the previous period.

Table 31.1
Taiwan's Exports to and Export Dependency on China via Hong Kong, 1981—1996 (in millions of U.S. dollars)

Year	Exports to China via Hong Kong [1]	Exports to Hong Kong F.O.B. [2]	Hong Kong's imports from Taiwan C.I.F. [3]	Taiwan's statistical discrepancy [4]=[2]-[3]	Estimated Taiwan's exports to China [5]=[1]+[4]	Taiwan's imports from China [6]	Estimated total Taiwan-China trade [7]=[5]+[6]	Taiwan's total exports [8]	Taiwan's export dependency on China [9]=[5]/[8]
1981	384.2	1,897.0	1,896.4	0.6	384.8	75.2	460.0	22,611.2	1.7
1982	194.5	1,565.3	1,570.1	-4.8	194.5	84.0	278.5	22,204.3	0.9
1983	157.8	1,643.6	1,600.0	43.6	201.4	89.9	291.3	25,122.7	0.8
1984	425.5	2,087.1	2,217.5	-130.4	425.5	127.8	553.3	30,456.4	1.4
1985	986.8	2,539.7	2,682.4	-142.7	968.8	115.9	1,102.7	30,725.7	3.2
1986	811.3	2,921.3	3,072.8	-151.5	811.3	144.2	955.5	39,861.5	2.0
1987	1,226.5	4,123.3	4,275.1	-151.8	1,226.5	288.9	1,515.4	53,087.7	2.3
1988	2,242.2	5,587.1	5,682.4	-95.3	2,242.2	478.7	2,720.9	60,667.4	3.7
1989	2,896.5	7,042.3	6,606.9	435.4	3,331.9	586.9	3,918.8	66,301.0	5.0
1990	3,278.3	8,556.2	7,439.9	1,116.3	4,394.6	765.4	5,160.0	67,214.4	6.5
1991	4,667.2	12,431.3	9,605.0	2,826.3	7,493.5	1,125.9	8,619.4	76,178.3	9.8
1992	6,287.9	15,416.0	11,156.3	4,259.7	10,547.6	1,119.0	11,666.6	81,479.7	13.0
1993	7,585.4	18,454.9	12,047.2	6,407.7	13,993.1	1,103.6	15,096.7	84,945.9	16.5
1994	8,517.2	21,263.0	13,757.7	7,505.3	16,002.5	1,858.7	17,881.2	93,060.0	17.2
1995	9,882.8	26,123.6	16,572.6	9,555.0	19,433.8	3,091.4	22,525.2	111,690.0	17.4
1996	9,717.6	26,804.8	15,795.1	11,009.0	20,727.3	3,059.8	23,787.1	115,950.0	17.9

Source: Mainland AffairsCouncil (1997: 24, 56).

Notes: When [4] is a negative number, [1] becomes [5]; [2] is based on Taiwan customs statistics; for [6], the pre- 1993 series is based on Hong Kong customs statistics; figures for 1994 and 1995 are based on Taiwan custom statistics. F.O.B. stands for free on board (i.e., price on departure); C.I. stands for cost, insurance, and freight (i.e., price on arrival)

The post-1997 period is likely to witness the GCEC evolving into a more fully and symmetrically integrated regional economic system, which will be facilitated further by the return of Macau to China in 1999. The loss of Hong Kong's political autonomy may weaken its strong pivotal and intermediary role. The recent opening of direct transshipping across the Taiwan Strait, which is not directly related to the turnover of Hong Kong, may further diminish Hong Kong's role by creating a more direct and balanced relationship between China and Taiwan. On the other hand, Hong Kong's continued economic advantages (e.g., remaining a free port with a free trade policy, separate customs territory status, separate membership of the World Trade Organization [WTO]) will help preserve its important go-between role in facilitating trade and investment flows between China and Taiwan.

REGIONAL INTEGRATION THROUGH LINKED PRODUCTION

The GCEC provides an excellent case for studying how "regions are once again emerging as important foci of production and as repositories of specialized know-how of technological capability, even as the globalization of economic relationships proceeds apace" (Scott 1995: 59). I address this empirical question by examining the formation and consequences of partial and complete production systems within the GCEC. As Scott (1995: 52) and Storper and Harrison (1991: 411) define it, a production system involves a functional division of labor through a network of input-output linkages set in a relational context of power and decision making. Conceptualizing production systems as global commodity chains, Gereffi (1996) argues that such organizational variables as the types of production and distribution networks are very important in creating overlapping and at times conflicting regional divisions of labor and determining export growth and industrial upgrading in East Asia. These perspectives point to the need to disaggregate the macro-level trade and investment ties binding the GCEC into its underlying production systems at the mezzo- and micro-levels. Since the China-Hong Kong and China-Taiwan dyads constitute the bulk of the production systems in the GCEC, they will be examined separately.

"Hong Kong as the Shop Window; China as the Factory Floor"

The production systems straddling the China-Hong Kong border began to form in the late 1970s when the experimental opening of Guangdong and Fujian Provinces triggered the relocation of Hong Kong's labor-intensive processing and assembling operations over the border. Fifteen years later, a distinctive, large-scale division of labor between Hong Kong and China, especially southern China, has been established, which is figuratively described by the title of this subsection (cited by Overholt 1993: 183). No aggregative statistics illustrate this restructuring more dramatically than the reduction of Hong Kong's manufacturing employment from over 1 million in the peak years of the 1980s to 600,000 in 1992 and to 400,000 today (Chen and Ho 1994: 37), and the creation of 4 million jobs in Hong Kong-invested factories in Guangdong alone, which account for 90% of all Hong Kong-invested factories in China as a whole (Luk 1995: 48).

Industry-level change provides a more specific avenue for examining the structure and consequences of the production systems linking China and Hong Kong. Over 90% of Hong Kong's toy industry, for example, has been shifted to China (HKTDC January 1997). An obvious outcome of this cross-border production is the slight decline of Hong Kong's domestic exports, and the substantial increase in re-exports of Chinese origin. Industry-specific developments demonstrate the striking features of the new manufacturing economy in Hong Kong, and point to its prospects beyond 1997.

Take the textile/clothing industry as an integrated example. Having moved the lion's share of the cottonspinning, cutting, dyeing, and sewing operations to China, the remaining firms in Hong Kong have been aggressively upgrading themselves at the upstream or input-supplying phase of the production process. Of the four mills still in operation (compared to over 30 at the peak of the industry in the 1970s), the largest miller—Central Textiles—invested $6.5 million in 1995 on the latest new machinery, including an almost fully automated spinner, which would work up to 10 times faster than the older ring-spindle machines (HKTDC November 1996: 4). To increase higher value-added production, the clothing companies left in Hong Kong have adopted modern technology such as CAD/CAM in pattern grading, marker making, and automatic cutting. Hong Kong's clothing industry has also moved to "Quick Response" as a flexible mode of operation which relies on modern production and communication technologies so as to supply the right products in precise quantities just in time.

In addition to upgrading its traditional labor-intensive industries, Hong Kong has begun to launch its knowledge-intensive, high-tech industries, especially the information technology (IT) industry, which had not been expected to emerge in a labor-intensive manufacturing center like Hong Kong. Two significant developments have occurred. First, endorsed and promoted by the Industry and Technology Development Council (ITDC), the Hong Kong government's highest-ranking advisory body on industry and technology development, Hong Kong is building its first ever science park, which will provide a focal point for both larger-scale companies and small fledgling high-tech ventures in IT, engineering materials, and drugs based on human genes. This government impetus, which is unusual for Hong Kong's free market economy, will foster the process of technological deepening in Hong Kong beyond 1997 (HKTDC March 1996). Second, the Hong Kong Productivity Council (HKPC) is working with China's State Science and Technology Commission and Ministry of Electronics Industry on a research project (funded by the ITDC) to help develop Hong Kong into the "Silicon Valley" of Southeast Asia. This project brings together the comparative and complementary advantages of China (e.g., a low-cost production base, a sizable market for IT products and services) and of Hong Kong (e.g., availability of venture capital funds, the entrepreneurial spirit of local industrialists and investors) (HKTDC September 1996). If successful, this high-tech cooperation between China and Hong Kong may lead to a new cross-border production system which could render the China-Hong Kong economic relationship more mutually beneficial after 1997.

"Keeping the Roots (in Taiwan); Letting the Branches and Leaves Grow (in China)"

Available survey data on industries and firms allow a more detailed and refined analysis of the production systems linking the China-Taiwan dyad in the GCEC. Although Taiwanese investment began to trickle into China through Hong Kong in the early 1980s, it swelled into a torrent from the late 1980s through the mid-1990s. A major push behind the flow of Taiwanese capital into China came in 1991 when Taiwan's Ministry of Economic Affairs (MOEA) began to promote a cross-Strait division of labor in which core factories would continue to be based in Taiwan ("keeping the roots planted") and peripheral (supporting) factories were allowed to be set up in China ("letting the branches and leaves grow").

Table 31.2 presents data from a survey that captures the main features of the emerging production systems linking Taiwan and China in four labor-intensive industry clusters. It is not surprising that a larger share of the more capital-intensive electrical equipment industry cluster maintained production after investing in China than the two more labor-intensive clusters (row 2). As expected, the largest proportion of the surveyed firms kept high-value-added production in Taiwan (3c). Firms in the electrical equipment and footwear industry clusters moved the more labor-intensive, low-value-added phases of the production to their mainland factories, and shipped them back to Taiwan for final assembling and finishing (3b). That 25.8% and 33% of all the surveyed firms chose 3a and 3e, respectively, suggests a certain degree of balance between the core and peripheral factories. The mother companies in Taiwan maintained control over the more profitable, skill-intensive phases of production such as order receiving and marketing (4a), although 27.8% of the firms shifted the connected processes of order receiving, manufacturing, and exporting to their subsidiary factories on the mainland (4b). While a much larger proportion of the firms in the footwear and textile industry clusters kept the financial functions of profit remittance, budgeting, and capital allocation in Taiwan (5a and 5b), a certain proportion of the firms in all industries maintained separate and independent accounting systems at the core and peripheral factories.

The evidence shows that in a similar way to the cross-border production involving China and Hong Kong, Taiwanese firms with investment in China have become more focused on high-value-added production, and kept the more lucrative segments of the production in Taiwan. The data also show that the peripheral factories of Taiwanese companies on the mainland were involved to a certain extent in finishing products and receiving orders. These organizational characteristics of the cross-border production systems have important implications for the export competitiveness of a range of industries and firms in China, Hong Kong, and Taiwan.

Table 31.2
The Division of Labor Between Taiwan's Mother Companies and Their Factories on Mainland China, by Industry Cluster, 1992

Industry Cluster / Survey Items	Total Sample	Electrical equipment, vehicles, metals	Footwear, daily necessities, bamboo & paper prods.	Textiles, garments, toys, plastic and leather products	Other industries
1. Number of firms (N)	140	33	25	44	38
2. Continuing production in Taiwan after setting up factories in China (%)	69.1	86.2	61.9	55.0	75.8
3. If continuing production in Taiwan, relationship with mainland factories:					
a. Taiwan makes parts or semiprocess, mainland assembles and/or finishes	25.8	26.9	22.2	23.1	29.6
b. Mainland makes parts or semiprocess, Taiwan assembles and/or finishes	29.9	34.6	44.4	26.9	18.5
c. Both finish products, but Taiwan makes high-value-added products (%)	67.0	69.2	72.2	61.5	66.7
d. Both finish products, but mainland makes high-value-added products (%)	7.2	3.8	5.6	3.8	14.8
e. Both factories finish identical products (%)	33.0	30.8	33.3	34.6	33.3
f. Both factories finish unrelated products (%)	17.5	23.1	11.1	15.4	18.5
4. Marketing and exports of products made by mainland factories:					
a. Taiwan receives orders; mainland manufactures and exports (%)	92.1	92.9	90.9	97.7	84.4
b. Mainland received orders; Taiwan manufactures and exports (%)	27.8	21.4	22.7	22.7	43.7
c. Taiwan receives orders; mainland makes parts and semiprocesses; Taiwan finishes and exports (%)	16.7	28.6	13.6	9.1	18.8
5. Capital circulation					
a. Taiwan companies and mainland factories have separate and independent accounting systems (%)	42.7	46.7	36.4	33.3	54.5
b. Mainland factories' sales and profits are remitted to Taiwan for budgeting and capital allocation (%)	52.4	46.7	54.5	59.0	48.5
c. Mainland factories' keep partial sales and profit revenues, and remit the rest to Taiwan; Taiwan companies provide financial assistance only when needed by mainland factories (%)	16.1	16.7	22.7	15.4	12.1

Source: Adapted from Kao et al. (1994: 177–178)

CLIMBING THE TECHNOLOGY LADDER

The evolution of the GCEC into a cross-national regional economy has created cross-border production networks which in turn have reshaped the relative competitive positions of China, Hong Kong, and Taiwan in technology development and exports.

Hong Kong's largest domestic exporter, the textile and clothing industry, lost some competitive advantage to China as its share of Hong Kong's total exports dropped from 40.9% in 1986 to 39.4% in 1990 and to 33% in 1994. Another major exporter, Hong Kong's footwear industry, also suffered from China's competitive advantage in lower production (mainly labor) costs as its share of Hong Kong's total exports declined from 0.8% in 1986 to 0.4% in 1990. In comparison, Hong Kong's more capital- and technology-intensive electronics and watch and clock industries sustained or even enhanced their export competitiveness: while the electronics exports' share of Hong Kong's total rose slightly from 6.2% in 1986 to 6.3% in 1990, the watches and clocks' share grew from 7.5% to 8.3% during the same period (Lin 1994). The changes in Hong Kong's export profile suggest that China has largely taken over Hong Kong's former position as a major exporter of labor-intensive, low-technology manufactured products.

Since the 1980s the decline of Taiwan's labor-intensive exports has accelerated, while the increase in its capital and technology-intensive exports has quickened. The balance between the high and low degrees of capital-intensive exports in 1982 changed to a large disparity in favor of high-capital-intensive exports in 1992. During 1982–1992, high-technology-intensive exports rose more than 10 percentage points, while low-technology-intensive exports dropped 17 percentage points (Kao et al. 1994).

The competitive positions of China's and Taiwan's specific labor-intensive exports to the U.S. market have changed abruptly. China's share of boys' shorts in textile products rose from 7.8% in 1990 to 38.1% in 1991, while Taiwan's share dropped from 60.5% to 35.7%. For children's shoes in the rubber and plastics market, China's share rose from 48.5% in 1990 to 65% in 1991, while Taiwan's share fell from 40.5% to 21.1% (Wang 1992). As recently as 1986, Taiwan had as many as 1,200 shoe factories, with a total output value of $3.9 billion. Even in 1989, Taiwan exported 1.1 billion pairs of shoes worth $3.9 billion. By 1994, over 800 shoe factories had been moved to China and another 100 elsewhere. In 1994, Taiwan exported only 108 million pairs of shoes (17% of the 1989 figure) valued at $876 million (a negative annual growth rate of 74% since 1989). The total output value of Taiwan's shoe production in 1994, however, rose to $5.2 billion. These statistics clearly show that Taiwan's footwear industry, especially shoe-making, has lost its export competitiveness to China on the lower rungs of world markets, but has gained competitive advantage at the higher end of these markets.

Taiwan's changing export position further illustrates simultaneous downloading and upgrading and the coexistence of "sunset" and "sunrise" industries that result from wide and deep investment and manufacturing linkages in a regional economic network. As Taiwan lost exports from the "sunset" industries or industry segments that have been downloaded to China, it has upgraded the "sunrise" industries that have either maintained or enhanced their export capacity. Taiwan's gain in capital- and technology-intensive exports, however, may be taking place at the expense of further weakening its labor-intensive exports. Yet between Taiwan and China, the shift in exports appears to be a win-win situation as both economies have become more specialized in manufacturing more competitive exports based on their comparative and complementary advantages.

In addition to making its industries more competitive in terms of composition and exports, Taiwan has been accelerating the development of high-tech industries. Although this process may not be directly associated with the GCEC, it has unfolded as another dimension of the overall industrial transition that has been fostered by the GCEC and spurred by regional competition with the other newly industrializing economies. Two distinctive yet related developments highlight Taiwan's going high-tech. First, following the successful model of its first science-based industrial park at Hsinchu (established in 1980), Taiwan's government has set up the second such park, in Tainan in southern Taiwan. This new park broke ground in 1996 and is expected to accommodate 30 next-generation (12-inch) integrated circuit (IC) wafer fabrication facilities in the next few years. Intended to spread the IC manufacturing now converging in northern Taiwan to the southern part of the island, the plan is reported to have attracted investment interests from Taiwan's top IC manufacturers such as Taiwan Semiconductor Manufacturing Corp. (TSMC), which has recently pledged NT$400 million to building a manufacturing hub project in the Tainan park (see IDIC February and April 1997).

Second, irrespective of the eventual success of the new high-tech park, IC wafer manufacturing appears to have become the focal point for Taiwan's high-tech sector. Besides the seven 8-inch wafer plants currently in operation, three more are expected to enter production by the end of 1997, boosting Taiwan's IC production by 30,000 units a month. By the year 2000, Taiwan is expected to have 22 8-inch wafer plants, which will account for 10% of the world's semiconductor market as opposed to the current 3% (IDIC March 1997: 2). Although this ambitious plan may be unrealistic and eventually unfulfilled, it represents the frontier of Taiwan's ongoing industrial transition. And it might not even have been unveiled, had there not been some pressure from the lost competitiveness of the labor-intensive, low-tech industries.

The shifts in Hong Kong's, Taiwan's, and China's export composition provide strong evidence in support of a staged model of technology learning, climbing, and upgrading (see Figure 31.1). Both Hong Kong and Taiwan have traversed the temporal trajectory of technological progress through exporting more capital- and technology-intensive products since the 1950s and 1960s. They have done it because their export-oriented industries and firms have moved from simple processing and assembling (P&A) in the beginning to original equipment manufacturing (OEM) and eventually to some original brand manufacturing (OBM) such as Acer Computers made in Taiwan. This process is characterized by the use of increasingly sophisticated process and product technologies, especially R&D. Will China travel the same path as Hong Kong and Taiwan have done? What opportunities and obstacles might move China along this path or make it stall or deviate from it? More specifically, will the complementary regional integration of China, Hong Kong, and Taiwan lock China into the trap of rigid specialization in labor-intensive, low-tech manufacturing? These questions will be a focal point for future comparative research that will examine the powerful role of technology in changing the export competitiveness of the various Chinese economies constituting the GCEC.

Figure 31.1
Latecomer Firms: Export-Led Learning from Behind the Technology Frontier

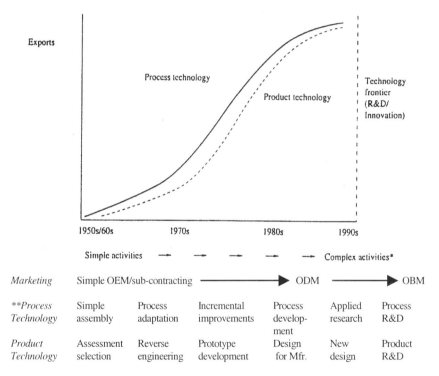

*No stages or linearity implied, but a general tendency to catch up cumulatively, through time with capabilities building systematically upon each other.
**Although it is useful to distinguish between process and product technology for analytical purposes, in practice the two are often inextricably entwined.

Source: Reprinted from *World Development, 23,* M. Hobday, "East Asian Latecomer Firms: Learning the Technology of Electronics." p. 1184, 1995, with permission from Elsevier Science.

REFERENCES

Axline, W. A. "Underdevelopment, Dependence, and Integration: The Politics of Regionalism in the Third World," *International Organization* 3, 1 (1977) 83–105.

Axline, W. A. "Comparative Case Studies of Regional Cooperation Among Developing Countries," in *The Political Economy of Regional Cooperation*. (W. A. Axline, ed.). London: Pinter Publishers, 1994, 7–33.

Chen, E. K. Y., and A. Ho. "Southern China Growth Triangle: An Overview," in *Growth Triangles in Asia: A New Approach to Regional Economic Cooperation* (M. Thant, M. Tang, and H. Kakazu, eds.). Oxford: Oxford University Press, 1994, 29–72.

Chen, X. "Taiwan Investments in China and Southeast Asia: 'Go West but Also Go South'," *Asian Survey* 36, 5 (May 1996) 447–467.

Christaller, W. "How I Discovered the Theory of Central Place: A Report About the Origin of Central Places," in *Man, Space, and Environment* (P. W. English and R. C. Mayfield, eds.). New York: Oxford University Press, 1972.

Drake, W. J. "Territoriality and Intangibility: Transborder Data Flows and National Sovereignty," in *Beyond National Sovereignty: International Communication in the 1990s* (K. Nordenstreug and H. I. Schiller, eds.). Norwood: Ablex Publishing Corporation, 1994.

Gereffi, G. "Commodity Chains and Regional Divisions of Labor in East Asia," *Journal of Asian Business* 12, 1 (1996) 75–112.

Hirschman, A. O. *The Strategy of Economic Development*. New Haven: Yale University Press, 1958.

HKTDC (Hong Kong Trade Development Council). *Hong Kong Trader* (various monthly issues, 1996, 1997).

Hobday, M. "East Asian Latecomer Firms: Learning the Technology of Electronics," *World Development* 23, 7 (1995) 1171–1193.

IDIC (Industrial Development and Investment Center). *Taiwan Industrial Panorama*. Ministry of Economic Affairs (various monthly issues, 1997).

Kan, C. Y. *The Emergence of the Golden Economic Triangle: Mainland China, Hong Kong and Taiwan* (in Chinese). Hong Kong: Lifework Press, 1994.

Kao, C. H. C., C.-C. S. Lin, C. Hsu, and W. Lin. *The Taiwan Investment Experience in Mainland China: A First-hand Report* (in Chinese). Taipei: Commonwealth Publishing Co., 1994.

Lin, J. "Hong Kong's Manufacturing System Toward the 21st Century and Its Policy," in *The Economic Cooperation of the Chinese Areas* (in Chinese) (R. Mei-jiao, ed.). Hong Kong: World Wide Publications, 1994, 220–256.

Luk, Y. F. *Hong Kong's Economic and Financial Future*. Washington, DC: The Center for Strategic and International Studies, 1995.

Mainland Affairs Council (Taiwan). *Monthly Statistics of Cross-Strait Economic Activities* (various monthly issues, 1997).

Myrdal, G. *Rich Lands and Poor*. New York: Harper and Brothers Publishers, 1957.

Oman, C. *Globalization and Regionalization: The Challenges for Developing Countries*. Paris: OECD Development Center, 1994.

Overholt, W. H. *The Rise of China*. New York: W.W. Norton & Company, 1993.

Perroux, F. "Economic Space: Theory and Applications," *Quarterly Journal of Economics* 64 (1950) 89–104.

Scott, A. J., "The Geographic Foundations of Industrial Performance," *Competition and Change* 1, 1 (1995) 51–66.

Simon, D. "The Question of Regions," in *Third World Regional Development: A Reappraisal* (D. Simon, ed.). London: Paul Chapman Publishing, Ltd., 1990, 3–23.

Storper, M., and B. Harrison. "Flexibility, Hierarchy and Regional Development: The Changing Structure of Industrial Production Systems and Their Forms of Governance in the 1990s," *Research Policy* 20 (1991) 407–422.

Tang, M., and M. Thant. "Growth Triangles: Conceptual and Operational Considerations," in *Growth Triangles in Asia: A New Approach to Regional Economic Cooperation* (M. Thant, M. Tang, and H. Kakazu, eds.). Oxford: Oxford University Press, 1994, 1–28.

Townsend, R. M. "Intermediation with Costly Bilateral Exchange," *Review of Economic Studies* 45 (1987) 417–425.

Wang, S. "Competition Between the Mainland and Taiwan for Major Export Markets" (in Chinese), *Taiwan, Hong Kong, Macao and Overseas Trade* 8 (1992) 11–17.

PART VI:
SUSTAINABILITY, ENVIRONMENT, AND BUSINESS: POLICY AND STRATEGIES

32

Sustainable Development: Concepts, Scenarios, and Strategies for R&D

Maria Laura Barreto, Heloisa V. Medina,
Carlos C. Peiter, and Roberto C. Villas Bôas

INTRODUCTION

This chapter presents the contemporary debate on the concept of sustainable development, also presenting two possible scenarios and their effects on technological research. This consideration is appropriate because the term "sustainable development" has been taken by the different groups of society—environmentalists, politicians, scientists, and others—that refer to it as an *a priori* concept. On the one hand, the frequent use of the term is important as it shows acceptance, but, on the other hand, the absence of a clear definition results in a lack of content. It is therefore often misunderstood as a mere environmental concern. This leads to serious consequences when we need to define policies for action because the lack of conceptual clarity leads to indecision regarding which path to follow. Concern with the concept of sustainable development is not merely academic; it is also related to practice, without which the term is empty and useless, reduced to a modernizing rhetorical resource.

The concern behind the proposition contained in this chapter is that the debate over sustainable development is receiving media attention and forcing a public opinion commitment that may lead to social pressures against changes in legislation which, in turn, may be translated into codes of behavior that are more restrictive of certain economic activities.

This chapter is divided into four parts: first, the concept according to international organizations; second, the theoretical concept; third, the scenarios; and finally, the implications of the scenarios for R&D.

THE CONCEPT ACCORDING TO INTERNATIONAL ORGANIZATIONS

Discussions about the concept of sustainable development go back to the 1970s when, during a United Nations Founex meeting, a new development option was outlined, incorporating "environmentally suitable strategies for fostering more equitable socioeconomic development" (Sachs 1993), known as ecodevelopment.

The 1972 Stockholm Declaration and the 1974 Cocoyoc Declaration reasserted the concept and the proposals of ecodevelopment. However, it was in 1980, in a document entitled "World Conservation Strategy," prepared by the International Union for the Conservation of Nature, that the phrase "sustainable development" was coined. The concept was closely followed by strategies of action for its implementation. In this sense it is not strictly speaking a theoretical concept, but an instrumental one.

It is in this guise that it is criticized, according to Baroni, by Khosla and Sunkel. Khosla criticizes the Stockholm Declaration because it establishes a

strategy confined to living resources, focused on the need to maintain genetic diversity, habitats and ecological processes, and incapable of dealing with controversial matters related to the international political and economic order, such as wars, armament, population and urbanization problems (Baroni 1992).

Sunkel makes a second criticism of the document to the effect that the strategy presented

essentially concentrated on the supply side, assuming that the structure and the level of demand were autonomous and independent variables, and ignoring the fact that 'if a style of sustainable development must be pursued, then both levels and particularly the demand structure, must be fundamentally changed (Baroni 1992).

A series of workshops were held and reports were produced by international organizations, as a way of giving substance to the term and to establish principles. Among the most important is the United Nations Environment Program (UNEP), which supported the document "World Conservation Strategy."

Finally, the World Commission on Environment and Development (WCED) took up the concept of sustainable development as being development that satisfies the needs of the present without jeopardizing the *abilities of future generations to satisfy their needs.*

It was also WCED that produced the first document to attempt to express the concept concretely: the Brundtland Report, presented at the UN General Assembly in 1987 (WCED 1990). The Report exhaustively defines so-called imperative strategies.

The great merit of this report seems to be the effort to make the concept of sustainable development operative, expressed in summary in the imperative strategies, as well as to seek to establish itself as a platform for international negotiations. For Baroni, the greatest criticism is that of referring to the withdrawal of the "requirement established originally in 1986 at the Ottawa Conference, regarding the need for equity and social justice for sustainable development" (1992).

THEORETICAL CONCEPT

In literature that attempts a more theoretical analysis, basically two concepts of sustainable development are found, according to the interpretation of development and sustainability or of the binomial development/environment.

For Baroni, for example, when seeking to define development and sustainability, the different and even contradictory concepts are quite clear. For Acselrad, by contrast, the differentiation of the concept of sustainable development emerges when the environmental crisis is interpreted. He says:

The first [meaning of the term] recognizes the market's inability to respect the environment's limits and proposes the creation of signaling elements that would make it possible to assure the continuity of the capitalist development model. The second line of interpretation sees the environmental crisis as a manifestation of a crisis in the capitalist development model and finds ways for overcoming it in the introduction of changes in the structure of power over natural resources (Acselrad 1993).

In fact, both authors share the same opinion about the existence of two ways of interpreting the term "sustainable development" and come to the same conclusions via complementary routes. As a matter of methodological choice, we use the differentiation of the term proposed by Baroni, because it will allow the two viewpoints to be distinguished more precisely.

In the first meaning of the term "sustainable development" means economic growth. For this viewpoint there is no contradiction between growth and sustainability because "governments concerned with long-term sustainability do not need to limit the growth of the economic product as long as they stabilize the consumption of aggregate natural resources," (Baroni 1992). Still adhering to this viewpoint, a more positive argument in favor of economic growth starts from the presupposition that poverty is largely responsible for environmental degradation. The elimination of poverty would be a condition for ecological sustainability and the role of economic growth in that process would be fundamental, with the proviso that the quality of such growth needs to be changed. Baroni says "It is argued...that economic growth is absolutely necessary for sustainable development" (1992). From this viewpoint, social objectives such as improving the quality of life and ending poverty are objectives of sustainable development, although based on an operational strategy of economic growth.

The term sustainability, for this meaning of sustainable development, is based on the ecological dimension, and not on social policy. Hence, it would be "the existence of ecological conditions necessary for providing support for human life at a specific level of well-being through future generations" (quoted in Baroni 1992). Acselrad, on this point, says:

Concerned with sustaining the basis of natural resources for future production, this concept proposes the introduction of a new environmental restriction.... Ignoring the conflict for control over natural resources, it seeks to create conditions for saving natural resources, without, however, considering the socio-political conditions that govern the control and the use of such resources (1993).

The same author considers that the concept of sustainability for this viewpoint has evolved from the simple intention to preserve natural resources to identify the environment as a capital in terms of the accounting system.

David Pearce elucidates this thinking: "Sustainable is the development which considers the expansion of environmental capital in proportion to population growth" and "Sustainable is the development that reinvests in the environment to assure its conservation and its recovery (Acselrad 1993).

Nature, which until then had provided the working capital (raw materials and input) and free services (water, soil, and air for disposing of waste), then begins to provide fixed capital elements—that is, those that need to be conserved throughout the productive cycle. This new viewpoint opted for appropriate economic concepts through analogy because in this case to consider the environment like other production factors, the elements of nature need to have an owner.

For Acselrad, the

diagnosis of this line of thought says that the roots of the environmental crisis are in the fact that capital considers the environment to be a free asset, and environmental damage as an externality...This system sanctions only what is the subject of private appropriation...All damage caused to the public interest is not expressed in prices. In this sense, the environmental crisis results from the inability of capitals to calculate the environmental damage that their activities generate....The solution would be to correct the shortsightedness of entrepreneurs and start seeing the environment as an economic asset, with a price....What is put forward as a solution...the "internalization of environmental costs" (1993).

The second meaning of the term "sustainable development" sets development against economic growth, considering sustainable growth to be a contradiction. For this meaning, development would therefore involve a socially desired phenomenon assuming two basic objectives: the end of poverty and better distribution of income, which are not very unlike the traditional declared goals of development. The so-called limitation of the ecosystem's ability to support, which appears in the two senses of development, has a special meaning for the latter, involving the redefinition and even limitation of humanity's consumption habits, which would be determined socially.

The term "sustainability," for this meaning of sustainable development, assumes the concept of ecological, political, and social sustainability; and in this sense, for both concepts, there is the matter of environmental rationality, strange because it is new, and which opposes or adds to traditional economic rationality. The difference lies in the way in which strategies should be established for attaining sustainability. In this meaning, what can be sustained, how, and for how long, are determined socially in a process of participation of society and even of social consensus. Hence, for the first meaning of the concept of sustainable development, the internalization of environmental costs depends upon the existence of basic instruments which make it possible to define parameters of the limits on the ecosystem's ability to support, and also the "measurement" and definition of the problems of environmental damage. However, for the second meaning, the lack of such instruments does not represent such an important technical-scientific barrier, to the point of significantly affecting either the debate or the definition of strategies for action.

For this latter viewpoint, the origin of the environmental crisis is precisely in the model of development and in the way of using nature that it implies. In capitalist logic, the use of natural resources is the result of the private economic reckoning of companies, which only consider those mercantile elements that are expressed through the price system—such as raw materials and land. The conditions and the global equilibrium of the environment are not considered. In this way, sustainable development would only be possible if the limits on the control of capital are put on the use of the environment, through predominantly political action. It can, therefore, be seen that sustainable development would involve change, or even societal rupture.

For Lélé, quoted by Baroni, the prevailing interpretation of sustainable development is

a form of societal change which, in addition to traditional development objectives, has the objective or the restriction of ecological sustainability. Obviously, this is not independent of other (traditional) objectives of development. Trade-offs normally have to be made between the extent and the rate at which ecological sustainability is attained *vis à vis* other objectives. In other cases, however, ecological sustainability and traditional development objectives (like satisfying basic needs) can mutually strengthen each other (1992).

SCENARIOS

Scenarios constitute methods of anticipation which indicate ways to future development[1]. In this text, two large-scale alternative scenarios are considered: Scenario I, inertial or tendential, which is characterized by the continuity of dominant present-day tendencies; and another of change or rupture, Scenario II, based on the discontinuity of present courses and wider-ranging transformations. The time horizon considered is the year 2015.

Scenario I

This scenario describes and explains the results of continuity and the expected evolution of the main tendencies noted at present with regard to factors that are critical in the future of materials.

The standards of consumption and production in the industrialized countries are undergoing constant pressure from environmental groups. As a consequence, there is a growing incorporation of new technologies that increase energy efficiency, intensive recycling of materials, and the substitution of scarce materials, especially rare metals, by abundant materials.

The tendential scenario adheres to the modernizing line of globalization of the economy, imposing a model of competitive insertion under the terms dictated by the "world class standard,"[2] in which the more dynamic sectors are those that are more technology-intensive.[3] Control over the new technical production paradigm (new ways of organizing production, greater flexibility in productive processes, and intensive use of new technologies) is exercised by the developed countries, to bypass environmental restrictions and to exploit their availability of natural resources. Technological innovation is seen only as a competitive factor—that is, technology is conceived as a strategic element to be incorporated into products so as to increase their penetration in world markets.

The market is the driver of the logic of the economic growth process, where external competitiveness is the basic indicator, with little emphasis on social problems. Political and socioeconomic inequalities are heightened between the developed and developing nations, as well as in the interior of the less-developed countries, where levels of absolute poverty get worse.

In international trade, there is a prevailing tendency toward greater liberalization with regard to industrialized products. With the new technical production paradigm, there is an accentuated reduction of trading in primary products and particularly nonrenewable products.

The internationalization of markets, however, heightens the concentration of trade among industrialized countries, strengthening integration in regional economic blocs, and increasing the North-South gap. Technological development, pressured by environmentalist interests, has a strong reducing impact on the exports of developing countries, which are mainly concentrated on primary products. Without foreign exchange revenues, these countries are not participating in the growth of world trade. Access of the countries of the South to new technologies is restricted on account of their strategic value in global competition.

Another significant aspect in this scenario is its environmental protection model. It differs, first, because it fosters a process of imposition, where the emphasis is put on the exercise of authority, particularly through regulatory mechanisms that inhibit and penalize. Second, in this scenario, the technological capability necessary for dealing with environmental problems remains virtually all concentrated in the developed countries, which make a point of exercising a monopoly over such technologies.

In this context, however, there is less possibility of environmental protection measures being really effective in the long term, since they are carried out "from the top down," without being essentially rooted in social participation and specific ecological awareness for each local context.

In connection with this model of environmental protection, it can be seen that there is a predominance of "cultural mimicry" distinguished by a dual reality: on one hand, a First World where consumption standards derive from a particular culture and are compatible with the rates of growth and of spatial distribution of their populations; and, on the other, a Third World that yearns after the First World's consumption standards. In this way, the First World's style of society and values are hegemonic and drive all the appraisals of level of development and quality of life, especially with regard to ecological issues. In the meantime, while the First World seeks to keep up its standard of living, with some incremental changes already under way in its consumption standards,[4] it tries to prevent the less-developed countries from attaining this standard of living, and to stop their economic exploitation of their natural resources, alleging that the environment must be preserved.

In this scenario, the state's role is marked by the fact that it is in the hands of two large interest groups: that of the big companies and, on a smaller scale, that of the international environmentalist movements. The financing criteria for sustainable development are dictated by international organizations.

The feasibility of the tendential scenario has its basis in the political-economic hegemony of the developed countries, which make their proposals prevail, dictating the rules and institutional models that govern international affairs. They preponderantly establish and benefit from pacts, blocs, and political and economic alliances among nations. This scenario represents the strengthening of "hypercolonialist" positions in the world context, the reduction of the sovereignty of states of the developing countries, and the strengthening of the role of transnational capital.

The political support for this model of international inequality is based on the performance of the industrialized countries. The new technical production paradigm applies, essentially, to the industrial sector which—because it saves labor—causes technological unemployment. To avoid the political and social consequences of unemployment, governments are obliged to direct their economies toward greater verticalization, fostering technologies that take advantage of their natural resources to use up the excess workers in primary production.

The peripheral countries are undergoing serious economic difficulties. Their exports are faced with the barriers of the regional blocs, their products are in sectors where there is declining demand, and they have no access to technology. As a result of the serious social tensions deriving from growing socioeconomic disparities and from the strengthening of ethnic movements, there is a proliferation of regional military conflicts, mainly involving peripheral countries. This political instability involves the industrialized nations which, through international organizations, are called upon to mediate in the conflicts and to receive their refugees.

In view of this situation of economic and social inequality and growing conflicts, migrations from the less-developed countries to the more-developed countries are growing in volume, increasing their unemployment problems.

In spite of the pressure from international organizations and environmentalist groups, the developing countries are doing little or nothing to alter their standards of natural resource exploitation, nor to restrain the degradation of the environment. The environmental consequences of the disorderly occupation of the rain forests, of atmospheric pollution caused by heavy industry and military conflicts, are worldwide. The environmental gains achieved in the industrialized countries are outweighed by the environmental degradation in the peripheral countries.

Scenario II

This scenario represents both a profound discontinuation of the historic tendency of competition and a global political, economic, and technological restructuring. These changes favor a more equitable and balanced international development.

In this scenario, the logic of the market coexists with the logic of its control and social satisfaction, in which the principles of reciprocity and redistribution in the conduct of economic and social activities are strengthened. Economic sustainability is provided by the pursuit of efficiency in macrosocial terms, and not only according to microeconomic profitability criteria. Growth rates are not as important as absolute indexes, but are rather seen as related to the social and spatial distribution of the population's quality of life. Economic and social development is fundamentally the result of production focused on raising living standards for the majority of the population and not to satisfy the sophisticated standards of those who are already part of the select consumer market. In this way, the market grows by absorbing those at present excluded.

In this scenario there is in fact a depolarization in the present pattern of North-South relations, as much in terms of trade as in financial and scientific-technological affairs. The more-developed societies are able to provide the less developed societies with funds to facilitate the acquisition of linguistic, educational, and professional attributes necessary to keep up with them and also to influence the redirecting of the "global development model."

Access to science and technology, as well as the development and redirecting of such knowledge, are the highlights of a new standard of international cooperation, particularly because of the need for new ways of using natural resources and of adapting to different ecosystems, in order to achieve environmentally healthy development.

In this scenario, there is a better balance of forces between the different players, while the dominant political force comes from civil society, which begins to have more access to information.

The environmental protection model is marked by encouragement and by making use of opportunities offered by a new paradigm that is globally rather than ecologically healthy, motivated primarily by endogenous ecological awareness adapted to each regional and local context in accordance with each other. It is this new awareness that alters market conditions which, in their turn, point to environmentally suitable productive practices. At the same time, the technical-scientific capability of developing regions in the environmental area is strengthened by the cooperation and financial support of the more developed regions.

In the context of Scenario II, competition is replaced by an interpretation of the concept of "competitiveness," which gains force in the relations it establishes with other concepts (equity and sustainability) and social values (democracy, human rights, and social participation). In this interpretation, competitiveness is not interchangeable with competition, but its status as a concept in its own right has yet to be clarified (Müller 1994).

The new productive paradigms may be described as a "partnership" between the developed regions and those that are developing technologically. Consumption and production standards will be profoundly altered in favor of energy efficiency, diversification of energy sources (especially renewable sources), greater durability of products, recycling of materials, the substitution of scarce materials by abundant materials, and by the miniaturization of components and products. In fact, scarce natural resources are already being progressively replaced by intellectual resources, with much benefit for the protection of the environment.

Finally, attention should be drawn to one aspect which might be called "unfinished" in Scenario II, the main difficulty that results from the generalization of promises of social participation, equity, and control of the environment. In the case of Scenario I, although the social security and unemployment aspects are not embodied, the actual continuity of the current economic rationale points to what should be done at micro and even sectoral levels. By contrast, in Scenario II, as the paradigmatic behavior of competition inclines toward a more complex competitiveness, the routes of this "world model-to-be" are less clear even at the micro and sectoral levels.

SCENARIOS AND IDEOLOGY

In the different approaches presented we can see that there is an ideological struggle to appropriate the term "sustainable development." At present, this struggle is not explicit, but it can be clearly observed in the various documents issued by international organizations and in the literature concerned with a conceptual definition. A deeper discussion of the aspects of the different definitions presented is usually avoided—especially at the level of international organizations—because an agglutinating concept of sustainable development is intended. The concept should bring together the various interests of the UN countries and not separate them, as the old and well-known Third World discourse or even the North/South dialogue did. This aspect could explain the lack of debate on the term "sustainable development."

It should be noted that the implementation of sustainable development will be possible only if it is global—at least according to one conception. Sustainable development would not be possible in one single country or in a single group of countries. Moreover, for those who adopt this conception, the wealthiest countries should contribute to the development of the poorest ones by providing them with financial, technological, and human resources, thus lowering their own consumption standards. All these objectives can be achieved only by means of a global political commitment. To reach this political commitment, the concept of sustainable development should unite interests rather than separate them. Actually, if anything can be considered uniting, it is not the concept of sustainable development, but its cause: the environmental crisis. Thus, sustainable development would be seen as a necessity, for the lack of an option before the imminent environmental crisis. Sustainable development would be the only way for rich and poor countries to avoid a crisis that would affect them equally. While ways to attain sustainable development or even the variables of sustainability are discussed, disagreements about the concept or the understanding of the term seem clear and even polarized, as can be observed in the "theoretical" and "senarios" sections of this chapter. The two scenarios may be viewed as shown in Table 32.1. The implications for the strategies of the respective R&D needs are shown in Table 32.2.

Table 32.1
Basic Dimensions of Scenarios I and II

SCENARIO I	*SCENARIO II*
Social dimension:	**Social dimension:**
▪ Social aspects are not stressed; growing inequalities of income between individuals and nations.	▪ Involving more equality in the distribution of income, property, and access to goods.
Economic dimension:	**Economic dimension:**
▪ Emphasis on international competitiveness imposed by the technical standards of production in the developed world.	▪ Macrosocial factors surpass microeconomic profitability as decision criteria, mainly for improving well-being and enhancing the value of work.
Ecological dimension:	**Ecological dimension:**
▪ Emphasis on preservation and recuperation of the physical environment through technology. Environmental costs enter decision-making criteria.	▪ Creative use of each ecosystem's potential; rational use/conservation of energy and natural resources; reduction of the volume of waste and pollution.
Political-institutional dimension:	**Political-institutional dimension:**
▪ "Global commitment" is imposed by the proposals, rules, and institutional models of the developed world.	▪ "Global commitment" is achieved by a new agreement between policy makers and domestic and international agents.
Cultural dimension:	**Cultural dimension:**
▪ Consumption standards in the developed world are maintained and used as a reference for the rest of the world, in spite of the modifications in way of life caused by environmental factors.	▪ The search for sustainable solutions is guided mainly by the increasing importance of local conditions, seeking better standards of living all over the world.

Table 32.2
Implications for R&D

SCENARIO I	*SCENARIO II*
Social dimension:	**Social dimension:**
▪ Emphasis on materials intended for increasingly selective and sophisticated markets, and highly specialized jobs.	▪ Emphasis on materials intended for meeting social needs and generating employment accompanied by training local labor.
Economic dimension:	**Economic dimension:**
▪ Emphasis on materials intended for increasing competitiveness in the external market.	▪ Emphasis on strategies for materials that produce a positive effect on earnings and employment.
Ecological dimension:	**Ecological dimension:**
▪ Emphasis on substituting materials that are scarce in the developed countries.	▪ Emphasis on materials based on renewable or abundant natural resources, according to local availability.
Political-institutional dimension:	**Political-institutional dimension:**
▪ Technological control by the developed countries.	▪ International technological cooperation in the materials field.
Cultural dimension:	**Cultural dimension:**
▪ Materials based on consumer standards imposed by the globalization process.	▪ Materials based on endogenous natural/mineral reality;
	▪ Materials based on endogenous S&T, business experience/capability;
	▪ Materials intended for local consumption standards.
Spatial dimension:	**Spatial dimension:**
▪ Spatial concentration of activities on materials.	▪ Materials that can be processed locally;
	▪ Regional coordination in materials;
	▪ Balance in the territorial distribution of activities.

NOTES

1. This section is an adapted extract from Medina et al. (1994).

2. John Sequeira (1990) defines world class standard as that which today's best business is able to achieve. In the current environment of "continual improvement" the better businesses are constantly redefining what is understood by "world class." That is, the parameters are always changing.

3. Such as, at present, the electroelectronic, aerospace, information technology, and communications sectors.

4. With regard to the environmental issue, the developed countries are still, with their development model, the greatest polluters. All the same, some attenuation of this process has already been felt since the 1970s, when the crisis of the model began at a world level, such as less-intensive use of natural materials, greater efficiency in the productive process with regard to raw materials and energy input in general, and greater emphasis on recycling of products and materials, among the principal changes. That is, the change began on the production side, precisely to prevent the level of material consumption of the developed societies being affected.

REFERENCES

Acselrad, H. "Desenvolvimento Sustentável: A Luta por um Conceito," *Revista Proposta* 56 (March 1993).

Baroni, M. "Ambigüidades e Deficiências do Conceito de Desenvolvimento Sustentado," *Revista de Administração de Empresa* 32, 2 (April/June 1992).

Medina, H. V., M. L. Barreto, I. C. Marques. "Desenvolvimento Sustentável e Microeletrônica no Brasil." Mimeo, 1994.

Müller, G. "A Competitividade como um Caleidoscópio," *São Paulo Perspectiva* 8, 1 (1994) 23–32.

Sachs, I. "Estratégias de Transição para o Século XXI," in *Desenvolvimento Sustentável* (M. Bursztyn, ed.). Rio de Janeiro, Brazil: Editora Brasiliense, 1993.

Sequeira, J. *Manufatura de Classe Mundial no Brasil: Um Estudo da Posição Competitiva*. Sao Paolo, Brazil: FIESP, 1990

World Commission on Environment and Development (WECD). *Sustainable Development. A Guide to Our Common Future*. Geneva: WECD, 1990

33

National Environmental Funds: A Financing Strategy for Environmental and Development Policies

Nicolas-Pierre Peltier

INTRODUCTION

If pollution reduction were profitable and if economies were perfectly following the market's invisible hand, environmental policies for industrial pollution abatement would be meaningless. With the increase of uncontrolled polluting industrial activities, various financing schemes aiming at supporting pollution abatement investments have been proposed in countries where private resources dedicated to environmental issues are not sufficient. Usually combined with a revenue-raising mechanism based on pollution taxes, they may offer a theoretically efficient redistribution mechanism in which the heaviest polluters are contributing more, and corporations willing to invest in clean technologies receive more assistance. In developing countries and transition economies where resources are scarce, they might even be the only available policy instrument to catalyze "green" financing. The issue then becomes relevant for aid agencies and development banks, as these existing financial mechanisms may become a way to channel external investments aiming at promoting more environmentally friendly activities. This chapter intends to promote recommendations for implementing this strategy for the environment, with a focus on pollution prevention. After acknowledging the importance of financing cleaner production activities, this chapter presents the model of environmental funds and describes other existing environmental funds from various economies, including the World Bank approach in China. On the basis of this experience, a policy design is proposed as well as recommendations for implementation and operation.

IMPORTANCE OF FINANCING CLEANER PRODUCTION

Cleaner Production: A Technological Approach from Reaction to Prevention

The history of environmental policies usually distinguishes several phases in approaching the reduction of pollution, from the promotion of the *dilution* of wastes, their *disposal*, their *treatment*, their *recycling*, and finally *source reduction*, also called *cleaner production* (CP). Following these various phases, a distinction should be made between the first four that are *reactive*—that is, trying to address the problem of wastes once these are generated, and the last one for which *prevention* measures are applied. While the so-called end-of-pipe approach remains widely used, preventive policies have been recently integrated in environmental statutes with generally good acceptance from the industry community. For environmentalists, the success of this new strategy depends on the reduction of pollution in the first place where it is generated, and for industry, on its capacity to reduce costs and risks and to identify new opportunities. Based on successful experiences in numerous industrial sectors, the United Nations Environment Program (UNEP) is defining cleaner production as "the continuous application of an integrated preventive environmental strategy applied to processes, products, and services to increase eco-efficiency and reduce risks to humans and the environment." As such, CP is intrinsically related to a successful management of technology and innovation.

Difficulties of Implementation

Although there exists considerable evidence of the superiority of CP compared to the reactive approach, many difficulties in implementation have been reported. These include in particular:

- *Rigidities to change:* As CP options target the production level inside the firm, enterprises may be reluctant to change an existing process, especially if there exists a technological risk. This is particularly observed with traditional industries such as breweries or tanneries. The separation inside the firm of the management of environmental and production competence may amplify the phenomenon of lack of *innovation* willingness.
- *Informational failures:* Small and medium-sized enterprises (SMEs) may not be aware of breakthrough innovations in the field of clean technologies. This penalizes the *diffusion* of existing clean technologies.
- *Externalities effects:* If national environmental policies do not provide a sufficient regulatory or economic incentive to internalize the social cost of pollution, certain

socially optimal clean technologies may not appear to be economically relevant to enterprises.

- *Market failures:* Although certain technologies may trigger sufficient economic benefits as well as significant environmental benefits, private capital may not exist for this type of investment. This is particularly observed in developing countries and transition economies.

The Role of Governments, Bilateral and Multilateral Aid Institutions

In order to successfully implement CP, governments should combine both the "command and control" regulatory approach and economic instruments to try to overcome the previously mentioned barriers. The proper environmental policy may vary with the level of economic development, especially when dealing with market failures, and certain instruments may be designed on a transitory basis only. International aid institutions as well as development banks have an important role to play in this regard, especially through the policy recommendations and the type of financial or technical assistance they provide to governments. The next section will describe an economic instrument that sometimes has been introduced by governments and aid institutions as a financial mechanism to promote CP.

DISCUSSION ON THE ENVIRONMENTAL FUND MODEL

National Environmental Funds and Externally Supported Revolving Funds

In several countries belonging to various types of economies, a possible approach has been to establish *national environmental funds* (NEFs), often handled by government-affiliated institutions. These economic instruments provide both a negative incentive to pollute through their sources of revenues usually based on effluent taxation, and a positive incentive to invest in pollution-abatement technologies by providing technical and financial assistance to enterprises willing to do so. Although this primary definition of funds as a redistribution mechanism for the environment is very simple, the primary design of NEFs may vary in numerous regards, including

- *The type of pollution that is targeted:* funds can be comprehensive or alternately media-specific (e.g., water agencies).
- *The type of financial assistance:* grants, soft loans or commercial loans.
- *The primary targets of NEFs:* local governments investing in publicly owned wastewater treatment plants, large enterprises, or SMEs (firms could also belong to a particular industrial sector).
- *The type of institution handling the fund:* governmental, autonomous but related to government, autonomous nongovernmental organization (NGO), or private.
- *The procedure to select enterprises and technologies:* this item is discussed later and allows us to define two large categories of funds.
- *The size of the fund:* starting from the small funds (less than US$1 million) to the largest ones (exceeding $100 million).

- *The sources of revenue:* earmarked revenues from pollution taxes (direct taxes), energy or raw material taxes (indirect taxes), general state budget, fines, and such.

In countries where such NEFs already exist, development banks and aid agencies have the possibility to use these financial mechanisms to channel financial assistance dedicated to industrial pollution abatement. Under this scheme are established *externally supported revolving funds* in which part of the capital is provided as a long-term loan, a grant, or is based on the so-called Debt-for-Nature Swaps, by the external lender or donor, and then used to provide short-term financial assistance to local enterprises. The external financial support is then being reused (it is "revolving") by disbursing subloans to enterprises, before being itself repaid in case of a loan. It should be noted that these environmental institutions are usually designed to overcome at least the "rigidities to change" and the "informational failure" barriers to CP by providing adequate technical assistance to enterprises. This chapter focuses hereafter on the relationship between funds and the two financial barriers: "externalities effects" and "market failures."

Examples of Application

The large variety in design is illustrated by numerous examples in industrialized countries, transition economies, and developing countries:

In industrialized countries, funds tend to be more media-specific, and with a focus on nonprofitable environmental investments. For example, France has designed six river basin agencies providing grants and soft loans to industries and municipalities for water pollution abatement. Denmark has set an NEF financed by an SO_2 tax based on fuel or electricity consumption to provide subsidies to industry for energy savings. The United States has established the Leaking Underground Storage Tank Fund, financed by gasoline taxes, to finance corrective actions at abandoned sites.

In many transition economies, the NEF model is usually comprehensive and with a focus on large investments for local governments and large enterprises. For example, the Polish NEF offers grants and soft loans to enterprises for air and water pollution control and is financed from air and water pollution charges. The Hungarian NEF has the same objectives but is financed from fuel taxes, a traffic transit fee, a European Commission PHARE grant, and pollution fines.

In developing countries, NEFs and externally revolving funds are widely used but very often promote nonprofitable environmental investments such as biodiversity or NGO activities rather than industrial pollution abatement. However, in countries such as China, NEFs financed by pollution charges and fines, and administered by local Environmental Protection Bureaus (EPBs), are disbursing grants and soft loans for industrial pollution control investments. In Ivory Coast, water charges and a sanitation tax finance a fund paying for the operation and maintenance of sewerage networks. Table 33.1 sums up some of these examples with their main characteristics.

Table 33.1
Examples of Existing National Environmental Funds and Externally Supported Revolving Funds in Various Economies

Country	Economy	Main features
Denmark *SO₂ tax*	Industrialized	*status:* governmental *revenues:* tax on consumption *disbursement:* subsidies to industry for energy savings
France *River Basin Agencies*	Industrialized	*status:* government affiliated *revenues:* effluent & water charges *disbursement:* grants & soft loans
United States *Leaking Underground Storage Tank Fund*	Industrialized	*status:* governmental *revenues:* gasoline taxes *disbursement:* grants
Hungary *National Environmental Fund*	Transition	*status:* ministry of environment *revenues:* taxes & PHARE grant *disbursement:* grants & soft loans
Poland *National Environmental Fund*	Transition	*status:* independent *revenues:* air & water pollution charges, water use charge *disbursement:* grants & soft loans
Russia *Regional Environmental Fund*	Transition	*status:* local government-affiliated *revenues:* pollution charges *disbursement:* grants
China *EPB funds*	Developing	*status:* governmental *revenues:* pollution charges *disbursement:* grants & soft loans
Colombia *Findeter*	Developing	*status:* independent *revenues:* banks' credits *disbursement:* loans to municipalities for water management
Ivory Coast *National Water Fund*	Developing	*status:* governmental *revenues:* water & effluent charges *disbursement:* grants

Sources: OECD and The World Bank.

Relevance as an Environmental Economic Instrument

Because environmental funds provide subsidies to enterprises, they have often been accused of being in contradiction with the "Polluter Pays Principle" (PPP). However, although NEFs do not always comply with a stricto sensu PPP because some polluters may receive more financial assistance than what they contribute to the fund, this policy model complies with the PPP at the macro level, as long as revenues are raised exclusively on polluters. In the case of externally supported revolving funds, the transitory use of external financial support is still in compliance with the 1974 OECD update of the PPP, which introduced a waiver "when (1) the subsidy does not introduce significant distortions in international trade and investment; (2) without the subsidy, affected industries would suffer severe difficulties; and (3) the subsidy is limited to a well-defined transition period adapted to the specific socioeconomic problems associated with the implementation of a country's environmental policy" (OECD 1993).

Another criticism of environmental funds addresses the earmarking of revenues. Earmarking is generally criticized by economists because the amount of revenues that is raised may not be used in the most cost-effective ways for society as a whole—that is, it could be more valuable to use revenues raised by environmental taxation for nonenvironmental purposes. However, there exist other advantages to earmarking, such as a better political acceptance of environmental taxes when these are introduced, an increase in transparency, or a more predictable financial management for the institution hosting the NEF. As a consequence, and because in most countries environmental financial needs are still unfulfilled, which reduces the risk of non-cost-effectiveness, earmarking is generally considered as a second-best and transitory solution when designing NEFs (OECD 1993). It should also be noted that most of the funds that exist in industrialized countries are based on the earmarking of taxation revenues for very specific environment-related uses.

Other theoretical advantages of NEFs are related to the traditional debate between regulatory and economic instruments. Although, under certain conditions, the command and control approach may encourage enterprises to innovate (Ashford 1985), economic instruments such as effluent taxes or emission allowance trading are usually more *dynamically efficient* because enterprises have a constant incentive to improve the environmental performance of technologies (Rajah and Smith 1992). Finally, recent literature has underlined a possible *double-dividend* effect of ecotaxation: effluent taxes may both correct market distortions by charging a polluter for the true opportunity cost of the resource being polluted, and be used to raise revenues instead of distortionary taxes.

In developing countries and transition economies, the use of existing NEFs as an indirect mechanism to channel financial support for the environment presents several theoretical advantages. First, it allows to build on existing competence toward a sustainable financing mechanism for the environment. Second, it contributes to the capacity-building of local environmental institutions and it reinforces or assists in improving existing environmental policies. Finally, it may help to target different types of enterprises (SMEs for example) and different types of technologies (less capital-intensive but requiring monitoring at the local level), which may improve the cost-effectiveness of many development projects that are too often exclusively focusing on large pollution control equipment.

EXISTENCE OF TWO DIFFERENT MECHANISMS FOR THE ENVIRONMENT IN VARIOUS ECONOMIES

Environmental Externalities and Market Failures

The first section of this chapter has mentioned two financial barriers to the implementation of CP that have been observed in industry. Economic instruments are usually introduced to internalize environmental externalities in order that the burden for the real cost of pollution that society as a whole is paying is borne by the polluter himself. Ecotaxes as they have been originally proposed by Pigou (1920), indirect taxes on energy and raw materials, or subsidies to finance pollution abatement may all theoretically achieve this goal. In a perfect world, these market mechanisms would convince enterprises to invest in the socially optimal environmentally friendly technologies, because these will be economically more interesting than the "old dirty" ones. From the double incentive they provide to industries, environmental funds may be used with this objective, provided that the taxation rate is set at the right level and that only the proper technologies are financed.

The second financial barrier to CP, which was previously identified, may also be overcome by using the environmental fund mechanism. Market failures arise in particular in developing countries and transition economies when the invisible hand is not working properly and when profitable clean technologies cannot be implemented from the lack of private funding. The concept of cleaner production sometimes has been defined by reference to these numerous technologies which are both profitable and environmentally friendly. National environmental funds could then be used to provide both the technical assistance that is necessary to the diffusion of clean technologies and to the promotion of innovation, and the financial resources (eventually with the financial incentive of a "soft loan" disbursement policy) that are necessary to correct market failures.

Selecting Enterprises and Technologies

Whatever their primary objectives are, all NEFs have to set up criteria to manage the portfolio of the CP projects they are supporting. This means that a precise policy has to be determined (1) to select *enterprises*; and (2) to select *technologies*. Moreover, for these two dimensions, both *economic* and *environmental* performances have to be taken into account. For example, the Fund Management Office (FMO) will have to decide between a nonprofitable, environmentally friendly technology in a profitable, average-polluting firm, and a profitable, less environmentally friendly technology in a less profitable, heavily-polluting enterprise.

These criteria can than be used to distinguish between the two types of funds that have been previously described. An "externality-internalizing" NEF will seek to focus first on the environmental performance of technologies and on heavy polluters, whatever their economic performance is. In that situation, taking into consideration the financial situation of firms would introduce distortions that could seriously hamper the primary objectives of the economic instrument. On the contrary, a "market failure-correcting" fund will be as concerned by the environment as it is by the economic performance of technologies. In an extreme situation in which the fund would act as a "green" private bank, the NEF would also consider the financial situation of firms (if the disbursement policy is based on loans, this may improve the default rate by reducing the financial risk).

Application to Several Existing Environmental Funds

These two dimensions are summed up in the matrix described by Figure 33.1. As previously mentioned, a green private bank will target the "corners" of the matrix, namely high economic returns technologies in profitable enterprises. In a more restrictive model, only the profitable heavy polluters would be concerned (left hand-side corners). A theoretical externality-internalizing NEF will focus on heavy polluters where most of the socially costly pollution is generated, and will finance high environmental impact technologies.

The French Water Basin Agency's model (one of them, the Agence de l'Eau Rhin-Meuse has been studied for the purpose of the research supporting this chapter) is a more restrictive version of the externality-internalizing NEF. For the purpose of saving public funding, the FMO chose to finance exclusively projects that are not profitable enough to justify private financing—that is, for which financial support is necessary to implement the technology. This presents the advantage that certain nonprofitable clean technologies, which would not be implemented otherwise, will be adopted by firms. On the other hand, it creates a distortion by discouraging firms to consider the environmental performance of profitable technologies.

Thus, the 3×3 enterprise/technology selection matrix allows us to identify the target of each NEF and to distinguish between the externality-internalizing and the market failure-correcting funds. It should be noted that the domains of the two types of funds are clearly different and even separated in certain cases (for instance for the French Water Basin fund and the Green Private Bank).

Figure 33.1
Positioning of Various NEFs on the 3×3 Enterprise/Technology Selection Matrix

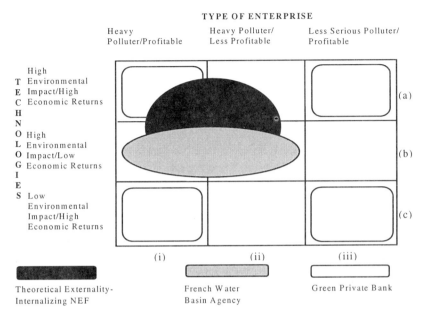

A DEVELOPMENT CASE STUDY AND RECOMMENDATIONS FOR FUTURE IMPLEMENTATION

The World Bank-Supported Revolving Funds in China

China is an example of a developing country in which a development bank (the World Bank) has been using an existing NEF model (administrated by the EPBs, which are the local bureaus of the National Agency for Environmental Protection) to establish externally supported revolving funds and to channel foreign funding for the environment. Since 1990, at least nine such funds have been established in China and two additional ones were under preparation in 1996 with the support of the UNEP Industry and Environment Office in Paris (Wijnants 1996). All of them have a relatively small size compared to the traditional World Bank projects, and they have been either focusing explicitly on assistance to "end-of-pipe" technologies or, on the contrary, to CP.

Compared to the initial governmental EPB fund model, the World Bank-supported revolving funds include a significant shift toward more private-oriented institutions. This is particularly illustrated by the choice of an autonomous status for the hosting institution, by a disbursement strategy based mostly on commercial loans, by the selection of profitable enterprises and of technologies with high economic returns, and by hosting institutions subcontracting the financial and technical appraisals of projects, instead of integrating "in-house" all the necessary competence as most governmental NEFs do. Such a strategy has several consequences, arising first from the choice of an indirect scheme using the EPB funds, and second from the shift toward more private-oriented institutions.

The use of existing EPB funds whose disbursement strategy is mainly based on grants or long-term soft loans may have caused difficulties to recovering commercial short-term subloans from enterprises. A default rate of 30% has been observed in the case of a World Bank-supported fund established in Tianjin (Peltier 1997), which is similar to what has been reported in the case of a typical EPB fund in the city of Jinan (Spofford 1996). Second, EPB funds, because of their relatively small size, tend to focus on small investments. As a result, after the fund was significantly enlarged by external financial assistance calculated to handle larger projects, and being possibly hampered by too-complex World Bank reviewing procedures, the FMO was unable to use a large proportion of the total fund resources.

The shift toward more private-oriented institutions may have resulted in an excessive focus on criteria selecting firstly profitable enterprises rather than top polluters. Moreover, in order to decrease the unused amount of revolving funds, incentives to raise the average project cost were given to FMOs. As a consequence, this has provoked the selection of more capital-intensive end-of-pipe technologies (wastewater treatment plants), even though the primary objective of funds were CP. Because end-of-pipe equipment is easier to identify and because it usually concerns larger enterprises with better ability to repay, it seems that the shift toward a more private-oriented fund model has distorted the selection of enterprises and of technologies. Such a strategy can, however, be justified from the point of view of development banks, by the importance of reducing the financial risk and having subloans to enterprises be effectively reimbursed.

Designing a Transitory Environmental Fund Model for Development

The experience of the World Bank in China illustrates the difficulties that can arise from trying to modify too rapidly the primary objectives of an environmental fund—that is, turning an "externality-internalizing" fund into a "market failure-correcting" one. Whatever the actual effectiveness of the EPB funds are, these financial mechanisms have been clearly designed with the intention of providing a negative incentive to pollute (through taxation) and a positive incentive to invest in pollution abatement, regardless of the economic performance of both enterprises and technologies. Thus, the domain they target in the selection matrix is exactly the one that corresponds to the theoretical externality-internalizing fund (see Figure 33.1).

On the contrary, the World Bank-supported revolving funds have been designed first because a true environmental market failure has been identified in China and it was necessary to remediate to it, and second under the constraints of a development bank, in particular the necessity to keep an acceptable financial risk. The "green private bank" model, based on the existing EPB funds but involving local private "agent" banks for the financial appraisal of projects, was therefore coherent with the objective of promoting the raising of private capital for profitable environmental investments. However, heavy polluters with more uncertain ability to repay are not targeted through such screening procedures, and thus a significant proportion of the heavy polluting enterprises are not concerned any more by the positive incentive of the original EPB fund to invest in pollution abatement.

In developing countries and transition economies where both categories of NEFs are needed, a focus on one kind of objectives exclusively, as it is observed through the World Bank experience in China, may become suboptimal and result in a partial solving of the lack of adequate financial resources for the environment. A possible solution could then involve cooperation between donors and lending institutions, in which donors would support government-affiliated funds aiming at correcting externalities, and development banks would offer long-term loans to more private institutions correcting market failures.

Another solution would be to define a transitory financial instrument, on the basis of an existing NEF such as an EPB fund, by mixing the two approaches with the possibility to turn into an "externality-internalizing" fund if market failures are disappearing and if the economic situation is improving. The domains that are targeted by such a fund are described by Figure 33.2. Under this approach, heaviest polluters with high ability to repay are targeted first as well as technologies with high environmental impact regardless of their economic returns. As a second priority, heavy polluters with less ability to repay would be selected, to support the implementation of technologies with high environmental impact and high profitability to reduce the risk of delayed repayments. Finally, if the default rate that is observed becomes unacceptable, less serious polluters with good ability to repay could also be selected.

Provided that accurate selection criteria are used by the FMO coherently with the domains identified in the selection matrix, such a financial mechanism may guarantee an acceptable financial risk to development banks, should internalize environmental externalities where they are most needed, and could also help to correct the most obvious market failures.

Improving the Model

From the Chinese experience, several recommendations concerning the financial parameters of the revolving fund model can be proposed to resolve the difficulties that have been observed in this particular case when establishing an externally supported revolving fund on the basis of an existing NEF.

- *Precisely evaluating the size of the fund*, based on the past practice from the existing NEF, in order to avoid a large unused proportion of resources. In the case of China, for example, an optimal external support to a large municipal EPB fund should range from $5 to $10 million.

- *Using a disbursement scheme based on soft loans* in order to keep an economic incentive for enterprises to request financial assistance instead of focusing exclusively on the correction of market failures by offering commercial loans. This can be achieved by mixing the external support (which can be disbursed at the commercial rate) with counterpart funding from pollution levies (used as grants or soft loans).

- *Avoiding a lower bounding of the amount of investments to be supported*; otherwise, it may favor the selection of capital-intensive technologies and introduce a non-cost-effective screening of potential investments (promoting end-of-pipe equipment or very specific CP approaches).

Figure 33.2
Positioning of the Proposed Transitory Financial Instrument on the 3×3 Enterprise/Technology Selection Matrix

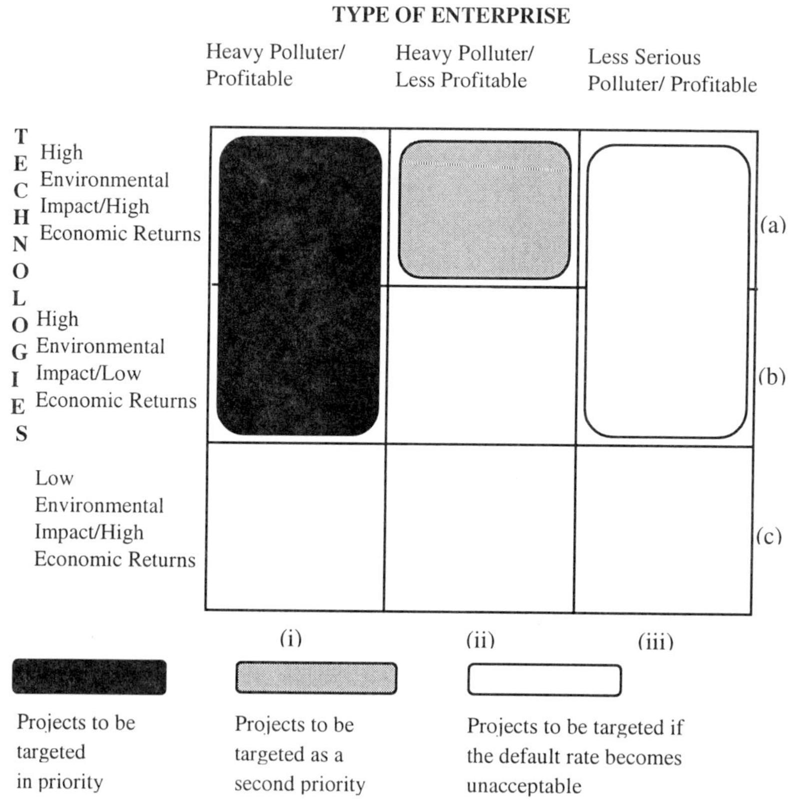

TYPE OF ENTERPRISE

| | Heavy Polluter/ Profitable | Heavy Polluter/ Less Profitable | Less Serious Polluter/ Profitable |

TECHNOLOGIES

High Environmental Impact/High Economic Returns (a)

High Environmental Impact/Low Economic Returns (b)

Low Environmental Impact/High Economic Returns (c)

(i)   (ii)   (iii)

Projects to be targeted in priority

Projects to be targeted as a second priority

Projects to be targeted if the default rate becomes unacceptable

In addition, reviewing and monitoring procedures could be simplified by using independent local resources. For example, in the case of China, primary reviewing of technologies could be done by a National Cleaner Production Center (NCPC) established with joint support from UNEP/UNIDO (United Nations Industrial Development Organization), or a similar institution. Furthermore, the disbursement policy for the external lending support could be redefined in order to match the requirements from the indirect funding mechanism of externally supported revolving funds.

Finally, if the size of the financial assistance to be given to an existing NEF is too small relative to the usual size of development projects, a larger amount could be disbursed by joining several local NEFs and having them compete one with another with a progressive ex-post disbursement scheme. The cash flow coming from reimbursement of subloans to enterprises should then be taken into account to determine what amount should be disbursed to a specific NEF. Such a policy would give an incentive to funds to increase their portfolio of technologies (and therefore reduce the unused resources) and to reduce the default rate. Figure 33.3 illustrates this "competition-based" disbursement scheme. By joining the preparation work necessary to establish the competing externally supported revolving funds, such an approach may as well significantly reduce the "transaction cost" of implementing these economic instruments.

Figure 33.3

A Competition-Based Disbursement Mechanism for an Externally Supported Revolving Fund Providing Assistance to Several NEFs

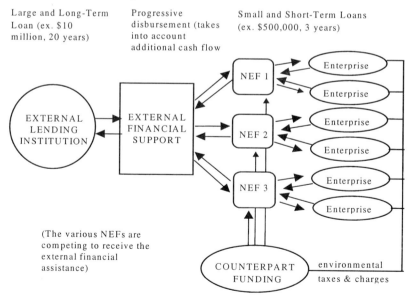

Dealing with this last issue, a practical difficulty observed is the high administrative cost of design and implementation for externally supported revolving funds. The traditional financing mechanisms used by most development banks are based on a "line-of-credit" approach in which only a few large top polluters are assisted to implement large-scale pollution abatement equipment. On the contrary, revolving funds still are considered innovative and their administrative cost of implementation may be particularly high, especially compared to the relatively small size of such projects. However, in countries such as China, with numerous NEFs existing at local levels, the diffusion of a particular model of revolving fund could be relatively inexpensive once the policy model is standardized.

CONCLUSION

This chapter has firstly defined the model of national environmental funds and of externally supported revolving funds as a relevant economic instrument to implement cleaner production in various economies. The model belongs to the economic approach to environmental problems in industry, and its political acceptability may vary depending on the level of governmental interventionism in the economy. Nevertheless, this policy instrument can also be found, with very specific objectives such as financing remediation at abandoned industrial sites, in very liberal economies. In most transition economies of Central and Eastern Europe, large-scale comprehensive NEF models can now be found.

However, the operating experience from existing funds shows the existence of two different models that can be distinguished by drawing their intervention domains on a 3×3 enterprise/technology selection matrix: the "externality-internalizing" NEF, on one hand, is usually governmentally managed and tends to give more weight to the environmental performance of technologies and enterprises than to their economic performance; the "market-failure-correcting" fund, on the other hand, is more "private-oriented" and focuses on the economic performance at least as much as on the environmental criteria.

Switching from one category to the other when, for example, implementing an externally supported revolving fund, may first create some operating difficulties for the fund and may, second, keep unsolved a significant proportion of the industrial sources of pollution. As a possible solution, this chapter proposes a transitory policy instrument that could be used by development banks operating in developing countries and transition economies. Some improvements of the model are also proposed, keeping the "market-based" characteristics of this financial mechanism.

Finally, it is important to remember that both categories of funds are based on a dual technological and financial competence. The selection matrix that has been proposed as a management and monitoring tool is an attempt to harmonize these two dimensions in order to optimize the design of NEFs as an environmental technology and policy instrument.

ACKNOWLEDGMENTS

This chapter is based on research done at the Massachusetts Institute of Technology, in the Technology and Policy Program, under the supervision of Nicholas A. Ashford, Professor of Technology and Policy. The evidence supporting this work was collected with the authorization of the World Bank Asia Information Service Center in Washington DC, the United Nations Environment Program's Industry and Environment Center in Paris, and the Agence de l'Eau Rhin-Meuse in Metz (France). The author was supported during the work described in this chapter by a *Lavoisier* scholarship from the French Ministry of Foreign Affairs.

REFERENCES

Ashford, N. et al. "Using Regulation to Change the Market for Innovation," *Harvard Environmental Law Review* 9, 2 (1985) 419–466.

Lovei, M. *Financing Pollution Abatement: Theory and Practice.* Washington, DC: World Bank Technical Report, 1995.

OECD. *Taxation and the Environment, Complementary Policies.* Paris: OECD, 1993.

Peltier, N.-P. "National Environmental Funds and Externally Supported Revolving Funds; Financing Schemes for Cleaner Production in Developing Countries: Experience from Various Economies and Recommendations for Future Application." Unpublished Master's Thesis, Cambridge: Massachusetts Institute of Technology, 1997.

Pigou, A. C. *The Economics of Welfare.* London: Macmillan, 1920.

Rajah, N., and S. Smith. "Taxes, Tax Expenditures, and Environmental Regulation," *Oxford Review of Economic Policy* 9, 4 (1992) 4–65.

Spofford, W. et al. *Assessment of the Regulatory Framework for Water Pollution Control in the Xiaoqing River Basin: A Case Study of Jinan Municipality.* Washington, DC: Resources for the Future Technical Report, 1996.

Wijnants, H. "World Bank-Assisted Pollution Control Funds in China," in *Proceedings of Workshop on the Use of Economic Instruments in Environmental Policies in China.* Paris: OECD, 1996.

34

Integrating Environmental Policy and Business Strategies: The Need for Innovative Management in Industry

Paulo C. Ferrão and Manuel V. Heitor

INTRODUCTION

The concern with the environment has been discussed for almost three decades and, at least in the last decade, has changed how industries operate and do business. The challenge which has been put forward consists of the development of a sustainable global economy, operating essentially in two dimensions, as extensively discussed in OECD (1996): (1) the quality of the environment to be sustained, mainly focused on the amelioration of the deleterious effects of industrial processes and on the optimization of the use of natural resources; and (2) the economy, which must continue to develop, but requiring what has been called a paradigm shift. However, for much of the developing world, sustainable development has begun to take on a different meaning and to incorporate two additional dimensions (e.g., Hart 1997): (1) social sustainability, which has been grounded in efforts to alleviate the grinding effects of poverty and is primarily focused on the need to guarantee total employment, besides the provision of basic services (sanitation, healthcare, education, and housing, mainly facilitating sustainable urban development); and (2) global trading, which is moving toward a more liberalized regime and may be affected by environmental policies in key areas, such as world trade, foreign investment, competition, and economic instruments, including ecotaxes. The challenge is to develop a sustainable global economy, which the planet is capable of supporting indefinitely.

Although much has been written on definitions of sustainability, the current trend among development practitioners is to include not just the biophysical world, but the whole range of natural and human resources in interaction with humans (Franks 1996). This calls for three different aspects to manage sustainable development: natural resources, institutional development, and appropriate management skills. This chapter focuses on the latter, although it is clear that public policy innovations and changes in individual consumption patterns will be required to move toward sustainability. However, corporations can and should lead the way, helping to shape public policy and driving change in consumers' behavior.

The practical consequence of the approach presented above has been described by Ehrenfeld and Howard (1996) as a four-stage evolution of corporate environmental management, in that companies should move from treating environmental management as a case-by-case problem-solving issue, to a regulatory compliance issue, to an issue for proactive management, and finally to a central element of business strategy. The challenge for companies, industry groups, and governments is to set goals that can be put into practice at the company level, but conform in a consistent fashion with the multidimensional and complex goals of sustainable development. It should be noted that the public awareness of sustainability has grown with the end of the cold war and the related swift advance of democracy in places where it was previously unknown and an even more rapid spread of market-based economies. Also, information and communication technologies have enhanced people's ability to exchange ideas and encouraged them to speak a greater voice for their interests.

The remainder of this chapter covers the following subject areas: the evolutionary patterns of corporate environmental behavior, as observed by many authors in recent years; the markets arising from increased environmental awareness; issues internal to corporations in terms of environmental responsibility; and finally the various issues facing companies today are brought together and presented as a related strategy to manage for the environment in order to achieve the transformation to a "green company."

BACKGROUND: THE EVOLUTION OF CORPORATE ENVIRONMENTAL MANAGEMENT

Historically, industry has played a minor role in setting environmental goals, while the environment was considered as a public good whose protection and development lay beyond the individual concerns of private business. However, in the last three decades, a shift has been observed in corporate environmental practices (e.g., Hunt and Auster 1990; Ehrenfeld and Howard 1996). First, at the most basic level, environmental protection is of no concern to corporate decision making and does not have a specific budget. Environmental goal-setting is virtually absent in this phase. Second, the firms consider regulations, but are not concerned with the environment itself. Third, firms believe that environmental protection has certain strategic advantages, as well as significant cost-reduction opportunities. This gives rise to proactive environmental management, in that the goals of the firm transcend mere compliance with government standards and encompass the voluntary establishment of stricter standards. By putting an economic value on pollution reduction, governments will entice companies to reduce pollution below the requirements in an attempt to increase profits. Fourth, the final stage of managing for the environment is reached when environmental concerns become a core strategic factor in corporate decision making in a manner that leads, rather than follows, public policy. The implication is that management for sustainability requires the incorporation of additional values and interests beyond those historically recognized by firms, as discussed by Angel and Berkhout (1996).

The practical result of the achievement of the fourth stage described above is that corporations now have to consider carefully not only their production processes, but also the use and disposal of their products in light of ecological market and regulatory processes (Steger 1996). Management of the complete life cycle of products will therefore become a requirement in many markets in the years to come.

While environmental programs that focus on a point in the product chain have resulted in resource conservation and pollution prevention, further advances will only be incremental in nature as long as the approach taken continues to separate all stages of economic activity, including product design, manufacture, use, and disposal. For example, when looking to reduce air emissions of a particular chemical associated with a product, the production plant is often not the only place to examine. Sometimes, larger and more cost-effective reductions can be found by analyzing emissions from transporting and distributing the product. A life-cycle approach captures the upstream environmental effects associated with raw material selection and use and effects from production processes and product distribution. It also reflects downstream effects associated with product use, recycling, and disposal, as discussed by Bohm and Walz (1996) and Angel and Berkhout (1996), among others. Figure 34.1 shows a schematic representation of potential environmental impacts across a typical product life cycle.

In this chapter, the life cycle is considered in terms of extended product responsibility, as an emerging principle to identify strategic opportunities for pollution prevention and resource conservation. It also addresses the underlying influence of consumer needs and preferences, government procurement, and the role played by these in the production and distribution chain. Under the principle of extended product responsibility, manufacturers, suppliers, users, and disposers of products share responsibility for the environmental effects of products and the waste stream.

Figure 34.1
Potential Environmental Impacts Across the Entire Product Life Cycle

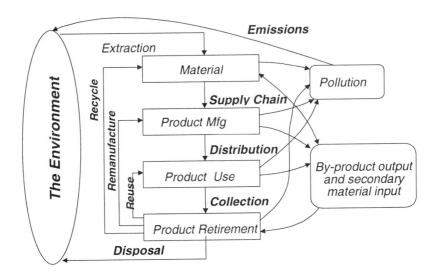

Source: Ehrenfeld and Howard (1996)

   Creating an innovative system of extended product responsibility would improve the current fragmented approach to waste reduction, resource conservation, and pollution prevention. When there are missing links in the chain of responsibility, waste and inefficiency result. Communities bear the greatest burden of the disposal of hazardous products. Similarly, decisions made upstream in the chain by suppliers can reduce manufacturers' emissions and wastes and improve profitability. Sharing responsibility implies not only understanding and communicating the environmental effects of product development, but also acting collectively to reduce them. By using a mix of regulatory and other incentives, information, education, and institutional support, this new system encourages individuals, governments, and corporations to recognize, understand, and act on the basis of their responsibility to advance sustainable objectives.
   Although this chapter does not address the issue of education for sustainability, it should be noted that this is not an add-on curriculum—that is, is not a new core subject like math or science. Instead, it involves an understanding of how each subject relates to environmental, economic, and social issues. Education for sustainability promotes high standards of achievement in all academic disciplines as well as an understanding of how these disciplines relate to each other and to the concepts of environmental quality, economic prosperity, and social equity.

THE MARKET AND THE ENVIRONMENT

It is known that a large proportion of OECD environmental policy is at present oriented toward the prevention and control of polluting activities. The "Polluter Pays Principle" (PPP), applied to manufacturers, started out in 1992 as an economic principle, but it has gradually evolved into more of an integrated environmental and economic principle, as environmental degradation also occurs through the overuse of natural resources during the life cycle of the product. This has led to calls for the extension of PPP to include the broader "User Pays Principle" (UPP), as discussed in OECD (1996), but it is clear that the importance of economic instruments remains relatively limited.

Much of the current interest in applying economic instruments to environmental objectives focuses on opportunities for gradually "greening" the tax system. This can occur in several ways, including imposing ecotaxes and reducing economic subsidies that harm the environment. It is clear that these instruments strongly affect company strategies based on short-term goals, but it is also clear that the use of "proactive" instruments that promote the development of long-term sustainability of industry also remains to be developed.

There are two ways in which markets can be restructured so as to ensure that environmental services enter into the market system more effectively. First, we could create markets in previously free services. This would require restriction of access to such resource services by charging entrance fees and changing property rights.

The second option would be to "modify" markets by centrally deciding the value of environmental services and ensuring that these values are incorporated into the prices of goods and services. This is known as the market-based incentives approach. It is to be contrasted with a direct ("command and control" or CAC) regulatory approach, which involves the setting of environmental standards (e.g., for air or water quality) enforced via legislation without the aid of market-based incentives.

Economists have long argued that the market-based incentives approach is more efficient than one based on CAC. The pollution control systems currently operating in industrialized countries, however, have been dominated by direct regulations. Thus, charges/taxes would enable a polluter to choose how to adjust to environmental quality standards. Polluters with high costs of abating pollution will prefer to pay the charge. Polluters with low abatement costs will prefer to install abatement equipment. By making abatement more attractive to "low cost" polluters than to "high cost" ones, charges tend to cut down the total costs of compliance.

The basic tenet of PPP is that the *price* of a good or service should fully reflect its *total production*, including the cost of *all* the resources used. Thus the use of air, water, or land for the emission discharge or disposal of wastes is as much a use of resources as are other labor and material inputs. The lack of proper prices for and the open access characteristic of, many environmental resources, means that there is a severe risk that overexploitation leading to eventual complete destruction will occur.

The PPP seeks to rectify this *market failure* by making polluters internalize the costs of use or degradation of environmental resources. The aim is to integrate use of the environment (including its waste assimilation capacity) into the economic sphere through the use of price signals and of economic instruments such as pollution taxes, charges, and permits. The use of regulation to internalize externalities is also, however, consistent with the PPP. Externalities are usually defined as unintentional side-effects of production and consumption that affect a third party either positively or negatively—for example, a factor that pollutes the surrounding local atmosphere creating an external cost.

The idea of internalization of externalities within a company leads to a definition of boundaries, as discussed by Curran (1996). These boundaries are based on the concept of *nested costs* in which life-cycle costs may be separated into two categories: internal (or company) costs and external (sometimes called *social or societal)* costs, as specified in Table 34.1. Internal company costs, in turn, fall into two major subcategories: (A) conventional and direct, and (B) less tangible, hidden, and indirect.

Table 34.1
Total Company Costs

Internal Costs			
A. Conventional Costs	**B. Indirect Costs**		**C. External Costs**
▪ Raw materials	▪ Environmental permits	**Product Life Cycle** →	Social costs related to:
▪ Process control	▪ Licensing		• Greenhouse effect
▪ Buildings	▪ Waste handling, storage and disposal		• Acidification
▪ Engineering	▪ Natural resource damage		• Wetland loss
▪ Design	▪ Worker compensation related to exposures to hazardous substances		As produced by the company's activities or the product use

Consider the case of automobile manufacturing. This industry relies on dozens of suppliers of various inputs, including coatings, plastics, steel, rubber, glass, and aluminum. Even the largest, most vertically integrated automakers, such as General Motors (GM), purchase materials and products from a large number of ancillary manufacturers. Making each of these materials and products requires some combination of raw material extraction, transport, and intermediate and final processing of materials before they reach GM for assembly in the motor vehicle. From an analytical perspective, one should include the emissions and associated health and environmental impacts of each of the ancillary industries within the life-cycle analysis of an automobile. Such a choice would mean, for example, that emissions from steel plants that supply the auto industry should be assigned to the final vehicle. Then as more intensive use of aluminum and plastic displaces steel in automobile manufacturing, one might track changes in the emissions inventory to determine how various emission types change in response to substitutions of materials.

For many of these ancillary industries, the pricing mechanism will capture environmental impacts. Control of air pollutants requires stack equipment; wastewater effluent may be handled by pretreatment facilities; and hazardous wastes must be properly labeled, transported, and disposed of. These activities incur capital and operating costs which, to some extent, are embodied in the prices that suppliers charge to the automaker. To the extent that this occurs, such costs find their way into vehicle prices, thereby signaling to buyers that purchase of a particular vehicle causes certain environmental impacts and that there are no external costs. All costs are internal, and the auto manufacturer accounts for upstream costs either directly or indirectly through the pricing mechanism.

It is clear that the real world does not operate so smoothly and precisely, as discussed, for example, by Ketola (1996) regarding two different oil companies, namely Shell and Texaco. The pricing mechanism is less than perfect and regulations do not cover all adverse impacts, which result from materials extraction, transport, and manufacture. Impacts such as residual health effects, wetlands loss, and climate change owing to greenhouse gas emissions exemplify costs to society at large which fall into area C of Table 34.1. They occur at all stages of the life cycle as a result of both direct activity (in our automobile example, final assembly of the vehicle) and ancillary industries (e.g., glass, aluminum, plastics manufacture) which supply the final product maker.

The question that may be raised at the corporate level concerns the reasons to incorporate external costs, and here we identify three specific aspects in order to promote environmental management as a central element of corporate business strategy:

1. Avoiding the "regulatory treadmill." An external cost today will be an internal cost tomorrow. While it is impossible to predict the exact direction of future regulations, the last decade clearly indicates that environmental policy is moving toward increasingly broader coverage in the materials and impacts it regulates. Impacts such as greenhouse gas emission and wetlands loss may see more stringent regulations in the next decade. This leads industry managers to consider such impacts and their associated costs in their planning and budging activities.
2. Internal competitiveness. An increasingly global economy means in many instances that the highest standards for product content (e.g., solvent- and heavy metal-free paper coating, unassembleable and recyclable automobiles) will increasingly dictate the global standard. To secure a place in an increasingly global marketplace, setting or exceeding the highest environmental standards is preferable to retrofitting designs and processes after they are put in place.
3. Accountability beyond responsibility. Business is under increasing pressure to protect the environment, regardless of whether formal regulatory mandates are in place. Environmentalists, investors, and the public at large are increasingly sensitive

to a firm's willingness to subscribe to voluntary codes of conduct and regularly and completely disclose its environmental performance in annual reports and through other reporting mechanisms. For many stockholders, the distinction between internal and external is an arbitrary one, and the firm ought to be held accountable for impacts even in the absence of regulation. This, many would argue, is the essence of environmental stewardship.

With this market-derived framework as background, we turn to a more detailed consideration of the corporate attitudes and management concepts which allow costing concepts to be operationalized. Then we will discuss the use of life-cycle analysis as a decision support tool in order to achieve environmentally based business strategies.

BUSINESS ENVIRONMENTAL RESPONSIBILITY

Based on the evolution of corporate environmental management just presented, and in terms of possible attitudes toward environmental responsibility, Fiksel (1996) estimates the distribution of worldwide companies among the three following categories.

1. 10 to 15% are at stage 1, characterized as *Problem Solving*. These companies have a traditional, reactive approach to environmental problems, and view regulatory compliance as a burdensome cost of doing business.
2. 70 to 80% are at stage 2, characterized as *Managing for Compliance*. These companies attempt to be more effective in coordinating their environmental management efforts, but are still mainly compliance-oriented.
3. 10 to 15% are at stage 3, characterized as *Managing for Assurance*. These companies are more far-seeing, and have adopted risk management as a rational means of balancing potential future environmental liabilities against costs. In this stage, companies are largely concerned with identifying and mitigating sources of risk that may result in financial liability. They will make risk control expenditures to the extent that the net marginal expenditure does not exceed the corresponding reduction in potential liability, although the latter is difficult to assess due to the presence of large uncertainties. This approach may lead companies to behave "beyond compliance"—for example, if there are significant residual risks associated with emissions that are exempt or below the regulatory threshold.

This classification scheme fails to capture the further evolution that has occurred within leading companies that have moved beyond stage 3. Although Ehrenfeld and Howard (1996) have considered a fourth stage of "managing for the environment," here we consider this stage in two steps, as follows:

- *Managing for Ecoefficiency*, including companies which have recognized that pollution prevention is more cost-effective than pollution control, and are seeking opportunities to be more environmentally efficient through waste minimization, resource reduction, and other approaches. In this stage, companies are largely concerned with identifying opportunities for improving efficiency while reducing waste and emissions. On a case-by-case basis, they will invest in pollution prevention opportunities to the extent that the net marginal expenditure does not exceed the corresponding savings in operating costs. Eventually they tend to reach a point of diminishing returns, where it is not cost-effective to continue reducing waste using existing technologies. However, new product and process technologies may change the economics to the point where zero-waste or closed-loop recycling is attainable.

- *Fully Integrated Management,* including companies that have adopted environmental quality as just one dimension of total quality to be managed in an integrated fashion. It is questionable whether any company today has truly attained this stage. However, many of the more enlightened stage 4 companies have established this as their ultimate vision, once their senior management has realized that environmental excellence is essential to profitability and competitive advantage.

Companies with a fully integrated vision are largely concerned with assuring stakeholder satisfaction and environmental quality over the complete life cycle of their products and facilities. This leads them to move from an opportunistic to a systematic approach, with environmental considerations factored into virtually all decisions. In fact, environmental decision making is no longer a separate exercise—it becomes an integral part of business decision making. These companies will invest in additional R&D, including environmental improvements, to the extent that they earn an adequate return on investment. If they are able to leverage their skills and technologies to profitably achieve superior quality, environmental or otherwise, this will constitute a competitive advantage.

INTEGRATED PRODUCT DEVELOPMENT FOR EXTENDED PRODUCT RESPONSIBILITY

It is well known that considerable uncertainty exists about the evolution of the variables which support environmental, economic, social, and trade policies due to their complexity and interrelationships, but we consider that the correct strategy is to recognize key elements of the linkage between the various processes in a way that considers all the environmental impacts of the whole life cycle of a product, as discussed by Pidgeon and Brown (1994) and Young and Rikhandsson (1996).

The Environmentally Oriented Value Chain

Management of the complete life cycle of a product covers mainly, but not exclusively, all the stages that the product goes through in the corporation, from production to disposal at the end of its service life, as represented schematically in Figure 34.1. Steger (1996) used the concept of the "value (creation) chain" developed by Porter (1985) in order to identify the corporation's strengths and weaknesses in the environmental area relative to competitors and to the financial benefits which result. Figure 34.2 shows the typical value-forming corporate activities which relate to the environment and comprise all the economic and technological activities which make a contribution to the manufacture of a product. The basic assumption of product life-cycle management is that it is integrated into an environment-oriented system of corporate management. All empirical research shows that if environmental protection is to be put into effect and credibly supported by top management, it is essential to anchor this objective at the normative level, by writing an environmental protection objective into the fundamental corporate principles. Although environmental protection is a matter for top management, the evidence is that the "top-down" principle must be augmented by "bottom-up" support. As Steger (1996) made clear, the relative environmental unfriendliness of a product can be reduced not just by a fundamentally new design, but also by changing the various value-intensive activities throughout the product life cycle.

Successful handling of the additional complexity brought by consideration of the entire value chain, represented in Figure 34.2, has been the subject of empirical research, but the main concern to express here is the need to extend the life cycle of a product beyond the value chain of a corporation. This is because the complementary tasks ranging from procurement to disposal have more intensive links with upstream and downstream value chains. In addition, opportunities arise for cooperation with competitors not only in take-back systems, but also in research. This calls for decision support tools which can be conveniently used to incorporate environmental aspects as corporate business strategies.

Decision Support Tools: From Assessment to Design

Motivated by the various types of costs mentioned above and environmental concerns, several techniques have been developed to exploit a very simple and intuitive concept: *every stage in the life cycle of a product or process has both costs and environmental impacts.* Some of these techniques, such as traditional life-cycle costing (LCC) as widely practiced in the defense industry, consider the direct capital and operating costs of producing, using, and maintaining products, or the analogous costs of operating manufacturing or service processes, but have typically ignored environmental costs. Other techniques, such as life-cycle assessment (LCA), assess the environmental impacts for all activities that take place over a product's full life cycle, "from cradle to grave," while linking these impacts to concrete business decisions (e.g., Angel and Berkhout 1996; Bohm and Walz 1996).

Figure 34.2
Environmentally Oriented Value Chain over the Complete Product Life Cycle

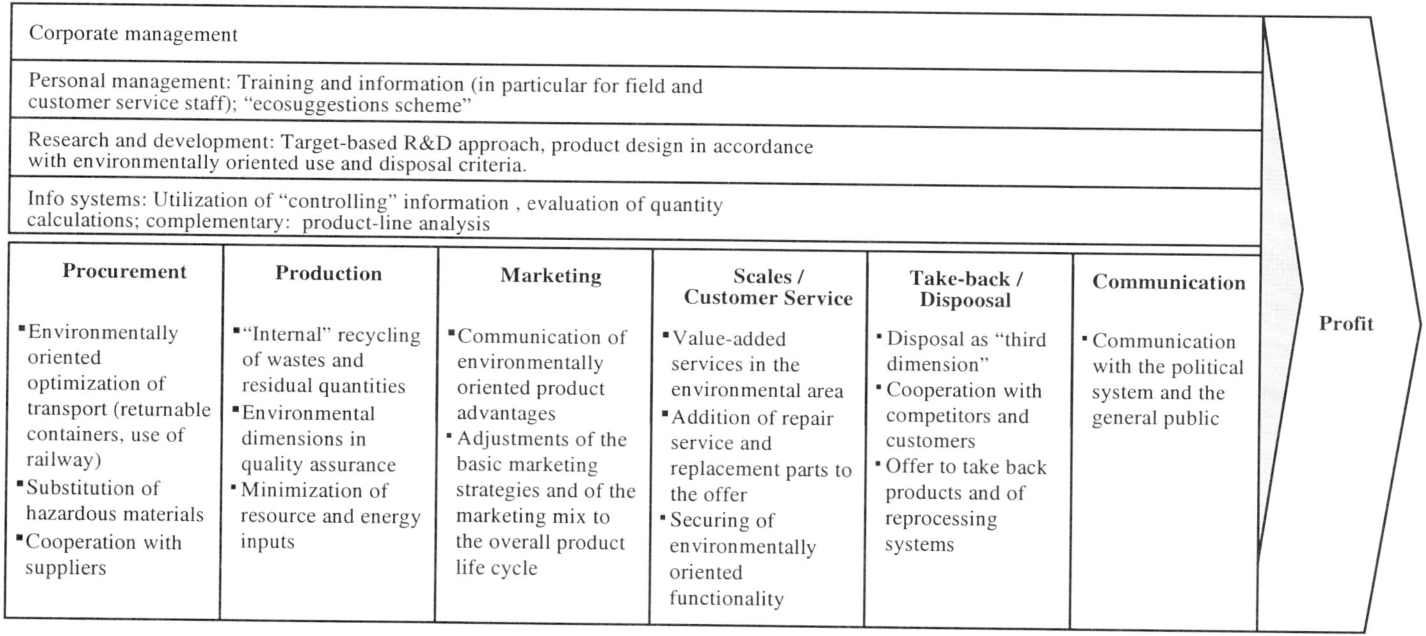

Corporate management

Personal management: Training and information (in particular for field and customer service staff); "ecosuggestions scheme"

Research and development: Target-based R&D approach, product design in accordance with environmentally oriented use and disposal criteria.

Info systems: Utilization of "controlling" information , evaluation of quantity calculations; complementary: product-line analysis

Procurement	Production	Marketing	Scales / Customer Service	Take-back / Disposal	Communication
▪ Environmentally oriented optimization of transport (returnable containers, use of railway) ▪ Substitution of hazardous materials ▪ Cooperation with suppliers	▪ "Internal" recycling of wastes and residual quantities ▪ Environmental dimensions in quality assurance ▪ Minimization of resource and energy inputs	▪ Communication of environmentally oriented product advantages ▪ Adjustments of the basic marketing strategies and of the marketing mix to the overall product life cycle	▪ Value-added services in the environmental area ▪ Addition of repair service and replacement parts to the offer ▪ Securing of environmentally oriented functionality	▪ Disposal as "third dimension" ▪ Cooperation with competitors and customers ▪ Offer to take back products and of reprocessing systems	▪ Communication with the political system and the general public

Profit

Source: Steger 1996.

It should be noted that the product life cycle is a representation of the flows of materials and energy that accompany a product from the primary production of materials used in its construction to its ultimate end-of-life disposal, including its potential reincarnation as recovered parts or materials. Life-cycle analysis is a systematic framework to identify and account for (inventory) all of these materials and energy flows and their embodied environmental impacts.

Figure 34.1 provides a schematic design of typical impacts, where each of the arrows in the figure is the energy use during each phase and flow. The production of by-products and the use of secondary inputs within each phase are also included with the "product" streams. These materials may come directly from the environment or go through various production sequences and represent other product life cycles, demonstrating the interconnectedness of economic activity. Pollution represents any output that is not used as an input to another product life cycle. The polluting materials may be treated or released directly into the environment.

From this analysis it is clear that LCA includes three main parts—inventory, assessment, and evaluation— which are described in detail in Bohm and Walz (1996). The evidence is that real and perceived environmental impacts of products vary a great deal. For car companies, LCAs are attractive because they provide a heuristic for making and communicating materials choice decisions. For chemical companies, LCAs provide a way of defending existing markets and expanding into new ones. In general, applications of life-cycle approaches therefore differ greatly from sector to sector and from company to company (Angel and Berkhout 1996). In addition, Schaltegger (1997) has shown recently that the technique can lead to economically inefficient results if not used properly and, in particular, in the absence of site-specific environmental management.

From Integrated Product Development to Innovative Management

The above discussion focused on the analysis of the entire life cycle of existing products; now we turn to the related implications for the design of "green products." As defined by Ehrenfeld and Howard (1996), design for the environment is a systematic process by which firms design products and processes in an environmentally conscious way based on industrial ecology principles.

In this context, integrated product development results from life-cycle thinking, and is to be adopted as *a process where all functional groups (e.g., engineering, manufacturing, marketing, etc.) are involved in a product life cycle participating as a team in the early understanding and resolution of key product development issues including quality, manufacturability, reliability, maintainability, environment, and safety.* Figure 34.3 represents the internal elements of life-cycle management, which should be considered in terms of the economics of product development. For example, in the automotive and electronics industry, it has been shown that up to 80% of product life-cycle costs are committed during the concept and preliminary design stages, and that the costs of design changes increase steeply as a product proceeds into full-scale development and prototyping. The practical implication of this leads to the use of concurrent engineering to examine a design from multiple perspectives and to anticipate potential problems or opportunities.

Figure 34.3
Internal Elements of Life Cycle Management

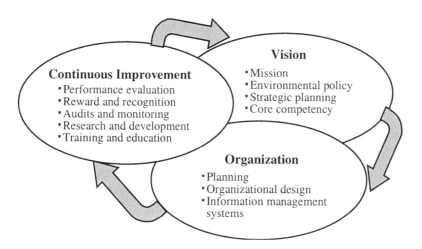

This analysis points up the need to develop innovative management practices which consider the ability to implement creative ideas and to move successfully in a new direction, taking the environment as a corporate business strategy. To date, the business logic for greening has been largely operational or technical, as described by Hart (1997): bottom-up pollution-prevention programs have enabled companies to make significant savings, in that greening has been framed in terms of risk reduction, reengineering, or cost cutting. Rarely is greening linked to strategy or technology development and, as a result, most companies fail to transform environmental opportunities into a major source of revenue growth.

A vision of sustainability for an industry or a company is like a road map to the future, showing the way products and services must evolve and what new competencies will be needed to get there. A contributing factor to achieve these objectives is the introduction of ecolabeling initiatives, which are to be integrated as a corporate business strategy, as represented schematically in Figure 34.4. While it remains unclear how consumers will respond to these initiatives, the potential marketing advantages of ecolabeling are of concern to companies and should be considered as drivers of new technologies. In this context, it is clear that a fully integrated environmental strategy should not only guide competency, but should also shape the companies, relationships with customers, suppliers, other companies, policy makers, and all its stakeholders. The policy is to use companies to change the way customers think by creating preferences for products and services consistent with sustainability.

Figure 34.4
A Vision for Strategic Management for the Environment

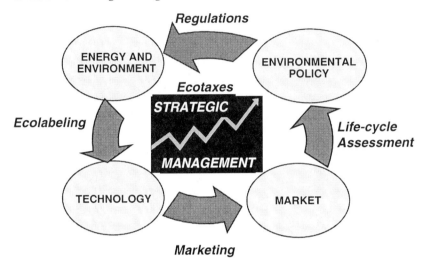

CONCLUSIONS

Significant opportunities exist to develop better tools for interpreting environmental data in policy-relevant terms, and, in this context, the use of environmental accounting should be performed in parallel to the life-cycle approach discussed in this chapter. In fact, the systematic use of *ex ante* tools, together with *ex post* analysis, should be included in long-term business strategies to achieve the current goal of sustainable development.

It is uncertain how effective the organizational changes which result from the management structure discussed here will be in practice, but it is likely that they will enjoy their greatest success in those countries where public awareness of environmental problems is high, particularly through systematic public participation.

REFERENCES

Angel, D. P., and D. P. Berkhout. "Building Sustainable Industries for Sustainable Societies," *Business Strategy and the Environment* 5 (1996) 127–136.

Bohm, E., and R. Walz. "Life-Cycle Analysis: A Methodology to Analyze Ecological Consequences Within a Technology Assessment Study," *International Journal of Technology Management Special Issue on Technology Assessment* 11, 516 (1996) 554–565.

Curran, M. A. *Environmental Life-Cycle Assessment.* New York: McGraw-Hill, 1996.

Ehrenfeld, J. R., and J. Howard. "Setting Environmental Goals: The View from Industries—A Review of Practices from the 1960s to the Present," in *Living Science and Technology to Society's Environmental Goals* (National Research Council). Washington, DC: National Academy Press, 1996, 281–325.

Fiksel, J. *Design For Environment.* New York: McGraw-Hill, 1996.

Franks, T. R. "Managing Sustainable Development: Definitions, Paradigms and Dimensions," *Sustainable Development* 4 (1996) 53–60.

Hart, S. L. "Strategies for a Sustainable World.," *Harvard Business Review* (January–February 1997) 67–76.

Hunt, C. B., and E. R. Auster. "Proactive Environmental Management: Avoiding the Toxic Trap," *Slower Management Review* 31, 2 (1990) 7–18.

Ketola, T. "Where Is Our Common Future? Directions: Second to the Right and Straight on Till Morning," *Sustainable Development* 4 (1996) 84–97.

OECD. *Integrating Environment and Economy: Progress in the 1990s.* Paris: OECD, 1996.

Pidgeon, S., and D. Brown. "The Role of Life Cycle Analysis in Environmental Management: A General Panorama or One of Several Useful Paradigms?" *Greener Management International* 7 (1994) 36–44.

Porter, M. *Competitive Advantage.* New York: The Free Press, 1985.

Schaltegger, S. "Economics of Life Cycle Assessment: Inefficiency of the Present Approach," *Business Strategy and Environment* 6 (1997) 1–8.

Steger, U. "Managerial Issues in Closing the Loop," *Business Strategy and the Environment* 5 (1996) 252–268.

Young, C. W., and P. M. Rikhandsson. "Environmental Performance Indicators for Business," *Eco-Management and Auditing* 3 (1996) 113–125.

35

Life Cycle Assessment:
A Tool for Innovation and Improved
Environmental Performance

Adisa Azapagic

INTRODUCTION

Life Cycle Assessment (LCA) is becoming an increasingly important decision-making aid in environmental system management. Its potential is being recognized not only by industry but also by policy makers and planners, educators, and others. LCA is still a relatively new environmental management tool and as such is associated with a number of unresolved issues and problems. Although the methodology is still developing, LCA has already found a number of different applications. Some of these include assessing the environmental impacts of consumer products such as packaging, washing machines, detergents, and the like. The results of these studies are aimed at helping consumers to choose a more environmentally friendly product from a group of equivalent products. Other applications are related to identifying possibilities for improvements in the environmental performance of an existing product or process or for the design of new ones. These studies are usually industry-specific and are mainly used for product or process innovation as well as for demonstrating the environmental progress of the company. Other examples include the use of LCA by governments for public policy making, for instance in ecolabeling schemes.

Although there is a growing need to adopt the LCA approach in environmental management and policy making, the wider acceptance and use of

LCA are still to come. In the meantime, a number of obstacles have to be overcome. Some of these, together with possible ways to deal with them, are discussed below.

LIFE CYCLE ASSESSMENT

Background

Life Cycle Assessment is a technique for assessing the environmental performance of a product, process, or activity "from cradle to grave"—that is, from extraction of raw materials to final disposal. Today's LCA originates from "net energy analysis" studies, which were first published in the 1970s (e.g., Boustead 1972; Sundstrom 1973) and considered only energy consumption over a life cycle of a product or process. Some later studies included wastes and emissions (Hunt and Franklin 1974; Ayres 1978; Boustead 1989), but none went further than simply quantifying materials and energy use. Most of the early LCA studies considered only packaging; it was only at the beginning of the 1990s that LCA started to be applied to different consumer products, as well as to chemicals, agricultural products, means of transportation, and so on. As the use and importance of LCA increased, it became apparent that a more standardized approach to conducting LCA studies was needed.

As a result, in 1990 the Society for Environmental Toxicology and Chemistry (SETAC) initiated activities to define and develop a general methodology for LCA. Soon afterward, the International Standards Organization (ISO) started similar work on developing principles and guidelines for LCA methodology. Although SETAC and ISO worked independently of each other, they reached a general consensus on the methodological framework. It is expected that both bodies will produce the final documents on an internationally standardized LCA methodology by the end of 1997.

Methodology

Life Cycle Assessment, as defined by SETAC, is "a process to evaluate the environmental burdens associated with a product, process, or activity by identifying and quantifying energy and materials used and wastes released to the environment; to assess the impact of those energy and material uses and releases to the environment; and to identify and evaluate opportunities to effect environmental improvements" (Consoli 1993). As illustrated in Figure 35.1, it follows the life cycle of a product, process, or activity from extraction of raw materials to final disposal, including manufacturing, transport, use, reuse, maintenance, and recycling.

Figure 35.1
Stages in the Life Cycle of a Product

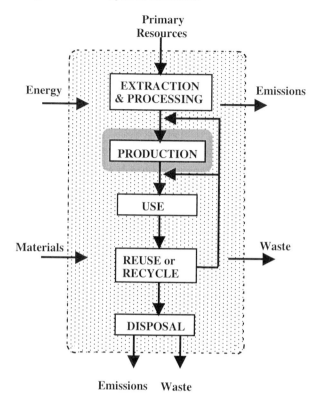

Emissions   Waste

The methodological framework for conducting LCA, as defined by both SETAC and ISO, comprises four main stages. The two approaches are compared below:

SETAC	ISO - 14040
1. Goal Definition and Scoping	Goal and Scope Definition (ISO 14041)
2. Inventory Analysis	Inventory Analysis (ISO 14041)
3. Impact Assessment	Impact Assessment (ISO 14042)
4. Improvement Assessment	Interpretation (ISO 14043)

Except for the final stage, which will be briefly commented on later in the text, the methodological frameworks are similar.

In the first stage, Definition of Goal and Scope, the boundaries of the system are defined. This stage also identifies the functional unit(s) of the system as a measure of the function that the system delivers. For instance, the function of packaging may be to store a certain amount of liquid. If different packaging is to be compared, then the comparison should be based on an equivalent function. Therefore, the functional unit can be defined as the amount of packaging needed to store a certain amount of liquid.

The second stage, Inventory Analysis, identifies and quantifies the environmental burdens associated with the system. This stage is in fact related to performing material and energy balances. The burdens are defined by resource consumption, emissions to air and water, and solid waste. Aggregation of the burdens into a smaller number of impact categories, such as global warming potential or acidification, is part of Classification, while evaluation of their potential impacts is part of the Characterization step within the Impact Assessment stage. Further aggregation of the impacts in the Valuation stage into a single environmental impact function, by attaching weights of importance to the impacts, is also a part of Impact Assessment. This is perhaps the most controversial stage of LCA, because it implies subjective value judgments in deciding on the importance of different impacts.

The final stage in the SETAC methodology is called Improvement Assessment and is aimed at identifying the possibilities for improving the performance of the system. ISO, on the other hand, decided to call this stage Interpretation and it is claimed to be quite different from Improvement Assessment (Saur 1997). The Interpretation stage is also aimed at improvements and innovations, but in addition covers the following steps: identification of major burdens and impacts, identification of stages in the life cycle that contribute most to those impacts, evaluation of these findings, sensitivity analysis, and final recommendations. In fact, this stage does not bring anything new to the methodology; it only redistributes the activities that used to be done within the other stages.

Uses of LCA

Because of its holistic approach to system analysis, LCA is becoming an increasingly important decision-making tool in environmental system management. Its main advantage over other, site-specific, methods for environmental analysis, such as Environmental Impact Assessment or Environmental Audit, lies in broadening the system boundaries to include all burdens and impacts in the life cycle of a product or a process, and not focusing on the emissions and wastes generated by the plant or manufacturing site only.

As an environmental management tool, LCA has two main objectives. Its first objective is to quantify and evaluate the environmental performance of a product or a process and so help decision-makers choose between alternative products or processes. Another objective of LCA is to provide a basis for assessing potential improvements of the environmental performance of an existing or newly designed system. This can be of particular importance to engineers and environmental managers, because it can advise them on how to modify or design a system in order to decrease its overall environmental impact.

LCA can thus be used both internally by a company or externally by industry, policy makers, planners, educators, and other stakeholders. If the results of LCA are to be used internally by a company then possible areas where LCA can be useful include, but are not limited to, the following:

- strategic planning or environmental strategy development;
- problem solving in the system;
- product and process design, innovation, improvement, and optimization;
- identification of opportunities for, and tracking of, environmental improvements.

External applications of LCAs include uses of LCA as a marketing tool, to support environmental labeling or claims, for educational or informational purposes, or to support policy decisions. Some of these uses of LCA are discussed in the rest of this chapter.

LCA FOR PRODUCT AND PROCESS INNOVATION

One of the internal applications of LCA is in product or process design and development. More recently, a new LCA-related tool called Life Cycle Product/Process Design (LCPD) has started to emerge as an extension of life cycle thinking and as an aid in product or process design and development. The LCPD procedure is outlined in Figure 35.2. Life Cycle Assessment is used throughout the development procedure, initially with reference to an existing product or process.

Figure 35.2
Life Cycle Product or Process Design as a Tool for Innovation

Once the main environmental impacts have been quantified, potential improvements are identified and the subsequent design is focused on these. Improvements are then achieved through the selection of materials and technologies so as to minimize environmental impacts but still to satisfy other parameters, such as technical performance, costs, legislation, and customers and suppliers. Once all of these requirements have been met, LCA is performed again to identify and quantify the improvements made. This whole process is iterative, with a continuous exchange of information between the stakeholders, and yields a number of possibilities for improvements.

Thus LCPD offers potential for technological innovation in the product or process concept and structure through the selection of the best material and process alternatives over the whole cycle. This can be of particular importance if placed in the context of EMAS standards of the European Union (EU) and EMS of ISO 14000, which require companies to have full knowledge of the environmental consequences of their actions, both on and off site.

LCA FOR IMPROVED ENVIRONMENTAL PERFORMANCE

Other uses of LCA include improving the environmental performance of an existing product or process. This is a part of the Improvement Assessment (or Interpretation) stage and involves identification, evaluation, and selection of the best available options. In many cases there will be a number of different improvement options and it may not always be obvious which one will provide the best solution for the system. It may also happen that it is not a single option but a combination of several that gives the optimum improvements in the system. The problem is further complicated by the number of different burdens or impacts that need to be considered simultaneously and by the fact that improvements in some of them will often mean deterioration in others. Therefore, to deal with this and other complexities often encountered in LCA, it is necessary to use system optimization in LCA as an aid for identifying optimum solutions for improvements. Because of the multiobjective nature of the problem in which optimum solutions are sought for a number of often conflicting objectives, it is necessary to use multiobjective optimization whereby the system is simultaneously optimized on a number of environmental objective functions subject to the constraints of the system (Azapagic 1996; Azapagic and Clift 1995, 1996a, 1996b).

In LCA terms, the objectives are defined by resource usages, emissions to air and water, and solid waste:

$$\text{Minimize} \qquad B_k = \sum_{i=1}^{I} bc_{k,i}\, x_i \qquad k=1,2,...,K \qquad (1)$$

where $bc_{k,i}$ is burden k from process or activity x_i. The objective functions can also be the environmental impacts:

Minimize $$E_l = \sum_{k=1}^{K} ec_{l,k} B_k \qquad l=1,2,...,L \qquad (2)$$

where $ec_{l,k}$ represents the relative contribution of burden B_k to impact E_l, as for instance defined by the "problem-oriented" approach to Impact Assessment (Heijungs 1992). In this approach, for example, Global Warming Potential (GWP) factors, ecl,k, for different greenhouse gases are expressed relative to the GWP of CO_2, which is therefore defined to be unity.

Depending on the goal of the study, the system can be optimized on either burdens or impacts. The optimization problem is then to simultaneously minimize the objective functions defined by equations (1) or (2) subject to the constraints on the system:

$$\sum_{i=1}^{I} a_{j,i} x_i \le A_j \text{ and } x_i \ge 0 \qquad j = 1,2,...,J; i = 1,2,...,I \qquad (3)$$

where $a_{j,i}$ is an input or output coefficient of an activity x_i. Definition of the optimization problem in LCA is therefore equivalent to that of the conventional optimization problem. However, the difference is that in the model formulation in the LCA context, the system boundary is extended to include all activities from extraction of raw materials to disposal. Furthermore, in conventional optimization the objective functions are normally defined as a measure of economic performance, while in the LCA context they are defined as environmental burdens or impacts.

The system can be optimized on all the environmental functions simultaneously in order to find a number of environmental optima of the system. It is important to emphasize at this point that objective functions do not have to be aggregated into a single environmental impact function by attaching weights to them to indicate their relative importance. In this way, the controversial issue of evaluating the importance of different environmental impacts is avoided. The environmental optima obtained define the multidimensional noninferior or Pareto surface. By definition, none of the objective functions at the Pareto optimum can be improved without worsening any other objective function. Therefore, some trade-offs between objective functions are necessary in order to reach the preferred optimum solution in a given situation. For example, if emissions of NOx and SO_2 are optimized simultaneously, the resulting Pareto optimum does not necessarily mean that these functions are at minimum achieved when the system is optimized on each of them separately. The Pareto optimum does, however, mean that the set of best possible options has been identified for a system in which both emissions should be reduced.

The value of multiobjective optimization in LCA therefore lies in offering a range of alternative solutions; they are all optimal, but the choice of the best one will depend on a range of technical, financial, environmental, and social factors. Consequently, system improvements cannot be carried out on the basis of environmental LCA only—these other factors have also to be incorporated into the model. Thus additional objective functions are identified and the system is optimized on all of them, yielding an n-dimensional Pareto surface with optimum solutions. An illustration of this approach is shown in Figure 35.3.

Figure 35.3
Multiobjective Optimization for Improved Overall Performance

ECONOMICS

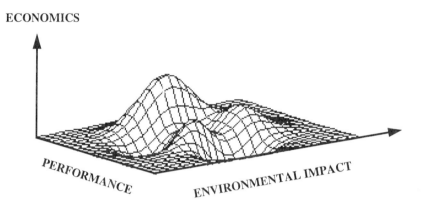

For a graphical representation, it is assumed that the system is to be optimized on three objective functions, defined as environmental impact, overall technical performance, and total costs to the system. A simultaneous optimization on these three objectives will result in a three-dimensional Pareto surface, which may take the shape of that shown in the figure. By definition, all solutions on the surface are optimal. By moving along the surface, the trade-offs between different objectives can be established in order to find out how much of one objective has to be given up to gain in the other. In this way, acceptable solutions, representing a compromise between conflicting objectives, can be found. This approach thus enables the choice of the Best Practicable Environmental Option (BPEO) and Best Available Technology Not Entailing Excessive Cost (BATNEEC).

The advantage of multiobjective over single-objective optimization is that the latter provides one solution only, which may be optimum but not appropriate for a particular situation. Decision-makers like to decide, and multiobjective optimization enables them to do so. It also helps them understand the gains and losses associated with each option, and therefore makes them more likely to compromise in conflicting situations. Thus, coupling LCA with multiobjective system optimization can simplify the decision-making process and so help overcome one of the many obstacles to wider use of LCA.

There are a number of optimization techniques that can be used in LCA. They include, for instance, linear programming (LP), mixed integer linear programming

(MILP), nonlinear programming (NLP), and mixed integer nonlinear programming (MINLP). The software for these techniques is readily available for personal computers and only basic programming knowledge is necessary to use them efficiently, which makes optimization techniques more accessible to LCA practitioners.

LCA FOR PUBLIC POLICY

LCA has so far mainly been used by companies for either internal or external purposes, as discussed above. Governments have, however, been much slower in adopting LCA as a tool in policy making. Although there have been a number of attempts to incorporate LCA in public decision making worldwide, LCA has still not become an integral part of this process. Many governments are still concentrating on reducing pollution to one medium only or from one life cycle stage only. One such example is the definition of the best practicable environmental option (BPEO), which is aimed at reducing pollution from a particular process or a plant without considering the effects of that option on the other parts of the life cycle. Recent research on the BPEO for the end-of-pipe removal of SO_2 (Vyzi and Azapagic 1997) and NOx emissions (Yates 1996) has demonstrated that this narrow thinking can only transfer burdens from one life cycle stage to another. These findings confirm that if a real BPEO is to be chosen, it has to be assessed in the LCA context.

There is thus an increasing need to incorporate LCA in strategic planning and governmental policy development. One of the attempts to integrate life cycle thinking into policy making is related to various ecolabeling schemes in a number of countries. Some examples include the German "Blue Angel" and the EU ecolabel, which are supposed to help consumers identify the most environmentally friendly product from a group of equivalent products. Similar schemes exist in other EU countries, such as France, The Netherlands, and the United Kingdom. However, although these schemes promote the use of LCA in decision making, they only concentrate on a limited number of burdens and usually on only one or two stages in a life cycle. Nevertheless, this approach is still preferred to single-issue considerations, related normally to the recyclability or biodegradability of the product.

Other policy-making schemes that are starting to use the LCA approach are related to taxation on pollution. One such example is France, which introduced a tax on CO_2 emissions based on the results of an LCA study (OECD 1995). The Belgian and Norwegian governments also have programs to introduce taxation on packaging, the information for which was obtained through life-cycle studies. In the United States, the use of LCA in policy making is being encouraged by the Environmental Protection Agency through various projects, such as establishment of subsidies or tax credits for alternative fuels and using a life-cycle approach in developing a maximum achievable control technology (MACT) standard under the 1990 Clean Air Act (Curran 1997).

Although these examples represent only the beginning of the application of LCA in policy making, they are nevertheless important developments because they demonstrate that governments are starting to consider broader, life cycle thinking and are prepared to integrate it into the decision-making process. However, before LCA is adopted more widely for the formulation of public policy, a number of obstacles have to be overcome. The first and probably the most important is the international standardization and harmonization of a methodology which would provide decision-makers with guidelines on how to perform an LCA study and how to interpret its results. The complexity of the LCA approach is another barrier to its adoption by nonspecialists, and guidelines on how to conduct simplified LCAs are therefore necessary. With the work of ISO and SETAC and their respective standards for the methodology, this problem may be resolved. The standards will also help to restore the credibility of LCA and reduce possibilities for misuse. This could help refute claims articulated by some groups that LCA is a useless or dangerous tool.

However, other problems still remain, one of them being data availability and reliability. There are many examples of LCA studies where unreliable and inaccurate data were used, which have contributed to damaging the credibility of the whole concept. This problem may be overcome by increased public awareness of the need to use LCA in policy making and by pressures on companies and industries, from both consumers and governments, to make the data available. This process has already started, which is demonstrated by a number of international databases available in the public domain.

Other kinds of obstacles are related to governments themselves and their ability and willingness to adopt an LCA approach in public policy making. The many layers of bureaucracy and office segmentation found in governmental organizations do not encourage the introduction of new approaches in decision making, particularly if these are seen to be complex and do not give the expected answers. In addition, lack of planning and failure to integrate other factors, such as economic and societal, of the decision-making process with environmental considerations is another major barrier to the adoption of LCA by governments (Christiansen et al. 1995).

It is therefore important to increase the awareness of the public and governments with respect to LCA. It is necessary to make these groups aware through a concerted educational process of the usefulness of LCA, as well as of its complexity and potential dangers and drawbacks. However, the effort to understand and adopt LCA has to be made by all the stakeholders involved, be they LCA practitioners, industry, public, or government. On the other hand, if this is to happen, then further development of LCA has to be directed toward meeting the needs and expectations of these stakeholders. It is only then that we can expect this approach to become fully integrated into the decision-making process.

CONCLUSIONS

There is a growing need to move away from narrow definitions and concepts in environmental system management. Life Cycle Assessment offers the potential to take on broader, life-cycle thinking and incorporate it into strategic planning and policy development. This is demonstrated by an increasing number of applications of LCA, by both industry and governments. Some examples include the use of LCA for design of new products or processes or improvements to existing ones. Others are related to various ecolabeling schemes or taxation on pollution.

However, before LCA can become a more widely used tool, a number of obstacles have to be overcome. It has to become more accessible and easier to use and understand. One of the ways to deal with this problem is through the standardization of the methodology and increased data availability. Other difficulties are related to the complexity of systems in LCA and its multiobjective nature. To aid the decision-making process, this chapter proposes to couple LCA with multiobjective optimization techniques, such as linear or nonlinear programming. The value of multiobjective optimization in LCA lies in offering a range of alternative solutions rather than a single solution which may be optimum but not appropriate for a particular situation. This approach also enables consideration of other factors, such as technical, financial, and societal, which should all be an integral part of decision making.

Finally, if LCA is to become integrated in public policy making, the concerted effort of all stakeholders is necessary. However, for this to happen, further development of LCA has to be directed towards enabling it to meet the needs and expectations of all stakeholders in the process.

NOMENCLATURE

A_j	Right-hand side coefficients in constraints
$a_{i,i}$	Input/output coefficients
B_k	Environmental burdens
$Bc_{k,i}$	Environmental burden coefficients
E_l	Environmental impacts
$ec_{l,k}$	Environmental impact coefficients
x_i	Activity (operation level of process)

REFERENCES

Ayres, R. U. "Process Classification for the Industrial Material Sector," Technical Report. New York: United Nations Statistical Office, 1978.

Azapagic, A. "Environmental System Analysis: The Use of Linear Programming in Life Cycle Assessment." Ph.D. thesis, University of Surrey, U.K., 1996.

Azapagic, A., and R. Clift. "Life Cycle Assessment and Linear Programming: Environmental Optimization of Product System," *Computer and Chemical Engineering* 19, Supplement (1995) 229–234.

Azapagic, A. & R. Clift. "Environmental Management of Product System: Application of Multi-objective Linear Programming to Life Cycle Assessment," *Proceedings of the 1996 ICheme Research Event* 2 (1996a) 558–560.

Azapagic, A., & R. Clift. "Application of Multi-objective Linear Programming to Environmental Process Optimization," AIChE 1996 Spring National Meeting, New Orleans, February 25–29, 1996b.

BPEO," *Proceedings of EngD Conference*, Guildford, U.K., (October 1996) 45/1–45/10.

Boustead, I. *The Milk Bottle*. Milton Keynes, U.K.: Open University Press, 1972.

Boustead, I. *Environmental Impact of the Major Beverage Packaging Systems: U.K. Data 1986 in Response to E.E.C. Directive 85/339*. London: INCPEN, 1989.

Christiansen, K., R. Heijungs, T. Rydberg, S.-O. Ryding, L. Sund, H. Wijnen, M. Vold, and O. J. Hanssen (eds.). "Application of Life Cycle Assessments," *Proceedings of the Workshop* held in Hankø, Norway, March 27–31, 1995.

Consoli, F. (ed.). *Guidelines for Life-Cycle Assessment: A "Code of Practice."* Brussels: SETAC, 1993.

Curran, M. A. "Life-Cycle Based Government Policies," *International Journal of Life Cycle Assessment* 2, 1 (1997) 39–43.

Heijungs, R. (ed.). *Environmental Life Cycle Assessment of Products: Background and Guide*. Leiden: MultiCopy, 1992.

Hunt, R. G., and W. E. Franklin. *Resources and Environmental Profile Analysis of 9 Beverage Container Alternatives*. Contract 68-01-1848. Washington, DC: Environmental Protection Agency, 1974.

OECD. *Life Cycle Summaries of OECD Countries*. January 1995.

Saur, K. "Life Cycle Interpretation: A Brand New Perspective?" *International Journal of Life Cycle Assessment* 2, 1 (1997) 8–10.

Sundstrom, G. "Investigation of the Energy Requirements from Raw Materials to Garbage Treatment for 4 Swedish Beer Packaging Alternatives." Report for Rigello Park AB, Sweden, 1973

Vyzi, E., and A. Azapagic. "Life Cycle Assessment as a Tool for Identifying the Best Practicable Environmental Option (BPEO)," University of Surrey, U.K, 1997.

Yates, T. "Life Cycle Analysis as a Tool for Assessing

36

Strategies for Social Sustainability Through the Built Environment: A Comparative Study

Teresa Valsassina Heitor

INTRODUCTION

Nowadays, demographic trends, including population, household sizes, and space occupancy, point toward ever-increasing pressures in urban areas. Cities and towns are thus growing in population, as well as spreading out and changing their structures into more widespread urban patterns, which synthesize and make visible a range of social, economic, and environmental conflicts. Social deprivation and isolation, unemployment, poor housing quality, inadequate urban service provision, high levels of environmental damage, and run-down housing estates in core areas are increasingly manifesting themselves as problems of urban concentrations, affecting especially, but not only, less privileged areas.

Consequently, urban areas are where the problems of the built environment most impact the quality of life of citizens who demand an understanding of its vulnerability in which social factors and dynamics of cities are integrated.

This chapter is organized as follows: it begins with the discussion of both the concept of social sustainability through the built environment and the processes for achieving and assessing it. In the following section a methodology to evaluate the degree of social sustainability of an urban area is proposed. The chapter continues with the review of two case studies. The main factors that contribute to the conception, formalization, and consolidation of these urban structures are identified and their morphological and configurational expressions are described and analyzed. It is concluded that there are some specific spatial attributes that contribute to bring vulnerability to built space. These variables interfere in the way space is used, affecting the capacity of social control and natural surveillability necessary for a socially sustainable urban setting.

"SOCIAL SUSTAINABILITY" THROUGH THE BUILT ENVIRONMENT

Whether we consider the massive physical and environmental challenge posed today by the renovation and restructuring of the existing cities or the continuing drive for economic growth, there can be little doubt that so-called traditional concerns about economic viability and the importance of meeting the needs of less-privileged social groups in cities have changed the agenda of those concerned with promoting urban change and growth.

Urban managers and planners, urban designers, and other pressure groups have nevertheless come to identify "sustainability" as a useful concept to guide strategy and to encourage national and local authorities to give prominence to long-term considerations—that is, to ensure that what is done now does not negatively affect what future generations may wish to do.

Sustainability has become an important goal in world society today. The United Nations Conference on Environment and Development (UNCED) in Rio de Janeiro put it on the agenda and the concept is now discussed and promoted worldwide. Although it is recognized within the literature as a contested concept, difficult to define in precise and concrete terms (see, for example, Shiva 1993; Myerson and Rydin 1996; Mathews 1996), it is most certainly concerned about excessive use of finite resources and the efficient management of the ecosystem (e.g., greenhouse gases, storm water pollution, efficient food production) as well as fundamental aspects of social equity and justice.

As far as urban managers and planners are concerned, sustainability is intended to reflect a policy and strategy for continued economic and social development without detriment to the urban environment, the quality of which continued urban activity and further development depend.

On a smaller scale and more pertinent to the role of the urban designer, sustainability has been adopted to reflect an attitude concerned with the social concept of design—that is, to provide qualified mechanisms for developing design strategies which will respond effectively to public demand. Based on the assumption that particular social processes or patterns of social life could have been embedded in the very spatial layouts of towns and cities, sustainability becomes about structuring town form to support security, health, and social responsibility while minimizing social malaise (Engwicht 1992; Hayward and McGlynn 1993; Hillier 1996; Klarkvist 1996). This means that the individual has choice but never at the expense of the collective, thus enabling as many of the citizens as possible to successfully determine the outcome of their daily lives in so far *as the layout of the town and the location of uses can assist*. In what follows, we refer this strategy, without loss of generality, as "social sustainability through the built environment."

But how do we achieve this in physical terms? The current debate on urban design strategies provides evidence of the need for an efficient long-term allocation of resources (both built environment and economic resources are included here) that maximizes social welfare by *optimizing the land use pattern with the way people use the space*.

It is generally agreed that utilization of space is concerned with an efficient use of space. Deserted spaces seldom contribute in an effective way to urban life, since lack of interaction and awareness in public areas can make spaces vulnerable and unsafe. Thus, investments in space should be well used by people. In other words, there is an attempt to establish some lawful relations between urban spatial form and social life. Such relations should permit us to approach the built environment in such a way that not only urban layouts with a great variety of patterns can be analyzed and compared in a purely formal sense by means of descriptive techniques, but the functional implications of different layout designs for social use could be also systematically accounted for against formal (spatial) analysis.

On the important question of how the relationship between urban spatial form and its various functions can be studied, such that the results from research may display more immediacy to design knowledge, there is an apparent lack of perspective and methods. Implicit in this question is the need to develop analytic instruments which permit: (1) a rigorous and nonarbitrary description of the urban form; and (2) a perspective for conceptualizing the relationship between urban spatial form and its social aspects.

There have been serious efforts made from architectural disciplines and social sciences to develop conceptual frameworks and methods for dealing with spatial analysis in the field of space-use (environment-behavior) studies. However, the architectural research has been mostly engaged in the detailed description and characterization of urban spatial form,[1] while the social sciences have almost always been concerned with studies characterized above all by their "environment-behavior relationship." As pointed out by Tecklenburg et. al. (1996), in practice one must admit that these studies, strong in describing and analyzing patterns of spatial behavior (or space use) and relating these to sociodemographic characteristics of populations, have been largely ineffectual when physical environment is conceived less as a social space and more as a physical setting or system.

However, this may not be true at all levels, and the Space Syntax methodology is a notable exception (Hillier and Hanson 1984). This approach provides us with a descriptive method of urban form as well as with a formulation of a perspective which accounts for the precise social implications of urban space (Hillier et al. 1987; Hillier 1989, 1996).

The Space Syntax method includes two basic parts: (1) a model postulating a settlement, or a part of a settlement, as an interfacing system; and (2) a set of concepts and techniques for representing the spatial structure of settlement plans. The essence of this method lies in that "it first establishes a way of dealing with the global physical structure of a settlement without losing sight of its local structure; and second—a function of the first—it establishes a method of describing space in such a way as to make its social origins and consequences a part of that description" (Hillier and Hanson 1984: 82).

The Space Syntax concept of urban spatial form can be summed up in two basic arguments: (1) the spatial configuration of an urban setting is considered capable of creating and sustaining a field of probable copresence and encounter, mainly through generating patterns of pedestrian movements; and (2) this spatial form-generated field of encounter has a describable structure, which varies with the syntax of the layout.

The fundamental Space Syntax assumption is that the distribution of pedestrian activity in urban areas is a function of the syntactic structure of layouts—that is, a spatial phenomenon (Hillier et al. 1987; Peponis 1989; Hillier and Penn 1992). Mainly through generating a field of pedestrian movement, the morphological structure of an urban setting is creating in its layout certain background conditions for people to encounter and become aware of one another's presence. Through this, people are able to naturally experience as well as participate in the every life taking place in a local area.

The results from the applications of research using Space Syntax theoretical framework are encouraging since they have demonstrated the potential for enhancing the quality of the environment for a range of situations. In particular it appears that many urban areas can be made more livable through designed circulation patterns that have the potential to improve levels of social interaction. Space Syntax techniques also permit the evaluation of the extent of functional capacity of a particular space embedded in the surrounding urban area. As Jones and Fanek (1997) stressed, if this is correct, such improvements may have further potential—that is, to create and modify an urban setting to more certainly achieve social sustainability.

Geographers, town planners, and traffic engineers also seem quite capable of describing the regional or urban environment in a way that satisfies their perceived need for objectivity. Traffic engineers in particular know how to describe road networks, classify the different types of roads, and relate the network to vehicular traffic. Even on a smaller scale, they are able to classify neighborhood road networks—for example, using graphlike representations. Geographical techniques for describing the urban environment may be quite sophisticated, but more often are derived from aggregate social behavior than from spatial behavior.[2] Batty and Longley (1987) describe urban forms using fractals, while White et al. (1993) use cellular automata to model urban land use dynamics. Batty and Longley describe the form of the urban boundary, with the intention of providing means for describing the formation of areas and urban edges, in the case using the fractal dimension, which may then be related to the way space is used. Similarly, the spatial patterns of land use, as described by White et al. are also a product of social uses.

ANALYTICAL STRATEGY FOR ASSESSING SOCIAL SUSTAINABILITY THROUGH THE BUILT ENVIRONMENT

In this section, an analytical strategy developed to describe urban spatial layouts is proposed, such that a description of their structural properties may in turn enable us to address the social implications of specific land use functions. This is based on the interaction between spatial analysis techniques and the study of the relation of the use of space.

The overall strategy proposed is one which, through a rigorous analysis of the physical objects (urban settings) seeks to establish links between design intentions, urban layout (land use), and patterns of space use. Figure 36.1 is a conceptual model sketched to illustrate this strategy.

Figure 36.1
Model Describing the Proposed Strategy

The box on the left side of the diagram refers to the various intentions and strategies underlying a plan design. The middle box is concerned with the end product of a design—that is, the complex spatial layout or configuration. The box on the right demonstrates the desirable or actual effect of a plan design.

The main considerations underlying this method for case study analysis are based on the belief that research should contribute to the generation of knowledge, which will both display general theoretical significance for related fields of study, and still have immediate implications for design practice. The case studies are presented in a way to make the structures and functional implications of these patterns both abstractly comparable and available to intuition. The procedure adopted is illustrated in Figure 36.2 and is described below.

Figure 36.2
Diagram Model Sketched to Illustrate the Procedures Adopted

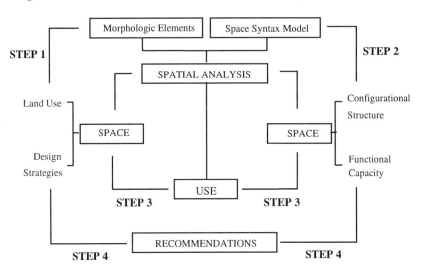

Step 1.   The spatial layout is subjected to a detailed description and characterization by means of its land use pattern. It is first represented as a collection of land uses or functions and described according to its basic morphologic elements and the way they are disposed and articulated in relation to each other. Thus, the plan is first represented as being composed by a built area and an open area and then characterized according to its design strategies.

Step 2:   Space Syntax concepts and techniques are applied. The configurational properties are identified and used to describe the abstract structure of the site plan and to assess its functional capacities by linking it to data collected in step 1. The implications which such structure may have for patterns of space use are subjected to due interpretation.

Step 3:   Spatial properties revealed by configurational analysis are used as a guide for making systematic observations of the way spaces are used. Data obtained from such observations are statistically correlated to the spatial properties (assessed in steps 1 and 2) to search for any pattern of covariation, thereby deciding if and how far the functional implications of the spatial layout can be captured by spatial analysis.

Step 4:   Based on the previous analysis, recommendations are made. When corrections to the urban layout are suggested, the proposed layout should be tested again following step 2.

The characterization of the spatial layout in terms of principal land uses is intended to provide an intermediate level which mediates between two different plan representations: a neighborhood plan, which exists as a set of graphic data and which could be understood only in those terms, and a plan which is basically represented by a set of numerical measures resulting from step 2.

The description of spatial structure using Space Syntax methodology is based on the axial map, which is a graphical transformation of the layout. This object allows a quantitative analysis of the spatial layout. It is a planar connected configuration consisting of the fewest longest straight lines covering all public urban spaces. These lines correspond to the image of physical and visual continuity tested by those who are static or in movement in the system.

The axial maps provide information about the pattern of connections between spaces (the way lines are distributed on the plane) and the connections of each space to all other spaces (the intersections of each line with all other lines). No information is given on areas and distances. To make an axial map suitable for computation, it must be converted into a graph in which vertices correspond to lines. Two vertices are adjacent if and only if the corresponding lines of the axial map intersect.

The axial map produces two types of output: numeric data in the form of line numbers, which is the basis for deriving measures of properties of the configuration, and graphic data in the form of core maps, which produce a standard understanding of the spatial pattern of the area. The key variable in axial analysis is depth (also referred to as integration). The depth or integration of a space v is the sum of distances from v to all other spaces. Put simply, integration measures the degree of accessibility of a space in relation to all others that constitute the spatial structure. It is designated as a global property. The way spaces are related to adjacent ones is expressed by local measures: connectivity measures the number of spaces directly connected to a space; control value expresses the degree to which a space is better or worse connected to other spaces in the system; and radius 3 refers to the (local) accessibility of a space considering the spaces disposed around a diameter comprising 6 spaces.

Space Syntax analysis is applied using the Axman software developed at the Bartlett School of Architecture and Planning, University College, London.

CASE STUDY

The objects of this case study were two housing areas—sites A and B—in the east side of the city of Lisbon. These areas were located in Chelas, an area of 510 hectares (approximately 1,080 acres), planned during the 1960s for 55,300 inhabitants. They were intended to serve different uses, although their major function was residential for a variety of income groups. The plan was subdivided into six major areas, each one with its own identity integrated by a multilevel center of dominant architectural form. Figure 36.3 illustrates the location of the selected study area in relation to the city using an axial map, and Figure 36.4 is a schematic plan of Chelas with reference to the selected areas.

Both sites were developed during the 1970s. They were an attempt to break through the Athens chart zoning and architectural principles in that they sought to recover the notion of streets. The urban design concept of treating blocks as part of the structure of the layout and not as single entities is also present, as well as the idea of treating areas inside blocks as extensions of the street in organizational terms.

These sites are stigmatized by their social and economic appearance due to high proportions of low-income people, raised levels of unemployment, and the presence of minorities with strong cultural identities. They have also gained a strong negative impact in the face of their physical features. Although they have been built according to accepted space standards and modern amenities, they have deteriorated and are suffering from poor or dated construction quality, disrepair, graffiti, vandalism, rubbish dumping, and high crime levels, showing high levels of spatial vulnerability. These facts, together with the peripheral and quite segregated location of the estates in relation to the rest of the city, are generally pointed out as the main reasons for the stigmas they have developed.

Figure 36.3
Axial Map of Lisbon Illustrating the Study Area

CHELAS

Source: Kruger, Heitor, and Tostões 1996

Figure 36.4
Map of Chelas with Reference to Study Sites

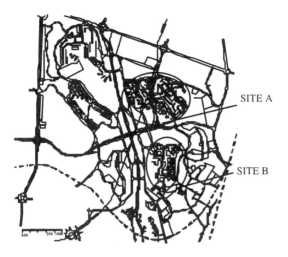

Source: Kruger, Heitor, and Tostões 1996

The main objective of the study was to understand the implications of the design layout in the way space is used, in particular considering the levels of environmental damage. The methodology outlined in the previous section was applied.

The urban layout was first decomposed in morphologic elements and described according to the design strategies: patterns of space accessibility (vehicular, nonvehicular; peripheral or internal to blocks); patterns of block permeability (unconstituted/blind facades versus constituted/direct dwellings or shops/indirect); patterns of block access (direct/indirect); and patterns of visibility (exposed versus enclosed/tunnel/recessed). Then it was submitted to a configurational description using Space Syntax methodology. The configurational structure was analyzed considering the area embedded in the surrounding urban fabric—Chelas—and on its own.

Space use was characterized with reference to the way people locate and informally consume space, as well as to physical marks denoting environmental damage. People, whether they were static or in movement, were recorded according to gender, age, activities, and relation to the site—that is, if they were residents or not. Environmental damage was classified according to four invariants previously tested by Coleman (1985): litter, graffiti, vandalism, and excrement.

Site A occupies a rather central position in the overall layout of Chelas, thus enhancing its spatial and functional integration in the surrounding urban fabric. As the spatial pattern is characterized by a high level of fragmentation, the potential integration of the site in a larger context is minimized. As observed in Figure 36.5, the most highly integrated spaces of Chelas only embrace the site without penetrating and reaching its internal parts. This contributes to the difficulty in crossing the peripheral spaces and reaching the interior of the site.

Figure 36.5
Corresponding Axial Graph with Reference to its Core Area

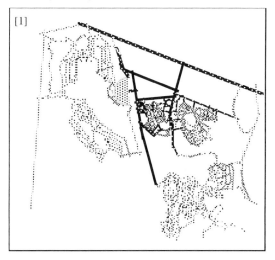

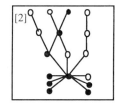

[1] Axial map of Chelas—integration "core" (25% of most integrated spaces)

[2] Axial-graph of Chelas
○ Integration "core" of Chelas
● Integration "core" of site A embedded in Chelas

[3] Axial map of site A (not integrated in Chelas)—Integration "core" (25% of most integrated spaces)

[4] Axial-graph of site A
● Integration "core" of site A
○ Spaces with "strong" local properties: high levels of connectivity (cl), control (c2), and radius 3 (R3).

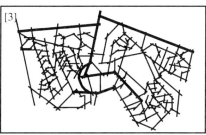

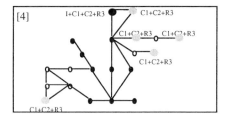

When analyzed on its own, the configuration suggests a structure mostly defined by global principles: the most globally integrated lines spread along the site are responsible for promoting the continuity of the grid. Locally integrated spaces tend to overlap the most globally integrated ones, giving shape to a central area with strong capacity to attract people and thus sustain commercial activities. Spaces with strong local properties surround this central area without being related to globally integrated spaces.

Site B occupies a marginal position in the overall layout of Chelas, without being penetrated by its most globally integrated spaces. The internal configurational structure shows a strong autonomy in relation to the surrounding grid. It works as a formally disconnected unit with a highly segmented internal core without the capacity to articulate the overall site fabric. Thus it functions as the summing of different parts with strong local properties.

It was verified that these spatial conditions interfere in the way space is used, by generalizing pedestrian resident activity to the overall area and restricting nonresident access to the more central and peripheral spaces. Strong levels of mixed activity occurred in the most integrated spaces, while globally segregated spaces with strong local properties tended to become territorialized. These spaces showed deficient rates of pedestrian occupancy due to low levels of interaction and awareness between people. Although environmental damage was spread around both sites, the study shows that it is related to some spatial and physical features—that is, to accessibility, visibility, block access, and permeability, as well as to the rates of pedestrian activity.

It was concluded that all the checked invariants, except casual litter, tended to occur in underutilized spaces or in spaces with low levels of pedestrian occupancy. In particular it was observed that: (1) spaces with high control over the adjacent ones but with restricted or asymmetric visibility—due to enclosure—or less surveillability—due to low supervision and lack of pedestrian activity for most of time—are highly vulnerable; (2) blocks with open doorways or public passageways also constitute a factor of vulnerability when they are spatially segregated and below the horizon of the public realm. The main findings are shown in Figure 36.6.

Following the conclusion that blocks with open doorways or public passageways constitute a main factor of vulnerability, some hipotetic scenarios without reference to these spaces were prepared and studied. This procedure has allow us to evaluate the role of these spaces on the configurational structure and their potential implications on the social life of the site.

Figure 36.6

Diagram Expressing the Main Findings Considering the Relations Between Space and Use

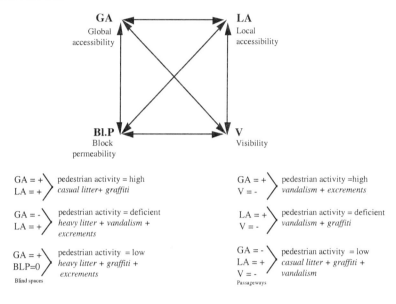

NOTES

1. Among those, Jane Jacobs is probably the first one in recent time who if not explicitly demonstrated then at least eloquently argued for such a possibility. She considers that the requisites for a reliable urban life are based on four generators—mix of primary uses, short blocks, diversity of building typologies, and high densities—all of which shall be at hand simultaneously (Jacobs 1961: 162–163). Despite its insightful comments on what might be the basic conditions for the creation of such generators by design intervention, Jacobs did not provide some descriptive method which otherwise may enable us to capture the exact implications which different urban settings have for patterns of space use.

2. One might say that dynamic space use at the geographical scale is static when observed from the frame of reference of the space use research at the scale of urban districts, plazas, and streets, down to the architectural scale of buildings and individual rooms. Because the static use at the geographical scale may be easier to describe than the dynamic use at a smaller scale, it might be easier to determine the properties of the environment to include in objective descriptions. Note that the case of vehicular traffic is somewhat exceptional in that it is itself a dynamic process, but because of its goal-oriented nature, may be easier to describe than other patterns of uses in urban environments.

REFERENCES

Batty, M. and P. A. Longley. "Fractal Based Description of Urban Form," *Environment, Planning and Design* 14 (1987) 123-124.

Coleman, A. *Utopia on Trial: Vision and Reality in Planned Housing.* London: Collier-Macmillan, 1985.

Engwich, D. *Towards an Eco-City: Calming the Traffic.* Sidney, Australia: Envirobook, 1992.

Hayard and McGlynn. *Making Better Places, Urban Design Know.* Oxford, U.K.: Butterworth Architecture, 1993.

Hillier, B. "The Architecture of Urban Object," *Ekistics* 56, 334 (1989) 5-21.

Hillier, B. *Space is the Machine.* Cambridge, U.K.: Cambridge University Press, 1996.

Hillier, B., R. Burdett, J. Peponis, and A. Penn. "Creating Life. Or Does Architecture Create Anything?" *Architecture and Behavior* 3, 3 (1987) 233–250.

Hillier, B., and J. Hanson. *The Social Logic of Space.* Cambridge: Cambridge University Press, 1984.

Hillier, B., and A. Penn. *The Mozart Estate Redesign Proposals.* Unit for Architectural Studies, Bartlett School of Architecture and Planning, 1992.

Hillier, B. and A. Penn. "Dense Civilisations: The Shape of Cities in the 21st Century," *Applied Energy* 43 (1992) 41-66.

Hillier, B., et al. *The Architecture of Maiden Lane Estate.* Unit for Architectural Studies, Bartlett School of Architecture and Planning, 1989a.

Jacobs, J. *The Death and Life of Great American Cities.* Harmondsworth, U.K.: Penguin Books, 1961.

Jones, M. A., and M. F. Fanek. "Crime in the Urban Environment," *Space Syntax First International Symposium Proceedings*, Vol. 2, 25.1-25.11. London, U.K.: Space Syntax Laboratory, The Bartlett School of Graduate Studies, 1997.

Klarkvist, B. "Spatial Properties of Urban Barriers: Analytical Tool for a Virtual Community." *Space Syntax First International Symposium Proceedings*, Vol. 2, 28.1-28.12. Space Syntax Laboratory, London, U.K.: The Bartlett School of Graduate Studies, 1997.

Mathews, J. "Social Progress and the Pursuit of Sustainable Urban Development in Buckingham" in *Environment Planning Sustainability* (S. Hatfiekd and B. Evans, eds.). Chichester, U.K.: John Wiley and Sons, 1996, 35-52.

Myerson, G. and Y. Rudin. "Sustainable Development: The Implications of the Global Debate for Land Use Planning In Buckingham" in *Environment Planning Sustainability* (S. Hatfield and B. Evans, eds.). Chichester, U.K.: John Wiley & Sons, 1996.

Peponis, J., C. Zimring, and Y. K. Choi. "Finding the Building in Wayfiding," *Environment and Behavior* 22, 5 (September 1989) 555–590.

Shiva, V. "Recovering the Real Meaning of Sustainability" in *The Environment in Question: Ethics and Global Issues* (D. Coopers and M. Palmers, eds.). London: Routledge, 1993.

Tecklenburg, J., Zacharias, T. and Heitor, T. "Spatial Analysis in Environment-Behaviour Studies: Topics and Trend*s,"* Invited Paper presented at the *Spatial Analysis Workshop: Evolving Environmental Ideas*, IAPS 14. Stockholm, Sweden: 1996.

White, R., Engelen, G. and I. Uljee. "The Use of Constrained Cellular Automatra for High Resolution Modeling of Urban Land Use," *Dynamics Geographical Analysis* (forthcoming).

PART VII:
BROADENING PERSPECTIVES

37

Art and Space

Jorge C. G. Calado

INTRODUCTION

The separation of the arts from the sciences is a convenience rather than a necessity. In Western culture, art, science, and technology have always formed an integrated whole—they stem from parallel concepts and a common, active impulse to create. This is, I think, even truer in Eastern civilizations. The arts and the sciences have influenced one another throughout the ages, beauty has permeated the framework of scientific theories and the design of machines, and scientific ideas and technological advances have been incorporated by artists into their creative processes. This underlying dialogue between art, crafts, science, and technology (engineering) is supported by the fact they all usually flourish at the same time. In the West, the great periods of scientific discovery and artistic creation were the Ancient Greek era (600 - 300 BC), the Ages of the Renaissance (14th - 17th centuries) and the Baroque (17th century), and the turn of the last century (1880–1930).

Beauty can be a criterion for judging scientific truth. When, in 1985, Harold Kroto and Richard Smalley submitted their paper on the structure of a "new" form of carbon, C_{60}, to *Nature*, they lacked the scientific evidence that the structure was, indeed, that of a soccer ball. But the solution was so beautiful that it had to be true (as it later turned out to be). They called it buckminsterfullerene, in homage to the great visionary architect and engineer, Buckminster Fuller (1895–1983). The extraordinary thing is that *Nature*, arguably the most prestigious scientific journal in the world, accepted it for publication on an aesthetic, rather than strictly scientific, basis. It is not the only case. Engineers often look for the most beautiful solutions because these are usually the most economical—some bridges, for instance, are true works of art.

To think of science and technology in aesthetic terms, to try to encompass in unifying vision both the sciences and the arts, is a scientific attitude. Science strives toward the universal. The ultimate purpose of physics is a grand unified theory of the four forces (electromagnetic, weak, strong, and gravitational)—even the development of a "theory of everything." I believe that what unites the arts and sciences is much more important than what separates them, and to show this is one of the aims of this chapter. Artists, too, are aware of this mutual identification, and the sculptor Naum Gabo (1890–1977) went as far as to say that "all the other constructions of our consciousness, be they scientific, philosophical, or technical, are but arts disguised in the specific forms of their peculiar disciplines" (Gabo 1962).

The art-science dialogue (or dualism) can be a powerful instrument of communication across the classic university-society and industry-society frontiers or interfaces. With the advent of modern physics (relativity and quantum mechanics) the capacity for visualization was lost. How to think of a four-dimensional universe in a three-dimensional space? How to understand mysterious dualisms (of the type *either/or* or *neither/nor*) like wave-particle, position-velocity, matter-energy? It should be noted that Eastern thought and culture are much more attuned to the paradoxes of the new physics. In July 1997, Nobel laureate John Polanyi, speaking at an international gathering of Nobel prize winners, "Great Science Under One Country Two Systems," at the City University of Hong Kong, said that "science, like politics, lives with paradoxes, and thrives on change." For Polanyi, the concept of "one matter, two descriptions" (particle and wave) was no more than a step toward greater understanding. The door remains open to further discoveries.

Science can be communicated through art and, likewise, aesthetic feelings can be expressed through science. The teaching of science and engineering is undoubtedly enriched by cross-references to similar developments in the arts. This was recognized by J. Bronowski (1908–1974) who once stated that "the layman's key to science is its unity with the arts" (Bronowski 1951). In this chapter, I will try to show how such an elemental concept as that of space is linked to crucial developments not only in the visual arts but also in literature, music, and architecture. (An elemental concept is, for me, one that follows St. Augustine's criterion—"If no one asks me, I know; if I wish to explain it to one that asketh, I know not.") For obvious reasons—lack of proper knowledge of Eastern culture—most of the examples will come from the history of Western art.

POINTS OF VIEW

Painting has traditionally been viewed as an art of space, and has also been widely used to explore the properties of space. The discovery of linear perspective by Leon Battista Alberti (1404–1472) in the 15th century, as well as the recovery of the foreshortening effect known to the Greeks, led to the introduction of the third dimension into the plane of painting. Paolo Uccello

(1397–1475) was, perhaps, the first painter to extensively use the new device of perspective to create the illusion of depth in his paintings, be it an interior, a landscape, or a battle scene. According to Giorgio Vasari (1511–1574), he would spend a lot of his time tackling the most difficult challenges, drawing from various angles in perspective complex shapes like "spheres showing seventy-two facets like diamonds, with shavings twisted round sticks on every facet and other oddities." (Donatello found these exercises useless, except for marquetry.) According to the same source (Vasari 1965), Paolo Uccello was fond of saying, "Oh, what a lovely thing this perspective is!" By the second half of the 15th century, an artist like Carlo Crivelli (active 1457–1493) could use linear perspective to dramatic effect, as in his painting of the *Annunciation with Saint Emidius*, shown in Figure 37.1.[1]

Rather than emphasizing the main theme of the painting—the *Annunciation*—perspective here is a distraction. The convergence of lines to a vanishing point leads our attention to what is happening in the busy streets of Ascoli Piceno (where the work was created). This is both a religious and a civic painting. On the bridge, in the middle distance, a gentleman reads a letter just arrived by pigeon post. St. Emidius, the patron saint of the city, seems engaged in conversation with the Announcing Archangel, Gabriel. Another important feature of the painting is the rectilinear path of the Holy Ghost (dove), made visible, like the track of any ionizing particle in a cloud-chamber, by a light beam. It amazes passers-by in the street (and a little boy, at the top of the steps, on the left). No wonder! Straight as a ruler, the ray of light plunges through a hole in the cornice of the Virgin's home to hit her, with laserlike precision, on the forehead. This is rectilinear geometry at its highest.

With linear perspective there is a clear distinction between the observer— the artist or a member of the public—and the world observed. The painter is not only immobile—frozen in space, so to speak—but also reduced to a single point. It is worth noting that Eastern art developed along different lines. The lack of interest in creating the illusion of depth led Chinese and Japanese artists to represent nature from the point of view of nature itself. A typical example is a handscroll of the Yodo river in Osaka, by the founder of the Maruyama school, Maruyama Okyo (1733–1795)—Figure 37.2. Both banks of the river are given equal prominence—the landscape is seen from the point of view of the river.

Another indication of this reversal of viewpoint—the observer seen by the observed—is the change in the hierarchy of scale and size with distance. In Eastern painting, figures and objects are sometimes smaller the closer they are to the viewer (because they are further away from the real subject of the picture.) Western painters influenced by Japanese art in the late 19th century used similar devices—the water lily ponds of Claude Monet (1840–1926) show bright flowers in middistance, in contrast with shapeless and fuzzy foregrounds.

Space in Oriental art is "treated as a linkage between pictorial motifs more than a means of suggesting intervening distance" (Singer 1995). Linkage is also a form of narration, and one could almost say that, unlike Western art, Eastern painting is more concerned with time than space. We still find hints of this aesthetic tradition in contemporary Japanese photography as, for instance, in the works of Hitoshi Namura (b.1945) and Hotaro Koyama (b. 1955)—Figure 37.3. Pointing the camera at a waterfall or a distant galaxy, the photographer animates the spirit of the things being photographed. In order to see the world, the artist must become the world.

When comparing linear perspective as developed in the West with the ambiguity of spatial relationships in Eastern art, relative space and the observer-observed dualism underlying Heisenberg's Uncertainty Principle come to mind. In the process of observation both the observer and the observed are transformed. When David Hockney (b. 1937) came to paint *Tian An Men Square and the Imperial Palace*—Figure 37.4—during his trip to China with Stephen Spender (1909–1995) in 1981, the result was both a view from above and a perspective lying down.

Here the Forbidden City of Imperial China and the modern Tien An Men Square appear as a kind of negative-positive pair—the former "haunted by the past and the dead," the latter "haunted by the future and the living" (Spender and Hockney 1982); one crammed with buildings, bridges, and monuments, the other void of the superfluous; full space as opposed to empty space, but both forming a harmonious unity, like the matching of the Yin and the Yang.

In literature the reversal of point of view is quite common. Novels are often told by one (or more) of their characters. In Percy Bysshe Shelley's *The Cloud* (1820) it is the cloud itself who composes the poem and tells its own story—the cycle of water:

I bring fresh showers for the thirsting flowers,
From the seas and the streams;

Shelley (1792–1822) was the prototype of the poet as scientist, and he wrote (in *A Defence of Poetry*) that "poetry comprehends all science." His poems are science by other means. In the *Tales of Scheherezade*—a meeting point of Eastern and Western imaginations—the fate of the protagonist is clearly connected to the flow of the narration.

NEWTON, ABSOLUTELY

With Isaac Newton (1642–1727) and his mechanics of the universe, the concepts of space (and time) became fixed and absolute—time flowed in one direction; space was immovable. With the publication of the *Principia Mathematica* (1686), phenomena as diverse as gravitation, collisions, pendulum oscillations, the firing of cannon balls, air friction, sound propagation, and so on, all came under the same umbrella of Newton's mechanics. Newton's ideas, which were to remain basically unchallenged for over two hundred years, were memorably enshrined by William Wordsworth (1770–1850) in two lines of *The Prelude* (1850)—"The marble index of a mind for ever / Voyaging through strange seas of Thought alone." The ideas of Nicolaus Copernicus (1473–1543) and Galileo (1564–1642) had been vindicated. How remote and senseless all those controversies about the heliocentric (as opposed to the geocentric) model seemed! A revolutionary work, the fresco painted in 1631 by Andrea Sacchi (1599–1661)—a contemporary of Galileo—on the ceiling of the Palazzo Barberini in Rome, remained ignored for over one hundred years. It is supposed to be an allegory to Divine Wisdom and she is given pride of place, sitting on a regal throne, center stage—Figure 37.5.

She is also depicted surrounded by her 12 attributes represented by appropriate constellations (Ariadne for nobility, Berenice for beauty, Lyre for suavity, Serpentarius or the serpent biting its own tail—the ancient symbol of chemistry—for eternity, etc.). However, the inescapable presence of the Earth in a peripheral position gives the game away. The painting should be read as a tribute to the Copernican view of the universe, and the central figure, bathed in sunlight, is Urania, the Greek goddess of Astronomy (Byard 1988).

With Newton's equations and the proportionality between force and acceleration (a definition of *mass*), the workings of the universe could be explained and the future predicted. Now, we not only knew how things were, we also knew what they would become. This omniscience was easily translated into literature. Like God or the wise physicist who could determine the path of anything that moves from first principles, the 19th-century novelist knew all there was to know about his or her characters. In carefully wrought novels, narration proceeded according to plan, with a beginning, a middle, and an end, and the narrator was a detached, omniscient being, capable of manipulating the characters and determining their destinies. This is what fiction meant! Also by the end of the 19th century physics was deemed a complete, finished science (and a boring one too, dealing with clocks, rulers, or falling rocks). Only two small problems (or "clouds" in the felicitous expression of Lord Kelvin) remained: that of the ether, and that of black body radiation. And what "little" problems these were, for the first ushered in relativity theory and the second quantum mechanics!

SPACE-TIME CONTINUUM

As so often happens, children were the first to intuit the *Zeitgeist*. Alice's land (in Lewis Carroll's *Alice's Adventures in Wonderland*, 1865) is a world of mutable frames of reference, an expanding universe where strange things happen. The novel is also about movement (it all starts with Alice's free fall down a deep well from the rabbit hole), with characters like the White Rabbit always running in a hurry. Alice's psychedelic changes—"I must be shutting up like a telescope!"—are consistent with time dilations and space contractions, as foreseen by the transformations of Hendrik Lorentz (1853–1928). All this was made clear by the famous drawings that Sir John Tenniel (1820–1914), a cartoonist for *Punch* magazine, created for *Alice's Adventures in Wonderland* (and *Through the Looking Glass*, 1871)—Figure 37.6.

Reading the Alice books one becomes aware that time and space might be linked, that they might even be one and the same thing. Relativity seems just around the corner. But then Charles Dodgson, a.k.a. Lewis Carroll (1832–1898), the creator of Alice, was a lecturer in mathematics at Christ Church, Oxford (and an eminent photographer). However, Albert Einstein's paper "On the Electrodynamics of Moving Bodies," which effectively introduced simultaneity and the theory of special relativity, appeared only in 1905 (Einstein 1905).

In matters of time and space, literature appeared to lead. In his first novel, *The Time Machine* (1895), billed as "a thrilling story about a journey into the far distant future" (this future was the year 802 701 AD), H. G. Wells (1866–1946) had the protagonist—the Time Traveller—explaining that "any real body must have extension in *four* directions: it must have Length, Breadth, Thickness, and—Duration." Later in the novel, the Time Traveller states emphatically, "There is no difference between Time and any of the three dimensions of Space except that our consciousness moves along it." Time had been regarded differently because it is unidirectional, like an arrow pointing to the future. But, as the Time Traveller was quick to remark, we cannot move freely in space, either. We can go backward and forward freely enough, but not up and down, for gravitation limits us there. Arthur Eddington, who was later the first to provide an experimental proof of the theory of relativity, called entropy "the arrow of time.") That time could be changed into space we knew already from wise Gurnemanz in Richard Wagner's *Parsifal* (1878), when he tells Parsifal that *Der Raum wird hier zur Zeit* (here time becomes space), and the scene gradually transforms itself from forest into temple. The famous physicist Hermann Minkowski (1864–1909) put it even more poetically when he declared, in an address to the 80th Assembly of German Natural Scientists and Physicians at Cologne (1908), that "Henceforth space by itself, and time by itself, are doomed to fade away into mere shadows, and only a kind of union of the two will preserve an independent reality." This new union is the space-time continuum that Einstein (1879–1955) expounded upon and that Edward Steichen (1879–1973) metaphorically photographed in 1920, in a famous series of still lives—Figure 37.7.

To the uninitiated, Steichen's photograph looks as complex and esoteric as Einstein's relativity. His Time-Space Continua are studies of both form and texture, with each object component evoking a function: a transparent glass globe (which also appears in the series *Triumph of the Egg* of the same year) standing for the Universe, anchored by an organic stump; a lever-like tool and a locksmith's diagrammatic board complete the mechanistic approach. Given a fulcrum, with the lever, the universe can be moved; understanding lies in a set of keys whose enigmatic patterns form an alphabet.

Even architecture, the most palpable of all the arts, moved into the fourth dimension. Erich Mendelsohn (1887–1953) seemed to echo the Time Traveller of H. G. Wells when he said that "objects do not exist in space, although they are spatially extensive." He built the Einstein Tower (1921)—a masterpiece of expressionist architecture—in the grounds of the Astrophysical Institute in Potsdam—Figure 37.8.

The tower is a building that is "in motion" through mass; space that turns into exploding matter. Guralnik (1979) would later comment that the Einstein Tower "represents the theory of relativity and the transitory nature of mass in the context of space and time—a mass moving in space in an unstable state in which the form remains incomplete." The title of Sigfried Giedion's Charles Eliot Norton Lectures at Harvard in 1938–1939 was *Space, Time and Architecture*. Later, under the same title, these lectures became a most influential book (Giedion 1941) (which, incidentally, completely—and unjustly—ignored the work of Mendelsohn).

THINKING IN THE CUBE

The equivalence of frameworks and the relativity of space-time found a revolutionary parallel in painting with the appearance, around 1908, of Cubism. Like many other important creations, it was a joint endeavor. Usually, teamwork is a prerogative of scientists (science is also a fertile field for simultaneous discovery.) With the rise of Impressionism there had been several cases of painters working together, on the same scene, side by side. Now with Cubism we had Georges Braque (1882–1963) and Pablo Picasso (1881–1973) stimulating one another and arriving at similar solutions. They were fond of painting musicians—the guitar player of Braque's *The Portuguese* (1911) or Picasso's *The Accordionist* (1911)—caught playing in the smoky atmosphere of Parisian cafés—Figure 37.9.

What these Cubist paintings usually show are fragments of newsprint, instruments broken down into their spare parts, disjointed bodies glimpsed from different vantage points at different times, a mustache here, a nail there, still lives that are anything but still, and the like. Cubism is about the restlessness of our vision and how things change within the blink of an eye.

Like Galileo and Newton before him, Picasso, in the words of Guillaume Apollinaire (1880–1918), "aggressively interrogated the universe." Like Galileo, he refused to recant. By restoring the supremacy of the straight line, long lost through the excesses of Impressionism, Cubism conferred a certain rectitude upon painting. Cubist pictures are structured in a crisscross of linear segments, forming a sort of haphazard marquetry. Cubism also restored volume (space) to the painter (and viewer) who are now free to exercise a multitude of points of view. The observer is no longer a fixed point, but a body of changeable volume that moves at will.

These conquests were later translated into photography by Hockney in the 1980s. His assemblages of snapshots are, in the words of the artist, "a critique of photography made in the medium of photography" (Arts Council of Great Britain 1983) . His main objection was that a photograph was a lifeless object that could not be looked at for a long time. His solution was to inject into photography what was missing—life, or rather, time, lived time. In his most successful composites, like that of *Gregory Waching the Snow Fall, Kyoto* (1983), there is a perfect match of form and content—the puzzle has been rearranged as to almost form a Japanese ideogram (Figure 37.10).

With Cubism, straight lines and planes, rather than curves and domes, became the order of the day. Marcel Proust (1871–1922), whose masterpiece is appropriately titled *À la recherche du temps perdu* (Remembrance of Things Past), wrote that "there is a plane geometry and a geometry of space. And for me the novel is not only plane psychology but psychology in space and time." Jean Cocteau (1889–1963), that most mercurial of artists, also wrote a cycle of Cubist-inspired poems, *Cocardes*, set to music by Francis Poulenc (1899–1963) in 1919. Theater, too, took a Cubist turn. The Italian playwright Luigi Pirandello (1867–1936) wrote several plays that reject the idea of the existence of an objective reality. In *Right You Are! (If You Think You Are)* (1917), each character presents his or her version of events and no two versions coincide. Truth is relative.

If the sphere had been, for 18th century visionaries and dreamers, the ideal shape, for the modernists of the 20th century perfection resided in the cube. It was the cube that gave its name to the new art of Braque and Picasso. Wells's Time Traveller denied the existence of an *instantaneous* cube precisely because its fourth dimension should be duration (time). In architecture, the cube became the symbol of the new International Style—the much derided skyscraping, brutalist box. Long before that, the Viennese architect Adolf Loos (1870–1933), one of the earlier advocates of "less is more" ("Ornament equals Crime," he wrote), had advised his students "to think in three dimensions, to think in the cube."

AND CURVES, TOO

Others, of course, looked at the round shape of the Earth and thought along curved lines. Mathematicians like Nicolay Lobachevski (1792–1856) and Georg Riemann (1826–1866) had developed *n*-dimensional spaces and non-Euclidean geometries based on curves. Later, Einstein would show (in the General Theory of Relativity, 1916) that even light rays, which Crivelli had depicted as the epitome of a straight line, would bend under the action of gravity (their frequency would also change). It was left to Arthur Eddington (1882–1944) to prove this experimentally in 1919, by studying the anomalous behavior in the motion of the planet Mercury during a solar eclipse. The observations were carried out on the (then) Portuguese island of Príncipe. The universe is, after all, a four-dimensional curved space.

A curious coincidence made the painter Pavel Tchelitchew (1898–1957) say in Kiev, in the same year of 1919, that "a line is never straight, but invariably curves spherically in conformance with the shape of our earthly globe," thus breaking away from the Cubist-Constructivist formula and plunging "into the curved line" (Tyler 1967). Many of Tchelitchew's paintings, in particular the series of heads that he created in the 1940s and 1950s, evoke this curved space, made of non-Cartesian lines—Figure 37.11.

SPEED

It was the breakdown of absolutes and certainties induced by modern physics at the beginning of the 20th century that led Emilio Marinetti (1876–1944) to proclaim, in the *First Futurist Manifesto* (1909), that "time and space died yesterday." What was very much alive was *speed*, which is the progress of space through time. Within the realm of the arts, movement became an obsession. A roaring car was, for Marinetti, more beautiful than the *Victory of Samothrace* in the Louvre! Where a photographer, Eadweard Muybridge (1830–1904), had led with his chronophotographs of humans and animals in motion—Figure 37.12—the futurist painters followed.

Like Giacomo Puccini (1858–1924), Gabriele D'Annunzio (1863–1938), and Benito Mussolini (1883–1945), the futurists were Italian, and like those three, obsessed with velocity. The most prominent was Giacomo Balla (1871–1958) whose *Dynamism of a Dog on a Leash* (1912), wagging tail and all, is as powerful a symbol of movement as Ford's Model T car (1908)—Figure 37.13.

This is a painting which, among other things, suggests that there is a space outside its frame. What is missing from the painting can be seen in a photograph, taken more or less at the same time, by a teenager who was also a photographic genius, J.-H. Lartigue (1894–1986)—Figure 37.14.

The boy Lartigue was one of the first photographers to capture anything that moves—people jumping and running, cars racing, buses speeding, kites and airplanes flying—and he did it with that fuzziness and distortion that we now equate with velocity.

Interestingly enough, Balla himself had had a premonition of the important role that Mercury's orbit would play in the establishment of the new physics. In 1914, five years before Eddington's observations and measurements, Balla watched, through his telescope, Mercury passing in front of the Sun. The event inspired him to paint a series of abstract canvases on celestial orbits. One of the most famous is *Mercury Passing in Front of the Sun Seen through a Telescope* (1914), where conical shapes compete with spheres and curves seem to obliterate straight lines—Figure 37.15.

A HANDFUL OF DUST

The greatest discovery about space at the turn of the century was the fact that matter, even the most solid and compact, is largely made of empty space. Matter is transparent to radiation, and indeed the study of the propagation of radio waves by Heinrich Hertz (1857–1894) and the discovery of X-rays in 1895 by Wilhelm Röntgen (1845–1923), played a crucial role in the emergence of modern physics. Radioactivity (1896) did the rest. On arriving at Cambridge in 1895 from his native New Zealand, Ernest Rutherford (1871–1937) devoted himself to the study of the transmission and reception of electromagnetic waves, amazed that they could pass through several thick walls, floors, and any other obstacles. Soon afterward he was sending, to his relatives back home, the latest novelty—rayograms (X-rays) of a frog, that eternal martyr of science. All these facts implied that matter was, like a sieve, full of holes.

In 1897, one year after the discovery of radioactivity by Antoine-Henri Becquerel (1852–1908) and two years after Röntgen stunned the world with his work on X-rays, H. G. Wells published the novel *The Invisible Man*. It was not a mere coincidence. Its protagonist, Griffin, had dropped medicine to take up

556 Science, Technology, and Innovation Policy

physics. *Light*, or rather optical density, fascinated him. Röntgen gets a mention in the novel. Matter, which physicists knew to be transparent to different types of radiation, could be rendered invisible. In fact, as Griffin explains, "the whole fabric of a man, except the red of his blood and the dark pigment of hair, are all made up of transparent, colorless tissue." Transparency could then be equated with nothingness, to no visible thing at all.

Science was, at last, catching up with fiction. The fact that matter is mostly made of empty space was consistent with Rutherford's planetary model for the atom (1910). The way atoms are packed together to form solid structures was analyzed by the Braggs, father and son (William, 1862–1942; Lawrence, 1890–1971), through their diffraction studies of crystals, carried out from 1912 onward. Electrons orbiting around a heavy nucleus, atoms occupying the vertices of a lattice structure, with nothing in between—this became the new view of matter. After reading Rutherford's account of atomic structure in 1911, Wassily Kandinsky (1866–1944) exclaimed: "The discovery hit me with frightful force, as if the end of the world had come. All things became transparent, without strength or certainty." The world was made up of a "handful of dust" (in the words of T. S. Eliot in *The Waste Land*, 1922). Or rather, to paraphrase the Portuguese writer Irene Lisboa, a handful of dust (atoms) and another of nothing (empty space). The historian of (Portuguese) science and poet António Gedeão (1906–1996) expressed it beautifully in his poem *World Machine* (1987):

> *The Universe is essentially made of nothing,*
> *Intervals, distances, holes, ethereal porosity.*
> *Empty space, in a word.*
> *The rest, is matter.*

For Gedeão, the void speaks and matter is simply silence—a reversal of *l'être et le néant*.

To a certain extent the future had been announced by that icon of modernism, the Eiffel Tower (1889). One of the earliest expressions of the interpenetration of inner and outer space (Giedion 1941), the Eiffel Tower functions both as a rayograph of a building—its visible metal structure as bare bones—and as a transmitter of radio waves, not unlike the old RKO Motion Pictures logo. Ascending or descending its spiral staircases, we are both inside and outside. Several photographers have explored this idea. In what was to become his most famous image, Marc Riboud (b. 1923) caught, in 1953, a worker in the process of painting the Tower—Figure 37.16. Framed by the iron structure, the painter becomes a tightrope walker doing his circus act.

More than a decade earlier (1939), Erwin Blumenfeld (1897–1969) had photographed the model Lisa Fonssagrives suspended in midair, butterfly-like, wearing a Lucien Lelong dress. Both are portraits on the edge of a precipice, taking advantage of the open structure of the Eiffel Tower.

The Tower's lightness and transparency are feats of engineering, made possible by the tensions of the metal grid. Matter under stress releases energy and becomes spirit. This, of course, is another triumph of relativity theory, perhaps the most widely known thanks to the appealing simplicity of Einstein's equation, $E = mc^2$. Sculptors followed suit and probed deep beneath the massive surface of things to reveal their flimsy interiors, void and vulnerable. Naum Gabo (1890–1977) was one of the first artists to define form in terms of empty space, rather than mass—Figure 37.17. He wrote:

Up to now sculptors have preferred mass and neglected or paid little attention to such an important component of mass as space ... we consider it as an absolute sculptural element ... I do not hesitate to affirm that the perception of space is a primary natural sense which belongs to the basic senses of our psychology."

His *Head of a Woman* (1917) has the spare elegance of a crystal structure— Figure 37.17.

Another example is provided by the "interior landscapes" that Tchelitchew produced in the mid to late 1940s—heads delineated with lines of light in dark space. His own self-portrait mask of 1929 (Figure 37.18) seemed to anticipate these creations—the head as a "wire basket," with only the sensory orifices (mouth and eyes) made opaque. Again, it is sculpture as X-ray and, in fact, its shadow looks uncannily like a skull.

Bringing the outside inside is one of painting's recurrent themes, particularly in landscape art. René Magritte (1898–1967) went through a period when, in his work, the outdoors was virtually indistinguishable from the indoors. In *Euclidian Walks* (1955) it is difficult to ascertain whether we are looking at a landscape through a glass pane, or at a painting on glass or canvas exactly matching the view from the window. The correct answer is, of course, neither, for the whole thing is a painting and an illusion—Figure 37.19.

FIELDS OF FORCE

There is yet another type of space that deserves attention—the space that lies between the work of art and the viewer. A painting, like any other work of art, creates a field of force, and energy flows both ways, between the observer and the work observed. To properly appreciate a work of art one often has to try out different points of view and experiment with increasing (or decreasing) distances from the work under observation. This is more than a matter of *scale*. Like a probe, the viewer measures, with his response, the intensity of the field in the space that separates her or him from the piece. This is obvious in paintings, like those of Bridget Riley, that play upon our mechanisms of perception. A good example is *Straight Curve* (1963)—Figure 37.20.

In her black and white paintings, Riley experimented with the juxtaposition of shapes, exploring active features like "contrast, irradiation and interaction" (Riley 1995), which are typical of the language of physics. Her works also evoke rhythm, modulation, and intensity—all qualities generally associated with music. When, in the late 1960s, Riley began to use color, it played the role of timbre. All the arts aspire to the condition of music, said Walter Pater (1839–

1894), and Riley's paintings come pretty close to possessing a musical essence. (Her admiration for Stravinsky and]is lectures on *The Poetics of Music* [1940] is well-known.)

Despite its abstractness, this painting generates sensations. These do not derive from any particular incident, but are triggered by the visual energy emanating from the picture plane. Space advances and recedes as if under the action of natural forces. A conflict between stability and disruption can be intuited; chaos is around the corner. The title of the painting is a statement, rather than a question. In its ambivalence, a straight curve sums up the difference between art and science.

NOTE

1. Due to copyright restrictions, the artistic figures referred to in this chapter have not been reproduced. A listing of the figures is provided below:

Figure 37.1: Carlo Crivelli, *Annunciation with St. Emidius*, 1486 (The National Gallery, London).

Figure 37.2: Maruyama Okyo, *The Two Banks of the Yodo River*, 1765 (Arc-en-Ciel Foundation, Tokyo).

Figure 37.3: Hitoshi Namura, *Spin and Gravity: For the Sea of Potalaka*, 1982–1984 (Publisher: Hara Museum of Contemporary Art, Tokyo; Distributor: Harry N. Adams, New York).

Figure 37.4: David Hockney, *Tian An Men Square and the Imperial Palace*, 1981 (Publisher: Harry N. Adams, New York).

Figure 37.5: Andrea Sacchi, *Allegory to Divine Wisdom*, 1631 (Palazzo Barberini, Rome).

Figure 37.6: Sir John Tenniel, Illustration for *Alice in Wonderland*, 1866.

Figure 37.7: Edward Steichen, *Time-Space Continuum*, 1920 .

Figure 37.8: Erich Mendelsohn, *Einstein Tower*, Potsdam, 1920–1924 (Publisher: The Architectural Press, London).

Figure 37.9: Pablo Picasso, *The Accordionist*, 1911 (The Solomon R. Guggenheim Museum, New York).

Figure 37.10: David Hockney, *Gregory Watching the Snow Fall*, Kyoto, 1983.

Figure 37.11: Pavel Tchelitchew, *Mercure,* 1956 (Collection of Mrs. L. B. Wescott).

Figure 37.12: Eadweard Muybridge, Plate from *The Human Figure in Motion*, 1887.

Figure 37.13: Giacomo Balla, *Dynamism of a Dog on a Leash*, 1912 (Albright-Knox Gallery, Buffalo, New York).

Figure 37.14: Jacques-Henri Lartigue, *Paris, Av. Du Bois de Boulogne*, 15 January 1911.

Figure 37.15: Giacomo Balla, *Mercury Passing in Front of the Sun Seen Through a Telescope*, 1914.

Figure 37.16: Marc Riboud, *Paris*, 1953.

Figure 37.17: Naum Gabo, *Head of a Woman*, 1917 (The Museum of Modern Art, New York).

Figure 37.18: Pavel Tchelitchew, *Self-Portrait Mask*, 1929 (Collection of Francis Sitwell).

Figure 37.19: René Magritte, *Euclidian Walks*, 1955 (Minneapolis Museum of Arts).

Figure 37.20: Bridget Riley, *Straight Curve*, 1963 (Collection of Victor Musgrave, Rye).

REFERENCES

Arts Council of Great Britain. *Hockney's Photographs*. London: Arts Council of Great Britain, 1983.

Bronowski, J. *The Common Sense of Science*. London: Heinemann, 1951.

Byard, M. M. "Galileo and the Artists," *History Today* (February 1988) 30–38.

Einstein, A. "On the Electrodynamics of Moving Bodies," *Ann. Physik* 17 (1905).

Gabo, N. *Of Divers Arts*. London: Faber and Faber, 1962.

Gabo, N. *Sculpture: Carving and Construction in Space*, 1937.

Gedeão, A. "Máquina do Mundo" (World Machine), from *Poesias Completas*, 10th ed. Lisbon: Livraria Sá da Costa Editora, 1987.

Giedion, S. *Space, Time and Architecture*. Cambridge: Harvard University Press, 1941.

Guralnik, N. *Erich Mendelsohn: Drawings of an Architect*. Tel Aviv: Tel Aviv Museum, 1979.

Riley, B. *Dialogues on Art*. London: Zwemmer, 1995.

Singer, R. T. *Photography and Beyond in Japan*. Tokyo: Hara Museum of Contemporary Art, 1995.

Spender, S., and D. Hockney *China Diary*. New York: Harry N. Abrams, 1982.

Tyler, P. *The Divine Comedy of Pavel Tchelitchew*. New York: Fleet Publishing Corp., 1967.

Vasari, G. *The Lives of the Artists*. Harmondsworth, U.K.:Penguin Books, 1965.

Index

About the Contributors

Daniele Archibugi is a director of the Italian National Research Council's headquarters and a Commissioner at the Authority for Public Utilities of the Rome Townhall. He is the co-author of *The Technological Specialization of Advanced Countries* (1992) and the co-editor of *Technology, Globalization and Economic Performance* (1997), *Trade, Growth and Technical Change* (1998), and *Innovation Policy in a Global Economy* (1998). He is in the editorial board of *Research Policy, Technological Forecasting and Social Change,* and *Technovation.*
Contact: archibu@www.isrds.rm.cnr.it

Suma S. Athreye is a Lecturer in International Business at the Manchester School of Management, Manchester, U.K. Her current research interests lie in the fields of technology markets and the use of foreign direct investment as an instrument of technological transfer, innovativeness and competition policy in the software sector, and comparative studies (intercountry, using firm-level data) on the determinants of innovative activity.
Contact: suma.Athreye@umist.ac.uk

Adisa Azapagic is a Lecturer in Environmental Technology at the University of Surrey in England. Her current research interest is focused on developing the methodology for life-cycle assessment and its application to corporate decision-making. Related research includes developing life-cycle product design techniques which offer the potential for technological innovation in product design through the selection of the best materials and process alternatives over the whole life cycle.
Contact: A.Azapagic@surrey.ac.uk

Maria Laura Barreto has been acting as researcher in environmental and natural resources legislation at CETEM, Center for Minerals Technology Laboratories of the Brazilian Research Council (CNPq). With a focused research interest in questions dealing with sustainable development, she has been very active in international projects and has published over 40 papers, articles, and book chapters. She is also a founder of AMLA, an association of Latin American mining lawyers and has been appointed the Deputy Coordinator of the Law Section of SGT 2 on Mercosul. Ms. Barreto is also an Adjunct Professor at the Law School of the Universidade Candido Mendes in Rio.

Jim Botkin is co-founder and president of InterClass—the International Corporate Learning Association—a consortium of 15 Fortune 500 companies from AT&T to Xerox exploring knowledge business, knowledge communities, and knowledge management. He is also a Senior Research Fellow at the IC^2 Institute, The University of Texas at Austin. Botkin's latest book, *Smart Business: Leveraging the Power and Potential of Knowledge Communities* will be published in 1999. In addition to *Monster Under the Bed* with Stan Davis, Jim is the author of *No Limits to Learning: A Report to the Club of Rome*, published in a dozen languages.
Contact: jbotkin@interclass.com

Jorge C. G. Calado is Professor of physical chemistry at Instituto Superior Técnico (IST, Lisbon). He has published over 150 papers in his field of research, the thermodynamics of molecular liquids. He is also a cultural critic for the Portuguese weekly newspaper *Expresso,* has curated several photography exhibitions at home and abroad, and was responsible for the creation of the National Collection of Photography in Portugal. Among his many interests are the relations and parallels between the arts and the sciences which are the subject of a course that he has taught at Cornell and IST, "The Art of Science."
Contact: jcalado@alfa.ist.utl.pt

João Caraça is the Director of the Science Department of the Calouste Gulbenkian Foundation in Lisbon. He is also Full Professor of Science and Technology Policy at the Instituto Superior de Economia e Gestão of the Universidade Técnica de Lisboa, where he coordinates the M.Sc. Course in Economics and Management of Science and Technology. He is Science Adviser to the president of the Portuguese Republic and has published 130 scientific papers. His interests are centered in science and technology policy and in prospective studies.
Contact: jcaraca@gulbenkian.pt

Pao-Long Chang is Professor at the Institute of Management Science, National Chiao Tung University. His research focuses on technology management and decision sciences. His work has been published in journals such as the *Journal of the Operational Research Society, Computers and Operations Research, Energy Policy, Energy Systems and Policy, Journal of Environmental Management, Technology Analysis and Strategic Management, Total Quality Management, Technovation, International Journal of Project Management,* and *International Journal of Technology Management.*

Xiangming Chen is an Associate Professor of Sociology at the University of Illinois at Chicago, a Research Fellow at IC^2 Institute, The University of Texas at Austin, and affiliated with the Center for East Asian Studies at the University of Chicago. He has been a recipient of the Chiang Ching-Kuo Foundation Chinese Studies

Postdoctoral Fellowship through the American Council of Learned Societies and the Social Science Research Council. He has published over 20 journal articles and book chapters on urban and economic development in China and East Asia.
Contact: xmchen@uic.edu

Po-Young Chu is a Professor of the Department of Management Science, National Chiao-Tung University. He has served as a management consultant to Philips Taiwan and Industrial Technology Research Institute. His major research interests include decision theory, technology management, and financial modeling.

Xuelin Chu is Professor of Management, Assistant Dean for International Affairs, and MBA Program Director at the School of Business and Management, University of Science and Technology of China (USTC) in Hefei. He is also a Senior Research Fellow of IC² Institute. His research interests are management of technology, human resource management, and strategic management. He has published many papers and books in Chinese, and translations both in English and Chinese. His books, *Modern Management Concepts* and *Management of Technology*, have been well received in China.
Contact: xlchu@ustc.ac.cn

Pedro Conceição is a Fulbright Scholar and Ph. D. candidate at the LBJ School of Public Affairs and a visiting scholar at IC² Institute, The University of Texas at Austin and lecturer at Instituto Superior Técnico. He co-authored a book in Portuguese (*New Ideas for the University*, 1998), contributed chapters to an edited book (*Inequality and Industrial Change: A Comparative Analysis*, forthcoming), and has published more than 15 papers in international refereed journals, including *Technological Forecasting and Social Change*, *Technovation*, and *The Review of Development Economics*.
Contact: pedroc@uts.cc.utexas.edu

Mark Dodgson is Professor of Management in the Australia Asia Management Centre at the Australian National University. His books include: *Effective Innovation Policy* (1996); *The Handbook of Industrial Innovation* (1994), *Technological Collaboration in Industry* (1993), *The Management of Technological Learning* (1991), *Technology Strategy and the Firm* (1989). He is at present completing *Technology and Innovation Management: An International Approach*. Dr. Dodgson is Editor-in-Chief of *R&D Enterprise—Asia Pacific*.
Contact: Mark.Dodgson@anu.edu.au

Rinaldo Evangelista is a researcher at the Institute for Studies on Scientific Research and Documentation of the Italian National Research Council, where he has worked in the areas of economics of technological change, technology and industrial organization, theory and measurement of innovation in manufacturing, and service sectors. He is the author of *Knowledge and Investment: The Sources of Innovation in Industry* (forthcoming) and has co-authored numerous papers in academic journals.
Contact: evangeli@www.isrds.rm.cnr.it

José Rui Felizardo is Director of ITEC, a non-profit technology institute, with specific competences in the area of technology and innovation auditing. His interests in current years include the development of the automobile industry in Portugal.
Contact: jrf@itec.pt

Paulo C. Ferrão is an Assistant Professor at Instituto Superior Técnico and is responsible for the course "Energy and Environment" of the Mechanical and Environmental Engineering courses. Professor Ferrão carries out fundamental research activities in Fluid Mechanics (particularly in Turbulent Combustion) and is interested in terms of their industrial applications, which has motivated his concerns in environmental management.
Contact: ferrao@dem.ist.utl.pt

James K. Galbraith is Professor at the Lyndon B. Johnson (LBJ) School of Public Affairs and the Department of Government, The University of Texas at Austin, where he teaches economics and other subjects. He is also a Senior Scholar with the Jerome Levy Economics Institute. His latest book is *Created Unequal: The Crisis in American Pay*, sponsored by the Twentieth Century Fund.
Contact: galbraith@mail.utexas.edu

David Gann is Professor of Technology Policy at the Science Policy Research Unit, University of Sussex, where he leads a research team on innovation in the built environment. He holds the IMI/Royal Academy of Engineering Chair in Innovative Manufacturing, with responsibility for approximately £2 million of research on the construction industries and other project-based industries.
Contact: d.gann@sussex.ac.uk

David V. Gibson, The Nadya Kozmetsky Scott Centennial Fellow, is Director of Global Programs at IC² Institute, The University of Texas at Austin. During 1999-2000 he served as a Fulbright Scholar under the Luso-American Educational Commission at Instituto Superior Técnico, in Lisbon. His research and publications focus on the strategic management of knowledge; cross-cultural communications, management, and technology transfer; the management and commercialization of technology; and the growth and impact of technopoleis or regional technology centers.
Contact: davidg@icc.utexas.edu

Zhicheng Guan is Vice President of Tsinghua University. His major research fields are in high voltage technology, high voltage insulation, composite insulators, electrical environment technology, and high voltage measurement. One of his major achievements, "Study on the Discharge Along the Surface of Polluted Insulators," won the First Class Award of Science and Technology Improvement and Development granted by the State Education Commission.

Manuel V. Heitor is Professor of Mechanical Engineering at the Instituto Superior Técnico in Lisbon, and is the director of the Center for Innovation, Technology and Policy Research, IN+. He is co-editor of several books and author of several scientific papers in the area of energy and environment and related fields, with emphasis on fluid mechanics and combustion. His current interests include the management of technology and the development of engineering and innovation policies, including higher education policies. He is a Senior Research Fellow of IC² Institute, The University of Texas at Austin. In this context, he has led the Organizing Committee of a series of International Conferences on "Technology Policy and Innovation," which began in 1997 in Macau. He is on the editorial board of *Technological Forecasting and Social Change*, and has published several papers in international refereed journals, including *Technological Forecasting and Social Change*, *Technovation*, *Higher Education Policy* and *Science and Public Policy*.
Contact: mheitor@dem.ist.utl.pt

Teresa Valsassina Heitor teaches Architecture and is Assistant Professor at the Civil Engineering Department, Instituto Superior Técnico (Lisbon). Her interests are in environmental-behavior issues and design education. Her research has centered on spatial analysis and postoccupancy evaluation mainly on the development of the participatory evaluation process.
Contact: teresa@civil.ist.utl.pt

Simona Iammarino is a researcher at the Italian National Institute of Statistics in Rome. She collaborates both with the Italian Institute for Foreign Affairs, Rome, and with Prof. J. A. Cantwell, University of Reading (U.K.), working in the field of multinational corporations and regional systems of innovation in the E.U.

Annamária Inzelt is Founding Director of IKU Innovation Research Centre. She is the Hungarian representative at OECD NESTI group, member of the National Committee for Technological Development, and holder of Hungarian Széchenyi Professorship. She is author and editor of several books, chapters, and articles. Her most recent publications are: "Transformation Role of FDI in R&D: Analysis Based on a Databank," in *Quantitative Studies for S&T Policy in Economies in Transition* (1998), *Technology Transfer: From Invention to Innovation* (1998), *Biotechnology Audit of Hungary* (with T. Reiss and U. Bross), (1998), and "Institutional Transfer in a Post-Socialist Country" (1998).
Contact: ainzelt@iku.omikk.hu

Vai Pan Iu is an associate professor in civil engineering and Dean of the Faculty of Science and Technology at the University of Macau. His research interests are on finite elements, nonlinear structural vibrations and computational techniques.

Dylan Jones-Evans is Director of the Welsh Enterprise Institute and Professor of Entrepreneurship and Small Business Management. His main research interests are the development of world-class small firms, the study of fast-growth entrepreneurial businesses, and increasing linkages between academia and industry. In the last four years, he has attracted over £1 million in grants from a variety of sources to examine these issues, and has published over 50 different articles on these subjects in refereed journals, academic books, and international conference proceedings. He is currently Chairman of Community Enterprise Wales, and has been involved in a number of initiatives to support small business development in Wales.

Junmo Kim is a researcher with the Institute of Governmental Studies at Korea University, where he also serves as a lecturer. His research areas include wage analysis, industrial, and technology policy. His recent works include papers on industrial and occupational wage analysis of Korea and analysis of Asian aerospace industry.
Contact: junmokim@unitel.co.kr

Yasuo Konishi works with the United Nations Industrial Development Organization (UNIDO) in Vienna, Austria.

Otto C. C. Lin is Vice President-Research & Development at the Hong Kong University of Science and Technology. He has received many awards and honors, and served on the boards of many organizations. His research interests include polymer materials for electronic applications, management of technology, and national competitiveness.

Fernando Machado works with the United Nations Industrial Development Organization (UNIDO) in Vienna, Austria.

Mauricio de Maria y Campos works with the United Nations Industrial Development Organization (UNIDO) in Vienna, Austria.

Rui Martins has been on the academic staff of the Instituto Superior Técnico since October 1980; since 1992 he has been on leave from IST and is currently with the University of Macau (UM) occupying the position of Visiting Full Professor in the Faculty of Science and Technology. He was the Dean of FST/UM from 1994 to 1997 and since 1997 he has been the Vice-Rector of UM.
Contact: rtorpm@umac.mo

Heloisa V. Medina works with the Centro de Tecnologia Mineral, Conselho Nacional de Desenvolvimento Científico e Tecnológico in Rio de Janeiro, Brazil.

Pedro Oliveira is pursuing doctoral studies in the field of Operations Management and Strategy at the University of North Carolina at Chapel Hill. He has conducted several works on research and development policy, focusing on policies for the protection of intellectual property and for the exploitation of research results.
Contact: oliveirp@icarus.bschool.unc.edu

Carlos C. Peiter is Head of the Studies and Development Department. He is co-author of the book *Sustainable Development and Advanced Materials: The Brazilian Case*, edited by R. C. Villas Bôas.

Nicolas-Pierre Peltier is working for the French Ministry of Industry, where he is responsible for industrial development in a regional bureau.
Contact: Nicolas.PELTIER@industrie.gouv.fr

Giulio Perani is researcher at the Italian National Statistics Institute where he is responsible for the production of statistics on technological innovation in business firms. He is a political scientist with research experience in the analysis of the research and technological innovation issues as well as technology policies in Italy and other countries.

Maria Rosaria Prisco has been collaborating, since 1991, with the Institute for Studies on Scientific Research and Documentation of National Research Council as researcher in the field of regional development in science and technology. In particular her work is concerned with the methodological aspects of measurement and evaluation of technological innovation at the local scale. She is currently working at the Instituto per lo Sviluppo della Formazione Professionale dei Lavoratori on the structural European policies for the less favored regions.
Contact: prisco@www.isrds.rm.cnr.it

Zhu Qin works in the Science and Technology Office of Tsinghua University, where she manages the State Key Laboratories and the National Engineering Research Centers. She manages cooperative projects between Tsinghua University and foreign universities and enterprises. Dr. Qin is also a section chief of the Science and Technology Office and assistant professor of Analysis Chemistry. She has published 12 papers which cover the field of research and development.

Fabio Rapiti is a researcher in the Italian National Statistic Institute.

Prasada Reddy is a Research Fellow at the Research Policy Institute, Lund University, Sweden. His research interests include globalization of R&D, global knowledge networks, innovation-based small and medium-sized enterprises, and intellectual property rights. He is involved in a research project on trade-related intellectual property rights and its implications for pharmaceutical and medicinal plants sectors. He is also the Convenor of the Working Group on Science and Technology for Development, the European Association of Development Research and Training Institutes.
Contact: prasada.reddy@fpi.lu.se

Nikolay Rogalev is a Senior Scientific Researcher at the Department of Boiler Installations and Ecology Energetics, serves as General Director of Science Park Izmaylovo of the Moscow Power Engineering Institute (MPEI), and as Senior Research Fellow of IC² Institute at The University of Texas at Austin where he serves

as a consultant on projects with the Commonwealth of Independent States (CIS). Dr. Rogalev also serves as an expert and special reviewer for the Ministry of Science and Technology, Ministry of Education, and CIS Association "Technopark" in technology management and technology transfer. He focuses his research efforts on technology transfer, management of technology, development of science parks and incubators, and energy and environmental control.
Contact: spark@aha.ru

Filipe Santos has been a member of the faculty at Instituto Superior Técnico since 1996. He is currently a Ph.D. student at Stanford University in the area of organizational studies applied to the management of technology.
Contact: fsantos@stanford.edu

Manuela Sarmento-Coelho teaches management science at Instituto Superior Técnico, where she is concluding a Ph.D. in Engineering and Industrial Management and is vice-president of the Portuguese Management Association.
Contact: sarmento@hidro1.ist.utl.pt

Syed Z. Shariq is Director of Research, KNEXUS (Knowledge: Networks, Exchange and Uses) Program, Institute for International Studies at Stanford University and a Senior Research Fellow of IC² Institute, The University of Texas at Austin. He has served as Senior Advisor for Information and Knowledge Management Technologies at NASA's Ames Research Center and has been the originator, principal founder, key implementor, and manager of the agency's commercialization initiatives. He was awarded the NASA Outstanding Leadership Medal, for "successful implementation of pathfinding innovations leading to new ways of science and technology commercialization" in 1993, the Group Achievement Award, for "the exceptional efforts and leadership in sharing NASA's technology wealth for the nation's economic benefit" in 1996, and the U.S. Technology Utilization Foundation's prestigious Life Time Achievement Award for Technology Transfer in 1996.
Contact: shariq1@stanford.edu

Alberto Silvani is Senior Researcher at National Research Council at the Institute for Studies on Scientific Research and Documentation. Since 1996, he has been involved in policy making activities, mainly in the field of evaluation and planning, as a scientific advisor with the Ministry for Universities and Scientific and Technological Research. His scientific activities cover many topics, including regional development in science and technology (policies and actions); science and technology indicators; technology transfer; policy evaluation and technology foresight; university-industry links, scientific infrastructures and technology parks; and international comparisons (R&D and RTD), particularly with less-favored regions. He is author or co-author of more than 50 papers on national and international journals.

Christopher E. Stiles is the Director of SOFTEX-Austin, a Brazilian software consortium representing 1,100 companies in 22 cities. He is also a Global Programs Research Associate at IC² Institute.

Robert S. Sullivan is Dean of the Kenan-Flagler Business School at the University of North Carolina at Chapel Hill. From 1995 to 1997 he was director of The University of Texas at Austin's IC² Institute where he also held the J. Marion West Chair for Constructive Capitalism in the university's Graduate School of Business.
Contact: rss@unc.edu

Tae Kyung Sung is Professor of Management Information Systems and Associate Dean of Sosung Institute for Advanced Studies at Kyonggi University, Korea. His research interests include information systems strategy, planning, management, and control; business innovation; knowledge management; technology transfer and commercialization, data mining, and virtual education. He has published in the *Journal of MIS Research, Journal of Industrial Studies, Technology Forecasting and Social Change, Journal of Management Education and Research, Management Science, Journal of Information Processing, International Business Review,* and other publications. He serves as Executive Vice President of Korean Management Information Society, editorial board member for the *Journal of MIS Research,* and Organizing Committee Chairman for 1998 KMIS International Conference.
Contact: tksung@kuic.kyonggi.ac.kr

Ming-Yung Ting is an Associate Professor in the Department of Information Management, Ming-Chuan University. His main interests are industrial innovation management, production management, organization theory, and BPR.

Nick Van Heck is at the Catholic University of Leuven (Belgium) as Research and Teaching Assistant in the domain of International Management and Strategy. His research has focused on (international) strategy and organization development, especially within the European context. In that context, he has written several in-depth business case studies—one of which won First Prize in the 1997 European Foundation for Management Development Case Competition in the category of European management—and has published articles on the management of the cross-border integration process in companies. He is involved in several consulting projects and management development programs on (international) strategy and organization issues in various companies and industries as a partner in Executive Learning Partnership (ELP), based in the Netherlands.
Contact: Nick.VanHeck@econ.kuleuven.ac.be

Francisco Veloso is a Ph.D. candidate in the Technology, Management, and Policy Program at the Massachusetts Institute of Technology. His research interests are economics of manufacturing and industrial development. In recent years, he has been studying the patterns of development of the automotive industry around the world, with a particular focus in late industrializing nations.
Contact: fveloso@MIT.EDU

Paul Verdin is Professor of Strategy and International Management at the Catholic University of Leuven (Belgium) and Visiting Professor at INSEAD (France). His earlier empirical research on industry and competence-based competition has been widely published. His current research focuses on the process of internationalization and integrating regional strategies and organizations particularly within Europe, for a variety of companies and industries facing substantial change in the wake of deregulation, technological innovation, and regional economic integration.

Roberto C. Villas Bôas is a Professor and Principal Researcher at CETEM, a R&D institution belonging to the Brazilian National Research Council. His main interests in R&D lie in developing sustainable development criteria for the extraction industries and performing research on robust engineering—mathematical technologies to assess robust quality to products and processes. Also he is Acting Chairman of the Advisory Committee for the International Materials Assessment and Application Centre of UNIDO, and the President of the Pan American Committee on Mining, Metallurgy, and Materials of the Pan American Union of Engineering Academies.
Contact: VILLASBOAS@cetem.gov.br

Bing Wang is a professor at the Tsinghua University Law School, Deputy Director of Tsinghua University's Science & Technology Office, and patent attorney at the Tsinghua University Patent Affairs Office. He has been involved in the management of research and development, intellectual property, and technology commercialization. He has published 64 papers and four books covering the fields of chemical engineering, intellectual property, and technology management.
Contact: wb-off@mail.tsinghua.edu.cn

Wu Yingjian works with the State Science and Technology Commission in Beijing, China.

Benjamin J. C. Yuan is at the National Chiao Tung University where he set up the first Institute of Technology Management in Taiwan. Most recently, he was a visiting scholar at the Center for Technology, Policy and Industrial Development at the Massachusetts Institute of Technology (1996–1997). His research interests focus on technology forecasting and assessment, project management, R&D performance evaluation, and incubators.